NANOCOMPOSITES WITH UNIQUE PROPERTIES AND APPLICATIONS IN MEDICINE AND INDUSTRY

Edited by **John Cuppoletti**

Nanocomposites with Unique Properties and Applications in Medicine and Industry
Edited by John Cuppoletti

Published by InTech
Janeza Trdine 9, 51000 Rijeka, Croatia

Publishing Process Manager Romina Krebel

Technical Editor Teodora Smiljanic

Cover Designer InTech Design Team

Additional hard copies can be obtained from orders@intechopen.com

Nanocomposites with Unique Properties and Applications in Medicine and Industry,
Edited by John Cuppoletti
p. cm.
ISBN 978-953-307-351-4

For EU product safety concerns:
IN TECH d.o.o., Prolaz Marije Krucifikse Kozulić 3, 51000 Rijeka, Croatia,
info@intechopen.com or visit our website at intechopen.com.

Meet the editor

Dr. Cuppoletti is Professor of Molecular and Cellular Physiology at the University of Cincinnati. (Ph.D. 1977 University of California, Davis). His research interests include pumps, carriers, and ion channels of the epithelia. In recent years, potassium and chloride channel regulation, pharmacology, and molecular engineering have become increasingly important tools with which to study the functional mechanisms of these proteins. More recently, as the X-ray crystal structures of these proteins have become available, it has become possible to combine these types of studies with computational chemistry approaches in attempts to understand how the structure relates to the function of these proteins. These transport proteins have been placed into synthetic membranes in order to obtain more detailed information regarding protein function that cannot be obtained from purely biological membranes. The natural and engineered proteins in these membranes are functional, and have led to the development of many interesting devices. Examples include drug screening devices, devices for identification of toxic bacterial products, and membranes for use in fuel cells.

Contents

Preface XI

Part 1 New Materials and Analytic Methods 1

Chapter 1 **On the Prediction of the Residual Behaviour of Impacted Composite Curved Panels 3**
Viot Philippe, Ballere Ludovic and Lataillade Jean-Luc

Chapter 2 **Fracture Toughness Determinations by Means of Indentation Fracture 21**
Enrique Rocha-Rangel

Chapter 3 **Techniques for Identification of Bending and Extensional Elastic Stiffness Matrices on Thin Composite Material Plates Based on Virtual Field Method (VFM): Theoretical and Numerical Aspects 39**
Fabiano Bianchini Batista and Éder Lima de Albuquerque

Chapter 4 **Analytical Research on Method for Applying Interfacial Fracture Mechanics to Evaluate Strength of Cementitious Adhesive Interfaces for Thin Structural Finish Details 67**
Tsugumichi Watanabe

Chapter 5 **Micromechanisms Controlling the Structural Evolution of Tribosystems 83**
Dmitry Lubimov and Kirill Dolgopolov

Chapter 6 **Damage Assessment of Short Glass Fiber Reinforced Polyester Composites: A Comparative Study 113**
Amar Patnaik, Sandhyarani Biswas, Ritesh Kaundal and Alok Satapathy

Chapter 7 **Review Fabrication of Functionally Graded Materials under a Centrifugal Force 133**
Yoshimi Watanabe and Hisashi Sato

Chapter 8 **Synthesis and Properties of Discontinouosly Reinforced Aluminum Matrix Composites 151**
Dusan Bozic and Biljana Dimcic

Chapter 9 **Modelling Reaction-to-fire of Polymer-based Composite Laminate 175**
Damien M. Marquis and Éric Guillaume

Chapter 10 **Production, Characterization, and Mechanical Evaluation of Dissimilar Metal/Ceramic Joints 205**
José Lemus-Ruiz, Leonel Ceja-Cárdenas, Egberto Bedolla-Becerril and Víctor H. López-Morelos

Chapter 11 **Measurement of Strain Distribution of Composite Materials by Electron Moiré Method 225**
Satoshi Kishimoto, Yoshihisa Tanaka, Kimiyoshi Naito and Yutaka Kagawa

Part 2 New Materials with Unique Properties 237

Chapter 12 **Joining of C_f/C and C_f/SiC Composites to Metals 239**
K. Mergia

Chapter 13 **Optical and Structural Studies of Binary Compounds by Explosive Laser Irradiation and Heat Treatment 267**
S. Kar

Part 3 Applications of New Materials 281

Chapter 14 **Development Liquid Rocket Engine of Small Thrust With Combustion Chamber from Carbon - Ceramic Composite Material 283**
Alexander A. Kozlov, Aleksey G. Vorobiev, Igor N. Borovik, Ivan S. Kazennov, Anton V. Lahin, Eugenie A. Bogachev and Anatoly N.Timofeev

Chapter 15 **New Routes to Recycle Scrap Tyres 293**
Xavier Colom, Xavier Cañavate, Pilar Casas and Fernando Carrillo

Chapter 16 **A Review of Thermoplastic Composites for Bipolar Plate Materials in PEM Fuel Cells 317**
Rungsima Yeetsorn, Michael W. Fowler and Costas Tzoganakis

Chapter 17 **High Voltage Electric Discharge Consolidation of Tungsten Carbide - Cobalt Powder 345**
Evgeny Grigoryev

Preface

This book contains chapters on nanocomposites for engineering hard materials for high performance aircraft, rocket and automobile use, using laser pulses to form metal coatings on glass and quartz, and also tungsten carbide-cobalt nanoparticles using high voltage discharges.

A major section of this book is largely devoted to chapters outlining and applying analytic methods needed for studies of nanocomposites. As such, this book will serve as good resource for such analytic methods.

Scrap tires nanocomposite particles for strengthening composites is one promising approach to recycling tires and preserving resources, and investgations into the use of electric fields to reduce friction can also help protect resouces including hydrocarbon lubricants. Some of these new composites and developments could, therefore, have a positive impact on the environment.

This book contains 17 chapters which have been grouped into three main parts:

1. *New materials and analytic methods*: This section is rich in analytic methods suitable for nanocomposites. Analytic methods include assessment of impact, studies of bending, damage assessment, models of reaction to fire, measurement of erosion wear and measurement of strain distribution.

2. *New materials with unique properties*: Studies on vibrations of composite plates and detailed analysis of methods of joining Cf/C and Cf/SiC to metals are presented.

3. *Applications of new materials:* Studies are presented on the development of new ceramic materials for rocket thrusters, new methods for preparation of tungsten carbide-cobalt nanoparticles and for the use of nanocomposites containing scrap tire particles.

I am pleased to have had the opportunity to work with the authors and to have served as editor of this book which expands composite materials research into so many exciting areas of development of materials, engineering, medicine and dental restoration.

The book contains a wide variety of studies from authors from all around the world. I would like to thank all the authors for their efforts in sending their best papers to the attention of audiences including students, scientists and engineers throughout the world. The world will benefit from their studies and insights. The new possibilities of the open access press bringing together such a diverse group and to disseminate widely on the web is revolutionary, and without the contributions of the group and the mechanism of InTech Open Access Publisher, this Book titled "Nanocomposites with Unique Properties and Applications in Medicine and Industry" would not be possible.

I also wish to acknowledge the help given by InTech Open Access Publisher, in particular Ms. Romina Krebel, for her assistance, guidance, patience and support.

John Cuppoletti, Ph.D.
Department of Molecular and Cellular Physiology
University of Cincinnati
Cincinnati OH
USA

Part 1

New Materials and Analytic Methods

1

On the Prediction of the Residual Behaviour of Impacted Composite Curved Panels

Viot Philippe, Ballere Ludovic and Lataillade Jean-Luc
Arts et Métiers -ParisTech, Institut de Mécanique et d'Ingénierie, UMR CNRS n° 5295, Esplanade des Arts et Métiers, F-33405 Talence France

1. Introduction

Composite materials are very often used in the aeronautical industry, because of their high specific strength, they are more appropriate for such applications than metals are. However, one disadvantage of such materials is the problem of detecting damage initiated by impact (e.g., dropping tools, collisions with foreign objects and other accidents), particularly when the reinforcement used is carbon fibre because may not be visible. Therefore, since it is difficult to avoid accidents, it is necessary to evaluate the effects of such damage on the residual resistance of the structure. This approach is related to the concept of the damage tolerance of structures. The structures considered here are filament-wound vessels subjected to high internal pressure loading and damage can be initiated in the carbon-epoxy shell during all their life cycle (manufacturing, storage, etc). In order to qualify the behaviour of these impacted structures, preliminary validation tests have to be done. However, these specific tests are generally very expensive and difficult to perform, especially when the structures' dimensions are large. An alternative way must be developed and a first one is to employ small-scale models.

The use of these reduced scale structures requires the identification of similitude models allowing the extrapolation of the small-scale model behaviour to the real structure. Although largely used in the case of homogeneous materials, such similitude techniques are not significantly developed for composite materials, mainly because of the interactive character of the different and multiscale damage mechanisms. As a first attempt, two scaling rule methods were developed based on a dimensional analysis using Buckingham's Pi theorem (Buckingham, 1914) or defined from dynamic equation of the system (Qian & Swanson, 1990). From these similitude models, some authors (Morton, 1988, Nettles et al., 1999) studied scale effects on composite structures taking into account the damage evolution during an impact but the gap was important between experimental results and predictions issued from scale models. For our study, in a preliminary phase of this research, scale models were evaluated (Viot et al., 2008) : a first approach consisted to apply similitude laws currently used on two scales (A and B) of composite structures . The purpose of this preliminary study was to predict the behaviour of the composite structure (scale A) from the knowledge of the response of the second scale model (scale B). It has been shown that existing similitude laws can be used to evaluate the elastic response of the two scales of composite structure but these models do not allow simulating the behaviour of the different scales when one of them is damaged ; it is due to non linearities.

For composite structures of large dimensions, an alternative and new approach of small-scale models must be developed since the experimental cost of impact study can be too expensive. Then, the main objective of our work is to predict the residual behaviour of impacted structures when the residual behaviour of small-scale structures is known. And because classical similitude laws cannot be used for damaged composite structures, another approach can be the use of a numerical model coupled with experimental data to predict the residual behaviour of impacted structures.

The proposed method is in three steps and must be applied on small scale panels to predict the behaviour of damaged vessels loaded by internal pressure. Before any numerical simulation, the analysis of the critical damage initiated during an accident must be quantified (point ⓿, figure 1). It is not the accident, the impact which is really important to qualify, even if it is interesting to know the impact conditions (mass, velocity, impactor's geometry, boundaries conditions...), but mostly the different kinds of damages (matrix cracks, fibre breakages or delamination) initiated during this impact which have to be precisely determined. And from the observations and analysis of the impacted composite microstructure, these damages must be classified from their critical effects on the performances of damaged structure. For vessel structures investigated, damages initiated during the impact were mainly delamination between carbon plies and fibre breakage. However, if delamination is a critical phenomenon for composite structures under bending load, this damage has not a drastic effect on vessel residual behaviour because the gap appearing between two delaminated plies decreases when the vessel is under the pressure and the propagation of delamination is then not effective. On the contrary, the breakage of fibres under tensile loading can obviously have a significant effect on the residual performance of the vessels.

From this preliminary study "identification of damage on real structure", the critical kind of damage was quantified and its effect must be experimentally and numerically evaluated on small scale structure. The main objective of the step 1 is then the development and the calibration of a numerical model, able to estimate the response of impacted small-scale structure: first, the critical damage has to be experimentally reproduced on small-scale structure by impact (working package❶, figure 1). Secondly, this damage must be controlled and precisely quantified (nature and size) at the scale of the composite microstructure (working package❷, figure 1). Finally, the residual behaviour of small-scale structure is estimated (working package ❸, figure 1) by imposing a state of loading close to the one imposed on the real structure (in order to initiate the propagation of the critical damage on small scale model in similar loading conditions than the ones imposed on real structure). This experimental approach is essential to calibrate the numerical model which has to be developed (working package ❹, figure 1), in taking into account the damage at the scale of the microstructure, in order to estimate the residual response of the small-scale structure.

The second step of this methodology is the evaluation of the performance of the numerical model. The same experimental study "impact – analysis of the damage- identification of the residual performance" is carried out on a second small-scale structure to obtain experimentally the effect of the damage on the residual behaviour of a second composite structure. This effect is also numerically evaluated in using the model developed during step 1. The performance of this model to predict the residual response of damaged composite structure is obtained from the comparison between experimental and numerical results obtained on the second small-scale structure (working package ❺, figure 1).

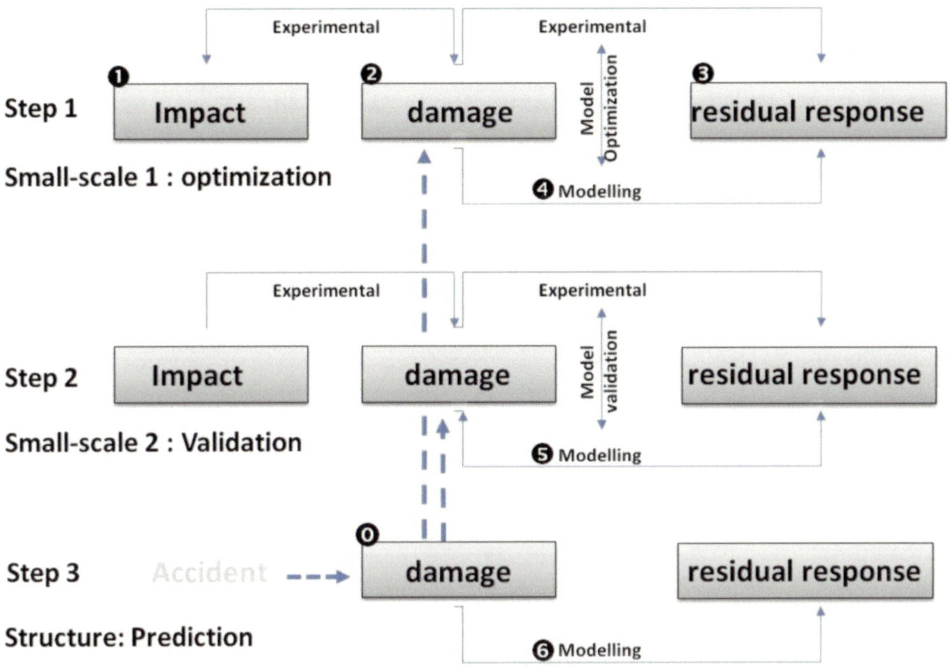

Fig. 1. Scheme of the multi-scale methodology

Finally the third step of the method is the prediction of the residual behaviour of the real structure. The critical damage identified on the real structure at the beginning of this methodology is implemented on the numerical simulation of the real structure. The residual behaviour of this structure can be then numerically estimated (working package ❻, figure 1).

This methodology was carried out for the study of the behaviour of impacted carbon-epoxy vessels under pressure. As an experimental study of damage tolerance using this type of structure is very expensive, the experiments were performed on curved panels extracted from tubes which had the same geometrical and mechanical properties as the vessels. The experimental procedure was carried out on these curved panels and the whole of the results were presented in a previous paper (Ballère et al., 2008): Firstly, the specimens were impacted to simulate an accident which can occur on such structures. Then, they were loaded in tension, according to their longitudinal axes, to reproduce the axial stresses caused by internal pressure being applied to the vessels' bottoms. The residual tensile strength was determined according to the initial damage states of the specimens.

The objective of this paper is to present a progressive failure analysis for the prediction of the residual properties of impacted curved specimens loaded in tension. First, the numerical results are compared with the experimental results obtained from undamaged specimens. Then, the damage observed experimentally is implemented numerically and the residual tensile strength is compared to the experimental results.

This methodology uses two scales of specimens. The first - close to the real scale - is used to validate the numerical modelling. The second- half the size of the first - is employed to highlight the mechanisms which have to be taken into account for the high-scale reduction of curved composite structures.

2. Numerical simulation

The modelling proposed in this study is based on a progressive failure analysis at the mesoscale (i.e., at the scale of the layer and the interface). Three steps are needed to build this model: i) choose a failure criterion; ii) choose damage kinetic; and iii) determine the consequences of the criterion activation on the elastic properties of the layer. Many approaches can be found in the literature to describe the progressive failure of a laminate, e.g., a state of the art approach was presented during the World Wide Failure Exercise (Kaddour et al., 2004). Different criteria are used in these approaches: for example, Ambur (Ambur et al., 2004) and Laurin (Laurin et al., 2007) use the Hashin-Rotem multi-criterion (Hashin and Rotem 1973); Bogetti (Bogetti, 2004) a 3-D maximum strain criterion and Zinoviev (Zinoviev, 2002) a maximum stress criterion. For this study, we have chosen the following maximum strain criterion.

2.1 Criterion: Maximum strain

The numbering of the orthotropic axes of the layer is shown in Figure 2a. The failure criterion is based on a damage variable, d_{ij}, defined as

$$\frac{\varepsilon_{ij}^{n}}{\varepsilon_{ij}^{R}} = d_{ij}^{n} \tag{1}$$

where i and j correspond to the orthotropic axes of the layer (i,j=1,...3), ε_{ij}^{n} is the component ij of the strain tensor at increment n and ε_{ij}^{R} its value to failure. The failure occurs when $d_{ij} \geq 1$. In this formulation, it should be noted that, for $\varepsilon_{ij} < 0$ (i.e., in compression), d_{ij} is always negative so that there is no failure in compression. This assumption can be justified here since this modelling is applied to the prediction of residual strength in tension. As soon as the failure occurs (i.e., $d_{ij} \geq 1$), the elastic modulus, E_{ij}, is reduced instantaneously to a residual value close to 0 (Figure 2b). This value is maintained regardless of the post-failure loading. This property avoids the healing of the damaged material. This approach proposes a degradation of the elastic properties of the layer in an independent way and the interface is considered non-damaging.
This progressive failure analysis has been implemented in the Finite Element Code, ZéBuLoN.

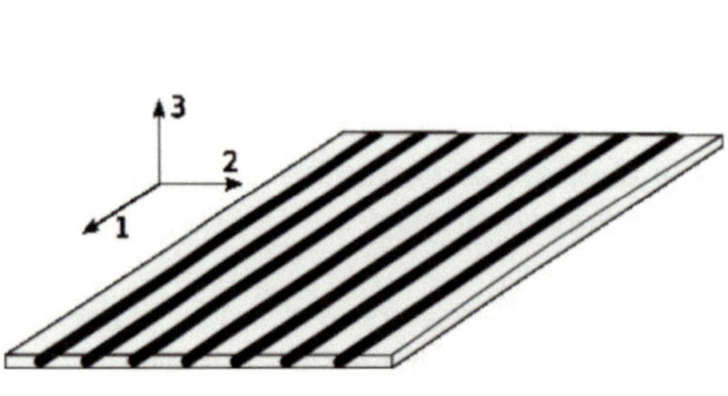

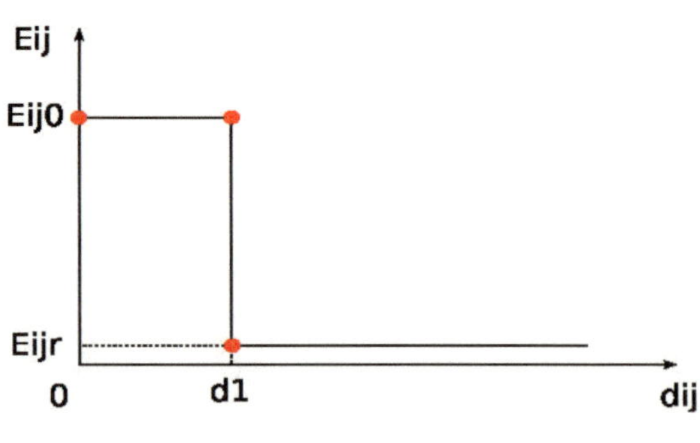

Fig. 2. a (left) Numbering of orthotropic axis, b (right) variation of elastic properties

2.2 Numerical simulation of curved panel

In order to validate this numerical approach, experimental tests were performed on two scales of composite curved panels. The methodology used and the results obtained for one of these scales of specimens (called «specimens Ø600») is presented in (Ballère et al., 2008). The first step in the numerical modelling is to check that the behaviour of an undamaged specimen is well-predicted.

2.2.1 Undamaged curved panel

The stacking sequence of the laminate used here is:

$$\text{Inner-}(90°)_2/[(\pm 20°)_2/(90°)_2]_3\}\text{-Outer}$$

For this modelling, the stacking sequence has been simplified and has been chosen to model the layers oriented at +20° and -20° independently (see Figures 3 and 4). For this scale of specimens, n is equal to 1.

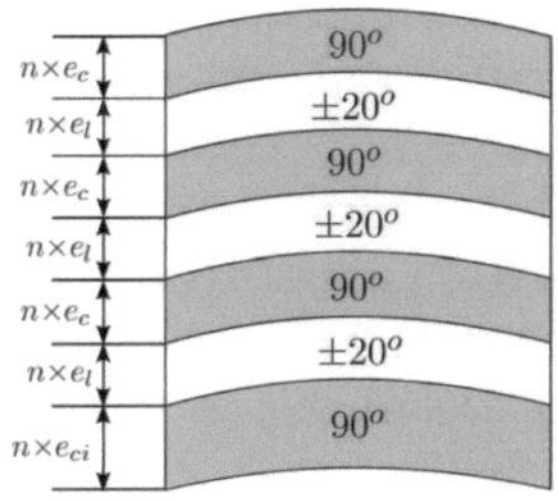

Fig. 3. Stacking sequence of the real structure

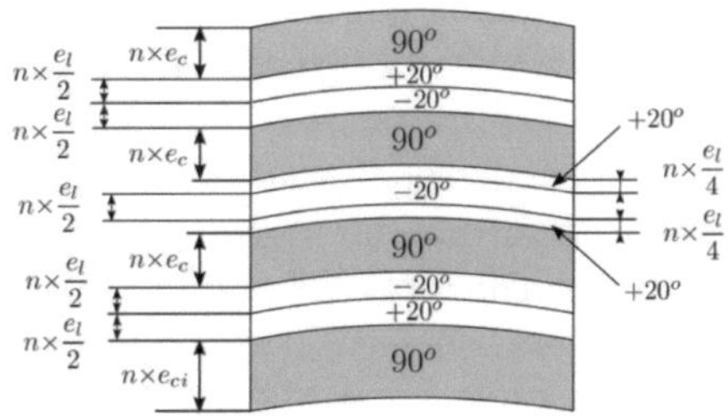

Fig. 4. Stacking sequence of the numerical specimen

The laminate is not symmetrical because the inner circumferential layer is thicker (e_{ci}}) than the others (e_c). Nevertheless, the choice of the stacking sequence for the numerical specimen was made in order to try to create a nearly symmetrical laminate according to the mid-plane so that the modelling would be easier. The elastic properties of the carbon/epoxy used are:

$E_1\,(GPa)$	$E_2\,(GPa)$	$G_{12}\,(GPa)$	ν_{12}	ν_{21}	$\sigma_1^R\,(MPa)$	$\sigma_2^R\,(MPa)$
165	7.1	3.9	0.39	0.015	2610	38

Table 1. Mechanical properties

The dimensions of the panels and the boundary conditions used in this numerical modelling are shown in Figure 5. They correspond to those used to perform the experimental tests. The radius of the curvature is 278 mm.

The influence of the element formulation for the failure prediction of these curve specimens was investigated in a previous study in which it was shown that the through-thickness displacement field is non-linear in tension. Therefore, in this research, an element denoted C3D20 (quadratic brick element) in ZeBuLoN (Carrère et al., 2009) has been chosen to model, through the thickness, each layer of a different orientation making it possible to detect the non-linearity of the displacement field.

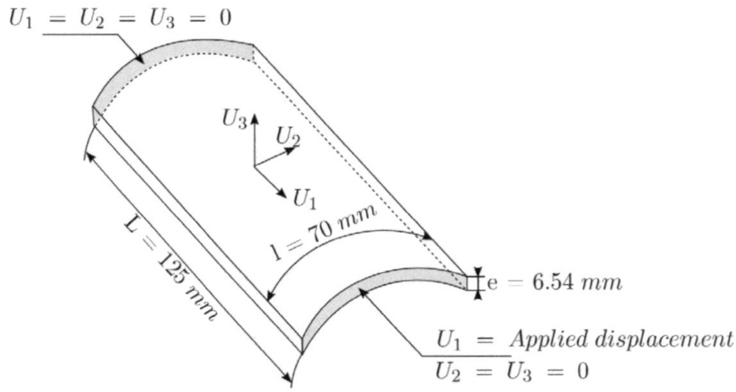

Fig. 5. Geometry and boundary conditions

2.2.2 Damaged curved panel ϕ 600

To implement the damage numerically, the typology of the damage mechanisms generated by impact had to be observed. This observation was undertaken during the experimental study (Ballère et al., 2008) and the results are summarized below.

For the impact tests, the specimen was clamped between two aluminium blocks and tightened with screws. It was fully supported on both surfaces except for a circular region of 30 mm in diameter in the centre corresponding to the impact zone. With this specimen-mounting device, classical damage mechanisms were observed. Delamination initiates and propagates during impact, even at low energy, but the delamination zone is always restricted in the centre because of the specimen-mounting device. Since impact energy is not fully dissipated by delamination, intra-laminar failures also occur (fibre failure, matrix cracking). Impacted specimens were loaded in quasi-static tension in order to evaluate their residual behaviour. It is well-known that the most prejudicial damage mechanism for laminates loaded in tension is fibre failure. For this reason, specific attention was paid to this phenomenon.

Table 2 presents some results of microscopic observations performed on specimens impacted with different impact energy levels. Each row is associated with a specific layer of the laminate. The columns of this table are ranked in order of increasing impact energy. The ≋ symbols indicate layers in which fibre breakages were observed. The number of layers damaged during the impact increased with the increase in the impact energy.

Structure		Damage1 (E = 22 J)	Damage2 (E = 30 J)	Damage3 (E = 38 J)	Damage4 (E = 71 J)
90°	Circ. 1	≋	≋	≋	≋
$\pm20^o$	Long. 1	≋	≋	≋	≋
90°	Circ. 2		≋		≋
$\pm20^o$	Long. 2			≋	≋
90°	Circ. 3				≋
$\pm20^o$	Long. 3				≋
90°	Circ. 4				≋

Table 2. Damage levels experimentally evaluated

In order to model the damage observed experimentally in a cylinder of 30 mm in diameter, an equivalent zone of the numerical specimen was defined (Figure 18). One or many layers can be damaged independently in this zone by decreasing all the elastic modulus E_{ij} to their residual value. This assumption can be justified since, at each time a fibre failure was observed in a layer, all the primary damage mechanisms (i.e., matrix cracking, fibre-matrix shear failures) were also observed. It is possible to suspect that the elastic properties of the layer decrease along all directions in the impact zone. In a first approach, the Poisson's ratio is not degraded.
All these observations were used to validate this modelling in the case of impacted specimens.

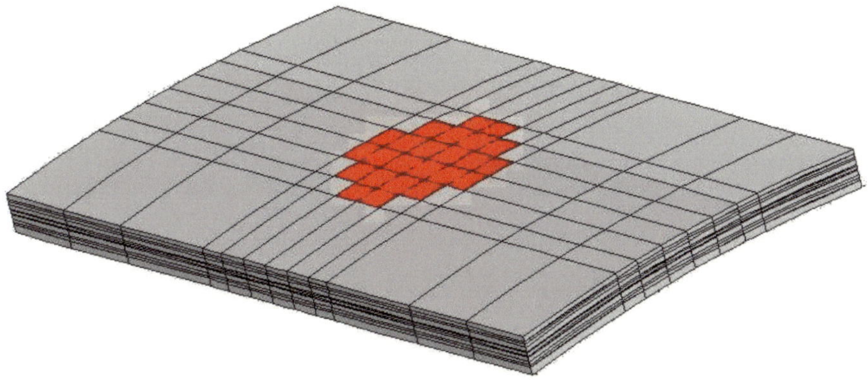

Fig. 6. Damage implementation

3. Model optimisation on a first small scale structure ϕ 600

3.1 Numerical results on non impacted panels: effect of the mesh size

Most progressive damage laws are very dependent on the mesh fineness. Therefore, two meshes of different element sizes were first considered in order to evaluate this effect (Figures 7 and 8). Mesh B consists of four times more elements in its surface than does mesh A. There is the same number of elements through the thickness in each mesh.

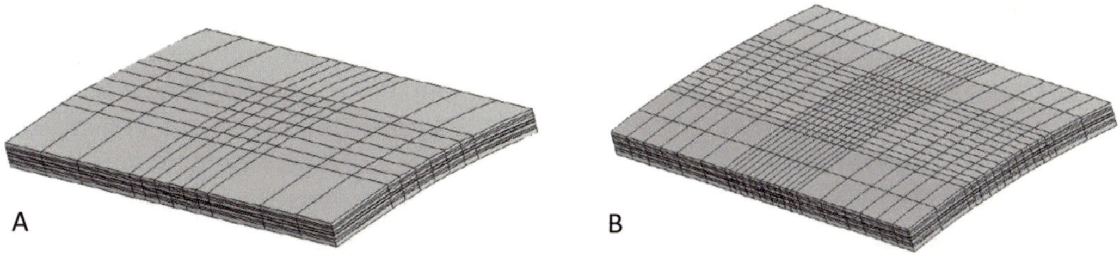

Fig. 7. Mesh A and Mesh B

Figure 8 shows a comparison between the stress-strain curves obtained using these two meshes. The horizontal line corresponds to the mean ultimate stress determined during experimental tests. Obviously, the two curves are similar in the first part of the loading, but a plateau occurs for them close to the experimental failure value and then, after this plateau, the main difference appears. For mesh A, the loading increases to reach a final failure value which is very far from the experimental value. Using mesh B, the final failure occurs just after this plateau with a stress value close to the experimental failure (3%). The decrease in the element size allows the failure value reached to be close to the experimental results. This can be explained by focusing on a zone of the graph located around the plateau (Figure 8).

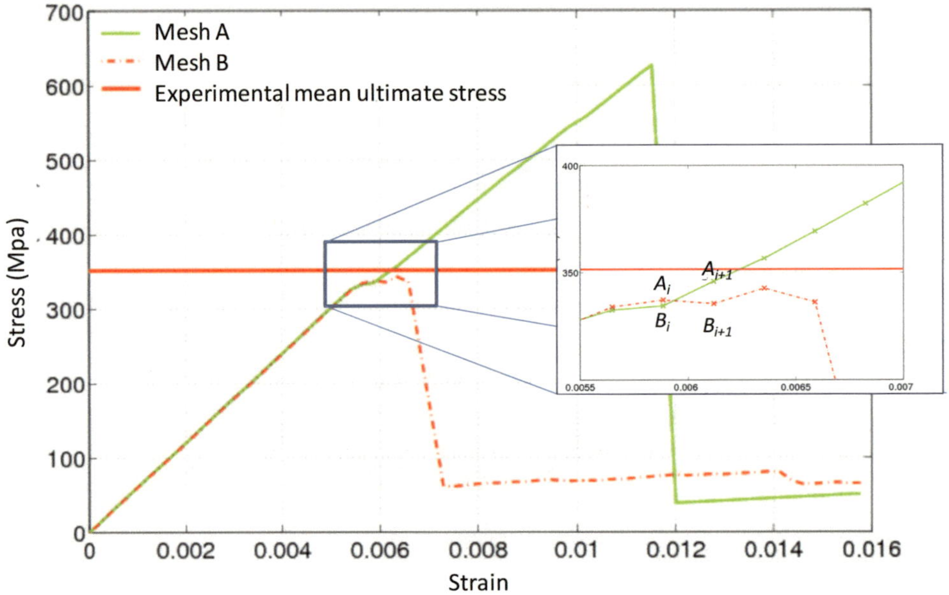

Fig. 8. Influence of the element size on the numerical stress vs. strain response

In order to identify the mechanisms which change according to the mesh fineness, attention has been paid to the criterion activation at particular points: i) points A_i and B_i, located just before the change of behaviour; and ii) points A_{i+1} and B_{i+1}, located at the next increment for mesh A and mesh B respectively. At points A_i and B_i, the criterion is highly activated by the damage variable d_{22}, (related to the orthoradial strain) in all the circumferential layers. The stress plateau observed for the two meshes is mainly due to this failure mode. For this level of loading, the damage variable, d_{33}, is also equal to 1 in many elements of the circumferential layers. Failures due to transverse shear stresses (damage variables d_{13} and d_{23}) also appear in all the layers of the laminate. The main difference between the behaviour obtained using these two meshes is in the detection of in-plane shear failures. The criterion is activated (i.e., d_{12}=1) in many elements with mesh B (Figure 9, left) but not in any element of the mesh A. The increase in the mesh fineness allows the detection of in-plane shear failures to be earlier.

This difference between the predictions from these two meshes is amplified at higher loading levels (i.e., points A_{i+1} and B_{i+1}). The criterion is still not activated with mesh A for in-plane shear stresses whereas there are many in-plane shear failures detected for mesh B in all the layers of the laminate and they are close to the free-edges of the specimen (figure 9, right). Experimentally, these failures lead to the delaminations observed post-mortem.

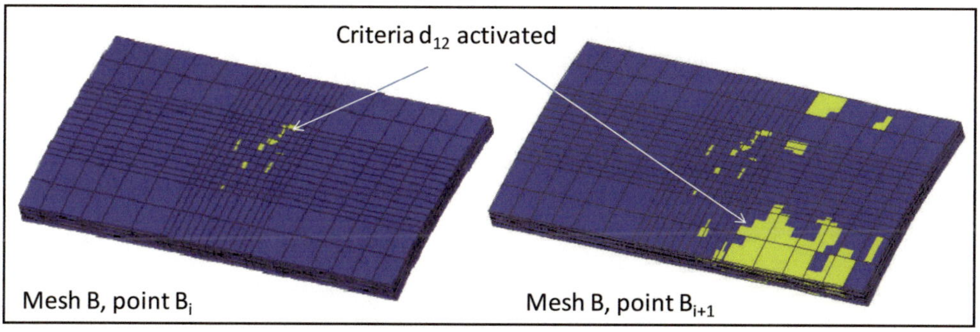

Fig. 9. Damage variable d_{12} calculated with mesh B at the point B_i (left) and at the point B_{i+1} (right)

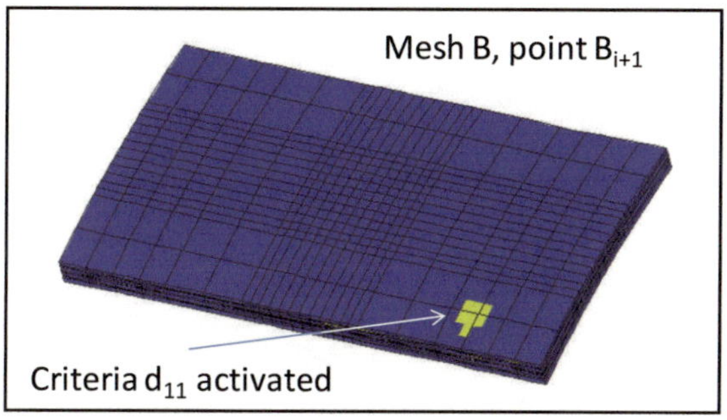

Fig. 10. Damage variable d_{11} calculated with mesh B at the point B_{i+1}

The existence of in-plane shear failures exhibited when using mesh B leads to failures in the fibre mode (i.e. damage variable d_{11}) in the longitudinal layers, as shown in Figure 10. This phenomenon leads to the global failure of the specimen. By decreasing the element size, it was possible to detect earlier the initiation of two damage mechanisms strongly prejudicial to the integrity of the specimens: in-plane shear failures and fibre breakages.

3.2 Numerical results on impacted panels

Since the proposed modelling was validated in the case of undamaged specimens, the next step was to use this model to predict the residual behaviour of impacted specimens. Each damage level presented in Table 2 was modelled and the residual tensile behaviour assessed. The stress-strain curves of the pre-damaged specimens are presented in Figure 11 and are compared with the response of the undamaged specimen (in using the mesh A).

Focusing on the slope of the first linear part of these curves, it appears that the more the pre-damage state is important, the more the slope decreases. This slope can be related to the homogenized elastic modulus of the specimens. Obviously, the damage generates decreasing stiffness. This phenomenon has been observed experimentally (Ballère et al., 2008) and could be analyzed in detail using this approach. Nevertheless, because the aim of this study is to predict the residual tensile strength of impacted specimens, particular attention was paid to the ultimate stress.

For each curve, the ultimate stress was extracted and then plotted versus the associated impact energy (figure 12a). Concerning an undamaged specimen, it was shown that an increase in mesh fineness leads to a global failure located at the stress plateau level. Thus, for this curve, the ultimate stress chosen was equal to this plateau value. These numerical predictions are in good agreement with the experimental results. Experimentally, there is a bi-linear evolution of the ultimate stress according to the impact energy. The impact energy yields (i.e., damage), when the ultimate stress starts to decrease. For this scale of specimens, this yield energy is equal to 40 J. This phenomenon is also observed in this numerical analysis since the ultimate stress obtained for the «damage 3» level is more or less the same as that for an undamaged specimen.

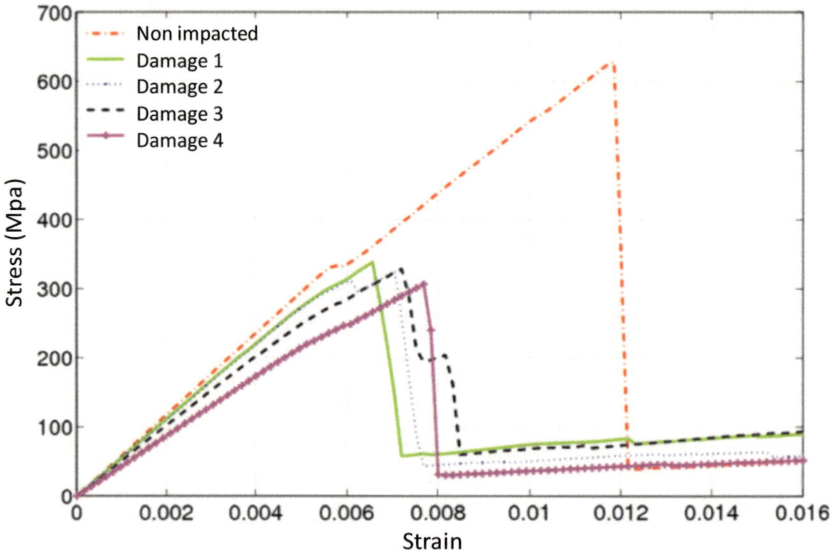

Fig. 11. Numerical stress vs. strain response for impacted specimens

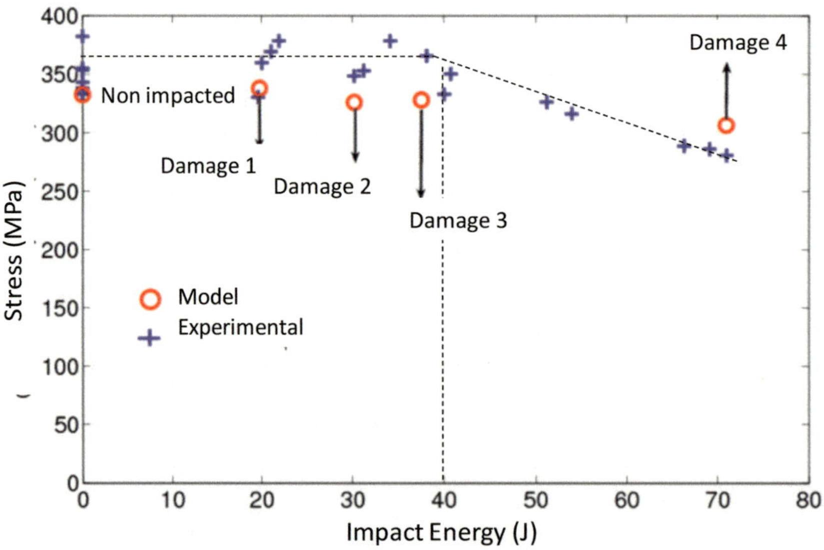

Fig. 12a. Comparison between model predictions and experimental results

It is interesting to note that an increase in the damage state of a specimen does not involve, systematically, a decrease in its ultimate stress. For example, the ultimate stress of a specimen damaged according to the «damage 3» level is almost equal to that of a specimen associated with the «damage 2» level (figure 12a). For instance, the «damage 2» level corresponds to the failure of two first circumferential layers (oriented at 90°) and one intermediate longitudinal layer (long. 1, oriented at ±20°, table 2) while the «damage 4» level is associated with the failure of the first circumferential layer and the two longitudinal layers (long. 1 and long. 2, table 2). For flat laminates, degradation of the longitudinal fibres is very harmful to the strength of the specimen in tension compared with degradation of the fibres oriented at 90°. It seems that, for these curved specimens, fibres oriented at 90° play a very significant role in their tensile strengths. Also, the influence of the damage organization through the thickness of the laminate has to be studied.

3.2.1 Influence of the damage on residual behaviour prediction

Experimentally, damage assessment was conducted using optical microscopy with a limited number of specimens. From the results, it was possible to correlate a residual tensile strength and an initial damage state of the specimen. Nevertheless, after the dynamic test, a variability of the damage could be suspected due to: i) dispersion of the properties of the different components; ii) variability in the mechanical properties of the specimens introduced by the manufacturing process; and iii) the boundary conditions used for the impact tests possibly being slightly different for each test. This damage variability was reflected in the experimentally observed strength dispersion. However, numerically, it cannot be taken into account without implementing a random damage variable.
In this approach, it was decided to quantify this strength variability by studying different cases of damage. »). Two new damage cases (called « Virtual damage A» and «Virtual damage B, table 3) were modelled to evaluate the residual behaviour of composite specimens if these kinds of degradation are imposed during an impact. The « virtual damage A» level was established from the «Damage 3» by adding the fibre breakage in the second circumferential layer. The « virtual damage B» was a damage level located between the «Damage A» level and the «Damage 4» level where five layers are damaged.

Structure		Dam.1	Dam. 2	Dam. 3	Virtual Dam. A	Virtual Dam. B	Dam. 4
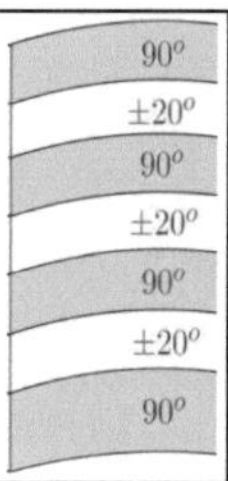	Circ. 1	≋	≋	≋	≋	≋	≋
	Long. 1	≋	≋	≋	≋	≋	≋
	Circ. 2		≋		≋	≋	≋
	Long. 2			≋	≋	≋	≋
	Circ. 3					≋	≋
	Long. 3						≋
	Circ. 4						≋

Table 4. Add of two damage level cases (A and B) for specimens ϕ600 mm

An increase in the initial damage state (from «Damage 3» to «Damage «A») obviously leads to a decrease in the ultimate stress (7%). This decrease is equal to 16% when one circumferential layer more than «Damage A» is damaged («Damage B»). For the damage B, the residual tensile strength is lower than the ultimate stress calculated for the damage 4 and experimentally obtained in any case of impact energy.

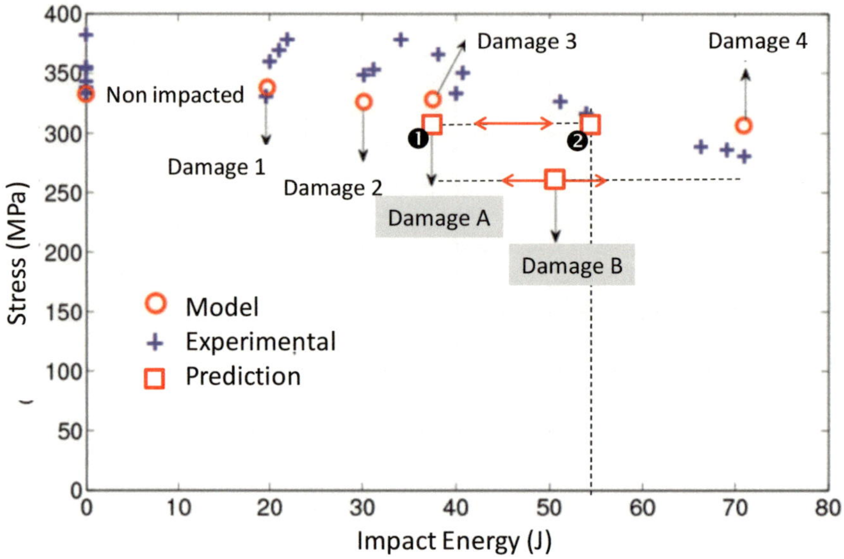

Fig. 12b. Comparison between model predictions for two new damages (A and B)

These two new results were plotted on the ultimate stress vs. impact energy diagram. It is clear that we cannot determine the level of impact energy necessary to obtain these two damage states A and B, we can just consider that the energy levels to reach these damages are in the range of the impact energies of the damage 3 and 4 (if we do not consider any variability of the composite specimen damage to the impact energy). For the damage A, the ultimate stress of which is really close the value numerically computed for the damage 3, it can be firstly assumed that the effect of supplementary damaged circumferential layers is not significant for the damage A (location❶, figure 12b). Or, in another way, the impact energy to obtain the damage A can be estimated to 55 J (location❷, figure 12b) by comparing the ultimate stress obtained experimentally to those one derived numerically.
In a first conclusion the used damage model is able to represent the residual resistance of a curved panel under tensile loading. However, it requires an identification of the damage induced during the impact. In the case of specimens Ø600, an increase in the number of damaged layers leads, beyond a specific threshold, to a global decrease in the residual strength. This decrease appears regardless of the type of damaged layers (i.e., circumferential or longitudinal). The shape of this decrease depends slightly on the location of the damaged layers through the thickness of the laminate. With the knowledge of the morphology and the size of the damage in the microstructure, the behaviour of larger specimen or larger structure previously damaged by an impact could be predicted. The second part of the paper presents the limits of this method and the difficulties which can appear if the scale of the specimen induces supplementary phenomena.

4. Prediction of the behaviour of impacted structures

The aim of the approach presented here is to predict the residual behaviour of specimens through the knowledge of the mechanical responses of small-scale specimens. In order to reveal the limits of the scale reduction in the case of curved composite structures, this approach was used to predict the residual behaviour of specimens which are half the size of specimens "Ø600".

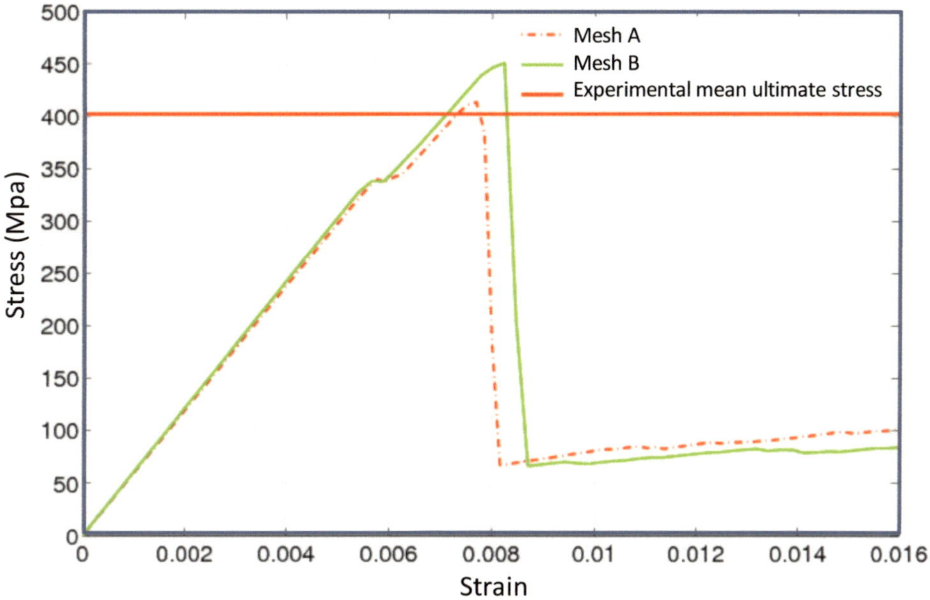

Fig. 13. Influence of the element size on the numerical stress vs. strain response

The dimensions of these small-scale specimens (denoted as specimens"Ø300") were chosen particularly to observe the influence of curvature and thickness on their mechanical responses. The radius of the curvature was divided by two (i.e., 139 mm). The thickness was determined by the stacking sequence of the laminate. The «ply-level scaling» technique has been chosen to design the stacking sequence of the small-scale specimens. Thus, the thickness of each layer of a different orientation was divided by two, thereby leading to a global thickness of the laminate equal to 3.27 mm. This technique allows the same stacking sequence to be maintained between each scale while only the layer thickness changes (i.e., in Figures 3 and 4, n=0.5). The width and length are the same for both scales.

The mesh used for this scale of specimens is equivalent to mesh A (Figure 6). Only the element thickness, which is related to the layer thickness, and the radius of the curvature are divided by two.

The influence of the mesh fineness was quantified using four more elements (equivalent to mesh B (Figure 7) for specimens "Ø600"). The stress-strain curves obtained using these two meshes are presented in Figure 13. For this scale of specimens, it seems that the numerical results are less dependent on the element size. A plateau appears in these two curves which, for the same reasons as mentioned in the case of specimens "Ø600", should be related to the global failure of the specimens. For specimens "Ø300", the damage mechanisms responsible for this plateau and, therefore, for the global failure, appear for the same stress value. In order to reduce computation times, mesh A was used for the modelling of the impacted specimens.

These small-scale specimens were tested experimentally and some of the results are presented in the next sub-section.

4.1 Numerical results on impacted panels ϕ 300

The different levels of damage modelled for this scale of specimens are presented in Table 5. They correspond to the damages observed experimentally.

Structure		Damage1 (E = 7 J)	Damage2 (E = 12 J)	Damage3 (E = 14.5 J)	Damage4 (E = 24 J)
90°	Circ. 1	≈	≈	≈	≈
±20°	Long. 1		≈	≈	≈
90°	Circ. 2	≈	≈	≈	≈
±20°	Long. 2		≈	≈	≈
90°	Circ. 3		≈		≈
±20°	Long. 3			≈	≈
90°	Circ. 4				≈

Table 5. Damage levels experimentally evaluated

Experimentally, a significant dispersion of the ultimate stress of the undamaged specimens is observed (figure 14). This dispersion decreases as soon as the specimens are pre-damaged by impact. Indeed, the defect initiated by the dynamic test restricts and governs the possible locations of failure initiation. Considering only one point, located at the mean ultimate stress for the undamaged specimens, it seems that there is a yield damage point from which the tensile strength starts to decrease.
The numerical predictions are in good agreement with the experimental results for the most-damaged specimens (i.e., «damage 4» level). They are located in the same area as the experimental dispersions for the undamaged specimens. Between these two damage levels, the numerical results do not match with those of the experiments. Moreover, the bi-linear evolution of the ultimate stress according to the impact energy cannot be predicted by this modelling.
It has been shown that this progressive failure analysis allows the residual tensile strength of one scale of curved specimens to be predicted. Unfortunately, it cannot be used for the prediction of the residual behaviour for smaller scale of specimens.

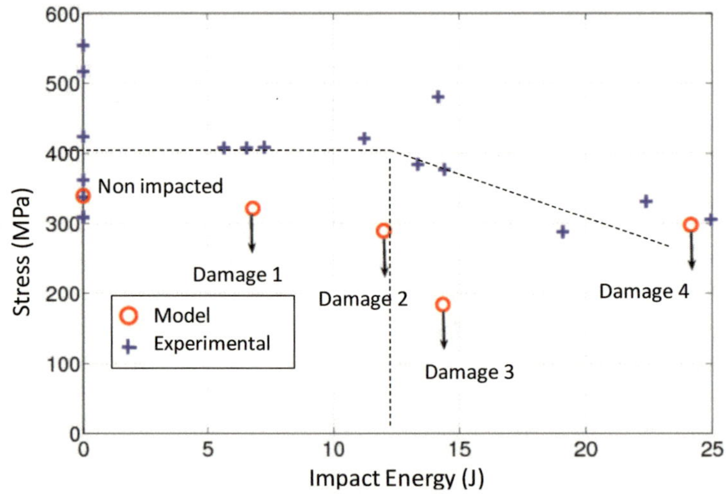

Fig. 14. Comparison between model predictions and experimental results

Because the numerical results are very different from the experimental ones in the case of small-scale specimens, a complementary study on the influence of the implemented damage on the ultimate stress was carried out. The results of this study are presented in the next section.

4.1.1 Influence of the damage on residual behaviour prediction

It was decided to quantify this strength variability by studying different cases of damage. This methodology was used for both scales of specimens starting with specimens "Ø300".

Structure		Dam.1	Virtual Dam. A	Dam. 2	Dam. 3	Virtual Dam. B	Dam. 4
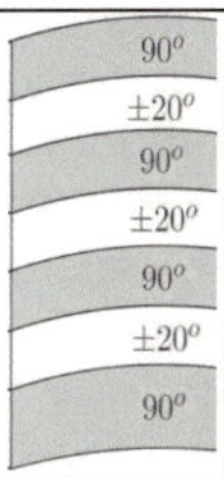	Circ. 1	≋	≋	≋	≋	≋	≋
	Long. 1		≋	≋	≋	≋	≋
	Circ. 2	≋	≋	≋	≋	≋	≋
	Long. 2			≋	≋	≋	≋
	Circ. 3			≋		≋	≋
	Long. 3				≋	≋	≋
	Circ. 4						≋

Table 4. Add of two damage level cases (A and B) for specimens ϕ300 mm

Thus, the «Damage A» level was created from the «Damage 1» level by adding the degradation of the first longitudinal layer (see Table 6). The «Damage B» level was established from the «Damage 3» level by adding the degradation of the third circumferential layer. These damage cases were chosen specifically to quantify the influence of one more failed layer on the ultimate stress of a specimen.

Figure 15 shows the ultimate stress evolution according to the impact energy (i.e., the initial damage state) from experimental and numerical analyses.

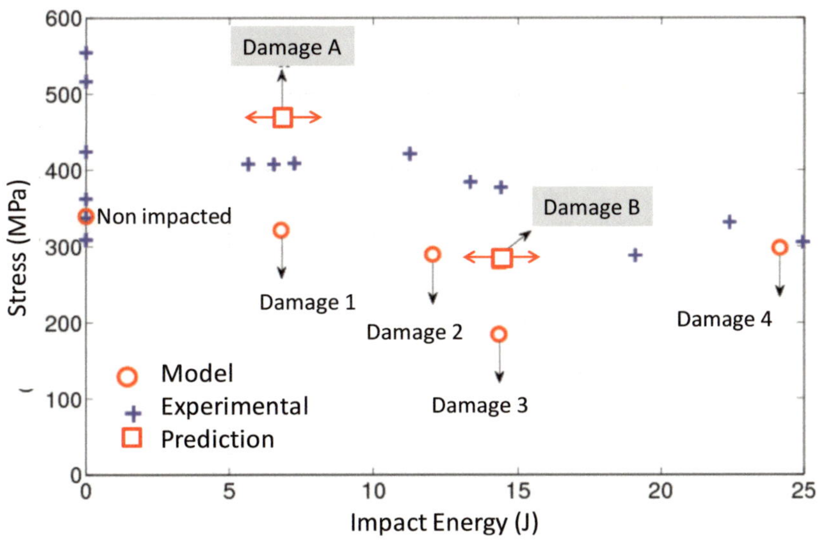

Fig. 15. Comparison between model predictions for two new damages (A and B)

The ultimate stress obtained in the case of «Damage A» is 46% higher than in the case of «Damage 1» while the damage level is more important. In the former case, the ultimate stress is also higher than that of the undamaged specimens. It seems that the degradation of all the layers located in the convex part of the curved specimen (i.e., above the mid-plane) increases the residual tensile strength. This is not true in the case of a partial degradation where only the circumferential layers are failed.
Moreover, increasing strength appears in the case of «Damage B». The ultimate stress is 60% higher than that obtained in «Damage 3» while «Damage B» considers one more circumferential layer failed. All these observations lead to a conclusion regarding the high sensitivity of the ultimate stress according to the damage organization through the thickness of these small-scale specimens ϕ300.
It seems that any changes in the radius of the curvature and/or the thickness of the damaged specimens modify the phenomena which occur when they are stressed in tension. Experimentally, during the tensile test, due to the lay-up and the boundary conditions, the specimen curvature tends to increase between the jaws [Ballère et al., 2008]. Circumferential layers are then progressively damaged according to the inter-fibre mode, thereby causing a change in the orientation of the fibres oriented at ±20° towards the loading direction (0°). This phenomenon is accentuated by the initial curvature of the specimen. In the case of a partial pre-damage, the type of layers damaged (circumferential or longitudinal), as well as their locations through the thickness of the specimen's boundary conditions, cause this change of orientation.
This phenomenon only exists because the interface between two layers of different orientations can be damaged. Looking at the delaminations observed after the global failure of specimens, the interface integrity strongly influences the residual tensile strength of curved panels. Because damage of the interface is not considered in this model, there are high variations in the ultimate stress in the case of specimens Ø300.

5. Conclusion

A progressive failure analysis, based on a maximum strain criterion, is presented in this paper. The aim of this approach is to predict the residual tensile behaviour of impacted curved panels. Applied on a scale of specimens (Ø600), it governs the progressive damage which appears at the layer level and leads to a residual tensile strength close to the experimental results. Using this approach, a scenario of failure was proposed and it was shown that the integrity of the circumferential layers (oriented at 90°) is very important for ensuring the tensile strength of the curved panels. Indeed, their failure leads to a change of a specimen's curvature between the jaws. This phenomenon alters the orientation of fibres oriented at ±20° towards the loading direction. High shear stresses then appear and lead to the global failure of the specimen.
This approach has been used for the prediction of the residual tensile strength of a second scale of specimens (Ø300), thinner and more curved than the previous ones. The numerical results are in good agreement with the experimental results in the cases of undamaged specimens and specimens fully damaged in the impact zone (i.e., all the layers are failed). But, for partially damaged specimens, the results do not match those of the experiments. Because of the higher curvatures of the specimens, the residual tensile strength seems to be very dependent on the organization of the damage through the specimen thickness.
This study indicates the need to pay attention to phenomena which can appear during a scale reduction in the case of composite curved structures. A significant decrease in the scale

of the specimens (i.e., a reduction in the thickness and the radius of the curvature) can change the nature of the main damage mechanisms responsible for the global failure of pre-impacted specimens. For slightly curved specimens, it has been shown that the damage which occurs at the interface has to be taken into account to accurately predict the residual tensile strength of pre-impacted specimens.

This possible degradation of interfaces could be conducted using cohesive elements (Elder et al., 2004, Pinho et al., 2006). These elements are implemented between two volumic elements representing two layers of different orientations. Their thickness is equal to 0 before loading. Choosing an appropriate behaviour law, they govern the displacement between two opposite nodes.

First results were obtained by taking into account the degradation of the interfaces. They show that the behaviour is different as soon as the interface is damaged. The change of the curvature, initiated by the failure of the circumferential layers, leads to high stresses which progressively damage the interface. The load transfer cannot take place anymore. This leads to criterion activation according to the fibre-mode in the longitudinal layers and the global failure of the specimen happens. These results confirm the fact that the interface damage has to be considered in order to reflect the physical mechanisms which lead to the failure of curved panels loaded in tension. Nevertheless, the computation time is very important when the cohesive elements are used with an intra-laminar progressive failure analysis. An alternative way has to be found to decrease this high computation time while maintaining the combined modelling of intra-laminar and inter-laminar damage mechanisms.

6. References

Ambur, D. R.; Jaunky, N.; Hilburger, M. & Dávila, C. G. Progressive failure analyses of compression-loaded composite curved panels with and without cutouts *Composite Structures*, 2004, *65*, 143 - 155

Ballère, L.; Viot, P.; Lataillade, J.-L.; Guillaumat, L. & Cloutet, S. Damage tolerance of impacted curved panels *International Journal of Impact Engineering*, 2009, *36*, 243 - 253

Bogetti, T. A.; Hoppel, C. P.; Harik, V. M.; Newill, J. F. & Burns, B. P. Hinton, M.; Kaddour, A. & Soden, P. *(Eds.)* Predicting the nonlinear response and progressive failure of composite laminates *Failure Criteria in Fibre-Reinforced-Polymer Composites, Elsevier*, 2004, 402 - 428

Buckingham, E. On physically similar systems; illustration of the use of dimensional equations. *Phys.Review*, Vol. 4, 1914

Carrere, N.; Rollet, Y.; Leroy, F.-H. & Maire, J.-F. Efficient structural computations with parameters uncertainty for composite applications, *Composites Science and Technology*, 2009, *69*, 1328 - 1333

Elder, D. J.; Thomson, R. S.; Nguyen, M. Q. & Scott, M. L. Review of delamination predictive methods for low speed impact of composite laminates *Composite Structures*, 2004, *66*, 677 - 683

Hashin, Z. and Rotem, A. A fatigue failure criterion for fibre reinforced materials. *Journal of Composite Materials*, 7:448–464, 1973.

Kaddour, A. S.; Hinton, M. J. &a. S. P. D. A comparison of the predictive capabilities of current failure theories for composite laminates: additional contributions *Composites Science and Technology*, 2004, *64*, 449 - 476

Laurin, F.; Carrère, N. & Maire, J.-F. A multiscale progressive failure approach for composite laminates based on thermodynamical viscoelastic and damage models *Composites Part A: Applied Science and Manufacturing,* 2007, *38,* 198 - 209

Pinho, S.; Iannucci, L. & Robinson, P. Formulation and implementation of decohesion elements in an explicit finite element code *Composites Part A: Applied Science and Manufacturing,* 2006, *37,* 778 - 789

Qian, Y. & Swanson S.R. An experimental study of scaling rules for impact damage in fibre composites. *J. of Composite Materials,* 24:559-570, May 1990.

Viot, P. ; Ballère, L. ; Guillaumat, L. & Lataillade., J.L. Scale effects on the response of composite structures under impact loading. *Engineering Fracture Mechanics Journal,* Vol 75/9 pp 2725-2736, 2008

Zinoviev, P. A.; Lebedeva, O. V. & Tairova, L. P. A coupled analysis of experimental and theoretical results on the deformation and failure of composite laminates under a state of plane stress. *Composites Science and Technology,* 2002, *62,* 1711 - 1723

2

Fracture Toughness Determinations by Means of Indentation Fracture

Enrique Rocha-Rangel
Universidad Politécnica de Victoria
México

1. Introduction

The assessment of fracture toughness (K_{IC}) on fragile materials such as ceramics or composites through conventional methods can be arduous. Recently, an alternative route referred to as the Indentation Fracture technique has been widely accepted with this purpose and extensively reported in literature (Weisbrod & Rittel, 2000; Plaza, 2003; Evans & Charles, 1976; Niihara et al., 1982). Different authors have derived math equations series as to fine tune and match with K_{IC} determination; those equations are based in the lineal mechanical fracture theory (Wang, 1996). The indentation fracture method and its application procedure are described in this chapter, whereas typical problems involved in the test are shown. Al_2O_3-based composites with different reinforced metals fabricated by both; liquid and solid pressureless sintering of an intensive mechanical mixture of powders were used as studied materials.
Ceramic materials have properties of great interest for various structural applications, specifically those that take advantage of their high hardness, chemical and thermal stability in addition to their high stiffness. However, their great fragility has severely limited their applications, although they have developed ceramic with reinforcement materials precisely to increase the toughness of the same (Miranda et al., 2006; Konopka & Szafran, 2006; Marci & Katarzyna, 2007; Travirskya et al., 2003; Sglavo, 1997). One of the macroscopic properties that characterize the fragility of a ceramic is the fracture toughness (K_{IC}). The fracture toughness describes the ease with which propagates a crack or defect in a material. This property can be assessed through various methods such as: Analytical solution, solution by numerical methods (finite element, boundary integral, etc.). Experimental methods such as: complianza, fotoelasticity, strain gauge, etc. and indirect methods such as: propagation of fatigue cracks, indentation, fractography, etc. The choice of method for determining the fracture toughness depends on the availability of time, resources and level of precision required for the application. In practice, measurements of K_{IC} require certain microstructural conditions on the material to allow propagation of cracks through it in a consistent manner. The strength of materials is governed by the known theory of Griffith, which relates the strength (S) with the size of the defect or crack (c) by $S = YK_{IC}/c^{1/2}$. This expression suggests the need to reduce the grain size and processing defects in the final microstructure to optimize the mechanical performance. Moreover, with increasing K_{IC}, resistance becomes less dependent on the size of the defect, thereby producing a more tolerant material to cracking. Due to high elastic modulus and low values of K_{IC} in brittle materials, achieving in them a stable crack growth is complicated and sometimes it is necessary sophisticated

measurement equipments and complex sample geometries (Wessel, 2004). The problem with applying these methods to evaluate K_{IC} is that required laborious procedures and only get one result by sample, being necessary multiple measurements to obtain reliable statistical results. In this sense many simple methods have been proposed to avoid these difficulties. One particularly attractive procedure due to its simplicity for routine evaluations of engineering materials is the indentation fracture (IF) method.
Although the IF method can only measure approaches of the values of K_{IC}, is a convenient technique for evaluation of many brittle engineering materials. This technique is based on normalized standards hardness tests (ASTM E1820 - 09e1 Standard Test Method for Measurement of Fracture Toughness, 2008 and ASTM C 1327-99, Standard Test Method for Fracture Toughness at Room Temperature of Advanced Ceramics, 1999. Assuming the presence of a preexisting, sharp, fatigue crack, the material fracture toughness values identified by this test method characterize its resistance to: (1) fracture of a stationary crack, (2) fracture after some stable tearing, (3) stable tearing onset, and (4) sustained stable tearing. This test method is particularly useful when the material response cannot be anticipated before the test, making reliable they obtained result.
These fracture toughness values may serve as a basis for structural flaw tolerance assessment. Awareness of differences that may exist between laboratory test and field conditions is required to make proper flaw tolerance assessment.
The test is relatively simple to implement and requires only a standard micro hardness tester. A small piece of material with a stress free surface and cracks is enough as test sample. The method, however, is not suitable for materials with values of K_{IC}, below 1 MPa $m^{1/2}$, significant ductility, large grain size and heterogeneous microstructures.

2. Antecedents

By the mid-60 began to apply empirically the concept of fracture mechanics to ceramic. The development of science has run parallel to the indentation techniques which helped to determine the resistance to penetration of the indenter in ceramics systematically. However, the most important development was the discovery of the transformation of the zirconia and the consequent increase in fracture toughness. From there they spent the previous design of the material and systematic study on the basis of the manufacture, characterization, testing and modeling. Another, important advance was the discovery that the addition of second phase substances such as fibers and spheroids particles improved mechanical properties such as fracture toughness (Konopka & Szafran, 2006; Travirskya et al., 2003; Bosch, 1990; Lieberthal & Kaplan; 2001). Finally, it started working with cermets, it means ceramic composites with second phase dispersed particles produced by particles processing or made directly by oxidation of metal.
However, progress in understanding the mechanisms of increasing fracture toughness has been slow for various reasons as the difference between the methods of measuring it, which raises questions about its usefulness. On the other hand, the most reliable methods of measurement tests require large numbers of samples which is not easy and possible in all ceramics.

2.1 Fracture definition

The fracture is the separation of a material under stress action. The fracture occurs by crack initiation and propagation (Figure 1). In this case the fault occurs with a small plastic deformation. The most important atomic mechanism is the breaking of atomic bonds due to the application of static loads. Although, theoretically the fracture can be caused by shearing forces, the majority of cases correspond to the application of normal, tension or bending stress.

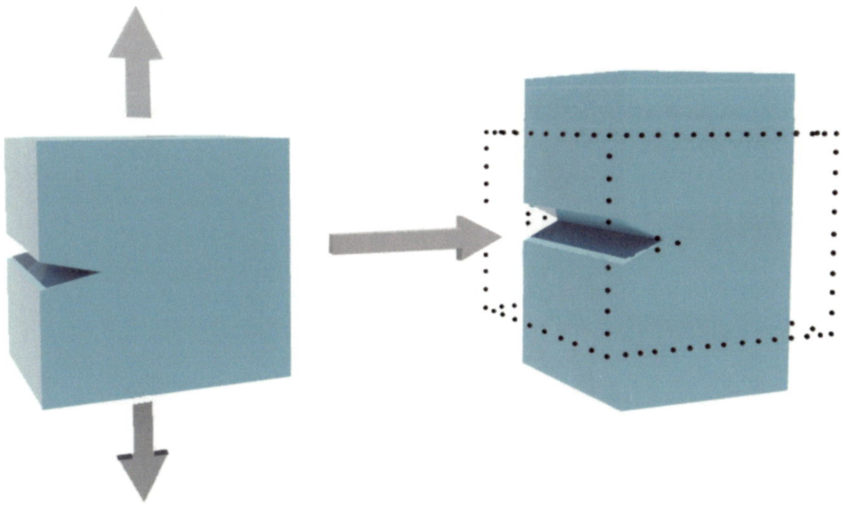

Fig. 1. Open crack and deformation of a body that is under tensile forces.

The normal stress is the most effective in breaking atomic bonds. The fracture can be ductile or brittle and involves a small or large consumption of energy, respectively (Figure 2). The comparison between both types of fracture shows always a sudden collapse in brittle fracture due to low energy absorption, originating from external stresses.

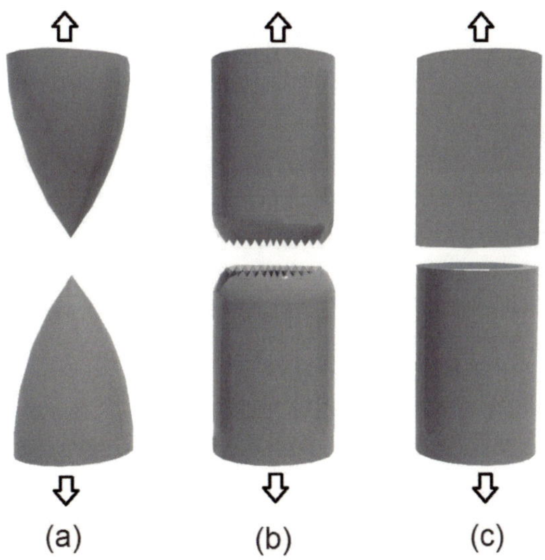

Fig. 2. Brittle vs. Ductile fracture (a) Very ductile, soft metals (e.g. Pb, Au) at room temperature, other metals, polymers, glasses at high temperature. (b) Moderately ductile fracture, typical metals (c) Brittle fracture, cold metals and ceramics.

2.2 Fracture and energy balance

It is generally conventionally considered that Griffith (1920) with his work: "The phenomena of rupture and flow in solids" was the first who introduced a scientific approach on the fracture in solids. The theory of linear elastic fracture mechanics began to be developed at that time. The Griffith fracture mechanics allowed to obtain a powerful criterion to predict crack propagation demonstrating to be generalizable to many types of materials and still remain valid.

Consider an infinite plate of unit thickness with a crack of length 2c is subjected to a tensile stress as shown in Figure 3.

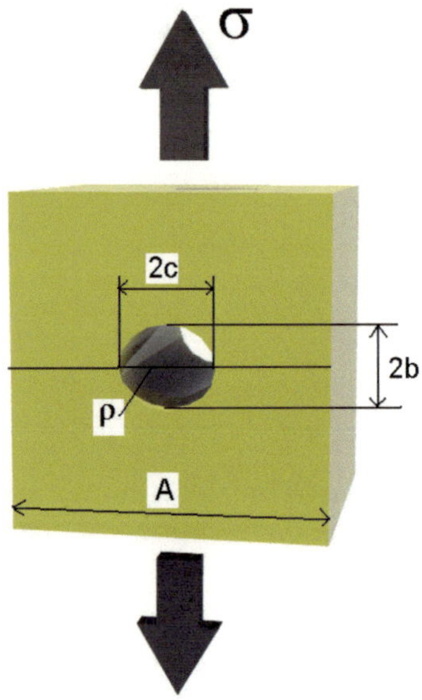

Fig. 3. Infinite plate of unit thickness with a crack of length 2c under tensile stress.

Griffith observed experimentally that small imperfections had a destructive influence on the materials much more than large and involved that in an energy balance not only care about the potential energy of external forces (W) and the stored elastic energy (UO), but another had not previously considered: the surface energy (US).

The surface energy (US) is needed to create new surfaces, incorporating work. For example, the blow and grow the surface of a soap bubble is required to make a job.

Correlated energies displayed are:

U - Total system energy

UO - Energy elastic plate with no cracks with applied forces (constant)

EU - elastic energy introduced to take place the opening of the crack

W - Work of external forces on the body

US - Energy associated with the "resistance" that opposes the material to the creation of new surfaces

Where:

$$U = U_0 + U_E - W + U_s \tag{1}$$

The parameters W and EU are associated with each other because they both promote the formation and propagation of the crack, while US represents the "resistance" that the material opposes to the creation of new surfaces, such as those generated by cracks.
Inglis proposed a solution that Griffith retums by the assumption that there is an atomic break caused by the crack, which is expected in brittle materials, such as ceramic. In a harmonic oscillator model (a_o is the atomic separation), once the springs have collapsed due to high normal stress the crack propagates in a perpendicular plane as shown in Figure 4.

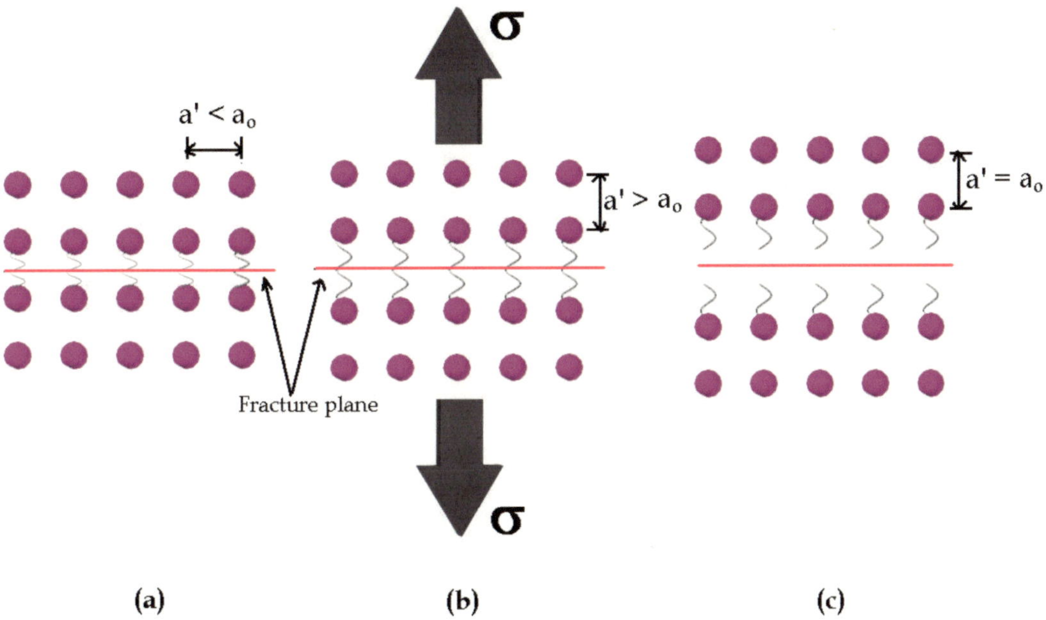

Fig. 4. Atomic break caused by the crack. a_o is the atomic separation, once the springs have collapsed due to high normal stress the crack propagates in a perpendicular plane.

Griffith's equation states:

$$U_e - W = -\frac{\pi\sigma^2 c^2}{E} \tag{2}$$

If it is consider this "elastic capacity" of material per unit area (dA) of the cracks generated, it can be defined a magnitude G.

$$G = \frac{d(U_E - W)}{dA} = \frac{d(U_E - W)}{2dc} \tag{3}$$

Where it is suppose a unitary depth of the crack and a length dc. The number 2 is because there are two surfaces created by crack.
The term "G" means the energy supplied per unit area as the elastic capacity of the body to create new surfaces (those of the cracks). It requires a continuous supply of stored elastic energy in the body to continue the propagation of a crack once initiated. From body parts with smaller loads comes the energy supply, although the main contribution is from the

stress concentration areas that cause local increase of the stress. Until several years ago, G was the parameter used to measure the "toughness" and then was replaced by K, the "intensity of stress". G may be considered (dimensionally) as a force (supplied by the body) per unit length of the crack, or force of crack propagation. Applying the equations (2) and (3) we have:

$$G = -\frac{\pi\sigma^2 c}{E} \tag{4}$$

Where:
σ - Tensile or bending stress of the material.
c - Semiaxis of the crack (assumed elliptical) in the direction perpendicular to the stress
E – Elasticity modulus
Inglis assumed that the difference between the elastic energy stored in the body and dissipated by the crack plus the external work exerted on the body by the applied normal stress is proportional to the elastic energy contained in a circle of radius c (2a) as can be seen in Figure 5.

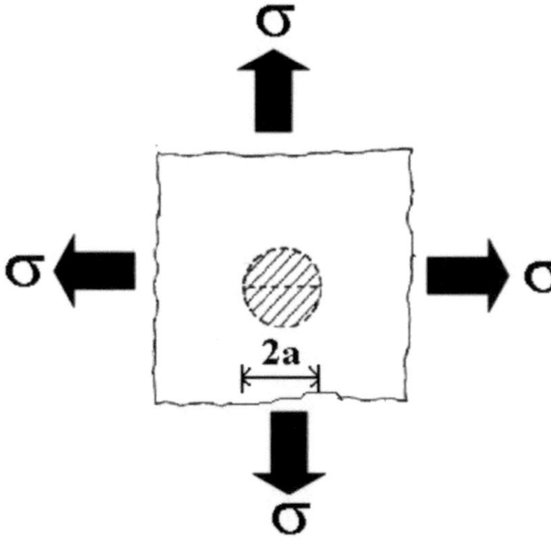

Fig. 5. Applied normal stress is proportional to the elastic energy contained in a circle of radius c (2a).

2.3 Cracking modes and the concept of fracture toughness (K_{IC})

The stress-intensity Factor (K) is a quantitative parameter of fracture toughness determining a maximum value of stress which may be applied to a specimen containing a crack of a certain length.
Depending on the direction of the specimen loading and the specimen thickness, three types of stress-intensity factors are used: K_{IC} K_{IIC} K_{IIIC}.
All the stress fields near cracks can be deduced from the three load shapes that cause the formation of three types of cracks, called "ways of cracking". As shown in Figure 6, the mode I is the crack opening by traction. The II is sliding on a plane and the third involves a lateral movement or "tearing" of the material.

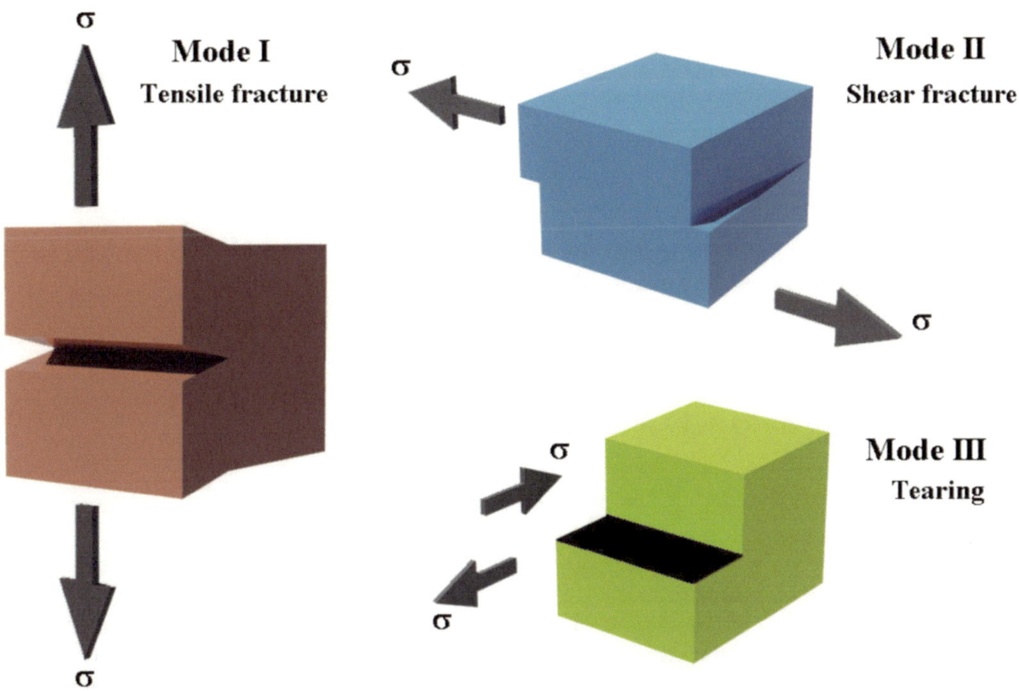

Fig. 6. Fracture modes.

The fracture toughness (K_{IC}) is a "threshold property" so that (theoretically) above or below face value there is or not crack propagation, respectively. Thus, from the standpoint of stress, this is the "stimulus" and the crack is the response to stimulation. The fracture toughness (or, ductility) enables an adequate redistribution of effort, as it is expected that local efforts have higher than average, and if local plastic flow would be possible for another part of the structure without stress bear so great, absorb the load. Hence, the importance in materials technology for finds procedures to increase K_{IC}. The fracture toughness depends on the elastic energy dissipated (Wang, 1996). Therefore, if the material has a high ability to dissipate elastic energy during crack propagation, without a catastrophic failure, we can say that the material has high fracture toughness.
Since the critical value of K is K_{IC} in the crack propagation mode I, one can expect a parabolic proportionality of K in relation to E, so that K increases with E, giving the equation 5. In fact in ceramics there is an erratic behavior and in some cases there was no simultaneous increase in K_{IC} and E. This may be due to the presence of other mechanisms such as microplasticity and the second phase substances (Wessel, 2004).

$$K_{IC} = \sqrt{2\gamma_s E} \quad (5)$$

The crack propagation occurs when G (ec.4) reaches a critical value Gc, which is equal to dU_S/dA that contains the term dissipation of energy by the formation of new surfaces in a material and is the "resistance" or opposition to it. Also known as R. Therefore, the crack propagation occurs when G becomes R, is to mean when is achieved and exceed a critical value. Obtaining a metastable equilibrium.

$$G_c = R \quad (6)$$

Assuming that σ_c and C_c are the stress and the half of crack length at the critical point respectively, we obtain:

$$R = \frac{\pi\sigma_c^2 c_c^2}{E} \quad (7)$$

In this case the crack propagates, but the increase in length is unstable and the total energy U [equation (1)] decreases. In contrast, if a stress is applied $\sigma 1 < \sigma$ cr then

$$G = \frac{\pi\sigma_1^2 c_{cr}}{E} \quad (8)$$

The crack does not propagate because

$$\frac{dU}{dc} < 0 \quad (9)$$

2.4 Use of hardness to characterize toughness and fragility (Correlations)

A systematic exposition of the expose above, allows to realize a conceptual connection with the measurement of another properties from hardness (H). It is possible to correlate the measurement of the hardness of the ceramic samples with different microstructural characteristics and properties such as fracture toughness, the extent of the crack and the elastic modulus. A possible sequence

1. The stress field is formed from:
   a. The applied load
   b. The stress field in the volume of the sample around the indented area that responds to conditions of elasticity (dependent of Young's modulus, E) or plastic flow of material (microplasticity) around the indented area. Since there are stresses when the load is removed, they are called residuals.
2. The maximum stress due to the application of the load in the volume under the area of indentation occurs at the interface between the elastic and plastic zone, which in turn creates microcracks that depend on the population of imperfections surface and nucleation mechanisms of the sliding planes of the material.
3. On the surface the indenter causes compression and not tension and acts not opposing to residual stresses.
4. When is retired the indenter the compression on the surface decreases to zero.
5. Residual stresses (which have no opposition and are of the order Hv/20) acting against the surface and cause radial cracks on the surface, visible under the microscope on slick surfaces. These are radial cracks.
6. The radial and meridional cracks combine to form semi-elliptical crack surface and their diameter is about twice the depth of the crack.
7. Radial cracks fully developed are in mechanical equilibrium and their dimensions are determined from the K_{IC}. Thus this let to measure the K_{IC}.

Lawn and other researchers (Marshall & Lawn, 1986) formulated the view that residual stresses do not contribute to a specific factor to the fracture toughness but instead affect the length 2c of the cracks (see Figure 7). The basic equations that determine this parameter and hardness are:

$$H_V = \frac{P}{\alpha a^2} \tag{10}$$

$$K_c = \frac{P}{\beta c^{2E}} \tag{11}$$

Where:
α, β - constants
a - half length of the indentation diagonal
c - half length of the crack generated by indentation
E – elasticity modulus
P - applied load in the sample for hardness testing
K_C - Fracture Toughness

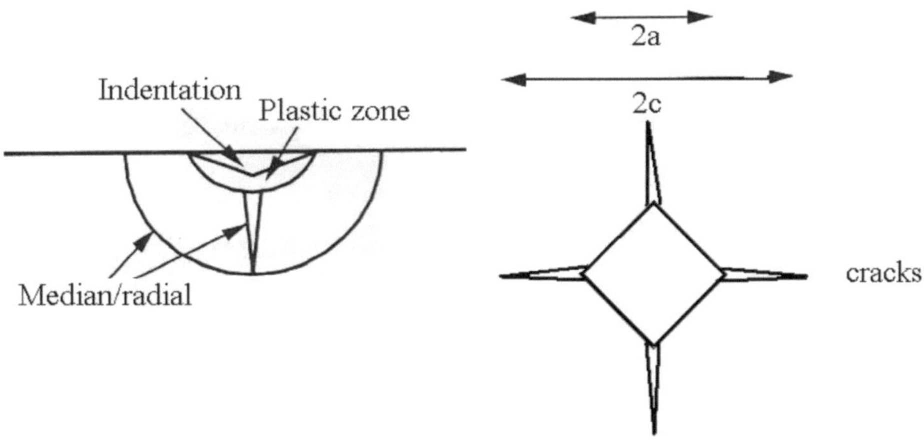

Fig. 7. Cracking around hardness indentation.

α depends on the geometry of the indenter and in the case of those are Vickers type its value is 2. On the other hand β is a constant but with more complex geometry that considers the effects of interaction between the free surface of the specimen and the cracks generated configurations. Its value is obtained experimentally.
When c >> a, the cracks centers can be considered at the point of the footprint of the apex of the indenter and the K_r contribution of residual stress to K_C is expressed as:

$$K_Y = \frac{\beta_T P}{C^{Z,2}} \tag{12}$$

In this equation βr is a constant proportional to $\sqrt{\frac{E}{H_V}}$. Since the ceramic with weak bonds such as: ionic or ionic-covalent, have lower values of E compared to those with covalent bonds, present also less hardness and less residual stresses too. The fracture toughness is also lower, as seen in all equations use "general" to be presented below.

When mechanical equilibrium is reached and no further propagation of cracks during the loading or after it was removed then:

$$K_\gamma = K_C = \frac{\beta_T P}{C^{1E}} = a_r \sqrt{\frac{E}{H_\lambda} \frac{P}{C^{1E}}} \tag{13}$$

Applying equation (10):

$$K_C = a_r \sqrt{\frac{E}{H_\lambda} \frac{H_\gamma a a^2}{C^{1E}}} = 0.032 \sqrt{\frac{E}{H\gamma} \frac{H\gamma a^2}{C^{1E}}} \tag{14}$$

Where $\alpha_r = 0.016 \pm 0.004$ y $\alpha = 2$ as was specified earlier.
All formulas are referring to this result, although some minor changes were made in the coefficient or in the exponent to adjust the results to the experimental values. The original formula is of Lawn.

$$K_C = 0{,}028 \sqrt{\frac{E}{H_V}} HVa^{\frac{1}{2}} (\frac{c}{a})^{-1{,}5} \qquad \text{(Lawn)(15)}$$

Another well-known formula is the (Evans and Charles, 1976).

$$K_{IC} = 0.16\,(c/a)^{-1.5}\,(Ha^{1/2}) \qquad \text{(Evans y Charles)(16)}$$

(Niihara, 1982) established that:

$$K_C = 0{,}067 \left(\frac{E}{H_V}\right)^{0{,}4} H_V \alpha^{0{,}5} \left(\frac{c}{a}\right)^{-1{,}5} \qquad \text{(Niiihara) (17)}$$

(Antis et al., 1981) did not replace the value of P as a function of HV and found this equation that is virtually the same of Lawn:

$$K_c = 0{,}016 \left(\frac{E}{H_V}\right)^{0{,}5} \frac{P}{c^{1{,}5}} \qquad \text{(Antis)(18)}$$

A final equation is that of Bhat, 2006)

$$K_{IC} = 1{,}59x10^5 E^{0{,}4} P^{0{,}6} a^{-0{,}7} \left(\frac{c}{a}\right)^{-0{,}36} \qquad \text{(Bhat)(19)}$$

In the equation of Bhat elasticity modulus E is expressed in Pa, the load P in grams and both a and c in microns. The result is given in MPa $m^{0.5}$ and it is assumed $c/a > 3$.

3. The indentation fracture (IF) method

As has be mentioned earlier the indentation fracture method, is derived from the experimental procedure commonly followed in hardness tests, it consists in relate the lengths of the cracks shown in Figure 8, growing in the corners of the Vickers indentation when a load (P) is applied, with the toughness of the material.

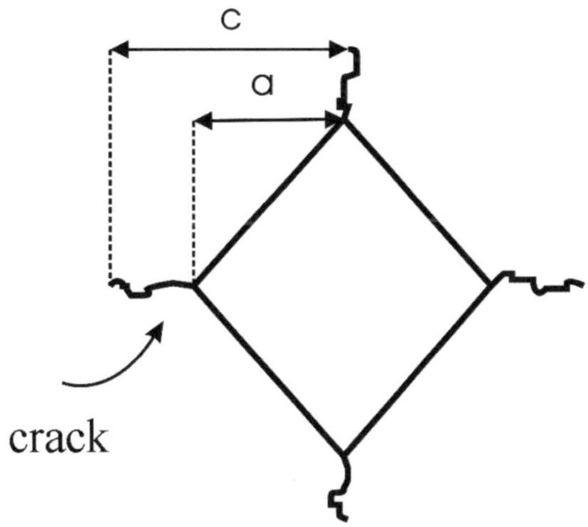

Fig. 8. Vickers indentation mark.

To calculate K_{IC} by this method they have developed a number of equations, some of which require the values of Young and Poisson modulus for their use in addition to the hardness test results. The equations are divided into two groups: empirical and experimental. One of the most used among the empirical group is the equation (16) proposed by Evans. While equation (17) proposed by Niihara is one of the most experimental frequently used

$$K_{IC} = 0.16\ (c/a)^{-1.5}\ (Ha^{1/2}) \qquad \text{(Evans y Charles) (16)}$$

$$K_C = 0,067\left(\frac{E}{H_V}\right)^{0,4} H_V \alpha^{0,5}\left(\frac{c}{a}\right)^{-1,5} \qquad \text{(Niihara) (17)}$$

Also:

$$H = 1.8P/a^2 \qquad (20)$$

Where:
K_{IC} = Fracture toughness (MPa m$^{1/2}$)
H = Vickers hardness (MPa)
E = Young modulus (MPa)
P = Test load in Vickers hardener (MPa)
c = Average length of the cracks obtained in the tips of the Vickers marks (microns)
a = Half average length of the diagonal of the Vickers marks (microns)

3.1 Applications of the IF method

With the idea to observe the application of the IF method investigations about the production of Al_2O_3-based composites with different reinforcement metals and intermetallics have been carried out. The synthesis of composites materials has been made by means of both; liquid and solid pressureless sintering of an intensive mechanical mixture of powders.

3.2 Experimental procedure

The starting raw material were powders of Al_2O_3, FeO, NiO, TiO_2 and ZrO_2 (99.9 %, 1 μm, Sigma, USA) and powders of aluminum, cobalt, copper, iron, molybdenum, nickel, titanium and zirconium (99.9 % purity, 1-2 μm, Aldrich, USA). For the composites reinforced with pure metals the amount of powders used was one that allowed obtaining Al_2O_3-based composites with 10 vol. % of the respective metal. For the composites reinforced with intermetallics, they were used as oxides of the respective metal for the in situ synthesis of the corresponding intermetallic phase. The sum of the starting materials; was fitted to the necessary amounts to form the products indicated in reaction (21) with 10 vol. % of each intermetallic phase.

$$Me_xO_y + Al \rightarrow Me_xAl_y + Al_2O_3 \quad (21)$$

Where: Me can be any of the next metals; Fe, Ni, Ti or Zr.

The processing and characterization of the composites were as follows: The weighted powders were put under a process of dry mix-milling at a speed of rotation of 300 rpm for 12 h, with the help of a horizontal mill (Cole Parmer, Labmill) using as milling elements balls of stabilized ZrO_2 (YSZ), the weight ratio of balls/powders was of 25:1. The powder mixtures were then fabricated into ten cylindrical samples of each composition with dimensions of 20 mm in diameter and 3 mm in thickness; this was done by uniaxial pressing of up to 200 MPa. The pressed samples were sintered in an electrical furnace (Carbolite, 1700) without the application of pressure at 1500 °C for 1 h in an inert atmosphere. The speeds of heating and cooling remained constant and were 10 $Kminute^{-1}$. The characterization of sintered products was carried out in the following way; the density was evaluated by the Archimedes' method, the hardness was measured with the help of Vickers indenter, the fracture toughness was determined by the method of fracture by indentation using the equation of Evans. Reported values are the average of ten measurements. The microstructures of the composites were observed with the help of a scanning electron microscope (SEM). The SEM was equipped with energy dispersive X-ray spectrometer (EDX) with which the phases present in the microstructure could be identified.

3.3 Results and discussion

3.3.1 Microstructure

Figure 9. Shows typical microstructures obtained by SEM of some of the composites investigated. Here, it can be seen fine and homogeneous microstructures, with the presence of two phases, on the basis of (EDX) analysis, it is deduced that the gray phase corresponds to the alumina matrix and the small white and brighter phase corresponds to the metallic reinforcement added to the ceramic matrix. The metallic phase is localized principally at intergranular positions. The main metallic particle size is on average 1 μm. In general all microstructures are fine, however the use of Ni, Ti and the corresponding intermetrallics help to obtain the finest microstructures in the composites. Judging from the trend disclosed by the Al_2O_3/intermetallic composites, it can be noticed that the microstructures have no cracks or pores, thus suggesting that the in situ formation of the intermetallics did not just occur, but in addition helped in the diffusion process in order to obtain well consolidated bodies. Image analysis performed on all the samples studied showed that the average volume fraction of the metallic phase in the composites was approximately 9.5%.

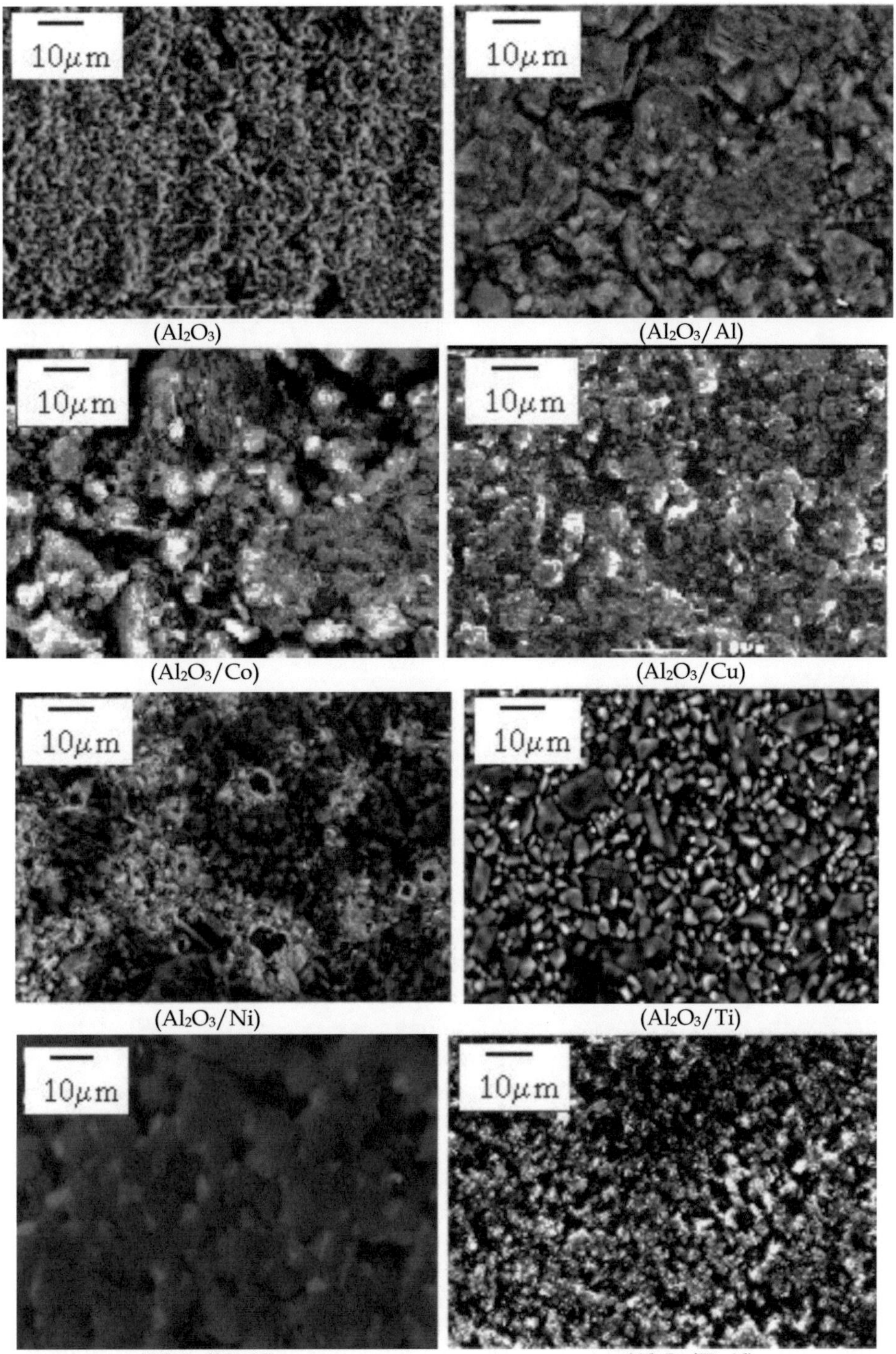

Fig. 9. Typical microstructures obtained by scanning electron microscopy of some of the composites investigated here.

The values of density, hardness and fracture toughness evaluated in the composite materials fabricated here are reported in Table 1. In this table also are reported the corresponding values for monolithic Al_2O_3 also processed here.

Reinforced Metal	ρ relative (%)	HV (GPa)	K_{IC} ($MPa{\cdot}m^{-1/2}$)
Al_2O_3	94.95 +/- 1.2	20.97 +/- 1.7	3.2 +/- 0.2
Al	89.01 +/- 0.88	18.62 +/- 1.3	4.1 +/- 0.1
Co	96.64 +/- 0.79	18.61 +/- 1.4	4.3 +/- 0.1
Cu	93.32 +/- 0.91	18.90 +/- 1.2	4.4 +/- 0.1
Fe	92.82 +/- 1.10	18.51 +/- 1.5	4.0 +/- 0.1
Mo	89.17 +/- 0.93	19.03 +/- 1.3	4.1 +/- 0.1
Ni	96.35 +/- 0.80	18.11 +/- 1.4	4.7 +/- 0.1
Ti	98.25 +/- 0.83	18.17 +/- 1.5	4.8 +/- 0.1
Zr	92.59 +/- 0.88	19.10 +/- 1.6	4.2 +/- 0.1
Fe_3Al	95.40 +/- 0.94	18.78 +/- 1.2	5.2 +/- 0.2
Ni_3Al	98.30 +/- 0.97	16.43 +/- 1.4	6.9 +/- 0.2
Ti_3Al	98.52 +/- 1.10	16.10 +/- 1.6	7.3 +/- 0.2
Zr_3Al	98.76 +/- 0.89	18.12 +/- 1.5	7.0 +/- 0.2

Table 1. Values of relative density, hardness and fracture toughness of the different Al_2O_3-based composites fabricated here.

3.3.2 Density

From this table it can be observed that the composite materials reinforced with Al, Cu, Fe, Mo and Zr display a lower relative density than monolithic Al_2O_3, whereas the composite materials reinforced with the other metals (Co, Ni and Ti) including all the intermetallics used, show better densification than monolithic Al_2O_3. The worst densifications were obtained in composites with Al and Mo, this may be due to the big differences in densities and melting points between these two metals in comparison with the corresponding values of monolithic Al_2O_3. This difference provokes poor diffusion during the sintering stage, leading to heterogeneous microstructures and in consequence bad densification of the products. For the cases were good densifications were obtained, as well as for the cases where in situ intermetallics were formed, in addition the reactions allowed some diffusion mechanisms to be activated during the process helping the densification of the products. The densification of the reinforced sample with titanium was very good, and it was equivalent to the densification obtained with the intermetallics. This was due probably to the close relation between the densities of titanium, the intermetallics and Al_2O_3, a situation than helps atomic movement during the sintering.

3.3.3 Hardness

With respect to the hardness results, from table 1 it can be seen that for all the systems monolithic Al_2O_3 is the hardest material. All the composite materials present hardness values between 18 and 19 GPa that are less than the almost 21 GPa reported for monolithic Al_2O_3. This is logical because a ceramic material has to be harder than the same ceramic material with the incorporation of ductile phases in its bulk volume.

3.3.4 Fracture toughness

From Table 1 and Figure 10 it can be observed that in all the composite cases the fracture toughness of monolithic Al_2O_3 was improved considerably, principally in composites reinforced with Ni and Ti and in all composites reinforced with intermetallic phases. The incorporation of ductile metal particles in the ceramic matrix enhances the fracture toughness due to plastic deformation of the metallic phase, which forms crack-bridging ligaments when a crack grows in the material under a tensile stress action. In other words, the energy absorbed for plastic deformation is unavailable for crack extension. Additionally, the deformed particles could bridge the faces of the crack wake, thereby exerting closure stresses, reducing the effect of the stress intensity at the crack tip (Ji & Yeomans, 2002; Lalande, 2003).

For the case of Al_2O_3/Ni system: nickel provides a liquid phase during the sintering stage that promotes diffusion and therefore densification of the composite On the other hand, Ni helps to refine the alumina microstructure by pinning its grain boundaries and thereby restraining the grain growth of alumina.

For the case of the Al_2O_3/Ti system: because the densities of titanium and alumina are very similar, Ti is well dispersed in the alumina matrix, forming a good homogeneous composite microstructure that promotes diffusion and densification, and as a consequence good toughening of the final material.

For the case of Al_2O_3/intermetallics systems: the use of intermetallics as reinforcement in Al_2O_3 gives an appreciable enhancement in the fracture toughness, this is due to the good ductility, low density and chemical compatibility of intermetallics with alumina. These factors help to obtain homogeneous microstructures with the formation of interfaces that allow the activation of different diffusion mechanisms thus improving the final density and then the mechanisms that improve the fracture toughness of the composites.

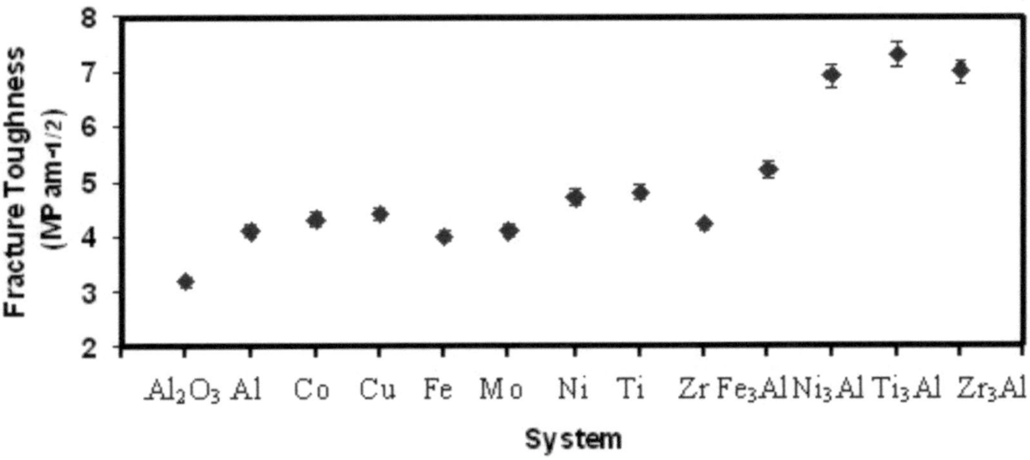

Fig. 10. Fracture toughness values measured for all the composites investigated.

Figures 11a and 11b show the fracture surface and the advance of a crack in an Al_2O_3/Ti reinforced composite. The fracture mode in figure 11a corresponds to microvoid coalescence as suggested by the dimple-like depressions that are typical of ceramic materials. From figure 11b it can be observed that the sample exhibits a mixed fracture mode, because metallic particles bridge the surface of the crack in the composite, but at the same time they can cause deflection of the crack. So the toughening mechanism in Al_2O_3/metal reinforced composites is due to crack bridging and crack deflection in this type of material. Steinbrech has reported that the improvement achievable in reinforced composites is governed by the mechanical properties of the ductile material, ligament diameter, volume fraction of the components, interfacial properties and the reaction products of the constituents (Steinbrech, 1992). This can explain the differences obtained in the fracture toughness of the materials investigated here.

(a) (b)

Fig. 11. (a) Fracture surface and (b) advanced of a crack in an Al_2O_3/Ti reinforced composite.

4. Conclusions

Al_2O_3-based composites reinforced with different metals have been fabricated by both; liquid and solid pressureless sintering of an intensive mechanical mixture of powders. By the use of ductile particles in a hard ceramic matrix, significant improvements in fracture toughness due to plastic deformation of the metallic phase has been obtained. However, there are metals that enhance the toughness of a ceramic better than others; these are those metals that have similar densities to alumina, because they help to obtain fine and homogeneous microstructures after sintering. From the fracture toughness measurements and microstructural observations, finally it can be commented that the toughening mechanism in Al_2O_3/metal reinforced composites is due to crack bridging and crack deflection.

Through fabricated composite-materials it was possible to analysis the application of the indentation fracture (IF) method for the determination in a simple and reliable way the fracture toughness of those materials. From the obtained results and with the help of bases given in the literature it can be commented that the main requirements to be met for samples that will be applied the IF methods for their fracture toughness determination are:

- Porosity fine and well distributed.
- Homogeneous microstructure.
- Good surface finish, free of residual stresses, pores and cracks.
- Parallel surfaces

5. Acknowledgements

Author would thank to the mixed foundation of promotion to the scientific and technological investigation from CONACyT-Tamaulipas Government, by the support offered for the accomplishment of the research work and its diffusion.

6. References

Anstis G. R., Chantikul P., Lawn B. R., and Marshall D. B. (1981). A Critical Evaluation of Indentation Techniques for Measuring Fracture Toughness: I, Direct Crack Measurements, *Journal of the American Ceramic Society*, Vol. 64, No. 9, (September 1981) pp. 533–538.

ASTM E1820 - 09e1 Standard Test Method for Measurement of Fracture Toughness. (2008) *Annual Book of ASTM Standards*, (2008).

ASTM C 1327-99, Standard Test Method for Fractute Toughness at Room Temperature of Advanced Ceramics (1999). *Annual Book of ASTM Standards*, (1999), pp. 14-02.

Bhat S. (2006). Fracture parameters estimation of alloy steel reinforced with maragin steel. *Fracture of Nano and engineering Materials and Structures, Proceedings of the 16th European Conference of Fracture*, Alexandroupolis Grecce, Edited by Gdoutos E. G. (July, 2006), pp.

Boch P., Chartier T. and Giry J. P. (1990). Zirconio Toughened Mullite / The Role of Circón Dissociation, *Ceram. Trans., Mullite and Mullite Matrix Composites*, edited by Somiya S., Davies R. F. and Pak J.A. Vol. 6, (1990), pp. 473-494.

Evans A. G. and Charles E. A. (1976). Fracture Toughness Determination by Indentation, *J. Am. Ceram. Soc.*, Vol. 59, (1976), pp. 371-372.

Griffith A. A. (1921). The Phenomena of Rupture and Flow in Solids, *Phil. Trans.* R. Soc. Lond. (January, 1921) pp. 163-198.

Ji Y. and Yeomans J. (2002). Processing and mechanical properties of Al_2O_3-5 vol.% Cr nanocompistes. *J. Eur. Ceram. Soc.*, Vol. 22, No. 12 (2002) pp. 1927-1930.

Konopka K. and Szafran M. (2006). Fabrication of Al_2O_3-Al composites by infiltration method and their characteristics. *J. Mater. Proc. Technol.*, Vol. 175, (2006), pp. 266-270.

Lalande J., Scheppokat S., Jansen R. and Claussen N. (2002). Toughening of alumina/zirconia ceramic composites with silver particles. *J. Eur. Ceram. Soc.*, Vol. 22, No. 13 (2002) pp. 2165-2168.

Lieberthal M. I. and Kaplan K. (2001). Processing and properties of Al_2O_3nanocomposites reinforced with sub-micron Ni and $NiAl_2O_4$. *Mater. Sci. Eng.*, Vol. A302, No. 1 (2001), pp. 83-91.

Marci C. and Katarzyna P. (2007). Processing, microstructure and mechanical properties of Al_2O_3-Cr nanocomposite *J. Eur. Ceram. Soc.*, Vol. 27. No. 2-3, (2007), pp. 1273-1277.

Marshall D. B. and Lawn B. R. (1986). Indentation of Brittle Materials", *Microindentation Techniques in Materials Science and Engineering, ASTM* Vol. 889, (1986), pp. 26-46.

Miranda Hernández J. G., Moreno Guerrero S., Soto Guzmán A. B. and Rocha Rangel E. (2006). Production and Characterization of Al_2O_3-Cu Composite Materials. *J. Ceram. Proc. Res.*, Vol. 7, No. 4, (2006), pp. 311-315.

Niihara K., Morena R. and Hasselman D. P. H. (1982). Evaluation of K_{IC} of Brittle Solids by The Indentation Method with Low Crack-To-Indentation Ratios, *J. Mater. Sci. Lett.*, Vol. 1, (1982), 1, pp. 13-16.

Plaza L. M. (2003). Determination of Uncertaines in Plane Tougness (K_{IC}) Testing, *Seminario Sobre la Evaluación de la Incertidumbre en Ensayos Mecánicos*, Universidad de la Rioja, España, (2003).

Steinbrech R. W. (1992). Toughening mechanisms for ceramics materials. *J. Eur. Ceram. Soc.* Vol. 10, No. 3 (1992) pp. 131-135.

Sglavo V. M., Marinob F., Zhang B. R. and Gialanella S. (1997). Ni_3Al intermetallic compound as second phase in Al_2O_3 ceramic composites. *Mater. Sci. Eng.*, Vol. A239-240, (1997), pp. 665-669.

Travirskya N., Gotmanb I. and Claussen N. (2003). Alumina-Ti aluminide interpenetrating composites: microstructure and mechanical properties. *Mater. Lett.*, Vol. 57, No. 22-23, (2003), pp. 3422-3424.

Wang C. H. (1996). Introduction to Fracture Mechanics, *DSTO Aeronautical and Maritime Research Laboratory*, (July, 1996), pp. 199-207.

Weisbrod G. and Rittel D. A. (2000). Method for Dynamic Fracture Toughness Determination Using Short Beams, *International Journal of Fracture*, Vol. 104, (2000), pp. 1-5.

Wessel J. K. (2004). *The Handbook of Advanced Materials*. John Wiley & Sons, USA (2004).

3

Techniques for Identification of Bending and Extensional Elastic Stiffness Matrices on Thin Composite Material Plates Based on Virtual Field Method (VFM): Theoretical and Numerical Aspects

Fabiano Bianchini Batista and Éder Lima de Albuquerque
Federal University of São João Del Rei
University of Brasília
Brazil

1. Introduction

Nowadays, the demand and the necessity for the use of materials with specific characteristics have increased in many engineering fields. Due to this necessity of making new materials, composite materials have been an alternative, or maybe the unique option, to attempt a large number of design requirements such as high strength-to-weight ratio, high resistance to mechanical shocks, chemical attacks, corrosion, and fatigue, that cannot be obtained only from the commonly used structural materials (metals, ceramics, polymers and wood). Because of this, their applications are present in the main industries such as aerospace, automotive, marine, and sportive

Thanks to their flexibility characteristics there are many combinations and arrangements and, consequently, constitutive properties, that are possible to be achieved. This particularity represents one of the main advantages of these materials. Nevertheless, some factors related to arrangements as the number of layers and the orientation of the fibers can introduce a behavior called anisotropy that, in the most of the cases, is not required. The anisotropy makes the structural analysis more complex due to increasing the number of independent variables, as for example, the number of bending and extensional elastic stiffness constants.

Recently, wide part of works presented by scientific literature whose goal is to identify constitutive parameters of materials (these being composites or not) is based in the called "inverse problems". Experimental data such as geometry, resultant forces and strain (or stress) fields are used as input data, and the unknown variables are the required constitutive parameters. In general, the solution is basically associated to two methods: iterative (or also called indirect methods) and non-iterative (or also called of direct methods). The first one is related to optimization problems where the design variables are constitutive parameters and the objective function represents, in general, a residue (or error) between experimental and numerical (generally obtained by finite elements) data. For the numerical simulations, it is considered structures that have the same geometrical characteristics and boundary conditions of the real ones. For each step, required parameters are checked out, and, the optimum represents the iteration whose residue (objective function) has its lowest value. Unlike indirect methods, the direct methods are ones where the required parameters are computed from the solution of constitutive equation(s) that are functions of these parameters.

In general, according to the type of experimental test, it is possible to separate the methods of elastic property identification in two categories: static (destructives and non-destructives) and dynamic methods (non-destructive), as shown in Fig.1. A large number of identification techniques that use data from these categories of tests have been proposed, especially ones dedicated to composite materials. It is possible to say that these techniques identify effective properties of the entire material. The way as each formulation is built, and, the adopted procedures and devices are the main differences among the many proposed methodologies.

Static tests with monotonic load are experimental tests that were more commonly used in the last years, and maybe the simplest ones, for this material property identification. Despite the simplicity of these tests, some aspects render them less attractive than dynamic tests, such as the fact of requiring a number of samples with fiber orientations according to standard norms, for example, American Society for Testing and Materials (ASTM), which, in the most of cases, are not in accordance to the real characteristics of the required material. Furthermore, some variables difficult of controlling during the tests can contribute to worsen the experimental results, e.g., the presence of non-uniform stress fields near the ends of the sample from the clamped boundary conditions. For these reasons, dynamic tests have been considered an interesting alternative. In general, they are tests that combine experimental data with numerical methods, and allow the identification of elastic constants from only one unique sample or even from composite material part. Sample is usually thin plate (that reflect Kirchhoff's hypotheses), cylindrical shell, or beam. In many cases, input data of the numerical methods are natural frequencies and/or mode shapes.

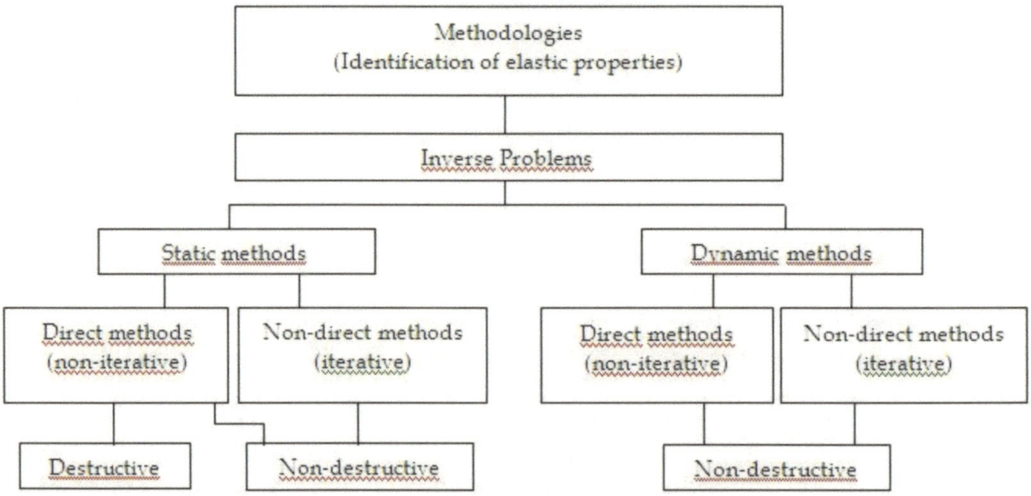

Fig. 1. Methodologies more used to identify elastic properties of materials.

Many authors have proposed to identify the elastic constants by iterative procedures adopting Rayleigh-Ritz and finite element as numerical methods. The difference among the identification techniques based on these iterative procedures is basically in the way as the optimization problem is formulated, for example, the type of adopted search method to find the minimal, the boundary conditions, the geometric characteristics of the samples, the type of anisotropy of the test material, the type of experimental devices, and the numerical method used to compute the mode shapes (or operational modes) with their respective frequencies (Deobald & Gibson, 1988; Pedersen & Frederiksen, 1992; Lai & Lau, 1993; Ayorinde & Gibson, 1995; Rikards & Chate, 1998; Ayorinde & Yu, 1999, 2005; Rikards et al., 1999; Bledzki et al., 1999; Hwang & Chang, 2000; Araujo et al., 2000; Chakraborty & Mukhopadhyay, 2000; Rikards et al., 2001; Lauwagie et al., 2003; Lauwagie et al., 2004; Lee & Kam, 2006; Cugnoni et al., 2007; Bruno et al., 2008; Pagnotta & Stigliano, 2008; Diveyev & Butiter, 2008a, 2008b).

In works that do not use iterative process, natural frequencies and mode shapes, or operational frequencies and modes, are input data of an algorithm based on the differential equation that governs the transversal vibration of sample in a specific direction and under specific boundary conditions (Gibson, 2000; Alfano & Pagnotta, 2007). In this methodology, it can be included the use of Virtual Fields Method, VFM (Grédia, 1996, 2004; Grédiac & Paris, 1996; Grédiac & Pierron, 1998, 2006; Pierron et al., 2000, 2007; Pierron & Grédiac, 2000; Grédiac et al., 1999a, 1999b, 2001, 2006; Giraudeau & Pierron, 2003, 2005; Chalal et al., 2006; Toussaint et al., 2006; Avril & Pierron, 2007; Pierron et al., 2007; Avril et al., 2008; Giraudeau et al., 2006). For VFM, weighting functions are the called virtual fields. Due to sensitivity to experimental errors and the presence of noise during the dynamic testing, it was proposed the use of specific virtual fields named "special virtual fields" (Grédiac et al., 2002a, 2002b, 2003). In order to decrease the noise contribution, the use of more accurate experimental modal analysis techniques or the application of some signal smoothing (or filtering) technique is mandatory.

The majority of works identifies only the bending stiffness matrix or directly the engineering elastic constants. However, the extensional elastic stiffness matrix is also needed to model composite materials under multi-axial loads. In general, these stiffness matrices are independent. The extensional stiffness matrix relates the in-plane resultant forces to the midplane strains, and, the bending stiffness matrix relates the resultant moments to the plate curvatures. In a laminate composite, if only the stacking sequence of layers is changed, the bending matrix is changed but the extensional matrix remains the same. In other words, different laminates can have different bending stiffness matrices and the same extensional stiffness matrix. It is not possible to obtain the extensional matrix from the bending matrix. It will be possible only if the stacking sequence of layers and their thickness are known and, also, if the material is the same for all lamina.

Sometimes, it is more convenient to use effective laminate engineering constants rather than the laminate stiffness. These effective laminated engineering constants may be easily obtained from the extensional elastic constants. However, due to difficulties on experimental in-plane modal analysis, such as the necessity of using specific devices to measure in-plane displacements and to excite high frequencies, the identification of extensional elastic stiffness constants using modal testing is less attractive. The main challenge to perform in-plane vibration testing is the excitation and measurement of only in-

plane and not out-of-plane vibration modes. Today there are some new techniques that are suitable for this kind of problems, for example, the excitation by piezoelectric (PZT) and measurements by digital image correlation.
In this chapter, a review about the VFM applied to compute bending elastic stiffness constants proposed by Grédiac & Paris, 1996 is presented. Furthermore, a formulation based on the VFM is proposed in order to identify the extensional elastic stiffness matrix of Kirchhoff's thin plates. The linear system of equations that provides the required elastic constants is obtained from differential equations that govern the forced vibration of anisotropic, symmetric and non-damped plates under in-plane loads. The common procedures to find the weak form (or integral form) of these equations are applied here. The correct choice of weighting functions (which are the virtual fields) and mode shapes represents a key characteristic to the accuracy of the results. Numerical simulations using anisotropic, orthotropic, quasi-isotropic plates are carried out to demonstrate the accuracy of the methodology.

2. Identification of elastic constants using VFM

2.1 Review of the Virtual Fields Method - VFM

The VFM has been developed for extracting constitutive parameters from full-field measurements and it is associated to problems of identification of parameters from constitutive equations. Two cases are clearly distinguished: constitutive equations depending linearly on the constitutive parameters and non-linear constitutive equations. The type of constitutive equations is chosen *a priori* for its relevancy and objective of it is to determine the parameters which govern the constitutive equations. The main difficulty comes from the fact that the measured displacement or strain components are generally not directly related to the unknown parameters (Grédiac et al., 2006), and no closed-form solution for the displacement, strain and stress fields is available.
Mathematically, the VFM is based on the principle of virtual work and can be written as:

$$-\int_V \sigma : \varepsilon^* dV + \int_{S_f} \mathbf{T}.\mathbf{u}^* dS + \int_V \mathbf{f}.\mathbf{u}^* dV = \int_V \rho\gamma\mathbf{u} dV \tag{1}$$

where V is volume of the solid, σ is the actual stress tensor, ε^* is the virtual strain tensor, $\mathbf{T}$ is the distribution vector of loading tractions acting on the boundary, S_f is the part of the solid boundary where the tractions are applied, $\mathbf{u}^*$ is the virtual displacement vector, $\mathbf{f}$ is the distribution of volume forces acting on V, ρ is the density and γ the acceleration. Eq. (1) is verified for any kinematically admissible virtual field ($\mathbf{u}^*$, ε^*). Kinematically admissible means that $\mathbf{u}^*$ must be continuous across the whole volume and it must be equal to the prescribed displacement on the boundary S_f where displacements are prescribed. Let's introduce the constitutive equations in the general case as:

$$\sigma = g(\varepsilon) \tag{2}$$

where g is a function of the actual strain. Thus, when constitutive equations are introduced and volume forces are disregarded, Eq. (1) can be rewritten as:

$$-\int_V g(\varepsilon) : \varepsilon^* dV + \int_{S_f} \mathbf{T}.\mathbf{u}^* dS = \int_V \rho\gamma\mathbf{u}^* dV \,. \tag{3}$$

It is possible to see in Eq. (3) that each virtual field originates a new equation involving the constitutive parameters. The VFM relies on this important property. It is a method based on setting virtual fields that provide a set of equations. This set of equations is used to extract the required unknown constitutive parameters. The correct choice of the virtual fields that combine to actual fields in Eq. (3) is the key issue of the method. Their number and their type depend on the nature of g in Eq. (3).

2.2 Review of the identification method of bending elastic stiffness matrix

The method proposed by Grédiac & Paris, 1996, consists of obtaining elastic constants based on the partial differential equation that governs the transversal vibration of an anisotropic thin plate (Kirchhoff's plate). This equation is given by:

$$D_{11}\frac{\partial^4 w}{\partial x^4}+4D_{16}\frac{\partial^4 w}{\partial x^3 \partial y}+2(D_{12}+2D_{66})\frac{\partial^4 w}{\partial x^2 \partial y^2}+4D_{26}\frac{\partial^4 w}{\partial x \partial y^3}+D_{22}\frac{\partial^4 w}{\partial y^4}=-\rho h\frac{\partial^2 w}{\partial t^2}, \quad (4)$$

where D_{ij} are thin plate bending stiffness constants; ρ is the mass density of the material; h is the plate thickness; x and y are coordinates of the plate; t is time; and $w(x,y,t)$ is the deflection function that represents the transversal displacement of a point of the plate at an instant t. Eq. (4) doesn't state the global equilibrium of the plate since the excitation force and damping are not considered. However, for many composite materials, as for example, aeronautic carbon epoxy tested in this work, the damping is low enough to disregard its contribution in the formulation. Besides, if the input data refer to resonant response of the plate, the work provided by the excitation is balanced by internal dissipation of the plate. A detailed discussion about when excitation and damping should be considered in Eq. (4) can be found in Giraudeau & Pierron, 2006.

After some mathematical manipulations in Eq. (4), Grédiac & Paris, 1996 obtained a linear system in which the unknown variables are the elastic constants. Briefly, the sequence of operations is as follows: (a) multiply both sides of Eq. (4) by an arbitrary weighting function; (b) integrate twice by parts along the plate domain; (c) eliminate the boundary integrals by applying the free-edge boundary conditions; (d) decompose the displacement function $w(x, y, t)$ as a product of the deflection amplitude Φ and $\sin(\omega t)$, where ω is the out-of-plane natural frequency of a particular mode shape of the plate; and (e) choose appropriate weighting functions and mode shapes to build the matrix of the linear system. At this point, as Grédiac & Paris, 1996 explain, the choice of mode shapes associated with the weighting function is extremely important for the accuracy of this method. Three particular modes are strongly dependent on the required coefficients D_{ij}: a twisting mode that strongly depends on terms D_{66}, D_{16}, and D_{26}; a bending mode along direction 1 that strongly depends on terms D_{11}, D_{12}, and D_{16}, and a bending mode along direction 2 that strongly depends on D_{22}, D_{12}, and D_{26}. These modes present smooth curvatures and are generally among the first modes, with lower frequencies. If these modes are not found, it is recommended to use modes that have similar shapes to them. Furthermore, they are modes that can be approximated by quadratic functions with constant curvatures: x^2, y^2, and xy. For this reason, these quadratic functions were the weighting functions chosen by Grédiac & Paris, 1996. Thus, using these previous quadratic-weighting functions, the following simplified system of equations can be obtained:

$$\begin{bmatrix} \cdots & \cdots & \cdots & \cdots & \cdots & \cdots \\ K_{xx}^{(j)} & 0 & K_{yy}^{(j)} & 0 & 2K_{xy}^{(j)} & 0 \\ 0 & K_{yy}^{(j)} & K_{xx}^{(j)} & 0 & 0 & 2K_{xy}^{(j)} \\ 0 & 0 & 0 & 2K_{xy}^{(j)} & K_{xx}^{(j)} & K_{yy}^{(j)} \\ \cdots & \cdots & \cdots & \cdots & \cdots & \cdots \\ K_{xx}^{(k)} & 0 & K_{yy}^{(k)} & 0 & 2K_{xy}^{(k)} & 0 \\ 0 & K_{yy}^{(k)} & K_{xx}^{(k)} & 0 & 0 & 2K_{xy}^{(k)} \\ 0 & 0 & 0 & 2K_{xy}^{(k)} & K_{xx}^{(k)} & K_{yy}^{(k)} \\ \cdots & \cdots & \cdots & \cdots & \cdots & \cdots \end{bmatrix} \begin{Bmatrix} D_{11} \\ D_{22} \\ D_{12} \\ D_{66} \\ D_{16} \\ D_{26} \end{Bmatrix} = -\frac{\rho h}{2} \begin{Bmatrix} \cdots \\ \omega_j^2 \int_S \Phi^{(j)} x^2 dS \\ \omega_j^2 \int_S \Phi^{(j)} y^2 dS \\ \omega_j^2 \int_S \Phi^{(j)} xy dS \\ \cdots \\ \omega_k^2 \int_S \Phi^{(k)} x^2 dS \\ \omega_k^2 \int_S \Phi^{(k)} y^2 dS \\ \omega_k^2 \int_S \Phi^{(k)} xy dS \\ \cdots \end{Bmatrix}, \tag{5}$$

where indices j and k represent a specific mode shape; and S is the plate domain. Elements of the matrix of Eq. (5) are given by:

$$K_{xx} = \int_S \frac{\partial^2 \Phi(x,y)}{\partial x^2} dS\,,\ K_{yy} = \int_S \frac{\partial^2 \Phi(x,y)}{\partial y^2} dS\,,\ K_{xy} = \int_S \frac{\partial^2 \Phi(x,y)}{\partial x \partial y} dS\,. \tag{6}$$

Eq. (5) can be represented in matrix form as:

$$\mathbf{KD} = \mathbf{C}\,, \tag{7}$$

where, considering L as the number of modes used in the linear system of equations, $\mathbf{K}$ is a $3L$ x 6 matrix, $\mathbf{D}$ is a 6 x 1 matrix, and $\mathbf{C}$ is a $3L$ x 1 matrix. As can be seen, Eq. (7) is an overdetermined system of equations. The solution can be found by least squares:

$$\mathbf{D} = \left(\mathbf{K}^T\mathbf{K}\right)^{-1}\left(\mathbf{K}^T\mathbf{C}\right). \tag{8}$$

2.3 Identification method of the extensional elastic stiffness matrix

In the general case of composite laminates, each lamina is assumed to have orthotropic material properties. After the assembly, the behavior can be anisotropic due to the interaction of different laminas. Considering a plate under plane state of stress and using Hooke's generalized law, stresses can be integrated over its thickness yielding the following force-deformation equations:

$$\begin{Bmatrix} \mathbf{N} \\ \cdots \\ \mathbf{M} \end{Bmatrix} = \begin{bmatrix} \mathbf{A} & \vdots & \mathbf{B} \\ \cdots & \vdots & \cdots \\ \mathbf{B} & \vdots & \mathbf{D} \end{bmatrix} \begin{Bmatrix} \boldsymbol{\varepsilon} \\ \cdots \\ \boldsymbol{\kappa} \end{Bmatrix}, \tag{9}$$

where $\mathbf{N}$ and $\mathbf{M}$ are vectors that contain normal forces and resultant moments, respectively, $\mathbf{A}$ is the extensional elastic stiffness matrix, $\mathbf{B}$ is the coupling elastic stiffness matrix ($\mathbf{B}$ is a null matrix in the case of a symmetric laminate), $\mathbf{D}$ is the bending elastic stiffness matrix, $\boldsymbol{\varepsilon}$ and $\boldsymbol{\kappa}$ are vectors that contain middle plane linear strains and rotations, respectively. Considering a symmetrical ($\mathbf{B}$ = [0]) and fully anisotropic laminate under free-edge in-plane vibration (the plate is not under bending) and using the equilibrium relations, the following equations can be written:

$$A_{11}\frac{\partial^2 u(x,y,t)}{\partial x^2}+2A_{16}\frac{\partial^2 u(x,y,t)}{\partial x\partial y}+A_{66}\frac{\partial^2 u(x,y,t)}{\partial y^2}+A_{16}\frac{\partial^2 v(x,y,t)}{\partial x^2}+ \\ +(A_{12}+A_{66})\frac{\partial^2 v(x,y,t)}{\partial x\partial y}+A_{26}\frac{\partial^2 v(x,y,t)}{\partial y^2}=\rho h\frac{\partial^2 u(x,y,t)}{\partial t^2}, \quad (10)$$

$$A_{16}\frac{\partial^2 u(x,y,t)}{\partial x^2}+(A_{12}+A_{66})\frac{\partial^2 u(x,y,t)}{\partial x\partial y}+A_{26}\frac{\partial^2 u(x,y,t)}{\partial y^2}+A_{66}\frac{\partial^2 v(x,y,t)}{\partial x^2}+ \\ +2A_{26}\frac{\partial^2 v(x,y,t)}{\partial x\partial y}+A_{22}\frac{\partial^2 v(x,y,t)}{\partial y^2}=\rho h\frac{\partial^2 v(x,y,t)}{\partial t^2} \quad (11)$$

where A_{ij} are the elements of matrix $\mathbf{A}$ (i,j = 1,2,6); ρ is the mass density; h is the plate thickness; x and y are the coordinates in the plate plane; t is the time, and $u(x,y,t)$ and $v(x,y,t)$ are functions that represent the displacements along x and y direction, respectively, of a point with coordinates (x,y) of the plate at an instant t. Multiplying Eq. (10) by a weighting function $W(x, y)$ and integrating along the domain of the plate, we can obtain:

$$\int_\Omega\left[A_{11}\frac{\partial^2 u(x,y,t)}{\partial x^2}+2A_{16}\frac{\partial^2 u(x,y,t)}{\partial x\partial y}+A_{66}\frac{\partial^2 u(x,y,t)}{\partial y^2}+A_{16}\frac{\partial^2 v(x,y,t)}{\partial x^2}+ \right. \\ \left. +(A_{12}+A_{66})\frac{\partial^2 v(x,y,t)}{\partial x\partial y}+A_{26}\frac{\partial^2 v(x,y,t)}{\partial y^2}\right]Wd\Omega=\rho h\int_\Omega\frac{\partial^2 u(x,y,t)}{\partial t^2}Wd\Omega \quad (12)$$

where Ω is the plate domain. Using chain rule, i.e.,

$$\frac{d}{dx}(f.g)=f\frac{dg}{dx}+g\frac{df}{dx}, \quad (13)$$

where f and g are any two continuous functions, and Green's theorem, i.e.,

$$\int_\Omega\frac{\partial f}{\partial x}d\Omega=\int_\Gamma fn_x dx, \quad (14)$$

where Γ is the boundary domain, and n_x is the component of the normal unity vector in directions x, the left hand side terms of Eq. (12) can be written as:

$$A_{11}\left(\int_\Gamma\frac{\partial u}{\partial x}Wn_x d\Gamma-\int_\Omega\frac{\partial u}{\partial x}\frac{\partial W}{\partial x}d\Omega\right)+2A_{16}\left(\int_\Gamma\frac{\partial u}{\partial y}Wn_x d\Gamma-\int_\Omega\frac{\partial u}{\partial y}\frac{\partial W}{\partial x}d\Omega\right)+ \\ +A_{66}\left(\int_\Gamma\frac{\partial u}{\partial y}Wn_y d\Gamma-\int_\Omega\frac{\partial u}{\partial y}\frac{\partial W}{\partial y}d\Omega\right)+A_{16}\left(\int_\Gamma\frac{\partial v}{\partial x}Wn_x d\Gamma-\int_\Omega\frac{\partial v}{\partial x}\frac{\partial W}{\partial x}d\Omega\right)+ \\ +A_{12}\left(\int_\Gamma\frac{\partial v}{\partial y}Wn_x d\Gamma-\int_\Omega\frac{\partial v}{\partial y}\frac{\partial W}{\partial x}d\Omega\right)+A_{66}\left(\int_\Gamma\frac{\partial v}{\partial y}Wn_x d\Gamma-\int_\Omega\frac{\partial v}{\partial y}\frac{\partial W}{\partial x}d\Omega\right)+ \\ +A_{26}\left(\int_\Gamma\frac{\partial v}{\partial y}Wn_y d\Gamma-\int_\Omega\frac{\partial v}{\partial y}\frac{\partial W}{\partial y}d\Omega\right)=\rho h\int_\Omega\frac{\partial^2 u}{\partial t^2}Wd\Omega \quad (15)$$

where n_y is the component of the normal unity vector in direction y. From the constitutive equation, one has:

$$\begin{Bmatrix} N_x \\ N_y \\ N_{xy} \end{Bmatrix} = \begin{bmatrix} A_{11} & A_{12} & A_{16} \\ A_{12} & A_{22} & A_{26} \\ A_{16} & A_{26} & A_{66} \end{bmatrix} \begin{Bmatrix} \varepsilon_x \\ \varepsilon_y \\ \gamma_{xy} \end{Bmatrix}, \tag{16}$$

where N_x and N_y are axial forces per unit length along directions x and y, respectively, N_{xy} is the shear force per unit length along plane xy, ε_x and ε_y are the middle surface axial strain along directions x and y, respectively, and γ_{xy} is the shear angular strain along plane xy. Rewriting Eq. (16), one obtains:

$$N_x = A_{11}\varepsilon_x + A_{12}\varepsilon_y + A_{16}\gamma_{xy} = A_{11}\frac{\partial u}{\partial x} + A_{12}\frac{\partial v}{\partial y} + A_{16}\left(\frac{\partial u}{\partial y} + \frac{\partial v}{\partial x}\right), \tag{17}$$

$$N_y = A_{12}\varepsilon_x + A_{22}\varepsilon_y + A_{26}\gamma_{xy} = A_{12}\frac{\partial u}{\partial x} + A_{22}\frac{\partial v}{\partial y} + A_{26}\left(\frac{\partial u}{\partial y} + \frac{\partial v}{\partial x}\right), \tag{18}$$

$$N_{xy} = A_{16}\varepsilon_x + A_{26}\varepsilon_y + A_{66}\gamma_{xy} = A_{16}\frac{\partial u}{\partial x} + A_{26}\frac{\partial v}{\partial y} + A_{66}\left(\frac{\partial u}{\partial y} + \frac{\partial v}{\partial x}\right). \tag{19}$$

Multiplying Eq. (17) by a weighting function W and by n_x, which is the x component of unit normal vector **n**, and integrating on the boundary Γ, one obtains:

$$A_{11}\int_\Gamma \frac{\partial u}{\partial x} W n_x d\Gamma + A_{12}\int_\Gamma \frac{\partial v}{\partial y} W n_x d\Gamma + A_{16}\left(\int_\Gamma \frac{\partial u}{\partial y} W n_x d\Gamma + \int_\Gamma \frac{\partial v}{\partial x} W n_x d\Gamma\right) = \int_\Gamma N_x W n_x d\Gamma. \tag{20}$$

Multiplying Eq. (19) by a weighting function W and by n_y, which is the y component of the unit normal vector **n**, and integrating on the boundary Γ, yields:

$$A_{16}\int_\Gamma \frac{\partial u}{\partial x} W n_y d\Gamma + A_{26}\int_\Gamma \frac{\partial v}{\partial y} W n_y d\Gamma + A_{66}\left(\int_\Gamma \frac{\partial u}{\partial y} W n_y d\Gamma + \int_\Gamma \frac{\partial v}{\partial x} W n_y d\Gamma\right) = \int_\Gamma N_{xy} W n_y d\Gamma. \tag{21}$$

Substituting Eqs. (20) and (21) into Eq. (15), and reorganizing the terms, one can write:

$$\begin{aligned} &\int_\Gamma N_x W n_x d\Gamma + \int_\Gamma N_{xy} W n_y d\Gamma - \int_\Omega \left[A_{11}\left(\frac{\partial u}{\partial x}\frac{\partial W}{\partial x}\right) + A_{16}\left(\frac{\partial u}{\partial y}\frac{\partial W}{\partial x} + \frac{\partial u}{\partial x}\frac{\partial W}{\partial y} + \frac{\partial v}{\partial x}\frac{\partial W}{\partial x}\right) + \right. \\ &\left. + A_{12}\left(\frac{\partial v}{\partial y}\frac{\partial W}{\partial x}\right) + A_{66}\left(\frac{\partial u}{\partial y}\frac{\partial W}{\partial y} + \frac{\partial v}{\partial x}\frac{\partial W}{\partial y}\right) + A_{26}\left(\frac{\partial v}{\partial y}\frac{\partial W}{\partial y}\right)\right] d\Omega = \rho h \int_\Omega \frac{\partial^2 u}{\partial t^2} W d\Omega. \end{aligned} \tag{22}$$

Now, if free-edge boundary conditions are considered, boundary integrals of Eq. (22) vanish. Considering that the plate is vibrating, functions u and v can be written as:

$$u(x,y,t)=U(x,y)\sin(\varpi t), \tag{23}$$

$$v(x,y,t)=V(x,y)\sin(\varpi t), \tag{24}$$

where ϖ is the in-plane natural frequency associated to any in-plane mode, and $U(x,y)$ and $V(x,y)$ are the displacement amplitudes along directions x and y, respectively, of a point with coordinate (x,y). In this sense, the amplitude is only a function of coordinates x and y. Eliminating the boundary integrals of Eq. (22), substituting Eqs. (23) and (24) into Eq. (22), and considering any mode j, one obtains:

$$\begin{aligned}\int_{\Omega}\Bigg[&A_{11}\left(\frac{\partial U^{(j)}}{\partial x}\frac{\partial W}{\partial x}\right)+A_{16}\left(\frac{\partial U^{(j)}}{\partial y}\frac{\partial W}{\partial x}+\frac{\partial U^{(j)}}{\partial x}\frac{\partial W}{\partial y}+\frac{\partial V^{(j)}}{\partial x}\frac{\partial W}{\partial x}\right)+A_{12}\left(\frac{\partial V^{(j)}}{\partial y}\frac{\partial W}{\partial x}\right)+\\ &+A_{66}\left(\frac{\partial U^{(j)}}{\partial y}\frac{\partial W}{\partial y}+\frac{\partial V^{(j)}}{\partial x}\frac{\partial W}{\partial y}\right)+A_{26}\left(\frac{\partial V^{(j)}}{\partial y}\frac{\partial W}{\partial y}\right)\Bigg]d\Omega=\rho h\varpi_j^2\int_{\Omega}U^{(j)}Wd\Omega\end{aligned}. \tag{25}$$

Now, if the same previous mathematical procedures used in Eq. (10) are used in Eq. (11), one obtains:

$$\begin{aligned}\int_{\Omega}\Bigg[&A_{16}\left(\frac{\partial U^{(j)}}{\partial x}\frac{\partial W}{\partial x}\right)+A_{12}\left(\frac{\partial U^{(j)}}{\partial x}\frac{\partial W}{\partial y}\right)+A_{66}\left(\frac{\partial U^{(j)}}{\partial y}\frac{\partial W}{\partial x}+\frac{\partial V^{(j)}}{\partial x}\frac{\partial W}{\partial x}\right)+A_{26}\left(\frac{\partial U^{(j)}}{\partial y}\frac{\partial W}{\partial y}\right)+\\ &+A_{26}\left(\frac{\partial V^{(j)}}{\partial y}\frac{\partial W}{\partial x}+\frac{\partial V^{(j)}}{\partial x}\frac{\partial W}{\partial y}\right)+A_{22}\left(\frac{\partial V^{(j)}}{\partial y}\frac{\partial W}{\partial y}\right)\Bigg]d\Omega=\rho h\varpi_j^2\int_{\Omega}V^{(j)}Wd\Omega\end{aligned}. \tag{26}$$

2.3.1 Choice of the weighting functions

Eqs. (25) and (26) are theoretically valid for isotropic, orthotropic, or anisotropic plates, provided that the laminate is symmetrical. As it can be seen, the function W is arbitrary since itself and its first order derivative is continuous in the domain Ω. Amplitudes $U(x,y)$ and $V(x,y)$, and frequencies ϖ_j are obtained from dynamic tests. Dimensions of the plate and parameters ρ and h can also be easily measured on the sample plate. Thus, the next steps are the choice of a suitable numerical method to compute the derivatives and integrals in Eqs. (25) and (26). Furthermore, suitable mode shapes and weighting functions should be chosen. In this work, finite differences and Gauss-Legendre numerical integration scheme are used to compute these derivatives and integrals, respectively. For numerical reasons, modes with several sign changes in the mode shape are avoided because their numerical derivatives and integrals are more sensitive to errors (Grédiac & Paris, 1996). Generally, first modes present more smooth curvatures and are, at the same time, easier to be obtained experimentally. It is worth noting that in-plane modal analysis presents much higher frequencies than transverse modal analysis (bending modes). This is because stiffness along direction x and y is much higher than stiffness along the transversal direction of the plate. For numerical reasons, smooth mode shapes associated with smooth weighting functions are preferred. For all these reasons, and in order to simplify Eqs. (25) and (26), the following group of weighting functions are proposed:

W(x, y) = x^2, which applied to Eqs. (25) and (26) provides the following integral equations:

$$2\int_{\Omega}\left[A_{11}\left(\frac{\partial U^{(j)}}{\partial x}x\right)+A_{16}\left(\frac{\partial U^{(j)}}{\partial y}x+\frac{\partial V^{(j)}}{\partial x}x\right)+A_{12}\left(\frac{\partial V^{(j)}}{\partial y}x\right)\right]d\Omega=\rho h\varpi_j^2\int_{\Omega}U^{(j)}x^2 d\Omega\,, \quad (27)$$

$$2\int_{\Omega}\left[A_{16}\left(\frac{\partial U^{(j)}}{\partial x}x\right)+A_{66}\left(\frac{\partial U^{(j)}}{\partial y}x+\frac{\partial V^{(j)}}{\partial x}x\right)+A_{26}\left(\frac{\partial V^{(j)}}{\partial y}x\right)\right]d\Omega=\rho h\varpi_j^2\int_{\Omega}V^{(j)}x^2 d\Omega\,. \quad (28)$$

W(x, y) = y^2, which applied to Eqs. (25) and (26) provides the following integral equations:

$$2\int_{\Omega}\left[A_{16}\left(\frac{\partial U^{(j)}}{\partial x}y\right)+A_{66}\left(\frac{\partial U^{(j)}}{\partial y}y+\frac{\partial V^{(j)}}{\partial x}y\right)+A_{26}\left(\frac{\partial V^{(j)}}{\partial y}y\right)\right]d\Omega=\rho h\varpi_j^2\int_{\Omega}U^{(j)}y^2 d\Omega\,, \quad (29)$$

$$2\int_{\Omega}\left[A_{12}\left(\frac{\partial U^{(j)}}{\partial x}y\right)+A_{26}\left(\frac{\partial U^{(j)}}{\partial y}y+\frac{\partial V^{(j)}}{\partial y}y\right)+A_{22}\left(\frac{\partial V^{(j)}}{\partial y}y\right)\right]d\Omega=\rho h\varpi_j^2\int_{\Omega}V^{(j)}y^2 d\Omega\,. \quad (30)$$

W(x, y) = 2xy, which applied to Eqs. (25) and (26) provides the following integral equations:

$$\begin{aligned}&2\int_{\Omega}\left[A_{11}\left(\frac{\partial U^{(j)}}{\partial x}y\right)+A_{16}\left(\frac{\partial U^{(j)}}{\partial y}y+\frac{\partial U^{(j)}}{\partial x}x+\frac{\partial V^{(j)}}{\partial x}y\right)+A_{12}\left(\frac{\partial V^{(j)}}{\partial y}y\right)+\right.\\&\left.+A_{66}\left(\frac{\partial U^{(j)}}{\partial y}x+\frac{\partial V^{(j)}}{\partial x}x\right)+A_{26}\left(\frac{\partial V^{(j)}}{\partial y}x\right)\right]d\Omega=\rho h\varpi_j^2\int_{\Omega}U^{(j)}(2xy)\,d\Omega\end{aligned}\,, \quad (31)$$

$$\begin{aligned}&2\int_{\Omega}\left[A_{16}\left(\frac{\partial U^{(j)}}{\partial x}y\right)+A_{66}\left(\frac{\partial U^{(j)}}{\partial y}y+\frac{\partial V^{(j)}}{\partial x}y\right)+A_{12}\left(\frac{\partial U^{(j)}}{\partial x}x\right)+\right.\\&\left.+A_{26}\left(\frac{\partial U^{(j)}}{\partial y}x+\frac{\partial V^{(j)}}{\partial y}y+\frac{\partial V^{(j)}}{\partial x}x\right)+A_{22}\left(\frac{\partial V^{(j)}}{\partial y}x\right)\right]d\Omega=\rho h\varpi_j^2\int_{\Omega}V^{(j)}(2xy)\,d\Omega\end{aligned} \quad (32)$$

Defining:

$$\int_{\Omega}\left(\frac{\partial U}{\partial x}x\right)d\Omega=K_{uxx}\,,\ \int_{\Omega}\left(\frac{\partial U}{\partial y}x\right)d\Omega=K_{uyx}\,,\ \int_{\Omega}\left(\frac{\partial U}{\partial x}y\right)d\Omega=K_{uxy}\,,\ \int_{\Omega}\left(\frac{\partial U}{\partial y}y\right)d\Omega=K_{uyy}\,,$$

$$\int_{\Omega}\left(\frac{\partial V}{\partial x}x\right)d\Omega=K_{vxx}\,,\ \int_{\Omega}\left(\frac{\partial V}{\partial y}x\right)d\Omega=K_{vyx}\,,\ \int_{\Omega}\left(\frac{\partial V}{\partial x}y\right)d\Omega=K_{vxy}\,,\ \int_{\Omega}\left(\frac{\partial V}{\partial y}y\right)d\Omega=K_{vyy}\,.$$

Eqs. (34)-(38) can be written as:

$$A_{11}K_{uxx}^{(j)}+A_{16}\left(K_{uyx}^{(j)}+K_{vxx}^{(j)}\right)+A_{12}K_{vyx}^{(j)}=\tfrac{1}{2}\rho h\varpi_j^2\int_{\Omega}U^{(j)}x^2 d\Omega\,, \quad (33)$$

$$A_{16}K_{uxx}^{(j)} + A_{66}\left(K_{uyx}^{(j)} + K_{vxx}^{(j)}\right) + A_{26}K_{vyx}^{(j)} = \frac{1}{2}\rho h\varpi_j^2 \int_\Omega V^{(j)} x^2 d\Omega, \quad (34)$$

$$A_{16}K_{uxy}^{(j)} + A_{66}\left(K_{uyy}^{(j)} + K_{vxy}^{(j)}\right) + A_{26}K_{vyy}^{(j)} = \frac{1}{2}\rho h\varpi_j^2 \int_\Omega U^{(j)} y^2 d\Omega, \quad (35)$$

$$A_{12}K_{uxy}^{(j)} + A_{26}\left(K_{uyy}^{(j)} + K_{vxy}^{(j)}\right) + A_{22}K_{vyy}^{(j)} = \frac{1}{2}\rho h\varpi_j^2 \int_\Omega V^{(j)} y^2 d\Omega, \quad (36)$$

$$\begin{aligned} &A_{11}K_{uxy}^{(j)} + A_{16}\left(K_{uyy}^{(j)} + K_{uxx}^{(j)} + K_{vxy}^{(j)}\right) + A_{12}K_{vyy}^{(j)} + \\ &+A_{66}\left(K_{uyx}^{(j)} + K_{vxx}^{(j)}\right) + A_{26}K_{vyx}^{(j)} = \frac{1}{2}\rho h\varpi_j^2 \int_\Omega U^{(j)}(2xy)\, d\Omega, \end{aligned} \quad (37)$$

$$\begin{aligned} &A_{16}K_{uxy}^{(j)} + A_{66}\left(K_{uyy}^{(j)} + K_{vxy}^{(j)}\right) + A_{12}K_{uxx}^{(j)} + A_{26}\left(K_{uyx}^{(j)} + K_{vyy}^{(j)} + K_{vxx}^{(j)}\right) + \\ &+A_{22}K_{vyx}^{(j)} = \frac{1}{2}\rho h\varpi_j^2 \int_\Omega V^{(j)}(2xy)\, d\Omega, \end{aligned} \quad (38)$$

or, in matrix form:

$$\begin{bmatrix} \cdots & \cdots & \cdots & \cdots & \cdots & \cdots \\ K_{uxx}^{(j)} & K_{vyx}^{(j)} & \left(K_{uyx}^{(j)} + K_{vxx}^{(j)}\right) & 0 & 0 & 0 \\ 0 & 0 & K_{uxx}^{(j)} & 0 & K_{vyx}^{(j)} & \left(K_{uyx}^{(j)} + K_{vxx}^{(j)}\right) \\ 0 & 0 & K_{uxy}^{(j)} & 0 & K_{vyy}^{(j)} & \left(K_{uyy}^{(j)} + K_{vxy}^{(j)}\right) \\ 0 & K_{uxy}^{(j)} & 0 & K_{vyy}^{(j)} & \left(K_{uyy}^{(j)} + K_{vxy}^{(j)}\right) & 0 \\ K_{uxy}^{(j)} & K_{vyy}^{(j)} & \left(K_{uyy}^{(j)} + K_{uxx}^{(j)} + K_{vxy}^{(j)}\right) & 0 & K_{vyx}^{(j)} & \left(K_{uyx}^{(j)} + K_{vxx}^{(j)}\right) \\ 0 & K_{uxx}^{(j)} & K_{uxy}^{(j)} & K_{vyx}^{(j)} & \left(K_{uyx}^{(j)} + K_{vyy}^{(j)} + K_{vxx}^{(j)}\right) & \left(K_{uyy}^{(j)} + K_{vxy}^{(j)}\right) \\ \cdots & \cdots & \cdots & \cdots & \cdots & \cdots \end{bmatrix} \begin{Bmatrix} A_{11} \\ A_{12} \\ A_{16} \\ A_{22} \\ A_{26} \\ A_{66} \end{Bmatrix} =$$

$$= \frac{1}{2}\rho h\varpi_j^2 \begin{Bmatrix} \cdots \\ \int_\Omega U^{(j)} x^2 d\Omega \\ \int_\Omega V^{(j)} x^2 d\Omega \\ \int_\Omega U^{(j)} y^2 d\Omega \\ \int_\Omega V^{(j)} y^2 d\Omega \\ \int_\Omega U^{(j)}(2xy)\, d\Omega \\ \int_\Omega V^{(j)}(2xy)\, d\Omega \\ \cdots \end{Bmatrix} \quad (39)$$

Eq. (39) can also be rewritten in a compact form as:

$$\mathbf{KA} = \mathbf{C}\,, \tag{40}$$

where, considering L modes, **K** is a $6L$ x 6 matrix in Eq. (40), **A** is a 6 x 1 matrix, and **C** is a $6L$ x 1 matrix in Eq. (40). Eq. (40) is an over determined system. Thus, the solution can be found by least squares:

$$\mathbf{A} = \left(\mathbf{K}^{T}\mathbf{K}\right)^{-1}\left(\mathbf{K}^{T}\mathbf{C}\right), \tag{41}$$

from where the extensional elastic constants A_{ij} are computed.

3. Results and comments

A commercial finite element code (ANSYS 11.0) was used to give particular mode shapes and their corresponding natural frequencies from both in-plane and out-of-plane numerical modal analysis. Element SHELL99 was used and plates under free-edge boundary conditions were considered.

To exemplify the method proposed by Grédiac and Paris (1996), it was used an anisotropic plate with dimensions 0.450 x 0.350 x 0.0021 m and density 1500 kg/m^3. It was used a laminate with 8 plies, [0 45 90 135]$_S$, and the following engineering elastic constants by ply: E_1=120 GPa (Young's module along the principal direction 1), E_2=10 GPa (Young's module along the principal direction 2), G_{12}=4.9 GPa (shear module along the plane 1-2), and ν_{12}=0.3 (Poisson's ratio along the plane 1-2). A mesh of 651 nodes was used. Fig. 2 shows the three modes used to identify the required properties. As can be seen, depending on type of material anisotropy, it is not possible to find all three modes necessary to apply the method. In this case, it is necessary to find the more approximated ones.

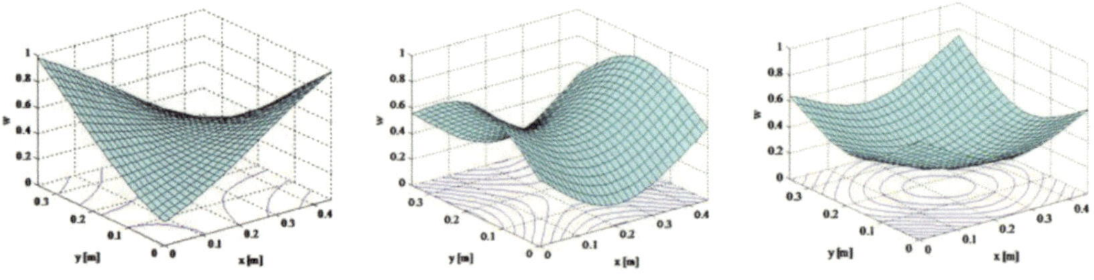

(a) (b) (c)

Fig. 2. Numerical modes obtained from Ansys: (a) Mode shape 1, (b) Mode shape 2, (c) Mode shape 3.

Table 1 shows the bending elastic stiffness constants computed using the engineering constants and the classical theory of laminates, and it also shows the errors computed after applying the identification method. As can be observed, the technique is able to find very satisfactory results when it is used the correct modes. The problem of this technique is the high sensitivity to noise presence because of second-order derivatives. More results and comments about this method can be found in Grédiac & Paris, 1996.

In order to verify the accuracy of the extensional elastic stiffness identification method, it was used six graphite/polymer symmetric laminated plates (Table 2): a fully anisotropic with all A_{ij} different from zero (i, j = 1, 2, and 6); a cross-ply orthotropic with $A_{11} = A_{22}$, $A_{16} = A_{26} = 0$, and $A_{11} - A_{12} \neq 2A_{66}$; a 0° unidirectional orthotropic; a 30° unidirectional orthotropic (generally orthotropic); a +30°/-30 angle-ply orthotropic; and a quasi-isotropic with $A_{11} = A_{22}$, $A_{16} = A_{26} = 0$, and $A_{11} - A_{12} = 2A_{66}$. These laminates have 8 plies with the following engineering elastic constants by ply: E_1=155 GPa, E_2=12.10 GPa, G_{12}=4.4 GPa, and ν_{12}=0.248. Dimensions considered were 0.450 x 0.350 x 0.003 m (x, y, and z plate coordinate axis, respectively), for the rectangular plate, 0.350 x 0.350 x 0.003 m, for the square plate, and, the density material was 1500 kg/m^3. The plates were modeled using a mesh with 651 nodes, for the rectangular plates, and with 441 nodes, for the square plate. The extensional elastic stiffness constants are shown in Table 2. The terms "Aniso", "Ortho" and "Quasi-iso" are simplifications of "Anisotropic", "Orthotropic" and "Quasi-isotropic", respectively. Only the first fifteen mode shapes were analyzed.

As can be seen in Table 2, the constants A_{12} and A_{66} for the orthotropic laminates I and II are equals. They also are equals to laminates with the same characteristics but with 90° unidirectional fibers. 0° and 90° unidirectional laminated plates are only different in relation to A_{11} and A_{22} terms. These are inverted: A_{11} term of the 0° laminate is equal to A_{22} term of the 90° laminate, and vice-verse. For the generally orthotropic laminate III and orthotropic laminate IV the difference are only the A_{16} and A_{26} constants: they are nulls for the laminate IV and non-nulls for the laminate III. For laminate III all extensional elastic constants are non-nulls, similar to fully anisotropic laminates, what, consequently, originates to full extensional elastic stiffness matrix.

Bending elastic constants	N x mm	Errors (%)
D_{11}	64363.9	0.02
D_{22}	24155.8	0.04
D_{12}	8875.1	0.02
D_{66}	10032.7	1.22
D_{16}	6019.6	0.63
D_{26}	6019.6	0.64

Table 1. Bending elastic stiffness constants of the tested anisotropic plate and the computed errors.

The key point of this technique of identification is related with the correct choice of mode shapes together to the weighting functions (virtual fields). The correct mode shapes are called here by "suitable modes" and the correct combination between these modes and the weighting functions are called by "suitable combinations". The identification of the suitable modes is not difficult, as it will be shown in the next topics. But, the suitable combinations are more difficult because they depend on the type and the geometry of the material. Fortunately, there are some aspects that help finding the best choice. Unlike the bending stiffness identification method originally proposed, for this method there are a lot of modes and suitable combinations that give satisfactory results.

A_{ij} 10^8 [N/m]	Aniso [90 0 0 45]$_S$	Ortho I [90 0 90 0]$_S$	Ortho II [0]$_{4S}$	Ortho III [30]$_{4S}$	Ortho IV [30 -30]$_{2S}$	Quasi-iso [90 45 0 -45]$_S$
A_{11}	2.7865	2.5186	4.6724	2.7840	2.7840	1.9776
A_{12}	0.3610	0.0905	0.0905	0.9020	0.9020	0.6315
A_{16}	0.2692	0	0	1.4012	0	0
A_{22}	1.7096	2.5186	0.3648	0.6301	0.6301	1.9776
A_{26}	0.2692	0	0	0.4641	0	0
A_{66}	0.4025	0.1320	0.1320	0.9436	0.9436	0.6730

Table 2. Laminated plates used for the verification of the method

[90 0 0 45]S anisotropic plate

Table 3 shows some errors computed for the anisotropic plate, rectangular and square. It was considered only the first fifteen in-plane modes shapes. Anisotropic plates, in general, give very satisfactory results using the combinations among suitable modes. This factor can be justified by the fact of these combinations be hardly involved with all required extensional elastic constants A_{ij}'s.

The numerical contribution of each mode to the computation of a specific constant cannot be jeopardized by numerical contribution of another mode during the solution of the system given by Eq. (40). The suitable modes are those that when associated with weighting functions do not null or give very low values for integrals of the right (**K** matrix) and/or left (**C** matrix) hand sides of Eq. (40). The suitable combinations are one composed by suitable modes and that give more accurate results. In the majority of the cases, combinations using a higher number of suitable modes can be suitable combinations. According to Table 3 is possible to see that using combinations with only two suitable modes very satisfactory results can be obtained. Satisfactory results would also be obtained using combinations with any modes since the number of suitable modes among all used modes is higher than non-suitable modes. But the accuracy of these results cannot be guaranteed for all combinations.

[0 90 0 90]$_S$ orthotropic plate (ortho I)

In general, for the orthotropic and isotropic materials is more difficult to find the suitable combinations when it is compared to fully anisotropic materials. It is necessary to take care to correctly identifying the combinations that give the best results. In these types of materials not all combinations are among suitable modes that can be considered as being suitable combinations. According to values found to terms of the **K** and **C** matrices, Eq. (40), and using combinations among suitable modes, it is possible to see the following types of systems:

For the rectangular plate:

Type 1:

$$\begin{bmatrix} I_1 & 0 & 0 & I_4 & 0 & 0 \\ 0 & 0 & I_3 & 0 & 0 & 0 \\ 0 & I_2 & 0 & 0 & I_4 & 0 \\ 0 & I_1 & 0 & 0 & I_4 & 0 \\ 0 & 0 & 0 & 0 & I_3 & 0 \\ 0 & 0 & I_3 & I_1 & 0 & I_4 \end{bmatrix} \begin{Bmatrix} A_{11} \\ A_{16} \\ A_{66} \\ A_{12} \\ A_{26} \\ A_{22} \end{Bmatrix} = \begin{Bmatrix} I_5 \\ I_6 \\ 0 \\ 0 \\ 0 \\ I_7 \end{Bmatrix}. \tag{42}$$

Type 2:

$$\begin{bmatrix} 0 & I_2 & 0 & 0 & 0 & 0 \\ 0 & I_1 & 0 & 0 & I_3 & 0 \\ I_1 & 0 & I_2 & I_3 & 0 & 0 \\ 0 & 0 & I_2 & 0 & 0 & 0 \\ 0 & 0 & 0 & I_1 & 0 & I_3 \\ 0 & I_1 & 0 & 0 & I_4 & 0 \end{bmatrix} \begin{Bmatrix} A_{11} \\ A_{16} \\ A_{66} \\ A_{12} \\ A_{26} \\ A_{22} \end{Bmatrix} = \begin{Bmatrix} 0 \\ 0 \\ I_5 \\ I_6 \\ I_7 \\ 0 \end{Bmatrix}. \tag{43}$$

Anisotropic rectangular plate - suitable modes: 2, 3, 6, 9, 10, 13, and 14						
Suitable combinations	Errors (%)					
	A_{11}	A_{12}	A_{22}	A_{16}	A_{26}	A_{66}
2-3-6-9	2.20	1.95	1.56	1.89	1.46	0.78
2-3-6	0.04	0.20	0.21	0.93	0.79	0.97
2-3-9	0.37	2.63	0.27	0.78	0.93	0.28
2-6-9	0.30	2.46	0.17	1.45	1.03	0.02
3-6-9	0.17	1.77	0.11	0.97	1.09	0.21
2-3	0.00	0.32	0.24	0.78	0.53	1.08
6-9	0.01	2.06	0.11	1.62	1.50	1.51
Anisotropic square plate - suitable modes: 2, 3, 6, 9, 10, 13, and 14						
Suitable combinations	Errors (%)					
	A_{11}	A_{12}	A_{22}	A_{16}	A_{26}	A_{66}
2-3-6-9	3.73	3.41	2.23	3.21	1.88	1.06
2-3-6	0.20	0.62	0.23	0.86	0.09	1.08
2-3-9	0.41	3.86	0.54	1.36	0.77	0.50
2-6-9	0.35	3.43	0.44	2.00	1.61	0.13
3-6-9	0.24	2.59	0.14	1.66	0.84	0.30
2-3	0.04	0.06	0.17	1.02	0.97	1.49
6-9	0.08	2.66	0.25	2.70	1.08	1.93

Table 3. Errors computed for the anisotropic plate using some suitable combinations.

where I_1, I_2, I_3, I_4, I_5, I_6, and I_7 are the integral values of Eq. (40). These integrals can be negatives or positives depending on strain direction and reference coordinate axis. As can be observed, for these two types of systems, Eq. (42) and Eq. (43), all elastic constants are involved, and, however, combinations associated to only one type of system can be suitable combinations and sufficient to give correct results. For this plate were found the following suitable modes: 2, 3, 6, 8, 11, and 14.

For the square plate: for this plate, the suitable modes are: 2, 3, 6, 7, 11, and 12. The systems of equations are full, even though of the additional terms, that are null in rectangular plate, to be low in this square plate. It is observed in this plate that for each suitable mode there is another identical but out-of-phase at 90°: modes 2 and 3, 6 and 7, and, 11 and 12.

Table 4 shows errors computed to some suitable combinations for these orthotropic plates. As can be seen, very satisfactory results can be obtained using correct combinations of modes. For this type of orthotropy, it can be more difficult to compute an accurate value for constant A_{12}. The identification of the suitable combinations is not so clear. As this technique of identification is associated to the solution of an equation system having different modal contributions, it is difficult to identify which modes compose a correct combination.

Ortho I rectangular plate – suitable modes: 2, 3, 6, 8, 11, and 14						
Suitable combinations	Errors (%)				Differences (N/m)	
	A_{11}	A_{12}	A_{22}	A_{66}	A_{16}	A_{26}
2-3-6-8-11-14	0.14	3.80	0.12	1.22	-2.7 x 10^4	0.8 x 10^4
2-3-6-8	0.11	2.90	0.24	0.74	-2.5 x 10^4	0.6 x 10^4
3-6-8-11	0.17	1.58	0.31	1.67	-1.4 x 10^4	0.5 x 10^4
2-8-11	0.17	0.29	0.23	1.46	-1.5 x 10^4	0.6 x 10^4
6-8-11	0.18	0.89	0.34	1.91	-4.8 x 10^4	1.0 x 10^4
3-8	0.02	5.75	0.91	0.13	-1.4 x 10^4	8.4 x 10^3
8-11	0.25	0.42	0.40	2.17	-2.6 x 10^4	1.1 x 10^4
Ortho I square plate – suitable modes: 2, 3, 6, 7, 11, and 12						
Suitable combinations	Errors (%)				Differences (N/m)	
	A_{11}	A_{12}	A_{22}	A_{66}	A_{16}	A_{26}
2-3-6-7-11-12	0.09	5.81	0.05	1.89	2.7 x 10^4	-1.6 x 10^4
2-3-6-11	0.10	5.09	0.06	1.92	-4.4 x 10^4	-8.7 x 10^4
3-6-7-12	0.38	2.33	0.08	2.92	6.4 x 10^4	-1.7 x 10^4
2-6-11	0.13	5.47	0.11	2.51	-1.0 x 10^5	-1.8 x 10^4
2-7-12	0.22	5.37	0.04	1.95	1.1 x 10^5	0.5 x 10^5
2-11	0.02	9.04	0.18	0.56	-1.3 x 10^4	-1.9 x 10^4
6-11	0.18	5.63	0.30	3.46	-1.1 x 10^5	-2.7 x 10^4

Table 4. Errors computed for the ortho I plate using some suitable combinations.

$[0]_{4S}$ orthotropic plate (ortho II)

For the rectangular plate: for this plate, modes 3, 10, 11, and 14 originate systems of type 1, Eq. (42), and modes 2, 5, 9, and 15 originate systems of type 2, Eq. (43).

For the square plate: for this plate, modes 4, 10, 12, and 14 originate systems of type 1, Eq. (51), and modes 2, 5, 9, and 13 originate systems of the type 2, Eq. (43).

Table 5 shows the errors computed to some suitable combinations for these orthotropic plates. Using correct combinations very satisfactory results can be obtained. Similar to ortho I plate, for this type of orthotropy, constant A_{12} is more difficult of being accurately computed.

Ortho II rectangular plate - suitable modes: 2, 3, 5, 9, 10, 11, 14, and 15						
Suitable combinations	Errors (%)				Differences (N/m)	
	A_{11}	A_{12}	A_{22}	A_{66}	A_{16}	A_{26}
2-3-5	2.77	6.06	0.08	0.38	2.4×10^4	0.5×10^4
11-14-15	1.14	2.15	0.92	1.15	2.3×10^5	0.1×10^5
2-3	2.36	0.57	0.19	0.72	1.7×10^4	0.1×10^4
2-10	2.10	0.82	0.60	2.42	-2.0×10^5	-0.4×10^5
2-14	2.04	5.22	0.37	1.12	-3.9×10^4	-1.2×10^4
11-14	1.64	2.14	0.87	0.98	-1.8×10^5	-0.1×10^4
Ortho II square plate - suitable modes: 2, 4, 5, 9, 10, 12, 13, and 14						
Suitable combinations	Errors (%)				Differences (N/m)	
	A_{11}	A_{12}	A_{22}	A_{66}	A_{16}	A_{26}
2-4-5	4.13	8.63	0.39	0.45	-3.5×10^4	-0.1×10^4
2-4-14	2.43	0.78	0.72	2.89	4.2×10^4	1.1×10^4
2-4	3.35	0.10	0.21	1.25	-3.2×10^4	-0.1×10^4
2-14	1.63	1.59	1.30	4.72	1.1×10^5	0.3×10^5
5-14	2.67	13.54	2.21	2.82	5.6×10^4	-1.0×10^4
9-14	2.12	62.05	0.85	3.89	1.1×10^5	0.2×10^5

Table 5. Errors computed for the ortho II plate using some suitable combinations.

$[30]_{4S}$ orthotropic plate (ortho III)

For these plates, rectangular and square the computed matrices **K** and **C**, Eq.(40), are full matrices, similar ones of the anisotropic plates. Thus, the majority of combinations among suitable modes are suitable combinations. Combinations that are not suitable present high errors for all constants, what, consequently, make them easy to be identified. Table 6 shows the errors computed to some suitable combinations. Using correct combinations very satisfactory results can be obtained.

Ortho III rectangular plate - suitable modes: 2, 3, 6, 8, 10, 12, and 13						
Suitable combinations	Errors (%)					
	A_{11}	A_{12}	A_{22}	A_{66}	A_{16}	A_{26}
3-6-12	0.25	0.51	0.10	1.05	0.69	0.99
8-10-12	4.26	5.42	3.67	3.43	4.60	5.62
2-3	0.85	1.72	1.22	1.39	0.10	0.96
3-6	0.63	0.32	0.04	0.22	0.53	0.14
6-12	0.71	1.06	0.05	1.39	1.19	1.30
10-13	4.42	2.05	0.71	4.07	5.21	0.63
Ortho III square plate - suitable modes: 2, 4, 6, 8, 9, 12, and 13						
Suitable combinations	Errors (%)				Differences (N/m)	
	A_{11}	A_{12}	A_{22}	A_{66}	A_{16}	A_{26}
2-4-6	2.62	3.94	3.18	0.59	2.02	3.23
2-4-8	1.20	2.40	1.64	1.27	0.11	1.19
2-4	2.35	1.11	0.02	5.46	4.26	2.90
2-6	0.76	2.76	3.52	0.01	0.53	2.90
2-13	1.75	2.75	1.51	1.92	0.33	1.25
12-13	3.08	2.41	2.63	0.53	1.85	2.54

Table 6. Errors computed for the ortho III plate using some suitable combinations.

[30 -30 30 -30]$_S$ orthotropic plate (ortho IV)

For the rectangular plate: it was found the following suitable modes: 2, 4, 5, 8, 11, 12 and 14. For this plate, the majority of combinations among the suitable modes are suitable combinations.

For the square plate: it was found the following suitable modes: 1, 4, 6, 9, 11, 12, and 14. Similar to rectangular plate, here the most of combinations among the suitable modes are suitable combinations.

Table 7 shows errors computed to some suitable combinations. Using correct combinations, very satisfactory results can be obtained.

Ortho IV rectangular plate – suitable modes: 2, 4, 5, 8, 11, 12, and 14						
Suitable combinations	Errors (%)				Differences (N/m)	
	A_{11}	A_{12}	A_{22}	A_{66}	A_{16}	A_{26}
2-4-5	1.52	2.15	1.28	1.54	-1.7 x 10^4	-0.6 x 10^4
2-4-8	1.50	2.09	0.83	1.14	1.4 x 10^4	0.9 x 10^4
5-11-12	0.27	0.09	0.04	1.11	-1.2 x 10^4	1.9 x 10^4
2-4	1.18	1.83	1.02	2.87	-1.2 x 10^4	-0.4 x 10^4
2-5	0.49	1.01	1.01	1.41	-1.8 x 10^4	-0.7 x 10^4
5-11	0.90	0.37	0.04	1.05	3.1 x 10^4	0.2 x 10^4
Ortho IV square plate – suitable modes: 1, 4, 6, 9, 11, 12, and 14						
Suitable combinations	Errors (%)				Differences (N/m)	
	A_{11}	A_{12}	A_{22}	A_{66}	A_{16}	A_{26}
1-6-12	0.16	0.30	0.13	3.07	1.7 x 10^4	2.5 x 10^4
6-9-12	0.18	0.35	0.30	2,68	1.3 x 10^4	4.8 x 10^4
9-12-14	0.18	1.87	0.40	0.68	-6.2 x 10^4	8.1 x 10^4
1-12	0.59	0.26	0.74	2.79	2.0 x 10^4	2.9 x 10^4
4-11	0.94	1.55	2.77	0.23	2.2 x 10^4	0.8 x 10^4
12-14	0.26	2.17	0.19	0.90	-9.7 x 10^4	9.4 x 10^4

Table 7. Errors computed for the ortho IV plate using some suitable combinations.

[90 45 0 -45]$_S$ quasi-isotropic plate

For the rectangular plate: it was found the following suitable modes: 1, 4, 6, 8, 11, 12, and 14. For this plate the most of combinations among the suitable modes are suitable combinations.

For the square plate: it was found the following suitable modes: 2, 3, 7, 8, 10, and 11. Similar to rectangular plate, here the majority of combinations among the suitable modes are suitable combinations.

Table 8 shows errors computed to some suitable combinations. Using correct combinations, very satisfactory results can be obtained.

Quasi-iso rectangular plate - suitable modes: 1, 4, 6, 8, 11, 12, and 14						
Suitable combinations	Errors (%)				Differences (N/m)	
	A_{11}	A_{12}	A_{22}	A_{66}	A_{16}	A_{26}
1-4-6	0.60	1.47	0.19	0.19	9.1 x 10^3	4.7 x 10^3
1-4-8	0.09	0.80	0.11	0.96	7.4 x 10^3	9.5 x 10^3
1-4-11	0.33	2.04	0.26	0.08	1.4 x 10^3	8.1 x 10^3
1-4	0.25	0.54	0.29	1.93	5.7 x 10^4	7.1 x 10^4
1-6	0.28	0.91	1.77	0.00	1.0 x 10^4	0.1 x 10^4
4-8	0.73	0.39	0.05	0.95	1.1 x 10^4	0.3 x 10^4
Quasi-iso square plate - suitable modes: 2, 3, 7, 8, 10, and 11						
Suitable combinations	Errors (%)				Differences (N/m)	
	A_{11}	A_{12}	A_{22}	A_{66}	A_{16}	A_{26}
2-3-7	0.37	1.33	0.35	0.91	4.7 x 10^5	4.7 x 10^5
2-3-8	0.32	1.36	0.33	0.97	-4.4 x 10^5	-4.4 x 10^5
2-10-11	0.07	3.29	0.45	0.81	-2.3 x 10^5	-1.1 x 10^5
2-3	0.33	0.79	0.34	2.68	2.2 x 10^4	1.3 x 10^4
3-7	0.56	1.73	0.43	0.80	6.1 x 10^5	5.2 x 10^5
8-11	0.89	3.05	0.06	1.34	1.1 x 10^5	-1.2 x 10^5

Table 8. Errors computed for the quasi-isotropic plate using some suitable combinations.

Figs. 3 to 14 show the fifteen first in-plane mode shapes to all the analyzed plates: rectangular and square geometry.

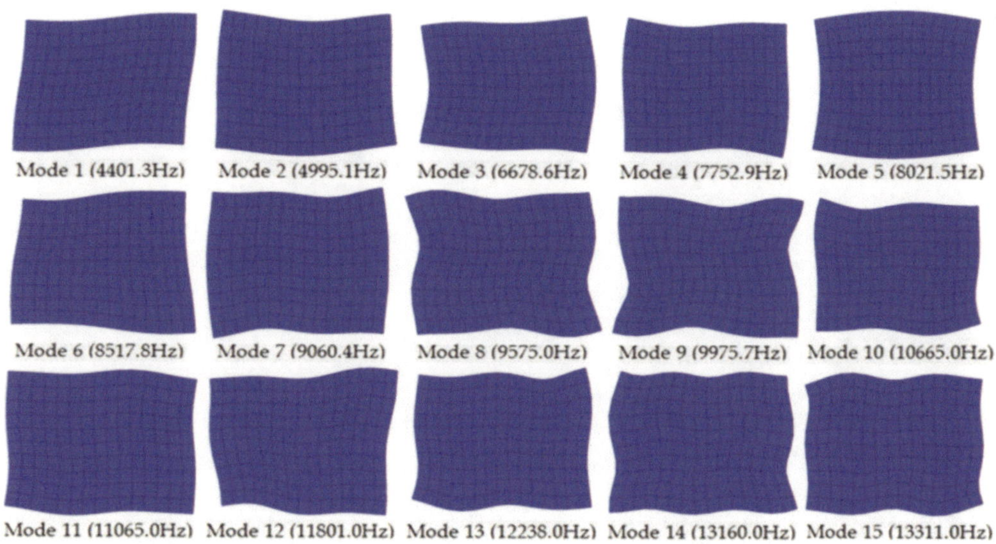

Fig. 3. Fifteen first in-plane mode shapes and natural frequencies to anisotropic rectangular plate.

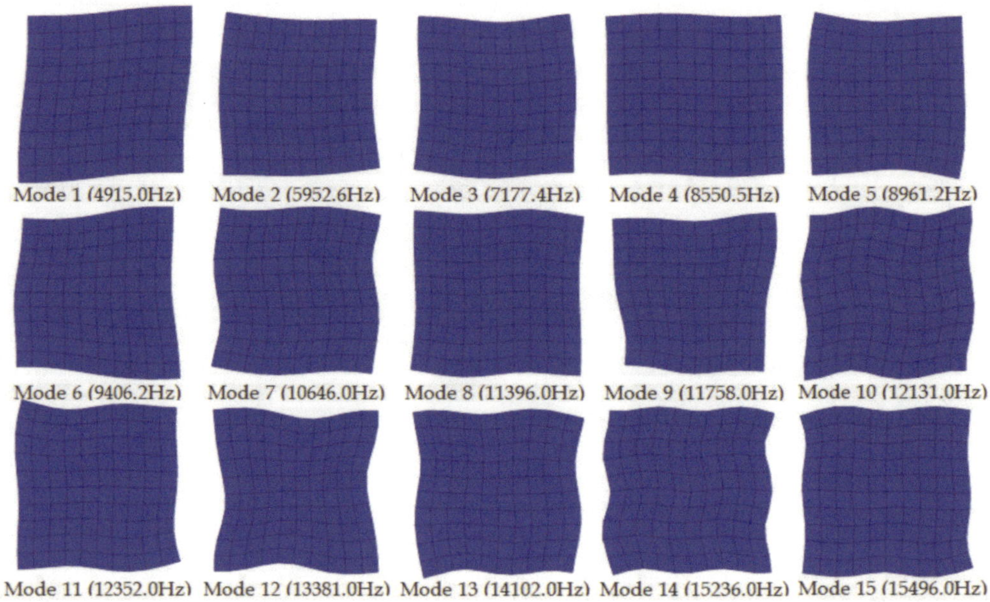

Fig. 4. Fifteen first in-plane mode shapes and natural frequencies to anisotropic square plate.

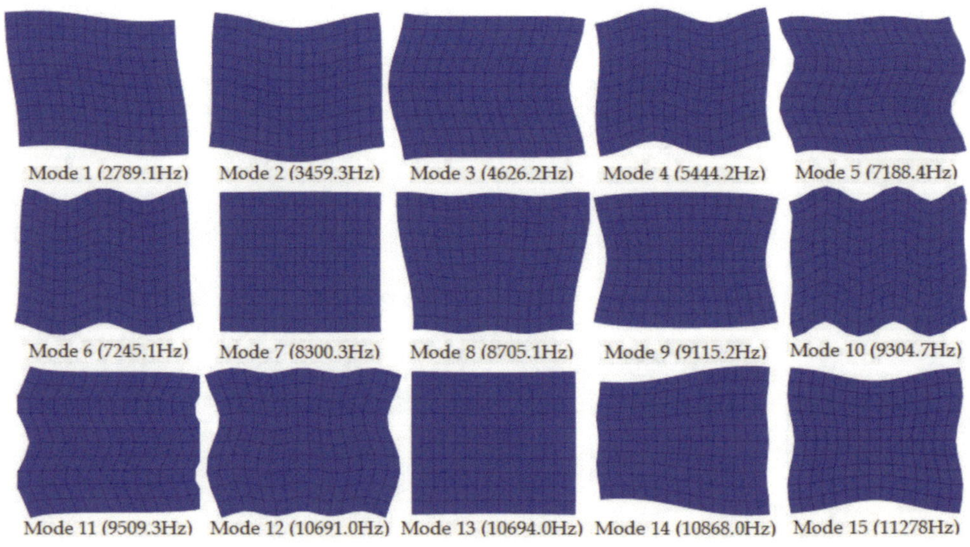

Fig. 5. Fifteen first in-plane mode shapes and natural frequencies to ortho I rectangular plate.

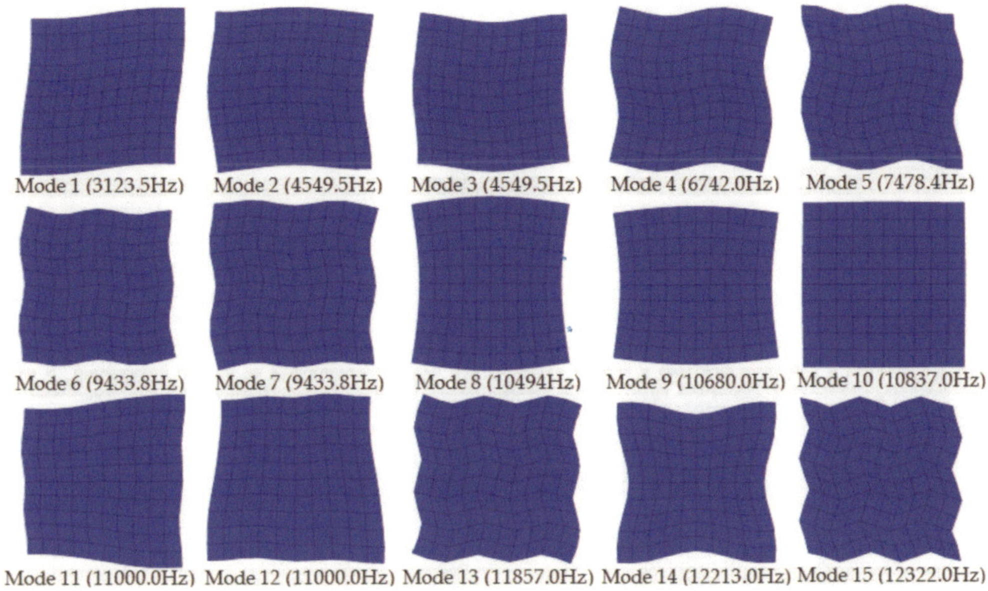

Fig. 6. Fifteen first in-plane mode shapes and natural frequencies to ortho I square plate.

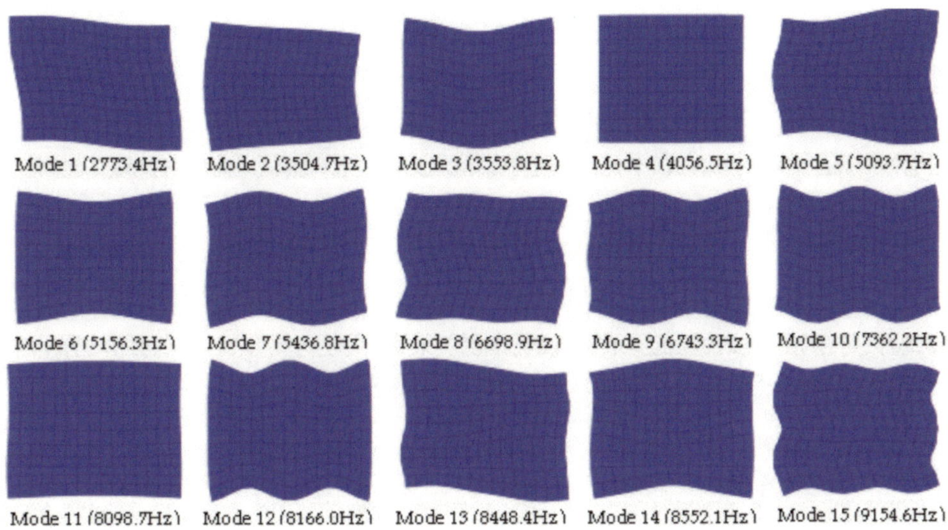

Fig. 7. Fifteen first in-plane mode shapes and natural frequencies to ortho II rectangular plate.

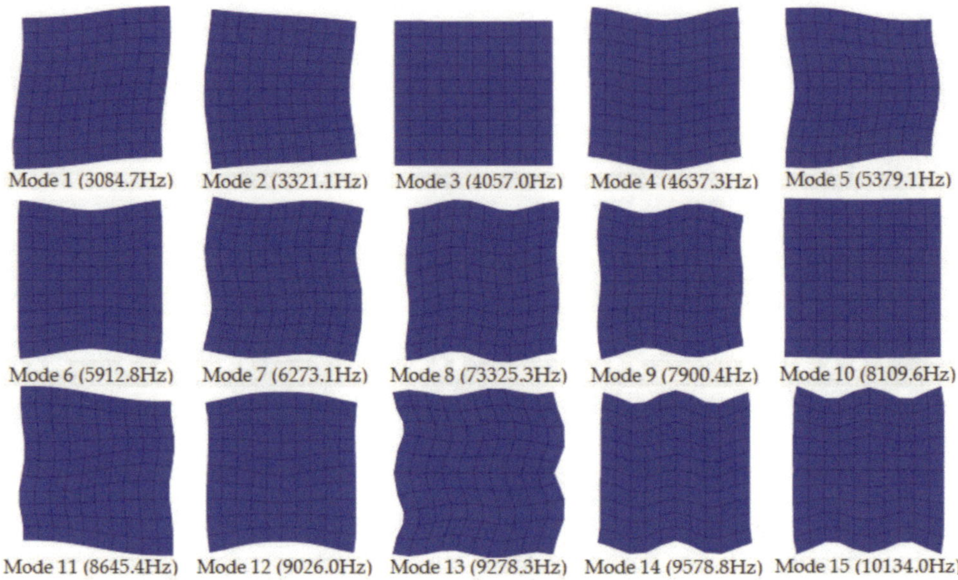

Fig. 8. Fifteen first in-plane mode shapes and natural frequencies to ortho II square plate.

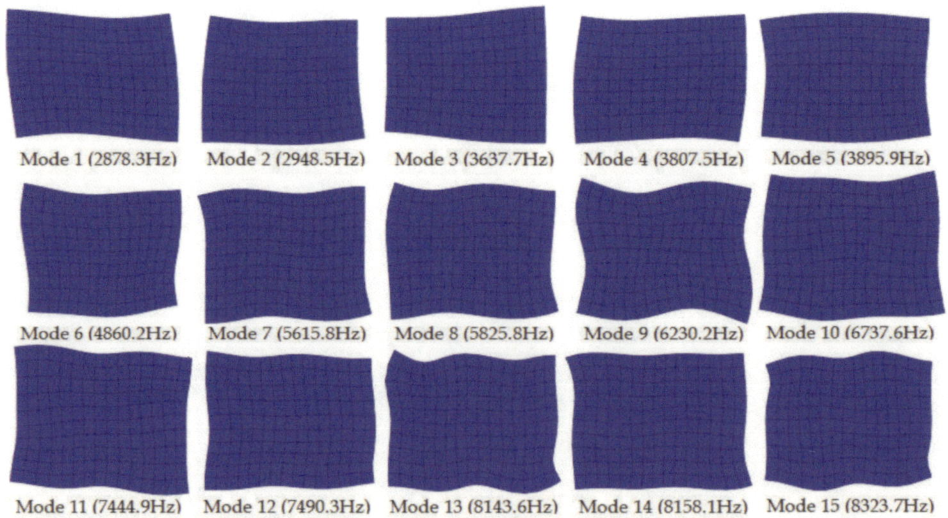

Fig. 9. Fifteen first in-plane mode shapes and natural frequencies to ortho III rectangular plate.

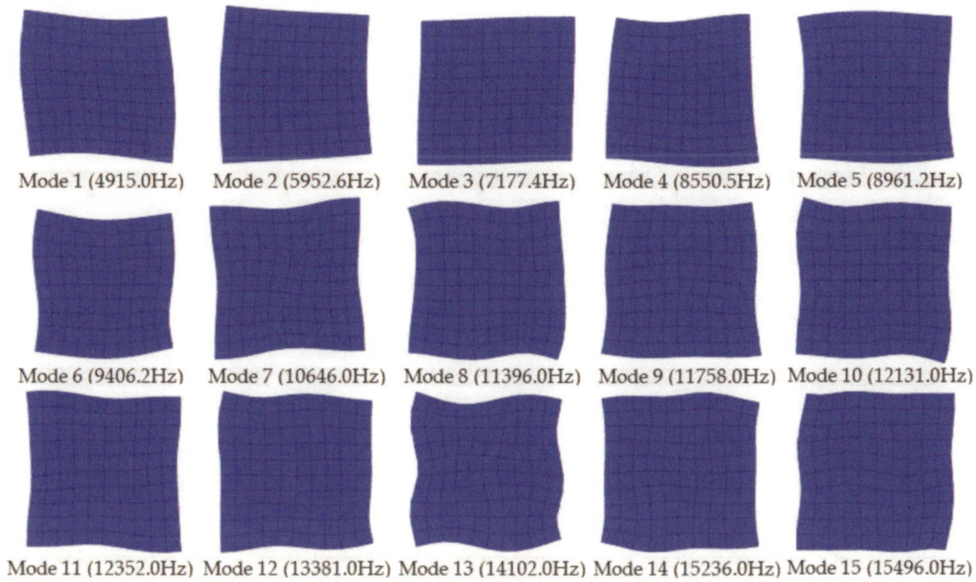

Fig. 10. Fifteen first in-plane mode shapes and natural frequencies to ortho III square plate.

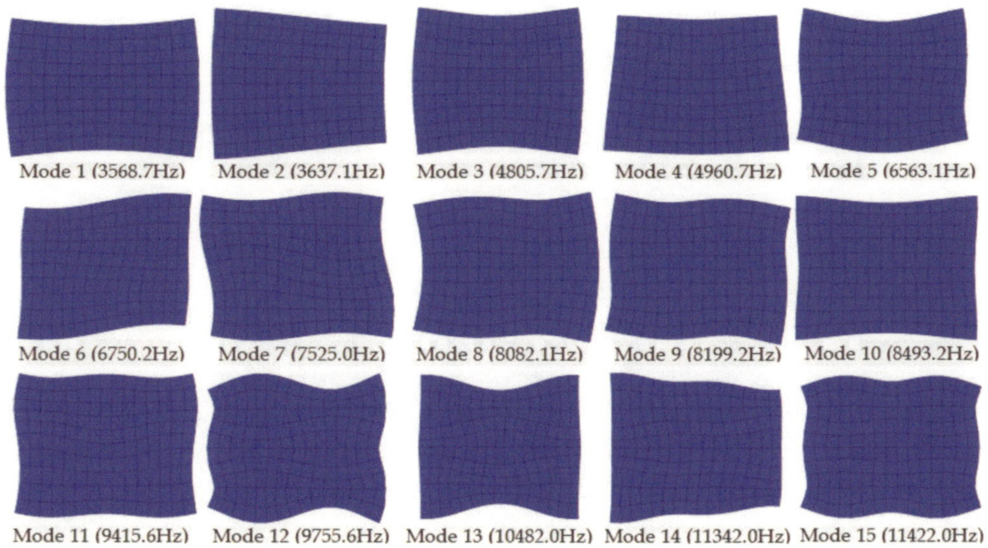

Fig. 11. Fifteen first in-plane mode shapes and natural frequencies to ortho IV rectangular plate.

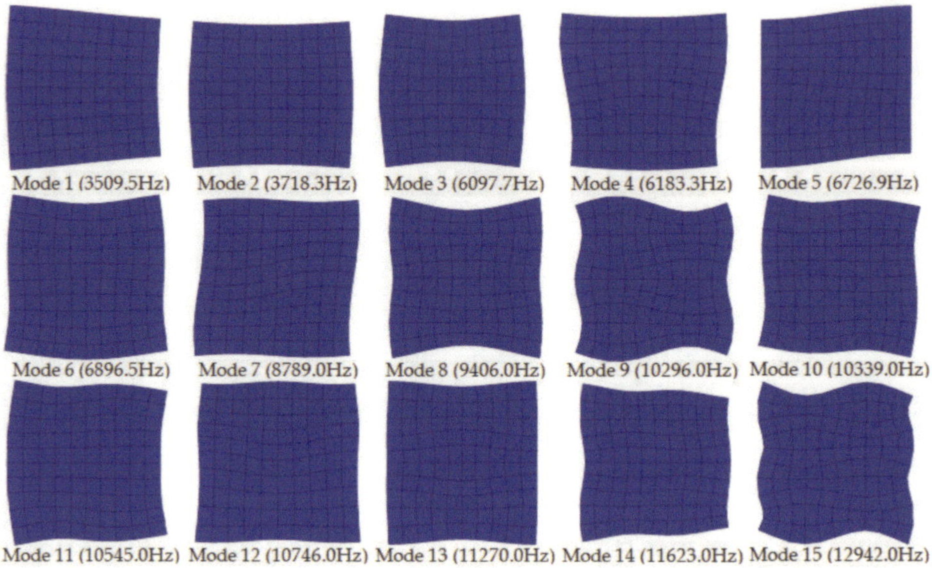

Fig. 12. Fifteen first in-plane mode shapes and natural frequencies to ortho IV square plate.

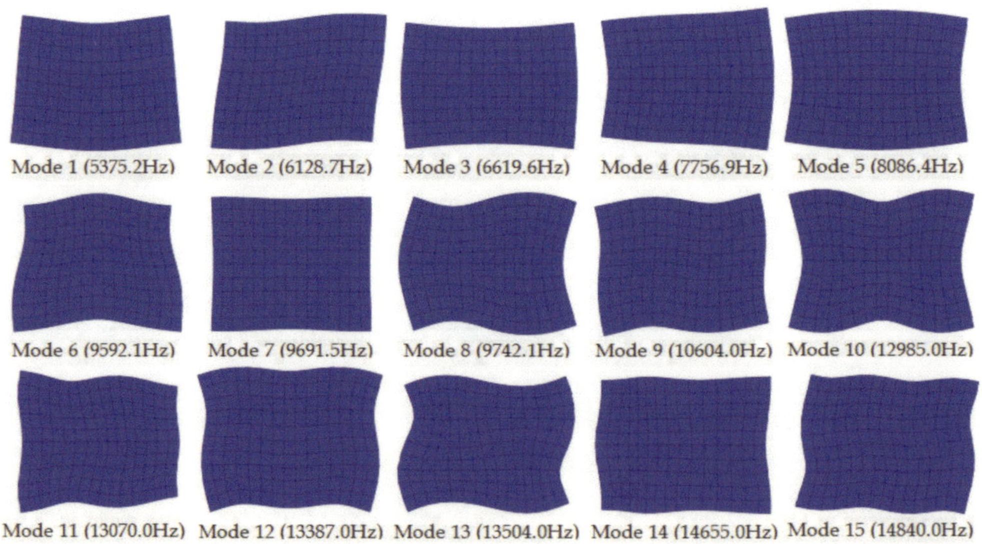

Fig. 13. Fifteen first in-plane mode shapes and natural frequencies to quasi-iso rectangular plate.

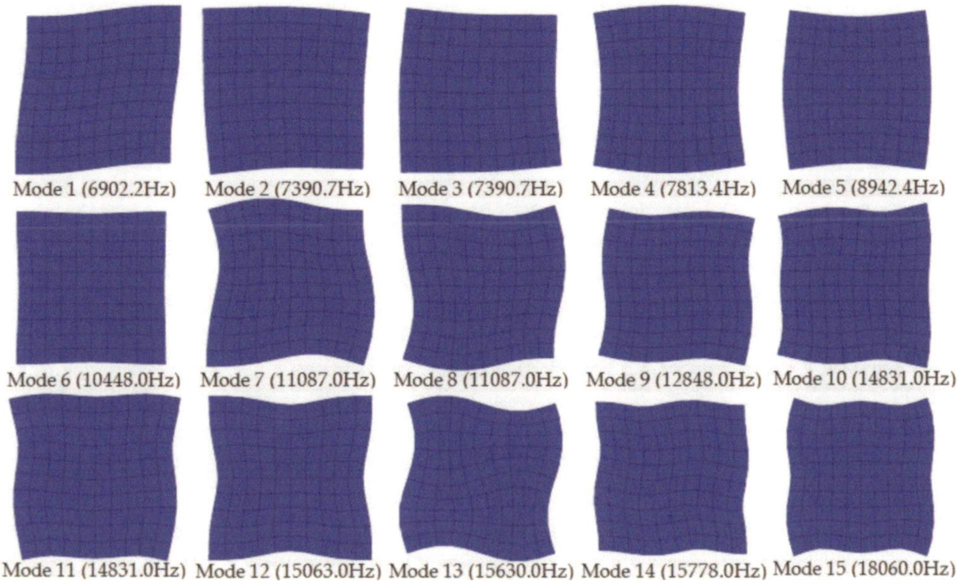

Fig. 14. Fifteen first in-plane mode shapes and natural frequencies to quasi-iso square plate.

4. Conclusions

The identification of elastic properties using VFM has shown to be a very efficient technique since the correct combinations among mode shapes and weighing functions are used. This factor is the key point to find the correct results. However, the identification of these suitable combinations is not so simple in some situations, mainly to the extensional elastic stiffness identification method. Fortunately, there are some characteristics that can help to find such combinations, as it was shown here. A great advantage of this method is related to the large number of possibilities to make combinations able to give very satisfactory results.

5. References

Alfano, M. & Pagnotta, L. (2007). A non-destructive technique for the elastic characterization of thin isotropic plates. *NDT Internatinal*, Vol. 40, pp. 112-120.

Araújo, A. L.; Mota Soares, C. M.; Moreira Freitas, M. J.; Pedersen, P. & Herskovits, J. (2000). Combined numerical-experimental model for the identification of mechanical properties of laminated structures. *Composite Structures*, Vol. 50, pp. 363-372.

Avril, S.; Huntley, J. M.; Pierron, F. & Steele, D. D. (2008). 3D heterogeneous stiffness identification using MRI and the virtual fields method. *IEEE Transactions on Medical Imaging*, Vol. 48, No. 4, pp. 479-494.

Avril, S. & Pierron, F. (2007). General framework for the identification of elastic constitutive parameters from full-field measurements. *International Journal of Solids and Structures*, Vol. 44, pp. 4978-5002.

Ayorinde, E. O. & Gibson, R. F. (1995). Improved method for in-situ elastic constants of isotropic and orthotropic composite materials using plate modal data with

trimodal and hexamodal Rayleigh formulations. *Journal of Vibration and Acoustics*, Vol. 117, pp. 180-186.
Ayorinde, E. O. & Yu, L. (1999). On the use of diagonal modes in the elastic identification of thin plates. *Journal of Vibration and Acoustics*, Vol. 121, pp. 33-40.
Ayorinde, E. O. & Yu, L. (2005). On the elastic characterization of composite plates with vibration data. *Journal of Sound and Vibration*, Vol. 283, pp. 243-262.
Bledzki, A. K.; Kessler, A.; Rikards, R. & Chate, A. (1999). Determination of elastic constants of glass/epoxy unidirectional laminates by the vibration testing of plates. *Composites Science and Technology*, 59:2015-2024.
Bruno, L.; Felice, G.; Pagnotta, L.; Poggialini, A. & Stigliano, G. (2008). Elastic characterization of orthotropic plates of any shape via static testing. *International Journal of Solids and Structures*, Vol. 45, pp. 908-920.
Chalal, H.; Avril, S.; Pierron, F. & Meraghni, F. (2006). Experimental identification of a nonlinear model for composites using the grid technique coupled to the virtual fields method. *Composite Part A: Applied Science and Manufacturing*, Vol. 37, No. 2, pp. 315-325.
Chakraborty, S. & Mukhopadhyay, M. (2000). Estimation of in-plane elastic parameters and stiffener geometry of stiffener plates. *Journal of Sound and Vibration*, Vol. 231, No. 1, pp. 99-124.
Cugnoni, J.; Gmür, T. & Schorderet, A. (2007). Inverse method based on modal analysis for charactering the constitutive properties of thick composite plates. *Computers and Structures*, Vol. 85, pp. 1310-1320.
Deobald, L. R. & Gibson, R. F. (1988). Determination of elastic constants of orthotropic plates by modal analysis technique. *Journal of Sound and Vibration*, Vol. 124, No. 2, pp. 269-283.
Diveyev, B. & Butiter., I. (2008). Identifying the elastic moduli of composite plates by using high-order theories. 1: theoretical approach. *Mechanics of Composite Materials*, Vol. 44, No. 1, pp. 25-36.
Diveyev, B.; Butiter., I. & Shcherbina, N. (2008). Identifying the elastic moduli of composite plates by using high-order theories. 2: theoretical-experimental approach. *Mechanics of Composite Materials*, 44(2):139-144.
Gibson, R. F. (2000). Modal vibration response measurements for characterization of composite materials and structures. *Composite Science and Technology*, Vol. 60, pp. 2769-2780.
Giraudeau, A.; Guo, B. & Pierron, F. (2006). Stiffness and damping identification from full-field measurements on vibration plates. *Experimental Mechanics*, Vol. 46, No. 6, pp. 777-787.
Giraudeau, A. & Pierron, F. (2003). Simultaneous identification of stiffness and damping properties of isotropic materials from forced vibration plates. *Comptes rendus Mécanique*, Vol. 331, No. 4, pp. 259-264.
Giraudeau, A. & Pierron, F. (2005). Identification of stiffness and damping properties of thin isotropic vibrating plates using the Virtual Fields Method. Theory and simulations. *Journal of Sound and Vibration*, Vol. 284, No.(3-5), pp. 757-781.
Grédiac, M. (1996). On the direct determination of invariant parameters governing the bending of anisotropic plates. *International Journal of Solids and Structures*, Vol. 33, pp. 39690-3982.

Grédiac, M. (2004). The use of full-field measurement methods in composite material characterization: interest and limitations. *Composites Part A*, Vol. 35, pp. 751-761.

Grédiac, M.; Auslender, F. & Pierron, F. (2001). Applying the virtual fields method to determine the through-thickness moduli of thick composites with a nonlinear shear response. *Composites Part A*: Applied Science and Manufacturing, Vol. 32, No. 12, pp. 1713-1725.

Grédiac, M.; Founier, N.; Paris, P. A. & Surrel, Y. (1999a). Direct identification of elastic constants of anisotropic plates by modal analysis: experiments and results. *Journal of Sound and Vibration*, Vol. 210, pp. 645-659.

Grédiac, M.; Founier, N.; Surrel, Y. & Pierron, F. (1999b). Direct measurement of invariant parameters of composite plates. *Journal of Composite Materials*, Vol. 33, pp. 1939-1965.

Grédiac, M. & Paris, P. A. (1996). Direct identification of elastic constants of anisotropic plates by modal analysis: theoretical and numerical aspects. *Journal of Sound and Vibration*, Vol. 195, pp. 401-415.

Grédiac, M. & Pierron, F. (1998). A T-shaped specimen for the direct characterization of orthotropic materials. International Journal for Numerical Methods in Engineering, Vol 41, pp. 293-309.

Grédiac, M. & Pierron, F. (2006). Applying the Virtual Fields Method to the identification of elasto-plastic constitutive parameters. *International Journal of Plasticity*, Vol. 22, pp. 602-627.

Grédiac, M.; Pierron, F.; Avril, S. & Toussaint, E. (2006). The virtual fields method for extracting constitutive parameters from full-field measurements: a review. *Strain: an International Journal for Experimental Mechanics*, Vol. 42, pp. 233-253.

Grédiac, M.; Toussaint, E. & Pierron, F. (2002a). Special virtual fields for direct determination of material parameters with the virtual fields method. 1-Principle and definition. *International Journal of Solids and Structures*, 39(10):2691-2705.

Grédiac, M.; Toussaint, E. & Pierron, F. (2002b). Special virtual fields for direct determination of material parameters with the virtual fields method. 2-Application to in-plane properties. *International Journal of Solids and Structures*, Vol. 39, No. 10, pp. 2707-2730.

Grédiac, M.; Toussaint, E. & Pierron, F. (2003). Special virtual fields for direct determination of material parameters with the virtual fields method. 2-Application to the bending rigidities of anisotropic plates. *International Journal of Solids and Structures*, Vol. 40, No. 10, pp. 2401-2419.

Hwang, S. & Chang, C. (2000). Determination of elastic constants of materials by vibration testing. *Composite Structures*, Vol. 49, pp. 193-190.

Lai, T. C. & Lau, T. C. (1993). Determination of elastic constants of a generally orthotropic plate by modal analysis. *International Journal of Analytical and Experimental Modal Analysis*, Vol. 8, pp. 15-33.

Lauwagie, T.; Sol, H.; Heylen, W. & Roebben, G. (2004). Determination of in-plane elastic properties of different layers of laminated plates by means of vibration testing and model updating. *Journal of Sound and Vibration*, Vol. 274, pp. 529-546.

Lauwagie, T.; Sol, H.; Roebben, G.; Heylen, W.; Shi, Y. & Van der Biest, O. (2003). Mixed numerical-experimental identification of elastic properties of orthotropic metal plates. NDT e International, Vol. 36, pp. 487-495.

Lee, C. R. & Kam, T. Y. (2006). Identification of mechanical properties of elastically restrained laminated composite plates using vibration data. *Journal of Sound and Vibration*, Vol. 295, pp. 999-1016.

Pagnotta, L. & Stigliano, G. (2008). Elastic characterization of isotropic plates of any shape via dynamic tests: theoretical aspects and numerical simulations. *Mechanics Research Communications*, Vol. 35, pp. 351-360.

Pedersen, P. & Frederiksen, P. S. (1992). Identification of orthotropic material moduli by a combined experimental / numerical method. Measurement, Vol. 10, pp. 113-118.

Pierron, F. & Grédiac, M. (2000). Identification of the through-thickness moduli of thick composite from whole-field measurements using the Iosipescu fixture. *Composites Part A: Applied Science and Manufacturing*, Vol. 31, No. 4, pp. 309-318.

Pierron, F.; Zhavoronok, S. & Grédiac, M. (2000). Identification of the through-thickness properties of thick laminated tubes using the virtual fields method. Internatinal Journal of Solids and Structures, Vol. 37, No. 32, pp. 4437-4453.

Pierron, F.; Vert, G.; Burguete, R.; Avril, S.; Rotinat, R. & Wisnom, M. (2007). Identification of the orthotropic elastic stiffness of composites with the virtual fields method: sensitivity study and experimental validation. Strain: an *International Journal for Experimental Mechanics*, Vol. 43, No. 3, pp. 250-259.

Reverdy, F. & Audoin, B. (2001). Ultrasonic measurement of elastic constants of anisotropic materials with laser source and laser receiver focused on the same interface. *Journal of Applied Physics*, Vol. 90, No. 9, pp. 4829-4835.

Rikards, R. & Chate, A. (1998). Identification of elastic properties of composites by method of planning of experiments. *Composite Structures*, Vol. 42, pp. 257-263.

Rikards, R.; Chate, A. & Gailis, G. (2001). Identification of elastic properties of laminates based on experiment design. *International Journal of Solids and Structures*, Vol. 38, pp. 5097-5115.

Rikards, R.; Chate, A.; Steinchen, W.; Kessler, A.& Bledzki, A. K. (1999). Method for identification of elastic properties of laminates based on experiment design. *Composites: Part B*, Vol. 30, pp. 279-289.

Toussaint, E.; Grédiac, M. & Pierron, F. (2006). The virtual fields method with piecewise virtual fields. *International Journal of Mechanical Sciences*, Vol. 48, pp. 256-264.

4

Analytical Research on Method for Applying Interfacial Fracture Mechanics to Evaluate Strength of Cementitious Adhesive Interfaces for Thin Structural Finish Details

Tsugumichi Watanabe
Osaka City University
Japan

1. Introduction

Concrete structures often incorporate large numbers of thin-section finishing components that are bonded or joined to the concrete with inorganic cement-type adhesives, such as cement mortar, when wet construction methods are used. These components, usually ceramic tile or other finished products, are emplaced by experienced workmen to create exterior finishing products using plastering cement mortar as adhesive. These exterior layers are typically 5–30 mm thick. Cement mortar finishing products are normally single-layer, while ceramic tile finishing products usually consist of two layers, with the tile emplaced atop adhesive cement mortar. One of the typical signs of decay for these components is delamination at the interface of the finishing product and the concrete structure. Delamination is usually attributed to high shear stresses and fractures at the interface caused by elastic and creep strain in the concrete due to the weight of the structure, and expansion-contraction strains caused by environmental variations.
Because of this, Japanese institutions have begun testing samples of thin sections attached to concrete with cement mortar using the method shown in Figure 1. An axial compressive load is then applied to exert a shear stress on the interface between the two materials. This approach is simple and provides a macroscopic measure of the shear resistance at the interface while the concrete is under compressive loading. An example of the test results obtained from ceramic tile finishing products is shown in Figure 2. This shows the how the strain on the lateral face of the tile varies with the mean stress at the loaded face of the concrete. Since the shear stress is concentrated at the edge of the interface between the finishing product and the underlying concrete, this is where delamination between the two bodies normally begins. Specifically, at the edge of the interface between the concrete and the cement mortar used as adhesive. Once the delamination propagates laterally out to the edges of the interface, where the strain is measured in these observations, the connection fails and the indicated strain terminates abruptly. These phenomena are often seen when separation has occurred due to extremely brittle fracture conditions at the interface. However, a method for evaluating the interface strength under the installed thin finishing components has yet to be developed.
It is possible that adhesive design technologies based on linear elastic interfacial fracture mechanics could prove useful for assessing the fracture strength of the cement-type

adhesive at the interface when it fails due to excessive brittleness, i.e., unstable fracture. If so, it will first be necessary to not only describe the stress field in the vicinity of the finishing product interface, but also to determine methods for calculating the stress singularity parameters governing interfacial fracture mechanics of the situation, and to subsequently validate the effectiveness of those parameters.

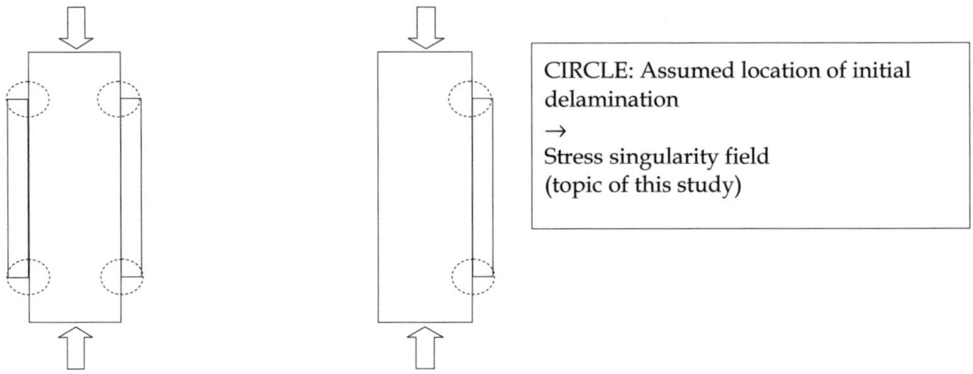

(a) Two-sided test (b) One-sided test

Fig. 1. Overview of test for improved deformation performance under compressive stress

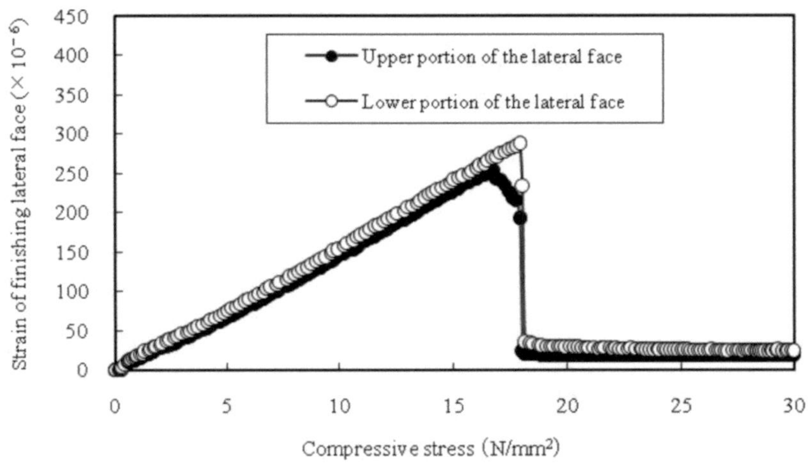

Fig. 2. Relationship between compressive stress and strains on the lateral face of a concrete sample

The purpose of this report is to propose a method for applying linear elastic interfacial fracture mechanics to problems related to the separation of cement-type materials as one method for testing simple shear stresses in ceramic tile, and to validate its practical applicability. The stress singularity parameters indicated by interfacial fracture mechanics can be handled as indicators of the interface strength for unstable fracture. If the proposed testing method is found to be practical, it could then be used in the same manner as the simple shear test when examining cement-type structure finishing products that are especially prone to separation. It may also be possible to extend these results to the design of the surface shapes that serve as structural interfaces in order to reduce the stresses acting on those interfaces.

2. Stress functions applied and their solutions

2.1 Stress singularity field at edge of the interface between dissimilar materials in a semi-infinite region

We will now re-draw the model of a notch in the edge of an adhesive interface between the components of a semi-infinite body composed of two dissimilar materials containing notches, which is a situation that gives rise to a stress field containing a singularity (Figure 3). It has previously been determined that the magnitude of stress can then be approximated by use of the logarithmic function shown in Eq. (1). Here, the parameters indicating the singularity are the order of stress singularity and the stress intensity factor. The order of stress singularity is determined by the notch angles θ_1 and θ_2 and the mechanical properties of the material. When both θ_1 and θ_2 are 180°, as occurs in the case of a crack, λ is 0.5, regardless of the material constants.

$$\sigma \propto 1/r^{\lambda} \tag{1}$$

Here, $\sigma(r)$ is the stress in the vicinity of the singularity point, r is the distance from the singularity point, and λ is the order of stress singularity.

However, Eq. (1) only provides the stress profile. It is still important to consider the stress intensity factor, which indicates the magnitude of the singularity stress field. In other words, the stress intensity factor is believed to be a function of material constants and of the shape of the bond including θ_1 and θ_2, but no unified theory of stress intensity factors that considers the bonded structures of differing materials has been formulated. Groth(1988),Hattori(1989, 1991) and Quaresimin(2006) expressed stress relationships as shown in Eq. (2) in order to apply Eq. (1) to practical circumstances. He expressed the stress relationship as shown in Eq. (2) for use in the practical applications of Eq. (1) and advocated using K as a generalized stress intensity factor when considering delamination caused by thermal stress loads. In this paper, we attempt to apply this simplified stress function to the simple shear stress test method shown in Figure 1.

$$\sigma(r) = K/r^{\lambda} \tag{2}$$

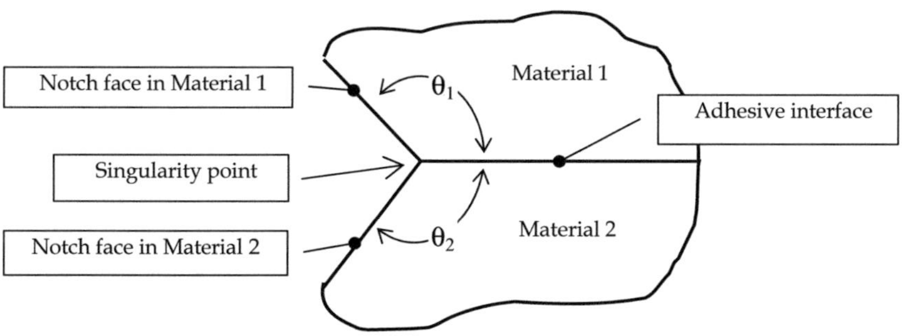

Groth(1988), Hattori(1989, 1991) and Quaresimin(2006)

Fig. 3. Model of interface between notch and dissimilar materials

2.2 Method for stress analysis

In this analytical model, the simple shear strength test sample is assumed to have the material composition and shape shown in Figure 1. The one-sided half-model shown in Figure 4 was used. The sample was assumed to have the mechanical properties found in actual tests. The stress applied by the loading plate was 10 N/mm^2 and the boundary condition between the loading plate and the concrete was assumed to be simple contact. A numerical analysis was performed on a two-dimensional elastic body under a plane strain conditions with the stress applied by a boundary element program. The minimum element size in the vicinity of the singularity was 0.2 mm; the material conditions are shown in Table 1. Compressive tests were carried out on concrete, cement mortar and ceramic tile and the elastic coefficients (longitudinal elasticity and Poisson's ratio) were found at 1/3 the maximum level for elastic behavior for concrete and cement mortar. Because it is difficult to measure Poisson's ratio in ceramic tile samples, this value was assumed to be 0.18.

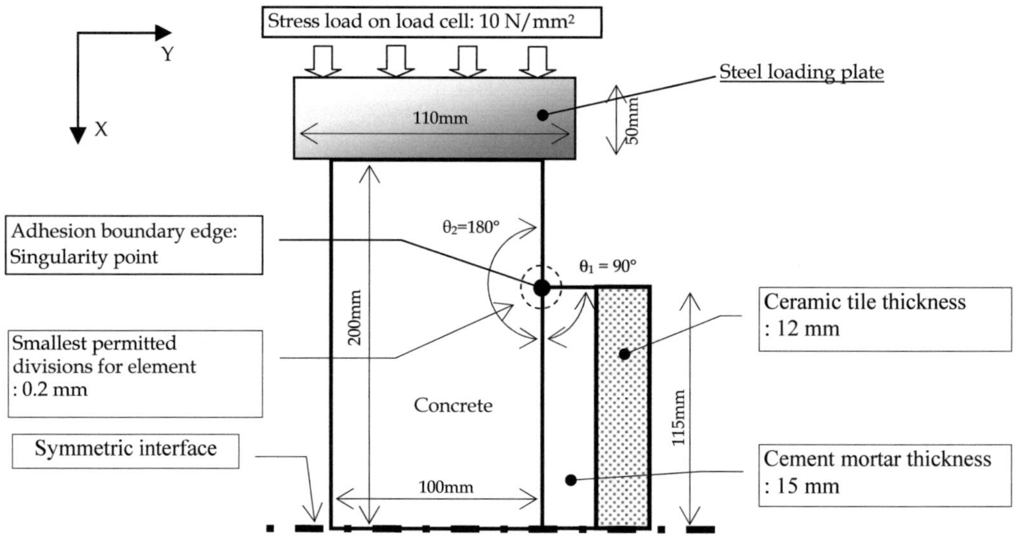

Fig. 4. Model for boundary element method

Concrete	Shape	• Width; 100 mm, Thickness; 100 mm, Length; 400 mm
	Properties	• Material: Ordinary concrete Young's modulus; 26700 N/mm^2, Poisson's ratio; 0.18
Cement mortar	Shape	• Width; 100 mm, Thickness; 15 mm, Length; 230 mm
	Properties	• Material: Ordinary mortar Young's modulus; 24000 N/mm^2, Poisson's ratio; 0.22
Ceramic tile	Shape	• Width; 100 mm, Thickness; 12 mm, Length; 230 mm
	Properties	• Material: Ordinary stoneware tile Young's modulus; 38100 N/mm^2, Poisson's ratio; 0.18

Table 1. Basic conditions for each layer

2.3 Stress field at the adhesive interface

A log-log chart of the absolute value of each interfacial stress versus the distance r (0.5–100 mm) from the leading edge of the interface is shown in Figure 5. It was found that σ_x (normal stress parallel to the interface) and τ (shear stress) varied in an approximately linear fashion within the range near the edge ($r \approx 0.5$ mm) to log (r) = 0.8 ($r \approx 7$ mm). This stress distribution indicates that it may be possible to express the stress function using Eq. (2). At distances over 7 mm, this relationship diverges from linearity, and ultimately, the singularity disappears. Meanwhile, σ_y (normal stress perpendicular to the interface) does not show any such linearity, even in the region quite close to the edge (within r = 0.5), and stress decreases rapidly, indicating disappearance of the singularity. These results indicate that incorrect values for the stress singularity parameters will be obtained by improper application of the linear relationship in Eq. (2) if calculations are carried out too close to the edge of the interface, or if the elements selected for numerical analysis are too coarse.

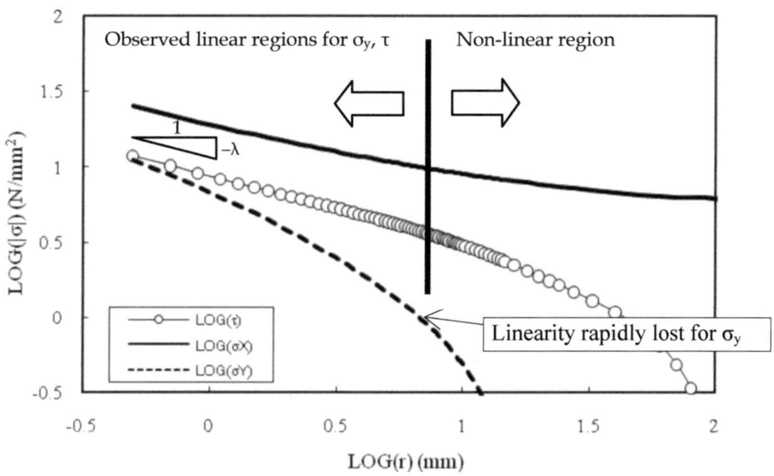

Fig. 5. Log-log graph of stress distribution (absolute values)

Two cracking modes are considered when assessing the risk of interface delamination in two-dimensional problems: Mode I (tensile stress-induced cracking) and Mode II (shear-induced cracking). These are induced by σ_y and τ. Here, σ_y is a negative stress, which acts to close cracks, so Mode I fractures cannot occur. Thus, a Mode II fracture caused by shear stress τ is the only mode assumed to act during this simple shear test. Therefore, the expressions for stress as it affects the stress singularity in the present study are exclusively τ, which has the chief influence over interface fracture here. The linear region of logarithmic stress in Figure 5 (up to $r \approx 7$ mm) was divided into three sub-regions. The solutions of further least-squares analyses for each portion are presented in Table 2. Each linearization showed a close resemblance to the order of the stress singularity, but varied from it by 3 to 7%. Thus, the piecewise linear data showed quite high linearity in each sub-region, but it is clear that the calculated values are vulnerable to several factors, including the choice of coarseness in the elements selected for the numerical analysis. This indicates that it is essential to develop an appropriate approach to these calculations. K′ is the apparent stress intensity factor found from the approximately linear segment where log (r) = 0.

Region	Distance r from edge (mm)	Log (r)	Apparent singularity order: λ (ratio with order near edge)	Apparent stress intensity factor K' (N mm$^{\lambda-2}$)
A	0.5–1.3	−0.3010 to 0.1139	0.445 (1.000)	8.57 (corr. coeff.: 0.9999)
B	1.3–3.1	0.1139 to 0.4914	0.414 (0.930)	8.52 (corr. coeff.: 1.0000)
C	3.1–6.9	0.4914 to 0.8388	0.458 (1.029)	8.95 (corr. coeff.: 0.9993)

Table 2. Results for piecewise linearization of shear stress

3. Calculation of the stress singularity parameters

3.1 Calculation of the order of stress singularity

Bogy(1971) employed the two-dimensional semi-infinite model shown in Figure 3, which incorporates Young's modulus and Poisson's ratio as E_1, v_1 and E_2, v_2 and notch angles θ_1, θ_2 for the respective materials. He derived theoretical values for λ from the roots **p** of the characteristic Eq. (3), which are used in Eq. (2). In this report, the stress singularity field occurring at the edge of the interface between the concrete and cement mortar layers in the finished structure is approximated with Eq. (1). It is proposed that the theoretical solutions to Bogy's characteristic equation be applied to the order of the stress singularity:

$$\mathbf{Q}(\theta_1,\theta_2,\alpha,\beta;p) = 0 \tag{3}$$

where the characteristic function Q and coefficients are determined by the following equations:

$$\mathbf{Q}(\theta_1,\theta_2,\alpha,\beta;p)= \\ =\beta^2\cdot A(\theta_1,\theta_2;p)+2\cdot\alpha\cdot\beta\cdot B(\theta_1,\theta_2;p)+\alpha^2\cdot C(\theta_1,\theta_2;p)+2\cdot\beta\cdot D(\theta_1,\theta_2;p)+2\cdot\alpha\cdot E(\theta_1,\theta_2;p)+F(\theta_1,\theta_2;p) \tag{4}$$

$$G_i = E_i \,/\, (2(1 + v_i)) \quad (i = 1,2) \tag{5}$$

$$m_i = \begin{cases} 4(1 - v_i);\ \text{plane strain} \\ 4/(1 + v_i);\ \text{plane stress} \end{cases} \quad (i = 1,2) \tag{6}$$

$$\alpha = (G_1\cdot m_2 - G_2\cdot m_2)/(G_1\cdot m_2 + G_2\cdot m_1) \tag{7}$$

$$\beta = (G_1(m_2 - 2) - G_2(m_1 - 2))/(G_1\cdot m_2 + G_2\cdot m_1) \tag{8}$$

If we use

$$\mathbf{H}(p,\theta) = \sin^2(p\theta)-P^2\sin^2(\theta) \tag{9}$$

then A–F in Eq. (4) are determined as follows:

$$\left.\begin{aligned}
\mathbf{A}(\theta_1,\theta_2;p) &= 4H(p,\theta_1)\cdot H(p,\theta_2)\\
\mathbf{B}(\theta_1,\theta_2;p) &= 2p^2\cdot\sin^2(\theta_1)\cdot H(p,\theta_2)+2p^2\cdot\sin^2(\theta_2)\cdot H(p,\theta_1)\\
\mathbf{C}(\theta_1,\theta_2;p) &= 4p^2\cdot(p^2-1)\cdot\sin^2(\theta_1)\cdot\sin^2(\theta_2)+H\{p,(\theta_1-\theta_2)\}\\
\mathbf{D}(\theta_1,\theta_2;p) &= 2p^2\{\sin^2(\theta_2)\cdot\sin^2(p\theta_1)-\sin^2(\theta_1)\cdot\sin^2(p\theta_2)\}\\
\mathbf{E}(\theta_1,\theta_2;p) &= -D(\theta_1,\theta_2;p)+H(p,\theta_1)-H(p,\theta_2)\\
\mathbf{F}(\theta_1,\theta_2;p) &= H\{p,(\theta_1+\theta_2)\}
\end{aligned}\right\}\quad(10)$$

The order of stress singularity λ is then found from the roots p of the above characteristic equation as follows:

$$\lambda = 1-p \quad (11)$$

where p is a real number, $0<p<1$, when the singularity occurs, and $p\geq1$ when no singularity occurs.

There are cases where p is complex, but a detailed analysis by Hein, Erdogan and others(1971) showed clearly that p is always real under the mechanical conditions imposed on the materials considered in the present study, as will be described below.

The solution to the characteristic equation using a singularity point case at the edge of a finish is shown in Figure 6. Here, notch angle θ_1 = 90° for the finish layer angle and θ_2 = 180°for the concrete adhesion angle under conditions that are otherwise identical to those for the simple shear test shown in Figure 1. Thus, the roots p are found by iteration until the characteristic equation Q on the left side of Eq. (3) is zero. This result indicates two values, p = 0.548, 0.925, yielding λ = 0.452, 0.075 in Eq. (11). As shown in Table 2, values for stress singularity λ differ depending on analytic region, and the log-log chart does not actually provide a linear solution because two orders of stress singularity λ exist. The two stress fields of these singularities are actually superimposed on each other to produce the overall field. If the stress function is simplified as in Eq. (2), one of these λ must be applied. Additionally, the order with the greater magnitude is believed to be dominant; this approach predicts that, of the two values for λ given above, 0.452 would be the appropriate value to employ. This theoretical value is very close to the value for the gradient found in Table 2, 0.445, confirming that under the present analytical conditions, the theoretically predicted value for the order of stress singularity is acceptable in the vicinity of the edge of the attachment interface.

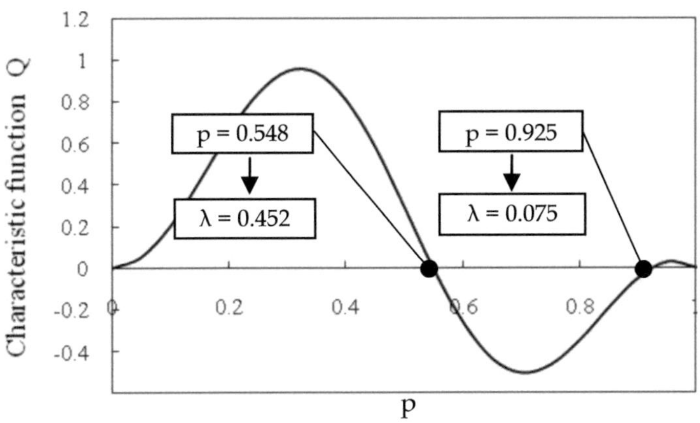

Fig. 6. Calculated characteristic function

3.2 Calculation of stress intensity factor at the edge of the attachment interface

The stress intensity factor at the singularity point is estimated by direct extrapolation from adjacent values. First, Eq. (2) is re-written as Eq. (12) in order to determine Kr from the distance r and the numerical solution for τ:

$$Kr(\tau,r) = \tau \cdot r^{\lambda} \tag{12}$$

where the larger of the two values found in Bogy's theoretical solution for order of stress singularity λ is provisionally employed. This will be examined for its suitability below. The results using Eq. (12) are shown in Figure 7. The value for Kr is nearly linear within the first 5 mm from the edge of the interface. The stress intensity factor within this range Kr_0 was estimated with the extrapolation formula in Eq. (13). This value was 8.52 N mm$^{\lambda-2}$, nearly the same as the apparent value given for K' in Table 2.

$$Kr_0 = \lim_{r \to 0} Kr(\tau,r) \tag{13}$$

When the stress intensity factor Kr_0 and order of stress singularity λ are calculated by the above method, the distribution of shear stress near the edge of the finish work is given by Eq. (14). The values for τ predicted by the numerical model and by Eq. (14) are compared in Figure 8; the extrapolation provided a good approximation of Kr_0 within the extrapolation region.

$$\tau(r) = Kr_0 / r^{\lambda} \tag{14}$$

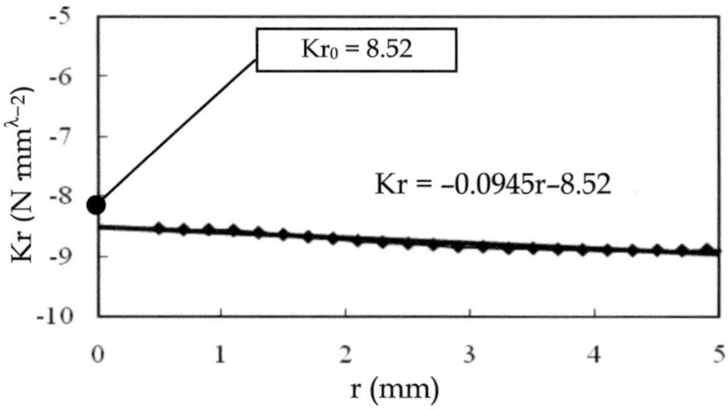

Fig. 7. Extrapolation of Kr_0 (theoretical value for λ)

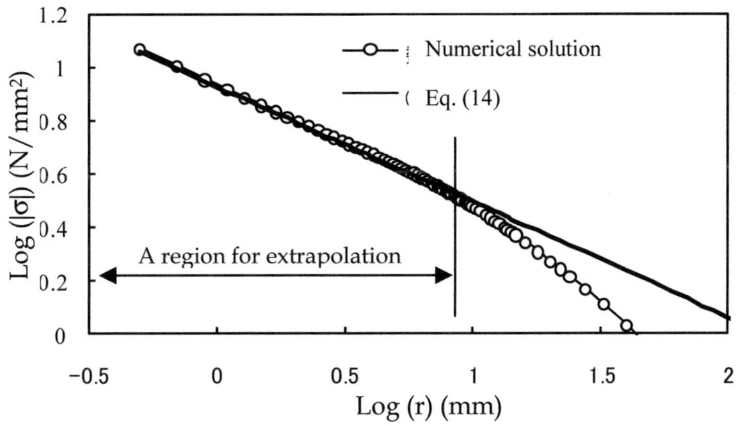

Fig. 8. Comparison of numerical solution with the solution yielded by the proposed method.

4. Parametric analysis of finishing conditions

4.1 Issues under investigation

This report examines the problem of whether the method for estimating the stress intensity factor by fitting the calculated value of the order of stress singularity in Eq. (13) to Eq. (2) is valid. There are two methods for calculating the order of stress singularity: finding the gradient with a linear approximation of the shear stress (the approximated solution) and the method proposed here, based on Bogy's theoretical analysis (the theoretical solution). The approximated solution provides an inaccurate value in the approximation region (Table 2) where there is a risk of influence on the estimate from the accuracy of the analytical method or from too-coarse elements. In contrast, the theoretical solution provides an unambiguous figure, based on the material constants and notch angles illustrated in Figure 3. This avoids the issues of the approximate approach. However, the material is thin near the edge of the adhesive interface and the situation here is very different from the semi-infinite region shown in Figure 3. For this reason, it is not clear whether the stress field takes the form predicted by the current theory. It must also be determined whether this theoretical solution should be fitted to Eq. (2) in situations where the theoretical solution provides multiple real-number solutions. Therefore, in the present study, this calculation method is evaluated by comparing the value for the order of stress singularity found by the theoretical approach to the approximated solution in an extremely narrow region near the singularity point, which is taken as the "true" value. The stress intensity factor is a very important parameter, used in the failure criteria for the adhesive interface and in other calculations. Therefore, the method for estimating this value is assessed in the same way as was previously described for the order of stress singularity. That is, by evaluating the approximated solution in close proximity to the singularity point and comparing this result to that provided by the approach proposed in this paper. The approximated solution was obtained in Region A, which is quite close to the singularity point, as shown in Table 2.

Both stress singularity parameters are believed to be strongly influenced by the geometrical conditions and mechanical characteristics of the finish material layer; thus, in the present study, the finish conditions are divided into three different series for separate parametric analyses.

4.2 Analytical models and basic conditions

The analytical conditions were defined as depth and length of the tile layer and mortar layer, as well as elastic coefficients. The typical real values of these parameters were used. The calculation methods are summarized in Figure 9. The analytical model was as shown in Figure 4, consisting of three layers: concrete, adhesive cement mortar, and ceramic tile. The model was one-sided and was symmetric about the horizontal center line. Two-dimensional plane strain analysis was used, assuming elastic media, and the boundary element method was employed. Since there was no influence on the stress singularity parameters from the loading, the same loading was used as in Figure 4, 10 N/mm^2. The basic conditions in each layer were also as shown in Table 1.

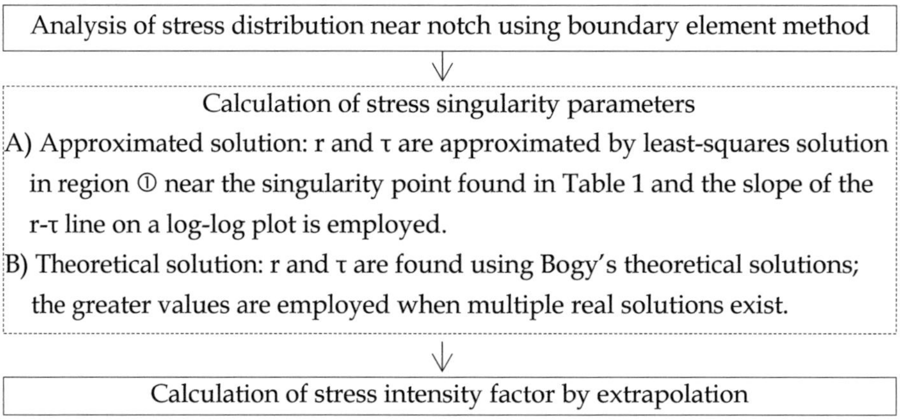

Fig. 9. Flow chart of calculations of stress singularity parameters

4.2 Scheme of analysis

The three series of analytical conditions are presented in Table 3.

In Series I, the mortar layer thickness and tile layer thickness were the analytical factors. Three mortar thicknesses and four integral multiples of the tile thickness (6 mm) were utilized.

In Series II, the analytical factor was finish layer length. The basic dimension was the length of a standard tile (227 mm). Four lengths were used, 0.25 to 1.5 times the basic dimension. The materials used in the mortar layer and tile layer were the same as in Series I.

In Series III, the relevant factors were the materials used for the mortar and tile layers. The basic conditions differed from those in Table 3. In anticipation of the future use of mortar with low Young's modulus, which has been under development in recent years, one of the two types of mortar in the mortar layer was assumed to have a low Young's modulus. The tile layer was also assumed to be one of two types, either stoneware finishing tile or porcelain finishing tile (both with Young's modulus of 44800 N/mm^2 and Poisson's ratio of 0.18).

Series	Layer	Conditions
I	Ceramic tile	Four thicknesses: 6–24 mm, Length: all standard length 227 mm Material: Stoneware tile
	Mortar	Three thicknesses: 5-15 mm, Length: same as tile Material: Ordinary mortar
II	Ceramic tile	Thickness: 12 mm Composition of tile and joints [Non-jointed configuration]: 4 lengths 1) 341 mm (1.5 × standard length) 2) 227 mm (standard length) 3) 114 mm (1/2 standard length) 4) 57 mm (1/4 standard length) [Jointed configuration]: 2 lengths 1) 2 tiles + 1 joint (tile 1/2 standard length, 1 joint 6 mm long in center) 2) 3 tiles + 1 joint (attachment length 354 mm) (tile 1/2 standard length, 2 joints 6 mm long in center) Material: Stoneware tile
	Mortar	Thickness: 15 mm, Length: same as tile Material: Ordinary mortar
III	Ceramic tile	Thickness: 12 mm, Length: all standard length 227 mm Material: 2 types, Stoneware tile and porcelain tile finishing 1) Stoneware tile 2) Porcelain tile (Young's modulus: 44800 N/mm^2;, Poisson's ratio: 0.18)
	Mortar	Two thicknesses: 5 and 15 mm, Length: same as tile Material: Ordinary mortar and light mortar 1) Ordinary mortar 2) Two types of light mortar (1/2 and 1/4 longitudinal Young's modulus of ordinary mortar)

* All widths 100 mm

Table 3. Analytical Conditions

4.3 Results of analysis for shear stress distribution

4.3.1 Series I

The results of this analysis are shown in Table 4. Almost no influence from the mortar or tile thickness was observed on the order of stress singularity in the approximated solution. The variation in the theoretical solution was small at 1-5%. The theoretical solution appears to be capable of predicting the order of stress singularity with adequate accuracy.

Variation in the calculated stress intensity factor versus finish layer thickness, the sum of mortar thickness and tile thickness is shown in Figure 10. The stress intensity factor increased gradually with tile thickness at all mortar layer thicknesses. This increase was most marked in the 5-mm mortar layer; above the finish thickness of 20 mm; however, it remained generally constant within the range 7–8 $N \cdot mm^{\lambda-2}$. No large variations were seen in the stress intensity factor under these analytical conditions. The stress intensity factor also showed extremely little variation with the method of calculation of stress singularity order.

Mortar thickness (mm)	Tile thickness (mm)	Stress singularity parameters by approximated solution		Stress singularity parameters by theoretical solution		Compared theoretical and approximated solutions	
		Order of stress singularity λ	Stress intensity factor*1 K	Order of stress singularity λ'	Stress intensity factor*1 K'	λ'/λ	K'/K
5	6	0.444	7.89	0.452	7.87	1.018	0.997
	12	0.437	8.59		8.54	1.034	0.994
	18	0.434	8.87		8.81	1.041	0.993
	24	0.431	8.89		8.89	1.049	1.000
10	6	0.450	8.18		8.18	1.004	1.000
	12	0.444	8.49		8.47	1.018	0.998
	18	0.442	8.63		8.60	1.023	0.997
	24	0.439	8.66		8.62	1.030	0.995
15	6	0.449	8.37		8.36	1.007	0.999
	12	0.445	8.51		8.52	1.016	1.001
	18	0.443	8.59		8.56	1.020	0.997
	24	0.441	8.58		8.55	1.025	0.997

*1) Unit: N mm$^{\lambda-2}$

Table 4. Analytical Results (Series I)

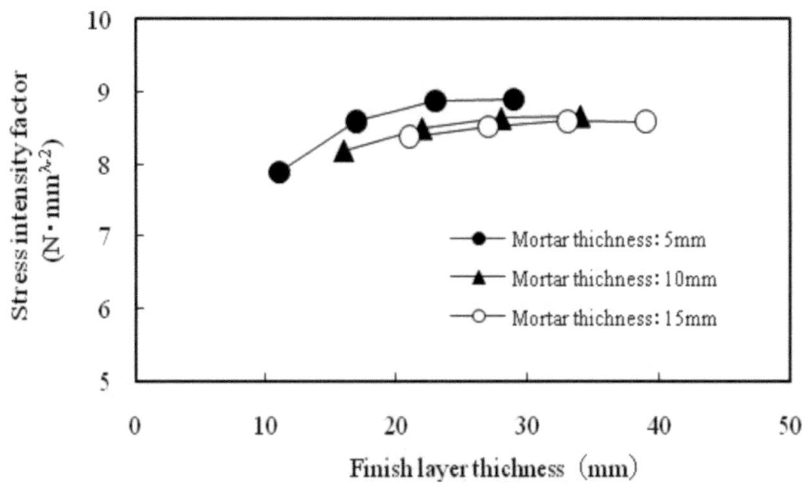

Fig. 10. Relationship between finish layer thickness and stress intensity factor (Series I)

4.3.2 Series II

The results were similar to those in Series I (see Table 5). No finish length influence was seen on order of stress singularity in the approximated solution method, which showed nearly the same values as the theoretical solution.

Mortar thickness (mm)	Tile length (times)	Stress singularity parameters by approximated solution		Stress singularity parameters by theoretical solution		Compared theoretical and approximated solutions	
		Order of stress singularity λ_a	Stress intensity factor*1 K_a	Order of stress singularity λ_b	Stress intensity factor*1 K_b	λ_B/λ_A	K_B/K_A
5	1.5	0.438	8.68	0.452	8.63	1.032	0.994
	1.0	0.437	8.64		8.54	1.032	0.988
	0.5	0.443	8.54		8.50	1.020	0.995
	0.25	0.438	7.24		7.20	1.032	0.994
15	1.5	0.445	8.61		8.58	1.016	0.996
	1.0	0.445	8.51		8.52	1.016	1.001
	0.5	0.446	8.33		8.31	1.013	0.998
	0.25	0.424	6.83		6.76	1.066	0.990

*1) Unit: N mm$^{\lambda-2}$

Table 5. Analytical Results (Series II)

Variation in the stress intensity factor with finish length is shown in Figure 11. The stress intensity factor was nearly constant at finish lengths over 100 mm, but was lower at the finish length of 57 mm. This indicates that when the finish material length is extremely small, the shear stresses operating on the interface are quite low, but at greater adhesion lengths, this low shear stress should not be assumed, as it could lead to dangerous underestimates. Despite this, when the finish length is greater than 100 mm, the obtained stress intensity
factor figures appear to be useful. It is also important to note that the differences between the approximated solution and the theoretical solution were quite small.

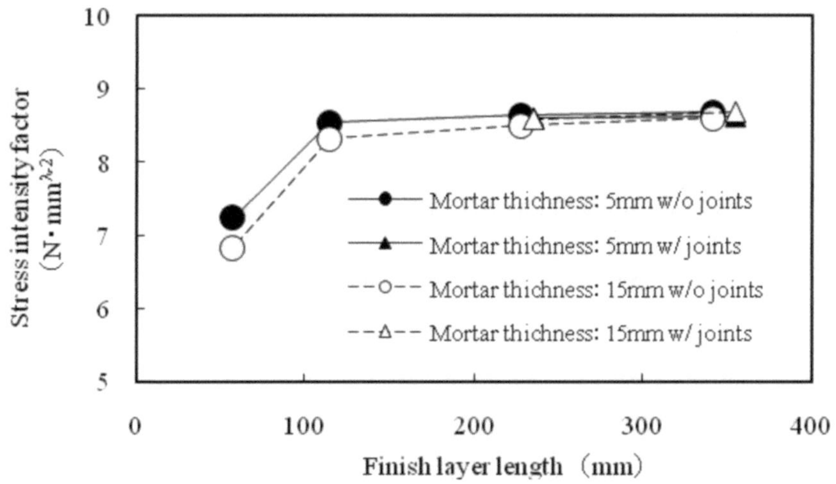

Fig. 11. Relationship between finish layer length and stress intensity factor (Series II)

4.3.3 Series III

The approximated solution indicated a strong effect of Young's modulus on the predicted order of stress singularity (see Table 6). That is, the lower the Young's modulus of the mortar, the lower the order of stress singularity. In addition, it was found that as the Young's modulus of the mortar decreased, the difference between the predictions of the approximated solution and the theoretical solution became larger; this difference also

Mortar material	Tile material	Stress singularity parameters by approximated solution		Stress singularity parameters by theoretical solution		Compared theoretical and approximated solutions	
		Order of stress singularity λ_A	Stress intensity factor*1 K_a	Order of stress singularity λ_b	Stress intensity factor*1 K_b	λ_B/λ_A	K_B/K_A
Ordinary	Stoneware	0.437	8.64	0.452	8.54	1.034	0.988
	Porcelain	0.433	8.80		8.73	1.044	0.992
Light (1/2)	Stoneware	0.372	6.24	0.400	6.16	1.075	0.987
	Porcelain	0.365	6.42		6.32	1.096	0.984
Light (1/4)	Stoneware	0.288	4.24	0.347	4.13	1.205	0.974
	Porcelain	0.275	4.36		4.21	1.262	0.966

*1) Unit: N·mmλ−2

Table 6. (1) Analytical Results (Series III ·Mortar thickness 5 mm)

increased with decreasing thicknesses of the mortar layer and increases in the Young's modulus of the tile. This is attributed to an increasing constraint on mortar deformation by the tile, which has a high Young's modulus that prevents the stress field from acting as predicted theoretically. Still, the discrepancy in order of stress singularity due to the calculation method remained within 5%. If this is the only parameter used to assess the risk of delamination, the method used to estimate the order of stress singularity will not be an issue.Variation in the predicted stress intensity factor with the ratio of Young's modulus (modulus of light mortar versus modulus of ordinary mortar) is shown in Figure 12. The tile material showed little influence, but reducing the ratio of Young's modulus lowered the ratio of stress intensity factor, resulting in a prediction of moderation in the resulting stress. This is a benefit of the development and use of lower-modulus mortars as a technique to prevent tile delamination. However, if used, it will also be necessary to compare the benefit of fracture toughness and other factors with that of the increased interface strength.

Lastly, in each of these series, the influence of the method of calculating order of stress singularity on the obtained values for stress intensity factor was minor. The extrapolation method proposed in the present study for finding the parameters was determined to provide acceptable accuracy across a wide range of finish attachment conditions.

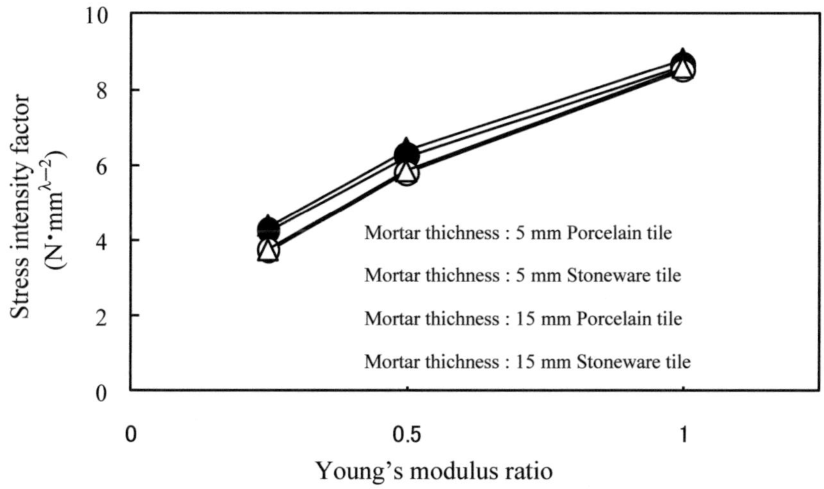

Fig. 12. Relationship between Young's modulus ratio and stress intensity factor (Series III)

5. Conclusion

A method was proposed for assessing the characteristics of the stress singularity field acting on the adhesive holding cement mortar structural finishing components on to concrete, and this method was analyzed. The method employed for testing finishing components was the simple shear test for ceramic tiles using axial compressive loads. The following results were obtained:

1. The various stress modes occurring during the simple shear test near the edge of the adhesive interface between concrete and mortar make up a stress field that contains a singularity point at its tip. The failure mode at the adhesive interface appears to be Mode II.
2. There are two real solutions to Bogy's theoretical solution. The stress field showed results that are fairly well approximated by the theoretical solution of greater magnitude, but in the strict sense, the stress field is complicated because it is influenced simultaneously by two orders of stress singularity.
3. Calculations of the stress intensity factor using an extrapolation method fitted to a generalized stress function revealed results that closely resembled those provided by numerical analysis.
4. The parametric analysis based on practical finishing conditions showed little influence of the geometry of the finish layer on the orders of stress singularity indicated by the approximated solution and the theoretical solution, but considerable influence of the Young's modulus ratio of mortar types. There are limits to the validity of the theoretical solution. The two solutions also provided consistent predictions for the stress intensity factor, regardless of the finish conditions. Thus, the approach proposed here was shown to provide useful data.
5. The thickness and length of the tile and mortar were also found to exert little influence on the stress singularity parameters within the range of practical dimensions. However, the Young's modulus ratio of mortar types did have a significant affect on these parameters. This analysis reproduced the stress moderating effect of low-modulus mortar.

6. References

Groth, H.L. (1988). Stress singularities and fracture at interface corners in bonded joints, *International Journal of Adhesion and Adhesive*, Vol.8, pp. 107-113, ISSN 0143-7496.

Bogy, D.B. (1971). Two edge bonded elastic wedges of different materials and wedge angles under surface tractions, *Journal of Applied Mechanics*, Vol.38, pp. 377-386, ISSN 0021-8936.

Hein, V.L. & Erdogan, F. (1971). Stress singularity in a two materials wedge, *International Journal of Fracture*, Vol.7, No.3, pp. 317-329, ISSN 0376-9429.

Quaresimin M & Ricotta M. (2006). Life prediction of bonded joints in composite materials, *International Journal of Fatigue*, Vol.28, pp.1166–76, ISSN 0142-1123.

Hattori T, Sakata S, Murakami G (1989). A stress singularity parameter approach for evaluating adhesive strength, *Journal of* Electronic Packaging ASME, Vol.111, pp.243–248, ISSN 1528-9044.

Hattori T (1991). A stress-singularity-parameter approach for evaluating the adhesive strength of single-lap joints, Jsme International Journal Series A, Vol.34–3, pp.326–331, ISSN 1347-5363.

5

Micromechanisms Controlling the Structural Evolution of Tribosystems

Dmitry Lubimov and Kirill Dolgopolov
"LIC" Ltd
Russia

1. Introduction

Development of scientific programs on the wear resistance is determined by economical significance of this issue for the development of productive potential of the world's countries. Deterioration is the main reason for removal of machinery and equipment from service; that is why controlling of the process of wear is the central core for such national scale issues as saving of non-renewable energy resources like hydrocarbon fuel and reducing the consumption of lubricants and structural materials. The condition of selective transfer in friction discovered by D.N. Garkunov and I.V. Kragelsky is the only effect theoretically admitting wearlessness (Garkunov, 2001). However, insufficiently developed physical and chemical basis for the theory of selective transfer hinders its wide deployment in engineering practices. The most controversial issue in the theory of selective transfer is considered to be the nature of products of chemical modification of friction surface. Traditional notions that are associated with the formation on a surface of either relatively simple inorganic compounds or tribopolymers are not sufficient to explain the unique tribotechnical characteristics in self-organizing friction systems. Considering the electronic structure of the rubbing metals, chemical properties of molecules of the active lubricant components, as well as the conditions of a frictional contact, one can expect that during friction of metals in the condition of selective transfer, in addition to normally expected products of the friction surface chemical modification, coordination compounds are generated. Those compounds, being more stable than tribopolymers and less stable than inorganic products of chemical modification can ensure a mass transfer process under the conditions of almost no-wear friction (Garkunov, 2001). Therefore, development of the fundamentals of the selective transfer theory and development of lubricants to put it into practice require considering the dynamics of physical and chemical processes and factors catalyzing the formation of complex compounds during friction, and their subsequent coordination on the surface. It is incontrovertible that with the help of additives it is possible to control almost all properties of lubricants. This being the case, oil acts as a carrier of the chemical reaction components, and a friction assembly act as a reactor, where processes are controlled not only by the composition of the lubricating medium and the nature of the rubbing surfaces, but also by the presence of external force action. In this connection, the analysis of the development trends for lubrication systems and existing advanced developments in chemmotology allows to highlight the following research directions in this research area:

a. decreasing the range of mineral lubricants replacing them with the highly compounded synthetic lubricants;
b. development of lubricants applied to the surface of a tribological conjunction once in the entire lifetime of a friction assembly;
c. development and creation of adaptive lubricating devices that represent an oil feeder controlled by a sensor responding to changes in key parameters of a friction pair;
d. application of centralized combined lubrication systems controlled by a microprocessor that will automatically change a lubrication mode either according to a preset program or depending on the operational performance of the friction assemblies controlled by sensors.

Active interference in electrochemical processes of a frictional contact is a very promising direction to prevent corrosion, to reduce oxidizing, mechanochemical and hydrogen wear, for instance, in metal-polymeric friction assemblies. It is known that the rate of electrochemical corrosion can be greatly reduced if a metal part is subjected to polarization. Polarization of a friction system pursues different objectives: to improve the wettability of the solid surface and adsorption of environment components. The friction system polarization method is that the potential of the system, with the help of an external source of polarization, is shifted to the area that is optimal for the friction process behavior, and then it is maintained at the required level during the work of a friction pair. The main difficulty in dealing with this kind of research is that the value of an electrode potential during friction may differ significantly from the electrode potential of the metal measured in static conditions (Goldade et al., 1993). In the first place, this is caused by mechanical cleaning of the surface from oxide films. In the second place, during friction secondary structures are generated shifting the potential to the positive or negative side. And, finally, the potential shift is affected by the plastic yield that leads to emergence of dislocated atoms in the crystal lattice of the metal. Being a fundamental characteristic of the electrical state of a "metal – electrolyte" interface, the electrode potential of the system can make a significant impact on the friction and wear of metals. The effect of polarization on the friction coefficient and surface damage of metals was discovered by Bowden and Young and further studied by other authors. Decreased friction factor and increased surface charge during polarization are associated with the increased wedge effect of the double electrical layer between the rubbing surfaces. Polymeric materials used in friction assemblies add some specific features to the electrochemical processes in the friction zone. During friction of metal-polymeric pairs, the polymeric components acquire the properties of surface-active substances, which substantially change the electrochemical activity of the metals (Pinchuk et al., 2004). The results of these studies are presented in details in the papers (Garkunov, 2001; Goldade et al., 1993; Kostetsky, 1980), where the influence of a friction system electrode potential on the adsorption, diffusion, oxidation, friction and wear of materials is noted. Unlike processes normally considered in electrochemistry, electrochemical processes during friction occur under the conditions of moving deformable discrete contacts of separate microroughnesses. This leads to generation of currents between the rubbing surfaces during friction (Ryzhikov & Dolgopolov, 2005). Roughnesses of two sliding bodies get elastic or plastic impacts; and consequently the equilibrium state corresponding to the minimal energy of the strained area is disturbed. Under the impacts, the microroughnesses oscillate with the frequency which is close to that of external action. Dynamic loads, causing a variable strain of the material in the mating parts contact, lead to emergence of an alternating magnetic flux in the strained layer. Changed magnetic flux induces an induction emf in the circuit generated by the mating parts. The electrical resistance between these parts conditioned by the properties of the oxide films and lubricant leads to a alternating potential difference in the contact area. If the body and counterbody are electrically isolated from each other with a dielectric layer (e.g., lubricant), the friction couple can be considered as a capacitor. When moving one body against the other, a point contact occurs due to existing surface roughness or other surface imperfections. When the point contacts diverge, the hydrodynamic effects occur like squeezing lubricant out of the friction area, generation of current-conducting seizure bridges from the products of erosion. Consequently, a kind of capacitor electric breakdown occurs, as a result of which a large amount of electrical energy accumulated by the friction surfaces is released. The lubricant and material of the surface layers are decomposed into the components of ionic, radical or ion-radical nature, transferring to the lubricating medium. In this specific short-circuited galvanic microelement, the potential probability is created of having on the friction surfaces the redox reactions of sedimentation of the lubricant medium's active components. Thus, during friction pair operation, an automatically functioning electrochemical mechanism of wear control appears. And this mechanism depends on such factors as the electromotive force of the galvanic cell, the shift of potentials during the circuit closing, the polarizability of the electrochemical circuit, and the rates of the electrode processes. Therefore, creation of alternative adaptive sources of polarization using design and technological features of a friction assembly itself, is of current importance. The main principle of operation of such adaptive lubricating device can be control of friction and wear processes via "external agents" - the force fields of different nature, which was named a "field effect" (Belyi, 1985; Shvedkov & Rovinsky, 1979). The "field effect" means a change of a friction coefficient and wear rate under the influence of electric, magnetic, thermal and radiation fields. Force fields are capable to change values of friction parameters by several times, and in some cases reduce them to zero ("the effect of abnormally low friction" (Shvedkov & Rovinsky, 1979)). Electromagnetic fields have a special place among the force fields (Garkunov, 2001; Goldade et al., 1993; Lubimov et al., 1992); the study of their influence on the performance of materials has occupied the minds of scientists since long ago. One of the problematic issues in reducing wear and friction of metals is the development of reliable and compact sources of electric polarization. The point is that the methods of polarization of a friction system by external sources have not found wide application in friction assemblies due to the bulky dimensions of the electrical sources, which significantly complicate the design of machines. However, today, in the century of rapidly developing microelectronics engineering, this problem can be perfectly solved. Based on the above said, we shall dwell on such aspect of the study of tribosystems as the capability to control their friction parameters directly in the conditions of dynamic interaction. This kind of work does not require interfering in the design features of a machine, which significantly reduces the cost and simplifies the potential possibility of its introduction into the production cycle of any engineering company.

2. Structural changes of tribounits

2.1 Third bodies of friction assemblies

According to present-day thinking, all triboeffects become evident on a macro scale, but the phenomena that cause and accompany them occur at a micro level. Friction is a combination of processes covering the thin surface layers of the contacting bodies. The work of the friction force equals to the energy of destruction of the interfacial bonds; about 99% of it turns into the heat that is spent for structural and thermal activation of the friction surfaces, and the remaining 1% of the energy is stored by a very small portion of surface and promotes activation of the surface bonds (Lubimov & Ryzhikov, 2001; Lubimov & Ryzhikov, 2006). Formation, growth and destruction of the interfacial bonds are determined by the nature of the contacting surfaces, the chemical processes occurring on them, and the stressed state of the surface layers caused by the loading conditions. I.V. Kragelsky coined the term "the third body" combining the interfacial bonds, the products of their destruction and the surface layers where the strains are localized. Later on, M. Godet developed the concept of the third body having expanded its meaning (Godet, 1984) that adds up to significant differences between the surface and bulk properties of the tribounit matter. The third bodies often have the form of film structures arising on the friction surfaces, which significantly affect the frictional characteristics of the friction assemblies. This phenomenon is especially evident when the effect of wearlessness or selective transfer takes place. In accordance with the model of Godet, the copper of the third body during selective transfer has a loose structure, different from a blocked one. It has a weak resistance to shear deformation and a low friction coefficient. These third bodies were named "servovite" films. Along with the servovite films, during a selective transfer the following structures that respectively belong to the third bodies are seen:

"Surfing" film - a molecular film structure consisting of associated coordination compounds;

Metal-cladding film – a protective film where a vacansion-dislocation deformation mechanism is partially realized;

"Dividal" film – a metal protective film formed as a result of discharge of metal ions in the contact area during friction;

"Nubial" film – a tribopolymer protective film consisting of tribopolymers chemisorbed on the active surface of the servovite film, which arise during destruction of lubricants (Polyakov, 1990).

Another type of third bodies appears on a friction surface during interaction of polymer or metal-polymer friction units. Such third bodies were named the friction transfer films; development of the theory of the friction transfer films is in many respects associated with the papers by V.A. Belyi and his school (Pogosyan & Oganesyan, 1986). In general case, during polymer-to-metal friction, as a rule, the transfer of the polymer onto the metal surface is observed. During the frictional transfer of the polymer, as a result of the molecular interaction, fine surface particles are separated and transferred to the active areas of the friction surfaces. During the transition to a steady friction state, in metal-polymer tribounits, actually the polymer rubs against polymer. When friction transfer films are generated, there are the following main types of adhesive interaction: adhesion of separate particles and a layer of particles, melt adhesion, wetting of the counterbody's surface, and film adhesion. Generation of the friction transfer films has been especially well studied during frictional interaction between polytetrafluorethylene (PTFE) and a steel counterbody. As it has been established (Nikolsky et al., 1988), during the interaction of PTFE with a metal surface, the PTFE molecules are decomposed, and during this process the fluorine atoms are coming off and are being replaced by the hydrogen atoms. Furthermore, generation of a chemical bond between the fluorine and the metal is observed, which determines the adhesion of the PTFE to the metal counterbody (Buckley, D. (1986). The friction transfer films are formed not only during the friction of metal-polymer tribounits, but also during the interaction of metal with graphite. The emergence of third bodies in the frictional contact area in most cases leads to reduced friction forces and wear levels of the tribounit materials (Sysoev et al., 1990). The main reason for such an effect is that the atoms and molecules of a third body passivate the surfaces of friction, thus reducing the adhesive component of the friction force. Intensity of the formation of third bodies depends on a host of aspects related to the fact that a friction assembly is a single tribosystem, the evolution of which is described via the laws of synergetics and thermodynamics of open systems (Garkunov, 2001).

2.2 Higher energy states of a tribosystem matter

As stated above, the energy released during friction in the form of heat and stored by the surface layer of the material, leads to structural and thermal activation of the surface. The structural-thermal activation means the origination and emergence at the surface of dislocations and point defects. Therefore, energy in tribosystems is degraded, its mechanical form is continuously transformed into the thermal one, and the entropy is constantly reproduced. The discovery of the wearlessness effect has made it possible to single out the general principle of evolution inherent to all friction systems, which consists in their thermodynamic "openness", ability to exchange energy with the external environment, and possibility to reduce entropy. On this criterion, a friction assembly by the level of self-organization approaches to biological systems (Kadolich et al., 2001; Pinchuk et al., 2007). The decreased entropy growth in an open thermodynamic system is determined by the interaction of two energy flows. The thermodynamic forces emerging in them have an influence not only within their own energy flow but also on the adjacent flow. In tribology, such flows are the mass transfer (diffusion) and the heat transfer, and their interaction determines entropy change and the friction assembly structure. American tribologist Buckley has associated a mass transfer with the processes of self-regeneration of friction assembly materials, which prevents the destruction provoked by the heat flow. In the first approximation, the influence of the flows on the tribosystem entropy looks as follows: the diffusion decreases the entropy of the friction assembly; the heat transfer, on the contrary, increases it. Based on the above said, the conclusion can be made that the decrease of tribosystem's entropy relates to the system's structural complication - structural adaptability, and consequently, is identical to the optimization of tribosystem's properties which, in thermodynamic terms, are expressed in minimized entropy growth, and in tribotechnical terms - in reduced wear and friction forces. The structural adaptability is preceded by the changes that have been recently named "tribomutation" (Voinov, 2010). Tribomutation is recrystallization of the thinnest surface layers, decreasing of the free surface energy, elimination of the boundaries of surface and near-surface crystallites, and annihilations of dislocations as they emerge at the surface. The well-known Soviet physicist Zeldovich has demonstrated that in the near-surface layers, the defects with a minimum elastic energy occur; resulting in the loss of elastic properties by the solid bodies, their becoming plastic and fluid, and their transition to a "liquid-like" state. During friction, such physical constants of the material as Young modulus, Poisson ratio, dynamic strength, and acoustic stiffness lose physical meaning. The described structural changes are explained by the action of the processes accompanying the structural and thermal activation of the surfaces matter of a tribounit. Though the percentage of the frictional energy reserved by the surface is small in an absolute value, but, if we take into account that it is accumulated by thin surface layers with a thickness of fractions of micrometer and calculate the density of this energy, then the received numbers can reach critical values for this aggregate state of the material. The surface is supersaturated with dislocations and point defects; its behavior obeys the hydrodynamic laws, and in some extreme cases - the gas dynamics laws. In this extreme case, the movement of dislocations, according to Kuhlmann – Wilsdorf, is described by the equation (Cahn, 1968):

$$\frac{\partial^2 W_z}{\partial x'^2} - \frac{\partial^2 W_z}{\partial y^2} = 0 \tag{1}$$

$x' = \dfrac{(x-\upsilon t)}{\beta}$; $\beta=(1-\dfrac{\upsilon^2}{c^2})^{\frac{1}{2}}$ - dimensionless parameter; c - velocity of elastic waves propagation in the matter; ϑ - velocity of dislocations propagation.
The energy of dislocation W ' moving with the velocity ϑ equals:

$$W' = \frac{W}{\beta} \tag{2}$$

W - energy of immobile dislocation.
When the dislocation velocity ϑ tends to "c", the parameter $\beta \to 0$ and, consequently, W' indefinably grows. Kuhlmann-Wilsdorf, by analogy with the theory of relativity, recorded the ratio for the W 'through the speed "c ":

$$W' = c^2 \times m' \tag{3}$$

$$m' = \frac{m}{\beta} \tag{4}$$

m - mass of the dislocation at rest, equal to $m = \rho \times b^2$; ρ - density of dislocation; b - Burgers vector.
Thus, the movement of dislocations is described by the equations identical to the equations of the special relativity theory (SRT). Therefore, in accordance with the postulates of the SRT, the specific effects predicted by this theory should occur in such systems: change of the space metric, increased mass and energy carried by the defects out to the surface, and the local time slowed down. The effect of these processes is determined by the relativistic parameter β. The value β depends on the speed "c", which, in the conditions of transition of the surface layers to the "liquid-like" state and then to the gaseous state, significantly decreases. Under the conditions of the frictional contact, the parameter β varies from 0.99 to 0.7, which makes the contribution of the relativistic correction to the equations of movement of dislocations very important. Consequently, a dislocation moving during friction can be regarded as a relativistic object, then according to the opinion of Nobel laureate L.D. Landau stated in the section "Relativistic Hydrodynamics" in the book "Gidrodinamika" ("Hydrodynamics") (Landau & Lifshitz, 2001): "The need for relativistic effects consideration ... may be associated not only with a high rate of macroscopic motion ... the equations (describing this motion (author's note)) also change significantly when this rate is not high, but high are the rates of the microscopic motions of the particles that constitute it". In other words, if a large "classical body" consists of small parts obeying in their movement the laws of the relativity theory, the entire body will be partly "relativistic". Hence, if the constituent

friction materials of a dislocation acquire relativistic properties due to structural and thermal activation, then the macro quantities of the tribounit matter start showing the special properties arising from the SRT equations. In this regard, it is very interesting to consider the time paradox that explains some anomalies in the physical and chemical processes of friction surfaces. The equations of the relativity theory do not make a distinction between the time directions; Einstein, as a matter of fact, stated that the past, the present and the future "are nothing but a human illusion devoid of physical sense..." At the same time, the equivalence of the future and the past leads to a number of paradoxes related to the violation of the causality principle, to obey which, the "time arrow" concept is introduced. The direction of the increasing entropy points which state in a couple of the adjacent events is the subsequent rather than the preceding. Many present-day physicists believe that the entropy is actually the arrow of time, which makes the time itself just a simple sequence of states. Time is a part of many physical laws, and it is quite possible that a change in the dimension of time can affect the behavior of various processes. In this connection, there are two interesting examples that illustrate the subsequent line of argument. The American theoretical physicist Feynman believed that an antielectron (positron) emerging in nuclear reactions is actually a normal electron moving backwards in time, and their positive and negative charges are nothing more than a direction indicator of time in which the particles make their way through the space-time at a given moment. The mathematical apparatus developed by Feynman has become a space-time interpretation of the quantum mechanics, for which he was awarded the Nobel Prize "for his fundamental contribution to the development of quantum electrodynamics, which had profound implications for particle physics" (Cholakov, 1986). Another leading theoretical physicist, the Englishman Hawking investigated changes in the entropy of a black hole system when a certain amount of substance gets to "beyond the event horizon". Since the latter is a measure of the entropy of a black hole, then, having become wider, it would increase its own entropy by exactly the same amount as was carried on itself by the substance falling into the black hole. But the substance has carried away into the black hole a part of the entropy which just equals to this increase, therefore the total entropy has not been changed, and its increment is zero. According to the "arrow of time" definition via entropy, Hawking has come to the conclusion about the stop of time in the black hole outskirts, relatively to an outside observer (Hawking & Penrose, 2007). This conclusion finds confirmation in astronomical observations of the movement of gas clouds in the areas of anticipated presence of black holes, where their speed is found to slow down. Based on this, we can assume that a tribosystem, in accordance with the definition given by Landau, can be classified as a relativistic object, representing an example of a relativistic solid state physics, which, according to Nobel laureate A. Heym, is "... a new scientific paradigm where quantum relativistic phenomena can be studied in normal laboratory conditions ..." (Morozov, 2008). And the decreased entropy increment and the related decrease of friction coefficients and wear rates are explained by slowing of the tribosystem's local time created by fast dislocations. An indirect example in favor of such an unusual interpretation of changes in anti-friction characteristics is the effect of anomalously low friction, opened by the Russian tribologist Silin (Svedkov & Rovinsky, 1979), where the friction factor was recorded to drop zero in the polyethylene-metal system when this friction pair was exposed to the radiation of α-particles flow and cooled down to the temperature of liquid nitrogen. Indeed, the exposure to low temperatures zeroes the entropy, and fast α-particles can be considered as "carriers of relativistic effects" into tribosystem. From a formal point of view, the zeroed entropy corresponds to the time stop within the tribosystem, which slows down the wear and decreases friction coefficients down to ultra low values. An extremely important factor influencing the change of tribosystems' surface states is the increased, due to relativistic effects, kinetic energy of dislocations which they pass to the matter as they emerge on the surface of the friction assembly. Given this adjustment to the energy reserved by the surface one can explain the transformations that happen with the aggregate state of the surface layer material. A discrete contact inherent to the rubbing surfaces leads to heavy stress fluctuations in the subsurface layers of the friction assembly at the initial moment of friction. Force and deformation processes in the spots of actual contact look like multiple short-time pulses the duration of which is limited by the time of the knock-on collision of the microroughness, ranging from 10^{-3} to 10^{-11}s. Under such loading conditions, the substance of the tribounit transforms to a high-energy state with significant concentrations of energy, nearing to the values leading to the phase transformations. A slightly increased energy flow associated with the relativistic increase of dislocation masses is capable to start the process of avalanche-like activation of the friction surfaces substance, the highest peak of which is the condition called triboplasma or magmaplasma (Garkunov, 2001; Heinicke, 1987). Triboplasma is an energetic bunch, into which the substance of the friction surface transfers. For a long time, triboplasma has been a hypothesis explaining many abnormal processes occurring during friction: mechanoemission by the friction surface of charged particles with energy of 5 eV; emission of electromagnetic radiation quantums, abnormal chemical activity, and many more. Since during friction the matter can not be fully ionized like it is observed to happen on stars or in a nuclear explosion, the triboplasma is classified as nonideal plasma.

In most studies, triboplasma is presented as a hypothetically postulated object, the properties of which are almost not described (Heinicke, 1987). In the works generalizing the previously published studies (Lubimov et al., 1998; Lubimov et al., 2007a; Lubimov et al., 2007b), reasonable assumptions are made about the properties of this aggregate state of substance. The listed papers are based on the study of the concentration profiles of the diffusion distribution of the matter in the surface layer of the friction materials, and the effect of the plasmatic states of the matter on this distribution. It was established that during friction the diffusion coefficients ($Д_э$) increase by tens compared with the diffusion coefficients typical for the solid state of matter. We have found the equation establishing the law of proportionality between the diffusion coefficient and the square root of the triboplasma temperature. This is typical for substances in a gaseous state, so it was assumed that triboplasma by its properties is close to gases (Lubimov et al., 2007b):

$$Д_э = \frac{2e}{h}\sqrt{\pi k} \times \frac{Q^*}{W}\sqrt{\frac{T}{n}} \tag{4}$$

T - temperature of triboplasma; n - average density of particles in the plasma, e - electron charge; $Q^* = kT + Q$ - value characterizing the kinetic energy of the thermal motion of the triboplasma particles; Q - heat released by friction; W – binding energy of the triboplasma components; h - Planck constant; k - Boltzmann constant.

Such a conclusion allows to simplify the discussion of the relaxation processes, applying the apparatus of the plasma kinetic theory and the wave physics. The value $\frac{Q^*}{W}$ is a characteristic for the degree of substance ionization as it shows by how many times the energy of the ionized plasma is greater than the energy of the same substance in a steady state. According to our evaluation, the ratio of $\frac{Q^*}{W}$ has an order of 10^{-7}. We have obtained the ratios allowing to evaluate the temperature, charge, mass and lifetime of triboplasma components (Lubimov et al., 2007a). Triboplasma lives for a very short time (less than 10^{-7} c) and has a very high local temperature (above 10^4 K) (Pinchuk, 2004; Lubimov et al., 1998). This is a higher excited state where the cohesive bonds in solids weaken considerably, and structural degradation processes occur associated with the emission of charged particles and photons of different wave lengths. In the gas plasma, in addition to the random thermal motion, particles are involved in the ordered processes of the so-called Langmuir waves having the plasma frequencies ω_p (Ginsburg, 1967). We have obtained the formula for the quantitative estimation of the ω_p value:

$$\omega_p = \frac{4\pi e^2}{h}\sqrt{\frac{k}{m}} \times \frac{Q^*}{W} \times \frac{\sqrt{T}}{Д_э} \tag{5}$$

m - electron mass.

The numerical solution of the equation (5) gives us the values of plasma frequencies in the gigahertz range of oscillations. Triboplasma as any high-energy metastable state is not stable and relaxes rapidly passing to the normal aggregate states of substance. This transition is a multistage process; in one of its intermediate stages, chemically very active postplasma states of substance are present, having a long lifetime. It is these long-lived postplasma states of the tribounit surface substance which cause the chemical processes to take place throughout the entire nominal contact area, not just in the spots of actual contact. Based on the plasma hypothesis, there exists a chemical model of "surfing" effect explaining the low values of the friction factor during a selective transfer. The latter process is associated with the time evolution of triboplasma's substance after it leaves the actual contact area and has step by step relaxation till the initial steady state. Complex compounds constituting the surfing film in the selective transfer mode and ensuring the minimal friction coefficient, are formed due to the tribochemical reactions associated with transformations occurring in triboplasma and postplasma phase (Garnovskyi, 1984). The modern day leading tribologists G. Heinicke, W. Ebeling, Belyi V.A., Garkunov D.N., Kostetsky, B.I., have pointed out the top-priority need to study tribochemical processes from the standpoint of plasma chemical substance transformations and the catalytic impact on them from physical fields.

2.3 Experimental study of triboplasma structural elements

Short lifetime of triboplasma and the fact that the friction contact can not be accessed with the analytical equipment make it difficult to study this aggregate state of the friction surface substance. For the purpose of this study, an experiment design was proposed aiming at separation of triboplasma in its "pure form". This plan was implemented using the experimental equipment representing a friction machine working in "shaft- bushing" contact mode with polytetrafluoroethylene - steel coupling. Shaft material is PTFE, bushing material is steel. The friction assembly is placed under a vacuum bell jar connected to a fore vacuum pump. The potential range from 0.5 to 200V can be applied to the friction assembly, and also a magnetic field is generated with an intensity of 10^4 $A \cdot m^{-1}$. Electromagnetic oscillations of different frequencies within the range from 0 to 10 GHz are superimposed on the friction assembly, and the characteristic signals are read out. In the steel counterbody (bushing) there are plasmatic traps keeping the friction substance in an active state (Fig. 1). Signal from a plasmatic trap is sent to a spectrum analyzer that resolves it into the spectrum representing a set of harmonics with individual frequencies. The friction torque is recorded by a change of the current in the winding of the machine electrical motor (similar to the Timken friction machine (US)). Frictional parameters were controlled at a sliding velocity of 1 m/s and a load of 10N, test time is 5 hours.

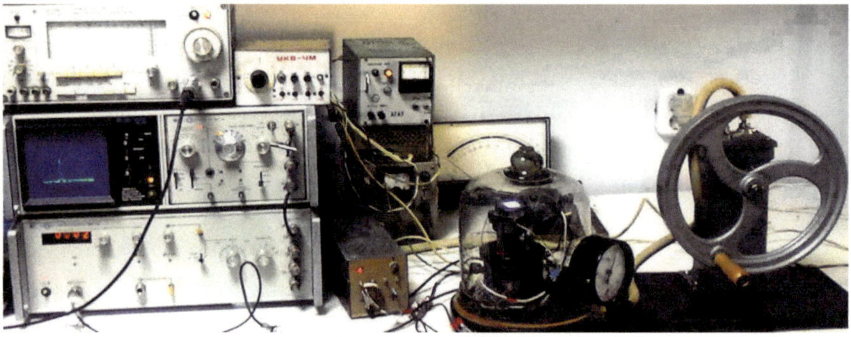

Fig. 1. Part of the laboratory system for the triboplasma study.

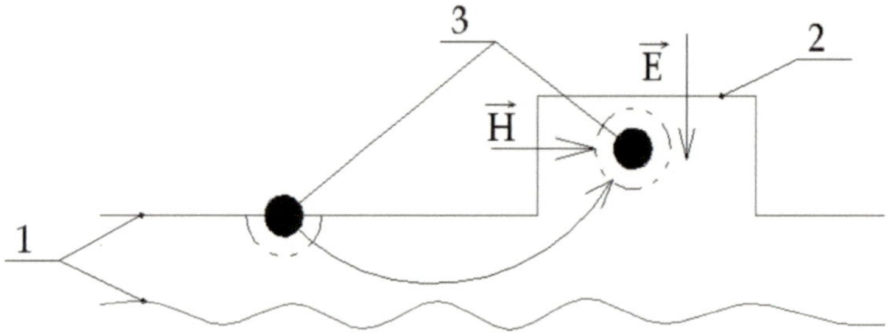

Fig. 2. Circuit diagram of a plasmatic trap: 1) friction surfaces; 2) plasmatic trap; 3) triboplasma particle. $\vec{H}$ and $\vec{E}$ are intensities of the magnetic and electric field.

We assumed that triboplasma is an electrically active state of substance, one of the fundamental properties of which is shielding the electrical potential applied to it. That is why its exposure to an outer electric or magnetic field discloses those oscillation spectrum harmonics which are directly connected with the triboplasma. Exposure to an outer force field discovers a response from the spectrum harmonics in the megahertz and gigahertz frequency range. Depending on alteration of the electromagnetic spectrum profiles recorded by the measuring equipment when the friction assembly is exposed to the electric and magnetic fields of different directions, we have established that the oscillators constituting the triboplasma are electrons and positively and negatively charged ions. The frequency values of the free oscillations of these oscillators split up into the two ranges: 10÷100 MHz and 1÷1.5 GHz. These match the theoretically found values of the plasmatic frequencies (5). We have discovered that in each of the mentioned range of frequencies (megahertz and gigahertz) there are two types of oscillators. For the gigahertz range, one of the oscillator types is electrons. While plotting the graph of the relation between the friction factor, the harmonic intensities of the electromagnetic spectrum (electrons and oppositely charged ions), and the test time, we have discovered that the positively charged ions of triboplasma decrease the friction factor, and the negatively charged particles (ions and electrons), on the contrary, facilitate its increase (Fig. 3). The decreased friction factor is seen to coincide with the increased harmonics corresponding to the positively charged particles.

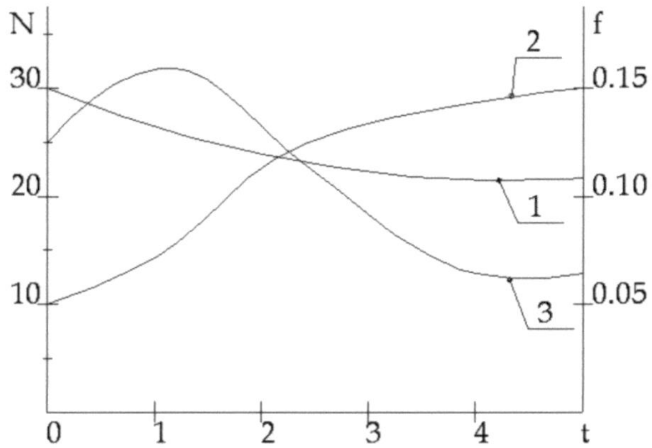

Fig. 3. Relation between the harmonic intensities N (in conditional units) of electromagnetic emission of tribloplasma components and the friction factor (f) of the "PTFE-steel" coupling, and the friction time (t, hours). 1) plot f(t); 2) plot N(t) for positively charged articles; 3) plot N(t) for negatively charged articles;

Based on the studies conducted, we can say that the exposure of the electrically active triboplasma components to an electromagnetic field of a certain direction and ripple frequency, leads to changes in the antifriction properties of a tribosystem. That is why, by changing the intensity of the electromagnetic field tuned to the resonance with the triboplasma components, it is possible to control the tribounit frictional parameters by intensifying the formation of lubricating structures on the contacting surfaces.

2.4 Impact of electromagnetic control of a frictional contact

It was shown that friction as a physical process generates electrically active responsive groups of triboplasma substance, which significantly influence the triboprocesses. Postplasmatic electrical states play a special role in that. They are complicated chemical structures carrying within them elements of free and bound charges, and also having extremely high chemical activity. They catalyze the formation of the "third bodies" driving a change in frictional characteristics of the friction assembly. In electrically conducting media, the influence of an electrical field upon the formation and growth of the thickness of the protective lubricating films on the friction surfaces, can be explained by the electrolytic transfer of the charges and substance. In dielectric media, which can be, for example, a lubricant or a friction transfer film of a metal-polymeric coupling, there are no free electrical charges in the form in which they are present in metals. Applying an electrical field of certain intensity to the frictional contact leads to redistribution of the charge densities in the constituting materials - a phenomenon of field polarization. The polarization is one of the fundamental properties of dielectrics, which consists in disequilibrium of charge distribution in the matter and appearance of a resultant moment $\vec{p}$ different from zero and proportional to the electric field intensity value $\vec{E}$:

$$\vec{p} = \alpha\vec{E} \tag{6}$$

α - substance polarizability, m^{-3}.

In the paper (Lubimov et al., 2009a) discussed the hydrocarbon lubricating media separating metal surfaces, it was proved that the interaction of an external electromagnetic field with a self electromagnetic field of the surface (induced by the frictional forces as a result of the dynamic contacting of the rubbing bodies), causes appearance of ponderomotive forces within the working layer of the friction assembly. These forces have a mechanical effect on the substance and cause to move the macrovolume of the friction substance. In the general case, the ponderomotive force value can be found from the ratio (7):

$$\vec{F} = \frac{m}{\mu}\alpha\omega\vec{S} \tag{7}$$

$\vec{F}$ - ponderomotive force; m - mass of the moving macrovolume; μ - molar mass of the moving substance; α - the substance polarizability; ω - frequency of the external, substance exciting factor; $\vec{S}$ - the Umov-Poynting vector.

Currents generated during friction, due to the contact discreteness, lead to appearance of electromagnetic fields in the gap between the rubbing bodies; that is why friction, as a physical process, per se generates forces of the electromagnetic nature. These forces, having orienting and structuring action on a very small amount of the lubricating medium near the surface of the solid body, facilitate emergence of ordered and stable states - multipoles, interfacing with the charged friction surface field. Under the electromagnetic field, the time limit of existence for the structurally activated substance of the gas triboplasma, increases due to non-dissipative electric drift of the plasma components in the electromagnetic field (Prokhorov, 1992). In other words, in the gas plasma, a weakly decaying electromagnetic wave is excited, which maintains the working layer substance in chemically active post-plasmatic state. Ponderomotive pressure exerted by the electromagnetic field upon the frictional substance depends on the properties of this substance (density, molecular mass, mass of polarizable volume), the field properties (field frequency and energy), and the contact geometry (gap size between the rubbing surfaces, dimensions of the metal surfaces). During the action of an electromagnetic field, favorable conditions are created for synthesis of the protective lubricating films covering the friction assembly surfaces. Our assumption is proved experimentally using the electromagnetic sanation method for the friction surface, during frictional interaction of the materials "epoxy-filled composite - zinc counterbody" (Lubimov et al., 2001). The amplitudes of the emergent high frequency harmonics of the electromagnetic oscillations spectrum depend upon the magnitude and direction of the electric field (Fig. 4). The recorded values of the harmonic amplitudes help to have a general idea of the quantity of the active particles that create oscillations in this frequency range, proved to be responsive to the external action.

From the plot one can see that with the applied potential (U, Volt) increased up to a certain value, the harmonic amplitudes increase correspondingly (ΔA). This is an evidence that more of the active centers (free charge states) response to the external action. The curve going parallel to the X-axis means that the mentioned charge states are present on the surface now not in the form of uncoordinated active centers, but in the form of multipoles constituting the "skeleton" of the protective lubricating structure - the "third body" - being formed on this surface. These multipoles passivate the solid body surface, and are electrically neutral formations; therefore, they do not react when exposed to an external field. Furthermore, the friction factor was recorded to drop during the transition from the negative to positive potential, which can serve as an indirect proof for the formation of the protective lubricating layer - the "third body".

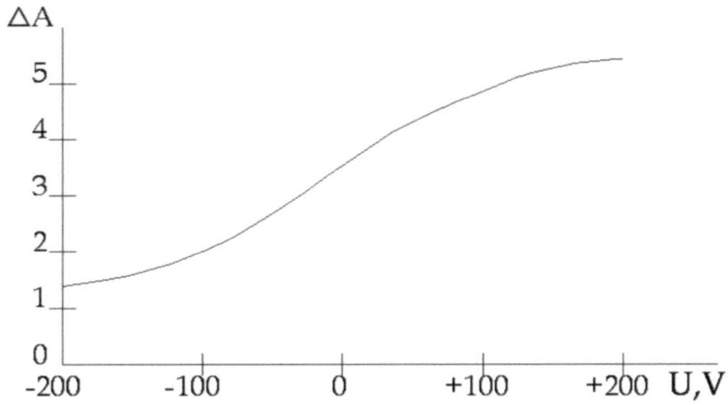

Fig. 4. Impact of the electric field magnitude and direction on the harmonic amplitudes change (ΔA) in a frequency range of 40 MGz for the friction pair "epoxy-filled composite - zinc counterbody".

Item No.	Potential applied to the metal counterbody, V	Friction Factor
1	-200	0,36
2	0	0,25
3	+200	0,12

Table 1. Change of frictional properties of the materials exposed to the electric field

The applied electric field influences not only the kinetics of the tribochemical reactions within the working layer, significantly increasing their speed, but also the orientation of components constituting the third body. At the same time, the frictional body polarized by the external electrical field, forms the lubricating film much more actively in comparison with the similar materials in the absence of the field (Fig. 5).

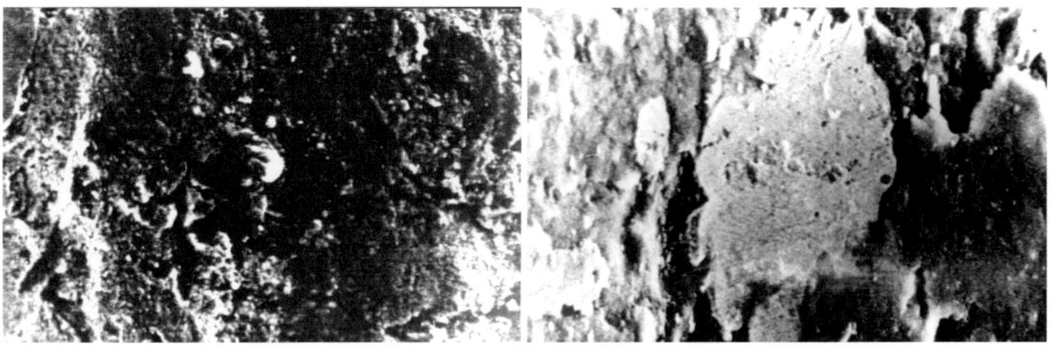

(a) (b)

Fig. 5. Photographs of friction paths in the coupling "PTFE - zinc counterbody" at the potentials applied to the metal counterbody: a) 0V; b) +100V.

As seen from the photos, when the field is present, the film structure is more homogeneous across the thickness, has no discontinuities and fully covers the zinc surface. On the right-hand photograph, the film has clearly seen new formations on the surface of the already existent layer. Besides that, the film, being developed under the action of the field, adheres to the metal surface more firmly. Such an intensive layer growth and visible increase of the third body volume filling the working layer, can be related to the better wetting of the metal surface by the polymer and higher adhesion of the polymer film. The electric potential in this example acts like the factor polarizing and maintaining the polarized state for the working volume matter. Molecule energy depends upon its position on the surface (Akhmatov, 1963). A polarized molecule (dipole and multipole) of a lubricating medium in the external force field of the surface, orients with regard to it so that its potential energy turns out to be minimal (8):

$$U(\alpha)=\frac{U_0\left(1-\cos(n\alpha)\right)}{2} \tag{8}$$

$$p(\alpha)d\alpha=C\times\exp\left(-\frac{U(\alpha)}{kT}\right) \tag{9}$$

U(α) - dipole potential energy; U_0 – potential barrier; n – dipole order of symmetry; α – inclination angle of the dipole symmetry axis to the surface; p(α) – relative probability of the molecule orientation angle; C - constant.

In accordance with the above formulas (8) and (9), applying an external electric field results in the multipole molecules tending to take up a position perpendicular to the metal surface, orienting along the electric field force lines. Electrical action on a polarized molecule, ion, or radical fixed with their active center on the friction surface can lead to the molecule precession. This fact was recorded by S.Z. Zaitsev during the study of water molecule behavior with the help of a field ion microscope, and was discussed in the paper (Lubimov et al., 1994). As the observations showed, in an electric field, the water molecule, structurally being a dipole, starts to precess and nutate, i.e. make movements similar to the movement of a gyroscope with one fixed point. Schematic sketch of this process is shown in Fig. 6.

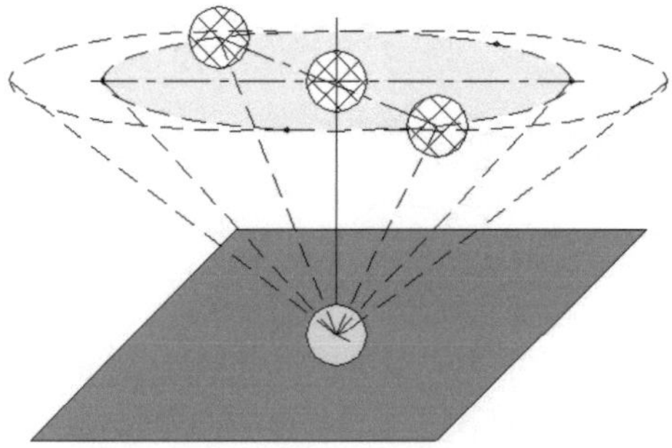

Fig. 6. Sketch of precession and nutation of a polarized molecule on the plane

There is none of the physically reasoned statement that would disprove the assumption that any other polar molecule, radical or ion ought to behave in an electric field in a similar way. For instance, during frictional interaction of polytetrafluoroethylene, the molecular chain is destructed with formation of its fragments that can be considered as radicals. Active ends of the molecular fragments make up a chemical bond with the frictional surfaces with energy of about 690 eV. Meanwhile, such a molecule tends to take up a perpendicular position, which is impeded by the frictional heat being emitted in the area of dynamic contact of the friction assembly, and by the intermolecular action (Nikolsky et al., 1988). Exposure to an external electric field changes the motion of a polytetrafluoroethylene molecule. As a result of precession and nutation processes, such a molecule becomes stable in its vertical position. The precession and nutation lead to the formation of an additional electron cloud, parallel to the friction surface plane. Presence of the Coulomb forces acting between the precessing molecules helps quite obviously to additionally organize their structural order, and resists to the disorganizing action from the temperature fields (Fig. 6). There being generated a structurally ordered molecular layer that is identical to a surfing film in terms of its action and structure.

3. Surfing films

The ultra low values of the friction coefficient recorded in the selective transfer mode are connected with appearance of surfing films on the friction paths. The "surfing" structures appear due to complex physical and chemical processes, and complexing processes become of a special significance among them (Polyakov, 1988). Complexing during friction occurs when the processes associated with triboplasma and postplasmatic transformation phase of the aggregate state of the tribounit surface layer matter, are directly involved. In the moment of closest approach of the rough edges of the conjugate bodies, the surface layer is cleaned from impurities, oxide films, and in the places of dislocation outcrops, resulting from the structural thermal activation, the surface atoms have additional bonds. As almost all metals are electron acceptors, and most additives or tribodestruction products of friction materials have electron donors within their structure, then the latter can act as ligands. This creates favorable conditions for the formation of complex compounds. In the wearlessness condition, it is typical that chemical bond occurs between lubricant ligands and metal surface atoms that have available bonds. In this case a particle on the surface can interact with several lattice atoms at once. Consequently, the total interaction will be very strong. The chemically active components of the lubricant - ligands - form metal-containing complexes with three-dimensional or two-dimensional structure at the dislocation outcrop areas or contact places. Formation of a coordination bond in a complex, leads to weakening or rupture of the metal atomic bonds in the complex with the metal atoms in the friction body lattice. The bond between the metal and complex becomes adsorptive. The complex is now capable to move across the surface and combine with other complexes in film islands (Lubimov & Ryzhikov, 2001). In further acts of contacting between the rough surfaces, the complexes may mechanically come off the surface transferring to the volume of the lubricant. This creates dynamic stability of the surfing structures, with which complex formation and decay processes run at the same speed. To minimize the frictional forces, the complex compound needs not only to develop on the friction surface but also to take a certain position. This positioning was named tribocoordination. Our studies conducted with a scanning tunnel microscope showed that the surfing films have quasi-crystalline structure, which is different from the surface structure where they appear. The important factor for the generation of a monomolecular and multimolecular highly oriented (coordinated) layer is the tribochemical interaction of the polar molecules with the solid body surface, which then provides easiest sliding of the lubricant on the boundary layer surface. This effect leads to the transition of the tribounit to the hydrodynamic friction condition with ultra low friction coefficient values inherent to it. A.S. Akhmatov has grounded the connection between the friction forces and structural organization of the lubricating layers' molecules. He has shown that ordered, pseudocrystalline structure of the lubricating layer adjacent to the solid surface is characterized by the friction modes transient to the hydrodynamic friction, and being accompanied with low values of the friction factor. For such a condition, it is typical to have oriented position of the filiform molecules. In this regard, specifically the surfing film's structure meets the friction moment minimization requirements most of all. Fig. 7 shows photographs of the friction surface conjugate with the epoxy-filled composite containing copper tetraethylthiuramdisulfide complex, which were made using a scanning tunnel microscope.

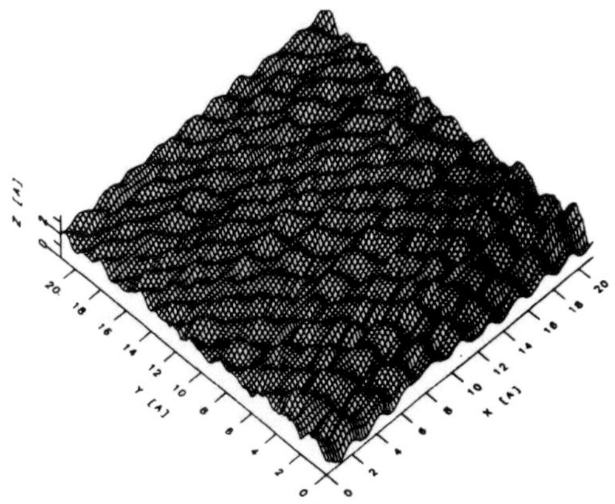

(a)

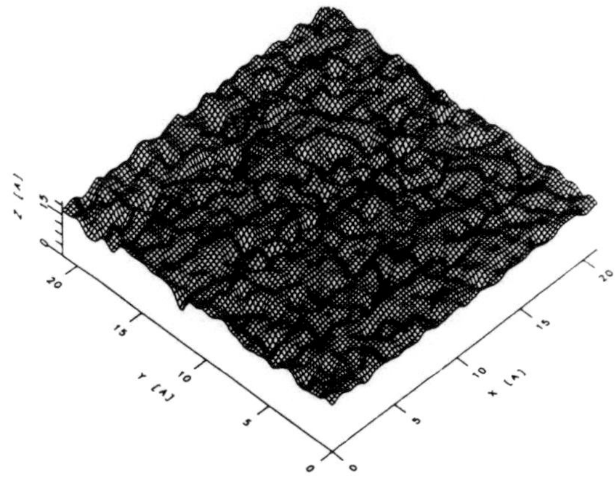

(b)

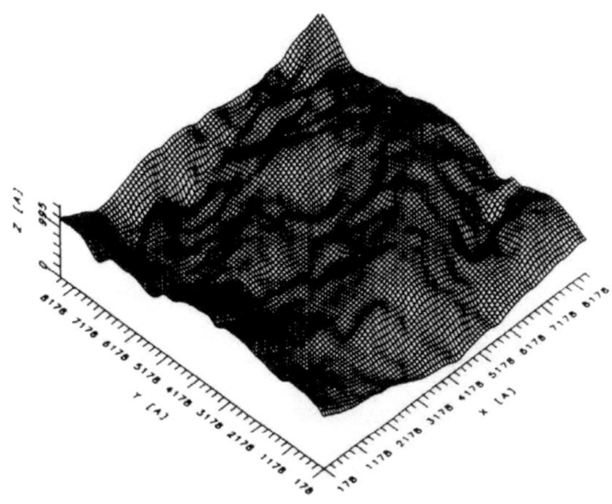

(c)

Fig. 7. Surfing film area made of copper complexes

The film from Fig. 7 is fixed on steel at a voltage of 0.1V and 0.5 mA current; one can see that it represents the ordered atomic lattice with a distance between the atoms of 1.9Å. Increased vertical scanning up to 15Å (Fig. 7b), also gives us the ordered atomic grid with 1.9Å distance between the atoms, which resembles a crystal lattice. When larger areas are used (Fig. 7c), apparently the topography can be described as chaotic. Obtaining the friction surface images with the scanning tunnel microscope was possible thanks to the small thickness of the oxide layer. The thickness of metal oxide on the rubbing specimens' surface did not increase (in spite of their contact with atmospheric oxygen), because the oxide layer was covered with the surfing film layer. That is exactly why the topography of surfaces shown in Fig. 6 reflects mainly the morphology of the surfing film. Images of different surface areas in different scales given for each specimen, apparently for the first time allow establishing the relation between the large-scale structure of the surfing film surface, and the structure of the same film at the atomic level. Thus, comparing the images from Fig. 6, the conclusion can be made that the ordered position of the surfing film atoms, evident at the 22×22Å area, leads to the formation of long molecular chains oriented mainly in one direction. While obtaining an image in a larger scale, separate molecules and especially atoms become invisible. And the topography itself now starts reflecting the sliding pattern of tribounit surfaces.

It can be taken for granted that the substances structured as surfing films, regardless whether they are liquid or plastically viscous in mass, with the same temperature and load, in the boundary state on the friction surface take on elastic forms, turning into a different quasi-crystalline aggregate state. Such lubricating layer has a higher stability, and is not squeezed out of the frictional contact area under external forces. Exposure to an electric field facilitates the emergence of a lubricating film on the friction surface, structural orientation of which resembles the tribo-coordinated surfing film, consisting of highly oriented molecules that have chemical bond with the metal surface. The orientation of such films is additionally maintained due to Coulomb forces, which provides them with additional stability and decreases the friction factor and material wear by many times.

4. Control of frictional processes with an electric field

The impact of an electric field action on a frictional contact for various materials of contacting bodies was studied using a side friction machine with the "plane - three pins" testing scheme. Schematic diagram for the friction assembly of the testing machine is shown in Fig. 8. The frictional parameters to be recorded are the linear wear and mass wear of the specimens, friction factor, bulk temperature of the specimens and lubricant.

The cartridge clip with the fixed indenters is held down to the counterbody at the set pressure. The specimens are loaded with the help of the lever mechanism. The machine design provides for the capability to alter the rotational speed of the indenters from 0.5 m/s to 1.2 m/s. The friction factor is determined based on the measured friction torque, recorded by the strain gages fixed on the beam 6 using the half-bridge circuit. The beam is connected to the friction assembly via the rigid constraint that prevents specimen 1 located in the metal cup 5 from turning. The cup 5 is installed on the thrust bearing. The friction assembly is electrically insulated from the machine frame with the help of the textolite pad 7. The friction torque is recorded by the pointer indicator of the instrument TMM-48. The uncertainty of such measurement is 1 - 1.5% which ensures high accuracy of the obtained results. Mass wear of the specimens is determined on the laboratory balance with 0.5 mg measurement error.

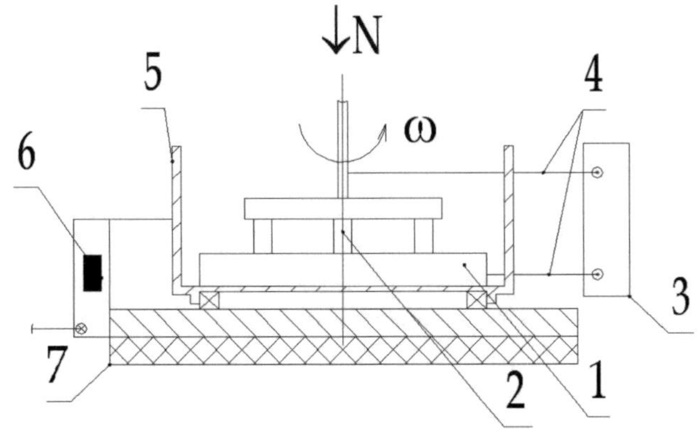

Fig. 8. Schematic diagram of the friction assembly to study the impact of an electric field upon the tribounit frictional parameters: 1) specimen; 2) counterbodies; 3) external source of electric energy; 4) electrodes providing the potential to the friction assembly; 5) testing cup for lubricants; 6) beam with resistance strain gages; 7) electrical insulating (dielectric) pad.

4.1 Application of an electric field to friction assemblies working in boundary lubrication conditions

Tests were carried out on the friction pair: "Bronze (БрО10Ц2, GOST 613-79) - Steel 45 (GOST 1050-88) in the lubricant medium - POLADYNE 10W-30 oil manufactured by Irvine, the US. Friction parameters: sliding velocity $\vartheta = 1$ m/s, contact pressure P = 0.5 MPa. A series of tests was conducted without an electric field, as well as tests where a positive potential was created on the steel counterbody, and then on the rubbing bronze indenters. The friction coefficient was selected as a measurand. The measurement results are shown in Fig. 9.

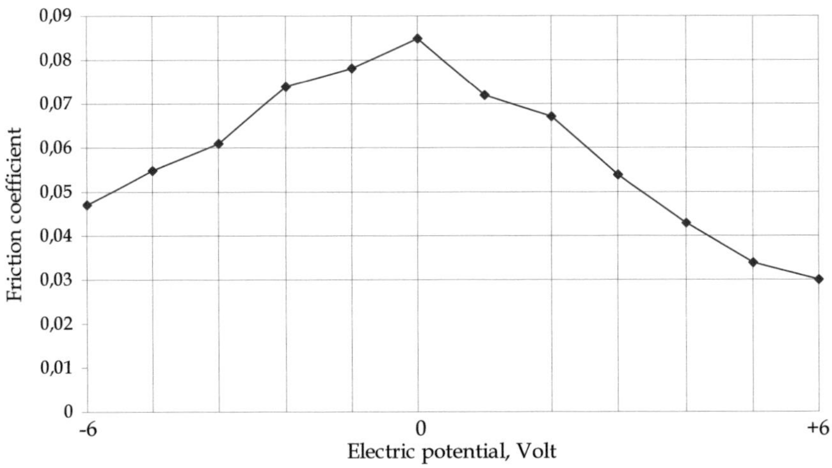

Fig. 9. Relation between the friction coefficient of the "bronze-steel" conjugation and the steel surface potential's value and sign

In our opinion, the significant decrease of the friction coefficient is driven by the intensifying effect of the electric field on the adhesive forces of the metal surface. The fact of the intensification of the adhesive forces under the action of the electric field was proved by experimental weighing (with an analytical balance) of the lubricant (adhesive), left on the steel counterbody surface after the friction. Relation of the adhesive mass and the magnitude and sign of the potential of the steel surface is shown in Figure 10. Along with insignificant amount of oil retained by the surface, it also has some amount of transferred bronze with the hardness that is smaller than that of steel.

Thus, the improved antifrictional characteristics observed in the tribounit are explained by the strengthened lubricant adhesion to the friction surfaces, which impedes its mechanical removal from the frictional contact area, and by the higher stability of the lubricating films. Also, higher chemical activity of the lubricant should be expected. Increased chemical activity of the present-day lubricants is one of the ways to achieve higher performance of friction assemblies. Oil's degree of activity can be visually monitored by the representative darkening after tests. Moreover, oil's reactivity can be evaluated by the decreased photo-emf and light flux, passing through the oil medium. The photo-emf measurement results for the tested oil samples, for each series, are shown in Fig. 11. It is evident that the oil taken after the tests where the positive potential had been applied to the steel surface is more chemically active. This can be explained by the fact that in this test, the recorded value of friction factor was minimal, as well as the frictional heating of the surfaces and the volume temperature of the lubricant. The oil has been depleted by the wear products and thermal decomposition products of its components. Visually it is lighter and therefore, weakens the light flux passing though it at a smaller extent.

Notable is total correlation of the dependencies shown in Fig. 9, 10, 11 which confirms the positive impact of the used electric field on the lubricant properties.

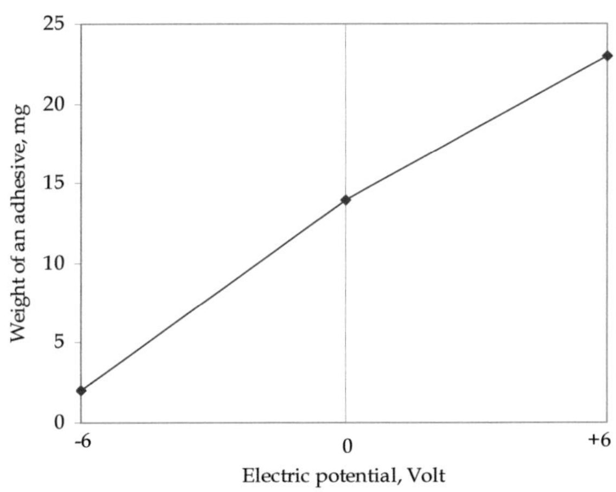

Fig. 10. Relation between the adhesive mass M (mg) and the magnitude and sign of the steel surface potential

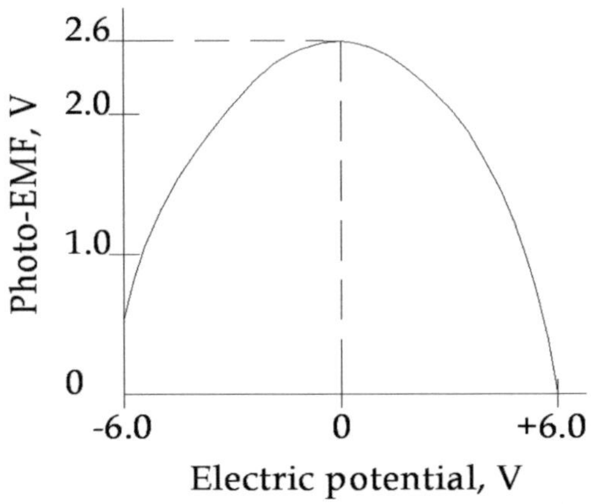

Fig. 11. Relation between the photo-emf and the magnitude and sign of the steel surface potential

4.2 Application of an electric field to metal-polymeric friction assemblies

Tests to evaluate the impact of an electric filed on metal-polymer conjugations with different combinations of polymer materials were carried out under the action of the load-and-speed factor ([pϑ]-factor) equal to 0.5 MPaxm/s, the test time was t = 2 hours. Intensity of wear of polymer materials is affected not only by the electrical field magnitude but by its direction, too (Fig. 12). This relates to the charge build-up on the metal surface in the metal-polymer friction pairs, and also to the fact that the metal surface has either cathodal or anodal polarization, depending on the rubbing polymer nature. Therefore, electric field application may lead to the surface depolarization, which in its turn will either increase or decrease the friction transfer of the polymer material.

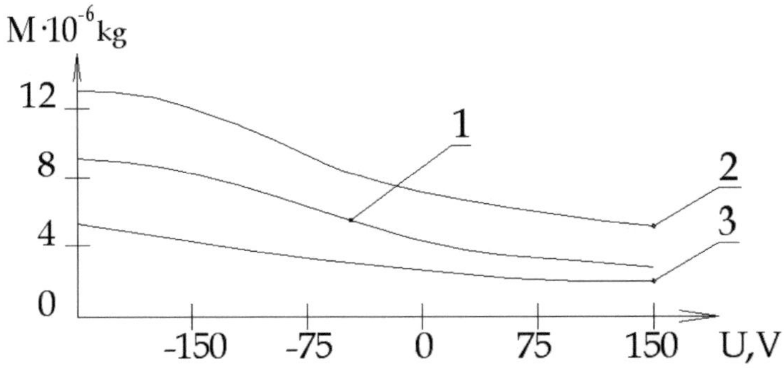

Fig. 12. Relation between the mass wear M and the magnitude and sign of the potential applied to the metal counterbody 1) polyethylene; 2) polypropylene; 3) polyamide 6.

More detailed research has determined that depending on the magnitude and direction of the external electric field, not only the extent of wear of the polymer material changes, but so do the wear conditions (Fig. 13), and the appearance of wear particles.

At the same time, it is evident that with the overall improvement of antifrictional properties of the metal-polymer tribounit, observed when a positive potential is applied to the metal counterbody, also the running-in time is observed to reduce. The plot of relation between the linear wear of the PTFE specimen and the test time for the friction pair "PTFE-steel" under an electric field, can serve as illustration to the previous statement (Fig. 14).

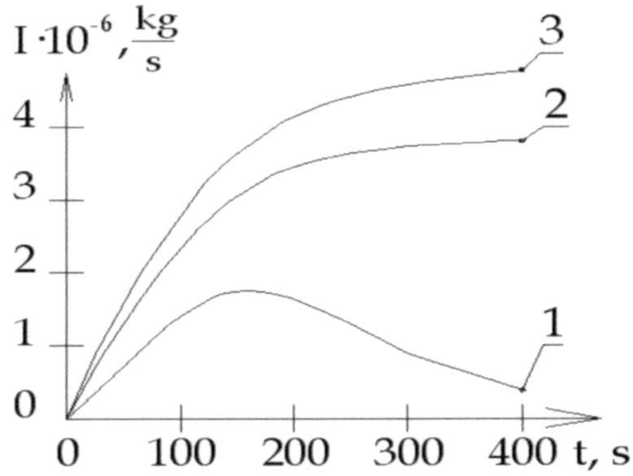

Fig. 13. Relation between the mass wear rate I and the friction time t for the friction pair "PTFE - copper": 1) +100V on metal; 2) 0V; 3) –100V on metal.

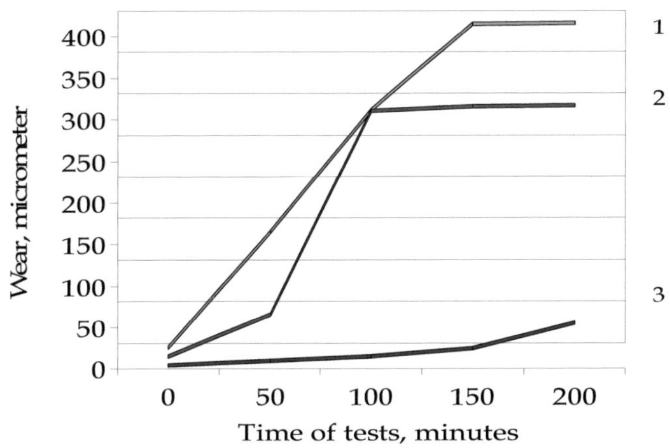

1) Heavy wear condition: -150V potential is applied to the steel counterbody; 2) Friction with no exposure to an electric field; 3) Minimal wear condition: +150V potential is applied to the steel counterbody.

Fig. 14. Relation between the running-in time for polymer material and the friction conditions.

It is evident that the superimposed electrical field is capable to not only decrease or increase the specimen wear by 6 times, but also to decrease the running-in time for the material. This is explained by the fact that, with a certain pattern of filed imposition on the metal countersurface, third bodies are increasingly generated which concurs with optimization of the frictional properties of the friction assembly and transfer to the steady state of wear. Therefore, the action of an electrical field leads to the improved tribotechnical parameters of friction assemblies. Characteristic feature of metal-polymer couplings is that it is possible, without fear of electrical breakdown, impose tens of volts on the friction assembly. As all negative issues associated with electroerosion damage of the surface taking place when an electrical breakdown occurs, depend not upon the magnitude of the applied voltage, but upon the breakdown current generated in the frictional contact. Having considered that, we developed and patented an electronic device for the dynamic control of friction and material wear processes, which was named "Electronic regulator of friction" (ERF).

4.3 Electronic regulator of friction

To control and regulate friction and wear processes of materials, it is necessary to provide a reference information channel which on the one hand, would transmit information about frictional processes, and on the other hand, would easily convert this information into a signal that controls the electrical field applied to the frictional contact area. The electronic friction regulator converts this reference signal into an electrical field with a certain voltage and direction. Frictional heating value of a friction pair can be selected as a reference signal. The heat Q emitted during friction is known to be a function of load N, friction coefficient f, sliding velocity ϑ, and friction time t:

$$Q = F(N; f; \vartheta; t) \tag{3}$$

In general terms, a change in tribological characteristics of a friction assembly reflects in the intensity of heating of the friction area. The less the contacting bodies are heated - the less apparent are the dilatation processes, the less is the bulk temperature of the lubricant, the better are the conditions to maintain the carrying capacity of the lubricating layer, stability of the lubricating protective films structure, and the less is the number of surface damages. Therefore, the increased heat release during friction may be identical to the deteriorated antifrictional parameters of a tribosystem. To prevent the wear of friction materials at high temperatures, surface active and chemically active additives are introduced into lubricants. Surface active additives ensure minimization of wear at relatively low temperatures – up to 300 ^{0}C, and chemically active additives decrease the wear and friction at higher temperatures – up to 800÷1200 ^{0}C. However, an additive added to a lubricant gradually deteriorates in the process of work, which entails its lower concentration in the friction area, and therefore the additive's modifying properties lower, too. Besides this, additional introduction of different types of additives into a lubrication material may disturb the thermodynamic, kinetic or mechanical compatibility of the base oil components. The maximum possible fuel savings (%), depending on an engine type and conditions of their operation, determined by the efficiency of the used motor oils and their additives, are shown in Table 2 (Matveevskyi et al., 1989):

Engine Type	Short Runs		Middle Runs	Long Runs
	Cold Start	Town Conditions		
Gasoline	7,5	6,0	2,2	1,8
Diesel	5,8	4,8	2,5	3,0

Table 2. Fuel savings for vehicle engines

Advantage of adaptive electronic lubricating devices over additives to lubricants is that you enable the electronic module only one time and then its positive action upon mechanisms' and machines' friction assemblies is continuous and unflagging, while the intensity of an additive's action decreases during operation process. The developed device in terms of its function may propose an alternative to quite a wide range of additives to lubricants, and has no restrictions related to the design and process features of friction assemblies. The device was developed for heavy loaded friction assemblies. The device was tested on the propulsion systems of passenger elevators, the electric drives of kettle mixers at lubricant factories, but the widest use this device has got in internal combustion engines. It was determined that this device connected to a power unit improves the power efficiency of engines for vehicles and drives of fixed machines. This happens due to the decreased friction forces in the mechanisms. Friction factors of the assemblies working in the boundary lubrication conditions decrease by 15%, and working with a lubricant - by up to 40%. Efficiency of the utilization of the Electronic Friction Regulator is determined by the load and speed performance of am operated friction assembly, as well as by the initial and maximal intensity of the electromagnetic field created by the device. The device itself has comparatively small dimensions (45 x 45 x 25 mm, its weight does not exceed 150 g). It is built as a stand-alone unit, and thanks to this it is adapted to the structures of machine's mechanisms without any design changes. At this stage of the accomplished work, the device is operable within the temperature range from minus 50 ^{0}C to plus 90 ^{0}C and has a service life no less than 15,000 hours. Application of an "ERF" in the friction assemblies of machines, at a certain intensity and direction of an electromagnetic field, will allow to:

- suppress the frictional electrification effect catalyzing development of corrosion and oxidation processes on the contacting surfaces, and metal anodic dissolution;
- improve lubricant adhesion to the surface being protected thus increasing the lubricant's protective functions;
- decrease a possibility for the hydrogen wear of the friction surface.

ERF

Fig. 15. General view of the test bed

Base Case												
Condition No.	Me, nm	Ne, kW	Gt, kg/hr	ge, kg/ kW-hr	ηe	ηm	Po, bar	Tdg, ⁰C	CO,%	CH, ppm	NO, ppm	CO_2, %
1	**2**	**3**	**4**	**5**	**6**	**7**	**8**	**9**	**10**	**11**	**12**	**13**
1	20,16	4,22	1,99	0,471	0,174	0,543	2,8	383	0,102	93	1563	11,61
2	40,32	8,44	2,81	0,333	0,246	0,714	2,7	425	0,073	112	2687	11,43
3	60,48	12,67	3,71	0,293	0,279	0,797	2,6	480	0,067	110	3046	11,93
4	80,63	16,89	4,84	0,286	0,286	0,847	2,5	560	0,643	114	2420	12,28
5	83,15	17,42	6,15	0,353	0,232	0,852	2,5	540	6,441	177	596	9,58
Final condition, with the Electronic Regulator of Friction in 5 motor hours												
1	**2**	**3**	**4**	**5**	**6**	**7**	**8**	**9**	**10**	**11**	**12**	**13**
1	20,16	4,22	1,87	0,443	0,185	0,543	2,5	395	0,096	76	1710	11,96
2	40,32	8,44	2,67	0,317	0,258	0,713	2,5	421	0,070	97	2812	11,69
3	60,48	12,67	3,43	0,271	0,302	0,797	2,4	460	0,062	105	3675	11,73
4	80,63	16,89	4,62	0,273	0,299	0,845	2,4	555	0,585	95	2556	12,43
5	84,67	17,73	5,98	0,337	0,243	0,854	2,3	535	6,664	155	554	9,39

Table 3. Load characteristic at n = 2000 rpm

Base case												
Condition No.	Me, nm	Ne, kW	Gt, kg/hr	ge, kg/ kW-hr	ηe	ηm	Po, bar	Tdg, ⁰C	CO,%	CH, ppm	NO, ppm	CO_2, %
1	**2**	**3**	**4**	**5**	**6**	**7**	**8**	**9**	**10**	**11**	**12**	**13**
1	20,33	6,39	2,96	0,463	0,177	0,503	2,8	502	0,213	82	2100	12,47
2	40,66	12,77	4,26	0,334	0,245	0,677	2,7	548	0,093	88	3487	12,56
3	60,98	19,16	5,38	0,281	0,291	0,765	2,6	596	0,089	97	3937	12,37
4	81,31	25,55	7,22	0,282	0,290	0,818	2,6	656	0,629	101	2981	13,12
5	97,58	30,65	9,81	0,320	0,256	0,848	2,8	610	7,171	189	602	9,95
Final condition, with the Electronic Regulator of Friction in 5 motor hours												
1	**2**	**3**	**4**	**5**	**6**	**7**	**8**	**9**	**10**	**11**	**12**	**13**
1	20,33	6,39	2,89	0,453	0,181	0,503	2,6	506	0,128	52	2211	12,51
2	40,66	12,77	4,10	0,321	0,255	0,677	2,6	537	0,097	73	3480	12,67
3	60,98	19,16	5,23	0,273	0,299	0,765	2,6	585	0,099	83	4013	12,61
4	81,31	25,55	6,90	0,270	0,303	0,818	2,6	650	0,785	90	2889	13,15
5	97,58	30,65	9,92	0,324	0,253	0,848	2,6	610	7,553	177	536	9,68

Table 4. Load characteristic at n = 3000 rpm

This entails the increased service lifetime of the machine mechanisms, as well as their better environmental performance. The device design and method of its connection to a car power plant are presented in the papers by (Lubimov et al., 2008; Lubimov et al. 2009b). There were conducted two series of tests of different designs of the device. Below are shown the device photograph and the results of its trials on the test bed simulating internal combustion engine's operation.

As you can see from the picture, the dimensions of the Electronic Regulator of Friction (blue colored device indicated with the arrow) attached to the machine body are small in comparison to the machine overall sizes.

Rotation frequency, rpm	Initial condition		Outcome	
	Closed throttle	Open throttle	Closed throttle	Open throttle
300	26,0	23,0	24,0	22,0
500	17,5	17,0	17,0	16,5
800	16,5	12,5	16,0	13,0
1000	16,5	12,5	16,0	12,5
1500	18,0	13,5	17,5	13,0
2000	19,0	14,0	18,0	14,0
2500	21,0	16,0	20,0	16,0
3000	24,0	18,5	21,0	17,5

Table 5. Mechanical loss moment, Nm

Effective Power	Fuel Consumption	Effective Efficiency	CO	CH	NO_x
0,840	4,344	4,621	18,90	19,05	- 6,53

Table 6. Averaged effects, %, against the engine base case condition according to the results of the first series of tests, considering the changes for all the conditions

Symbols in the table: n – rotation frequency for the engine crankshaft; Me – net troque; Ne – effective power; Gt – fuel consumption per hour; ge – specific fuel consumption; ηe – effective efficiency; ηm – mechanical efficiency; Po – oil pressure; Tdg – temperature of discharge gases; CO – carbon oxide content in the engine discharge gases; CO2 – carbon dioxide content in the engine discharge gases; NO – nitrogen oxide content in the engine discharge gases; CH – residual hydrocarbons content the engine discharge gases.

From the given data, one can see that the enabled device decreases the fuel consumption and mechanical loss moment; increases the power and efficiency of the machine engine, decreases harmful contamination in the discharge gases, as well as their temperature. The observed effects were recorded after 5 motor hours of the experimental unit operation. After the device had been disconnected and the simulation machine had been restarted, the controlled variables again decreased down to the values corresponding to the base case condition of the engine. The next series of tests were aimed to study the impact of an electromagnetic field with 60 MHz frequency pulse action on the engine performance. For this purpose, the design of the Electronic Regulator of Friction was modified with some slight increase of its dimensions. The results of the development tests of this design version at n = 3000 rpm, as well as the averaged effects are shown in Table 7 and summary table 8.

At this moment, we can not provide the definite explanation for the increased content of nitrogen oxides in the discharge gases. One of the draft versions of reasoning is the ability of an electromagnetic field to inhibit generation of soot depositions on the friction surfaces of the piston-cylinder group of an internal combustion engine. Under the action of an electric field the working elements of the car engine are cleaned. Indirectly, this is proved by photographs of oil filters taken from the machines during operational tests of the device

(Fig. 16). In the presented figure, the upper image is a new filter of the VAZ 21099 car that has run 1500 km with ERF, the lower image is the condition of the oil filter after the scheduled oil replacement after 10,000 km. The upper filter is filled with particulate depositions of the soot nature. It seems that under the action of an electromagnetic filed, the above mentioned depositions were removed from the cylinders' wall surfaces and retained by the oil filter.

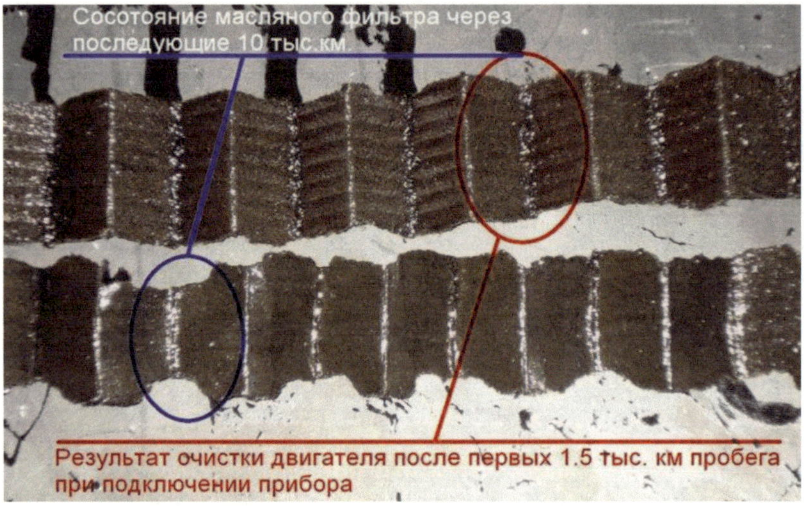

Fig. 16. Condition of the oil filters with the utilization of the ERF (above) and without it (below)

Base case												
Condition No.	Me, nm	Ne, kW	Gt, kg/hr	ge, kg/ kW-hr	ηe	ηm	Po, bar	Tdg, ⁰C	CO,%	CH, ppm	NO, ppm	CO_2, %
1	**2**	**3**	**4**	**5**	**6**	**7**	**8**	**9**	**10**	**11**	**12**	**13**
1	20,33	6,39	2,90	0,455	0,180	0,503	2,7	494	0,135	55	2059	12,60
2	40,66	12,77	4,19	0,328	0,250	0,677	2,7	536	0,092	78	3421	12,67
3	60,98	19,16	5,28	0,276	0,297	0,765	2,6	585	0,099	95	4017	12,71
4	81,31	25,55	6,98	0,273	0,299	0,818	2,7	650	0,644	97	2987	13,32
5	98,59	30,97	9,98	0,322	0,254	0,849	2,8	612	7,356	174	574	9,96
Final condition, with the Electronic Regulator of Friction (60 MHz) in 5 motor hours												
1	**2**	**3**	**4**	**5**	**6**	**7**	**8**	**9**	**10**	**11**	**12**	**13**
1	20,33	6,39	2,79	0,436	0,187	0,503	2,7	506	0,132	55	2255	12,56
2	40,66	12,77	3,95	0,309	0,265	0,677	2,6	538	0,095	74	3562	12,64
3	60,98	19,16	4,99	0,261	0,314	0,765	2,6	582	0,100	86	4066	12,59
1	**2**	**3**	**4**	**5**	**6**	**7**	**8**	**9**	**10**	**11**	**12**	**13**
4	81,31	25,55	6,68	0,262	0,313	0,818	2,6	661	0,675	87	2988	13,21
5	98,59	31,23	9,38	0,303	0,270	0,849	2,7	608	8,024	163	488	9,48

Table 7. Load characteristic at n = 3000 rpm

Effective Power	Fuel Consumption	Effective Efficiency	CO	CH	NOx
1,392	5,036	5,233	21,14	24,78	-4,43

Table 8. Averaged effects, %, against the engine base case condition according to the results of the second series of tests, considering the changes for all the conditions

The important feature of the developed device is that it can be used both for the improvement of the lubrication conditions of friction assemblies and for controlled tightening of their frictional parameters. This feature allows to use the device for faster running-in time of new machines and mechanisms in the conditions minimizing the initial wear, and it can be helpful for machine engineering companies to reduce time and energy consumption when testing their fabricated products.

5. Conclusion

Today, research of the fine mechanisms of friction which are responsible for change of materials' antifriction parameters, and the impact of external force fields (electric, magnetic) on them, are underway. At this stage, a thesis has been defended which discusses the main principles of transformation of a matter aggregate state during friction, and impact of electromagnetic fields on lubricating action of oils and greases in metal couplings. The design of the Electronic Friction Regulator is being improved, and there being checked possible ways of its connection to vehicle engines with the purpose of finding the best option providing the maximal improvement of vehicle/machine power plant performance. Also, the research is underway to study possible negative consequences from utilization of this device. However, over the three years of monitoring of the technical condition of the machines operated with the electronic device, no negative aspects from its utilization were found.

6. References

Akhmatov, A.S. (1963). *Molekulyarnaya fizika granichnogo treniya (Molecular Physics of Boundary Friction)*, Fiziko-matematicheskaya literatura, Moscow, USSR.

Buckley, D.H. (1986). *Poverkhnostnye javlenija pri adgezii i frikcionnom vzaimodejjstvii (Surface Effects in Adhesion, Friction, Wear and Lubrication)*, Mashinostroenie, Moscow, USSR.

Belyi, V.A. (1988). Problema sozdanija kompozicionnykh materialov i upravlenie ikh frikcionnymi svojjstvami (Creation of Composites and Control of their Frictional Properties). *Trenie i iznos*, Vol.3, No.3, (June 1988), pp. 384-396, ISSN 0202-4977.

Cahn, R. W. (1968). *Fizicheskoye metallovedenie (Physical Metallurgy)*, Vol.3, Mir, Moscow, USSR.

Cholakov V. (1986). *Nobelevskie premii. Uchenye i otkrytija (Nobel Prizes. Scientists and Discoveries)*, Mir, Moscow, Russia.

Garkunov, D.N. (2001). *Tribotekhnika, iznos i bezyznosnost' (Tribotechnics, Wear and Wealessness)*, Moscow Agricultural Academy, ISBN 5-94327-004-3, Moscow, Russia.

Garnovskyi, A.D.; Rjabukhin, Yu.I. & Kuzharov, A.S. (1984). Koordinacionnaja khimija (Coordination Chemistry). *Trenie i iznos*, Vol.5, No.6, (December 1984), pp. 1011-1033, ISSN 0202-4977.

Ginsburg, V.L. (1967). *Rasprostranenie ehlektromagnitnykh voln v plazme (Electromagnetic Wave Propagation in Plasma)*, Fiziko-matematicheskaja literatura, Moscow, USSR.
Godet, M. (1984). The body approach. A mechanical view of wear, *Wear*, Vol. 100, pp. 437-452, ISSN 0043-1648.
Goldade, V.A.; Struk, V.A. & Peseckyi, S.S. (1993). *Ingibitory iznashivanija metallopolimernykh system (Wear Inhibitors of Metal-polymeric Systems)*, Khimija, ISBN 5-7245-0876-1, Moscow, Russia.
Heinicke, G. (1987). *Tribokhimija (Tribochemistry)*, Mir, Moscow, USSR.
Hawking, S & Penrose, R. (2007). *Priroda prostranstva i vremeni (The Nature of Space and Time)*, Amfora, ISBN 978-5-367-00590-5, St.-Petersburg, Russia.
Kadolich, Zh.V.; Pinchuk, L.S. & Anisov, A.P. (2001). Issledovanie poverkhnosti trenija polimernykh implantantov tazobedrennykh sustavov (Studying Friction Surface of Polymeric Implants of Hip Joints), *Trenie i iznos*, Vol.22, No.1, (February 2001), pp. 78-83, ISSN 0202-4977.
Kostetsky, B.I. (1980). O roli vtorichnykh struktur v formirovanii mekhanizmov trenija, smazochnogo dejjstvija i iznashivanija (About the Role of Secondary Structures in Formation of Friction Mechanisms, Lubricating Action and Wear). *Trenie i iznos*, Vol.1, No.4, (August 1980), pp. 622-637, ISSN 0202-4977.
Landau, L.D. & Lifshitz E. M. (2001). *Teoreticheskaja fizika (Theoretical Physics)*, Vol.6 *"Gidrodinamika" (Hydrodynamics)*,Fiziko-matematicheskaja literatura, ISBN 5-9221-0121-8, Moscow, Russia.
Lubimov, D.N.; et al. (1990). Ehlektromagnitnye spektry i fiziko-khimija metallopolimernogo frikcionnogo kontakta (The Electromagnetic Spectra and Physicochemistry of a Metal-polymeric Frictional Contact). *Trenie i iznos*, Vol.11, No.6, (December 1990), pp. 1084-1086, ISSN 0202-4977.
Lubimov, D.N.; et al. (1992). Kinetika formirovanija i struktura smazochnykh sloev pri izbiratel'nom perenose v metallopolimernykh tribosoprjazhenijakh (The Kinetics of Formation and The Structure of Lubricating Layers during Selective Transfer in Metal-polymeric Tribounits). *Trenie i iznos*, Vol.13, No.3, (June 1992), pp. 496-500, ISSN 0202-4977.
Lubimov, D.N., Kozachenko P.N., Ivanov A.E. (1994). Ehlektropolevojj "serfing-ehffekt" (Electrofield "Surfing Effect"), *Sovershenstvovanie tekhniki, tekhnologii i problemy ehkologii proizvodstv* (Improvement of Technology and Machinery, Manufacturing Environmental Issues), Vol.3, pp. 66-68, Shakhty Technology Institute for Consumer Services, Shakhty, Russia.
Lubimov, D.N.; Levkin, V.V. & Ivanov, A.E. (1998). Nakhozhdenie kontaktnojj temperatury pri pomoshhi diffuzionno-ehnergeticheskikh predstavlenijj o trenii (Finding of Contact Temperature Using Diffusion and Energy Notions of Friction), *Oborudovanie i tekhnologii byta i uslug (Equipment and Technologies of Housekeeping and Services)*, Vol.29, pp. 104-106, Shakhty, Russia.
Lubimov, D.N & Ryzhikov, V.A. (2001). *Osnovy teorii trenija (Friction Theory Fundamentals)*, South Russian State Technical University, ISBN 5-88998-220-6, Novocherkassk, Russia.
Lubimov, D.N. & Vershinin, N.K. (2001). Vlijanie frikcionnogo vzaimodejjstvija poverkhnostejj metallopolimernogo tribosoprjazhenija na ikh ehlektricheskie i khimicheskie svojjstva. (Impact of Frictional Interaction of the Metal-polymeric

Tribounit Surfaces on Their Electrical and Chemical Properties). *Sostojanija i perspektivy razvitija vostochnogo Donbassa (Conditions and Prospects of Eastern Donbass Development)*, pp. 274-278, ISBN 5-88998-252-4.

Lubimov, D.N & Ryzhikov, V.A. (2006). *Fiziko-khimicheskie processy pri trenii (Physicochemical Processes in Friction)*, South Russian State Technical University, ISBN 5-88998-394-6, Novocherkassk, Russia.

Lubimov, D.N.; Kozhemyachenko, A.V. & Dolgopolov, K.N. (2007a). Vlijanie postplazmennykh processov na formirovanie smazochnykh plenok na poverkhnostjakh trenija (Influence of Postplasmatic Processes on the Formation of Lubricating Films on Friction Surfaces), In: *Bytovaja tekhnika, tekhnologija i tekhnologicheskoe oborudovanie predprijatijj servisa i mashinostroenija (Household Equipment, Technology and Process Equipment of Service and Engineering Companies)*, pp. 86-87, South Russia State University of Economics and Service, ISBN 978-5-9383-4-302-3, Shakhty, Russia.

Lubimov, D.N.; et al. (2007b). Issledovanie fizicheskikh svojjstv triboplazmy, kak osobogo agregatnogo sostojanija veshhestva (Study of the Physical Properties of Triboplasma As a Special Aggregate State of Matter), *Mezhdunarodnaja nauchnaja konferencija Problemy razvitija estestvennykh, tekhnicheskikh i social'nykh sistem (International Scientific Conference. Development Issues of Natural, Technical and Social Systems), Part 2*, pp.38-43, ISBN 5-88040-047-6, Anton, Taganrog, Russia, April, 2007.

Lubimov, D.N.; et al. (2008). Osnovnye principy i rezul'taty ispol'zovanija "Ehlektronnogo reguljatora trenija" dlja povyshenija ehkspluatacionnykh pokazatelejj dvigatelejj transportnykh sredstv (Main Principles and Results of Operation of the "Electronic Friction Regulator" to Imporve the Performance of Vehicle Engines), In: *Intellektika, logistika, sistemologija (Intellectics, Logistics, Systemology)*, pp. 146-149, Chelyabinsk Center of Scientific and Technical Information, ISBN 978-5-94218-081-2, Chelyabinsk, Russia.

Lubimov, D.N.; Dolgopolov, K.N. & Bai, N.M. (2009a). Ehlektrodinamika poljarizacii gorjuche-smazochnykh materialov ot vneshnego istochnika (The Electrodynamics of Polarization of Fuels and Lubricants from an External Source), In: *Ehkonomika i proizvodstvo (Economics and Production)*, pp. 124-130, Chelyabinsk Center of Scientific and Technical Information, ISBN 978-5-94218-088-1, Chelyabinsk, Russia.

Lubimov, D.N.; et al. (2009b). Ehlektronnyjj reguljator trenija (ustrojjstvo i rezul'taty stendovykh ispytanijj) (Electronic Friction Regulator. Design and Development Testing Results), In: *Ehkonomika i proizvodstvo (Economics and Production)*, pp.118-123, Chelyabinsk Center of Scientific and Technical Information, ISBN 978-5-94218-088-1, Chelyabinsk, Russia.

Matveevskyi et al. (1989). *Smazochnye materialy: Antifrikcionnye i protivoiznosnye svojjstva. Metody ispytanijj (Lubricants. Antifriction and Antiwear Properties. Testing Methods)*, Mashinostroenie, Moscow, USSR.

Morozov, S.V.; Novoselov, K.S. & A. Heym, A.K. (2008). Ehlektronnyjj transport v grafene (Electron Transport in Graphene), *Uspekhi fizicheskikh nauk*, Vol.178, No.7, pp. 776-780, ISSN: 0042-1294 (Print), 1996-6652 (Online).

Nikolsky, A.V.; Kozakov, A.T. & Kravchenko, V.I. (1988). Dinamika izmenenija khimicheskogo sostojanija poverkhnostejj trenija metallopolimernogo soprjazhenija v processe frikcionnogo vzaimodejjstvija (The Change Trend of the Friction Surface

Chemical Condition of a Metal-polymeric Tribounit during Frictional Interaction Process). *Trenie i iznos*, Vol.9, No.5, (October 1988), pp. 860-869, ISSN 0202-4977.

Pinchuk, L.S.; Goldade, V.A. & Makarevich, A.V. (2004). *Ingibirovannye plastiki (Inhibited Plastics)*, Metal-Polymer

Research Institute of National Academy of Science of Belarus, ISBN 985-647727-1, Gomel, Belarus.

Pinchuk, L.S. & Chernyakova, Yu.M. (2007). Sinovial'nyjj sustav kak "umnyjj" uzel trenija (Synovial Articulation as a "Smart Friction Assembly"), *Trenie i iznos*, Vol.28, No.4, (August 2007), pp. 410-417, ISSN 0202-4977.

Pogosyan, A.K. & Oganesyan, K.V. (1986). Javlenie frikcionnogo perenosa: osnovnye zakonomernosti i metody issledovanija (Phenomenon of Frictional Transfer: Main Patterns and Study Methods). *Trenie i iznos*, Vol.7, No.6, (December 1986), pp. 998-1007, ISSN 0202-4977.

Polyakov, A.A. (1988). Samoorganizacija struktury izbiratel'nogo perenosa (Self-organization of Selective Transfer Structure), In: *Dolgovechnost' trushhikhsja detalejj mashin (Durability of Rubbing Parts of Machines)*, Vol.3, pp. 45-95, Mashinostroenie, Moscow, USSR.

Polyakov, A.A. (1990). Terminy izbiratel'nogo perenosa (Selective Transfer Terms), In: *Dolgovechnost' trushhikhsja detalejj mashin (Durability of Rubbing Parts of Machines*, Vol.4, pp. 11-15, Mashinostroenie, Moscow, USSR.

Prokhorov, A.M. (1992). *Fizicheskaja ehnciklopedija (Physical Encyclopedia)*, Vol.3, Bol'shaja Rossijjskaja ehnciklopedija (Big Russian Encyclopedia), ISBN 5-85270-019-3, Moscow, Russia.

Ryzhikov V.A. & Dolgopolov K.N. (2005). Ehlektrokhimicheskie processy v uzlakh trenija (Electrochemical Processes in Friction Assemblies). *IV Mezhdunarodnaja nauchno-prakticheskaja konferencija Problemy sinergetiki v tribologii, triboehlektrokhimii, materialovedenii i mekhatronike (IV International Scientific Workshop Conference. Synergetics Problems in Tribology, Triboelectrochemistry, material science and mechatronics)*, pp. 7-11, South Russian State Technical University, ISBN 5-88998-629-5, Novocherkassk, Russia, November 4, 2005.

Sysoev, P.V.; Bliznets, M.M. & Pogosyan, A.K. (1990). *Antifrikcionnye ehpoksidnye kompozity v stankostroenii (Antifrictional Epoxy-filled Composites in Machine Engineering)*, Nauka i tekhnika, ISBN 5-343-00536-5, Minsk, Belarus.

Shvedkov, E.L. & Rovinsky, D.Ya. (1979). *Slovar'-spravochnik po treniju, iznosu i smazke detalejj mashin (Glossary for Friction, Wear, and Machine Parts Lubrication)*, Naukova dumka, Kiev, USSR.

Voinov, K.N. (2010). *Tribologija: mezhdunarodnaja ehnciklopedija. Istoricheskaja spravka, terminy, opredelenija (Tribology: International Encyclopedia (Background, Terms, Definitions)*, Vol.1, Anima, ISBN 978-5-9902064-2-7, St.-Petersburg, Russia.

6

Damage Assessment of Short Glass Fiber Reinforced Polyester Composites: A Comparative Study

Amar Patnaik[1], Sandhyarani Biswas[2], Ritesh Kaundal[1] and Alok Satapathy[2]
[1]*Department of Mechanical Engineering, National Institute of Technology, Hamirpur*
[2]*Department of Mechanical Engineering, National Institute of Technology, Rourkela*
India

1. Introduction

The main goals of many challenges to the utilization of by-product cement kiln dusts (CKDs) as partial replacement of Portland cement (PC) environmental protection agencies are to seek ways to minimize the dual problems of disposal and health hazards of these by-products. For many years, by-products such as fly-ash, silica fume and slag were considered as waste materials. These by-products have been successfully used in the construction industry as a portland cement substitute (Krazowski & Emery, 1981; Al-jabri et al., 2006). However, with the growing of industries the new and different types of by-products are being generated by various industries, which could have being a promising future for partial replacement of portland cement. CKDs typically fine powders that are generated during the cement manufacturing process, and then carried off in the flue gases then subsequently collected in electrostatic precipitators. The portion of CKDs that are not returned back to the cement manufacture process, or otherwise used beneficially, are placed in landfills. A limited number of studies have carried out the use of CKD as industrial wastes in fabrication of polymer composites. The effect of CKD on the compressive strength of cement paste was studied and the corrosion behavior of embedded reinforcement and observed that up to 5wt% of CKD of cement had less effect on cement paste strength and on reinforcement (El-Sayed et al., 1991). It has also reported that when CKD and blast furnace slag are added in proper ratio to ordinary cement, the compressive strength and the corrosion-resistance of the mix increase dramatically (Batis et al., 2002). It has also been reported that the use of CKD in concrete as a partial replacement of cement (Sri Ravindrarajah, 1982). The percentages of cement replacement by weight were in different percentage i.e. from 0, 25, 50, 75, and 100 in cement paste and 0, 15, 25, 35, and 45 in both 1:1.5:3 and 1:2:4 concretes. As CKD is a cementitious material and it showed good strength as compared with general cement. Therefore, particulate played a vital role in the development of commercially feasible fiber reinforced polymers. Not only they provide significant cost reductions but also certain fillers may improve processing, physical properties and improved the mechanical strength (Rothon, 2002). There are few, if any, publications exploring the effect of hard ceramic particles on the wear resistance (especially for erosion resistance) of composites. Further, there are few publications discussing the erosion wear mechanisms of elastomers and their composites. Based on the above study, computational simulation has provided an effective and economical approach to study wear problems. During past decades, there were a number of computational models proposed to simulate wear processes at macro/meso and atomic/nano-scales (Ling & Pan, 1986; Ludema & Bayer, 1991; Hsu et al., 1997; Komvopoulos & Choi, 1992). The most widely used macro/meso-scopic model is FEM (finite element method), which can be used to deal with elastic/plastic contact problems and failure processes in the contact region. The two-dimensional (2D) water drop impact model with DYNA3D code was established (Adler, 1995). The calculation did not involve any solid erodent. Despite this, to date there are no economically viable and environmentally acceptable solutions for effective utilization of the high-residue volumes that have developed. Current best practice is to contain the material within specially constructed landfill sites, called red mud ponds/stacks or bauxite residue disposal areas (BRDAs). Recently, the potential utilization of red mud filled glass/bamboo epoxy composites was explored and investigated its significant effect on the erosion wear rate of the resulting composites (Biswas and Satapathy, 2009, 2010). Few researchers also studied the solid particle erosion behavior of particulate filled polymer matrix composites and revealed that parameters such as impact velocity, impingement angle, fiber loading/filler content, erodent size, erodent temperature and stand-off distance etc. largely depends on the erosion behavior of polymer composites (Patnaik et al., 2010a, 2010b, 2010c, 2010d, 2010e, 2010f, 2010g).

In view of the above literature in this work, a non-linear finite element model has been proposed and compared the simulated results with experimental results. The utility of the new formulation is proven through its application onto a short glass fiber reinforced polyester based composites filled with cement by-pass dust as reinforcing particulates. This model is simple and flexible, which helps to gain insight into various wear processes. The eroded surfaces of these composites are analyzed with scanning electron microscopy (SEM), and the erosion wear mechanisms of the composites are investigated.

2. Experimental

2.1 Preparation of composites

Short glass fibers (elastic modulus of 72.5 GPa and possess a density of 2.59 gm/cc) of 6mm length are used to prepare the composites. The unsaturated isophthalic polyester resin (Elastic modulus 3.25GPa, density 1.35gm/cc) is manufactured by Ciba Geigy and locally supplied by Northern Polymers Ltd. New Delhi, India. The composite fabricated in two different sets. One having different fiber loading, with varying the weight fraction of fibers from 10wt% to 50wt% at an increment of 10wt% and the designation of individual composites are reported in Table 1. Secondly, cement by-pass dust (CBPD) particulate is mixed with short glass fiber reinforced polyester resin with three different percentages (0wt%, 10wt% and 20wt% of CBPD) and the mixture is poured into various moulds conforming to the requirements of various testing conditions and characterization standards. The entrapped air bubbles (if any) are removed carefully with a sliding roller and the mould is closed for curing at a temperature of 30^0C for 24 h at a constant pressure of 10 kg/cm^2.

2.2 Air-jet erosion tester

The erosion experiments are carried out as per ASTM G76 on the erosive wear test rig. The equipment was designed to feed abrasive particles into a high velocity air stream, which

Sl. No.	Designation of composites	Composition of composites
1.	PGFL-1	Polyester resin + 10wt% Short glass fiber
2.	PGFL-2	Polyester resin + 20wt% Short glass fiber
3.	PGFL-3	Polyester resin + 30wt% Short glass fiber
4.	PGFL-4	Polyester resin + 40wt% Short glass fiber
5.	PGFL-5	Polyester resin + 50wt% Short glass fiber

Table 1. Designation and composition of unfilled short glass fiber reinforced composites

propelled the particles against the specimen surface (Strzepa et al., 1993; Routbort et al., 1981). The erodent particles entrained in a stream of compressed air and accelerated down to a 65mm long brass nozzle with 3mm inside diameter to impact on a target material mounted on an angle fixture. The steady state erosion rates are measured as a function of the erodent particle size and the amount of erodent particle used in each experiment was measured with the help of double disc method. The steady state erosion rate was determined by weighing the sample after the end of each test. While the impingement angles ranges from 30° to 90° and the test duration was 20min for each run. The erodent used for this test was river silica sand particle of three different sizes, i.e. 250, 350 and 450μm. The sample was cleaned with a blast of compressed air before each weighing to remove all loosely adhering debris. The mass loss from the target was measured with an analytical balance of ±0.01mg accuracy. The erosion rate was measured in terms of mass of material removed per unit mass of erodent impacted. The process is repeated every 10 min till the erosion rate attains a constant value called steady-state-erosion-rate.

2.3 Finite element model

The erosive processes are simulated using an explicit dynamic code ANSYS/AUTO-DYN. The eight-node brick hexahedral elements with one integration point are used in the 3D simulation. The mesh is refined to standard cubic element at the impacted area in order to calculate the erosion rate. It has reported in literature that simulating a single particle was not sufficient and three or more particles were needed to simulate the erosion process (ElTobgy et al., 2005). In this study, 100 spherical particles were used so as to ensure the accuracy of the model. All of the particles strike the target center area at random locations. The "simultaneous group" was employed for computational efficiency and analytical organization. There are 10 groups which contain 100 particles aggregately in the proposed model. Every group has 10 particles which would impact the surface simultaneously and be followed by another simultaneous particles group, and so on. According to the researchers, the distance between any two particles' centers in the same group is no less than 0.6r (r is the radius of the particles) to avoid the damage interaction (Woytowitz and Richman, 1999). Figure 1 shows the schematic diagram of the prepared composites by using ANSYS/AUTO-DYN. The finite element model of the target material and simulated nozzle is shown in Figure 2. There are 164, 765 nodes and 147, 840 finite elements in the model. All of the bottom nodes of the target materials are fixed. For the particles, the rotation degrees of freedom are constrained. Generally, the erosion rate (gm/gm) was used to characterize the erosion performance of the target materials. The solution time is set to 1.3t10, where t10 is the time when the last particles group contacted to the target. In this study, the impact angle varies from 30° to 90° in increments of 15°. The impact velocity varies from 43 to 65 m/s.

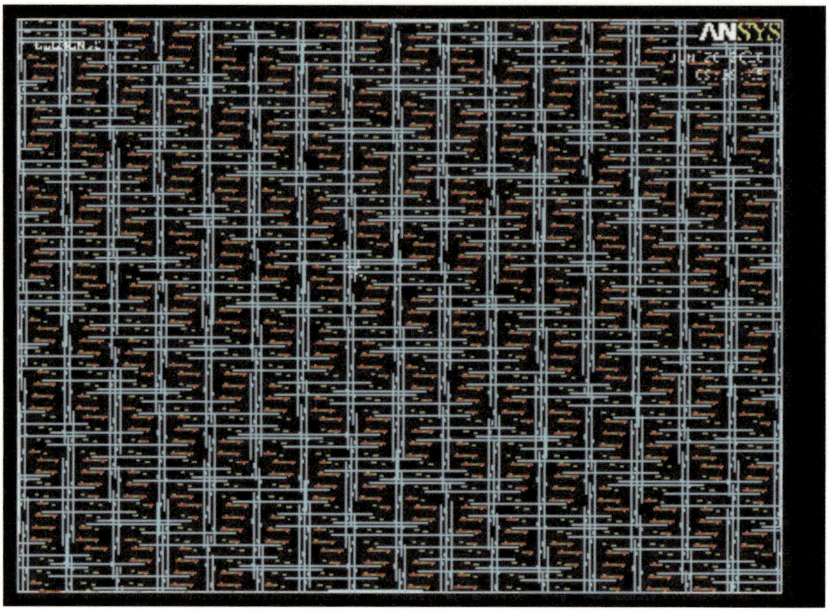

Fig. 1. Schematic diagram of the particulate filled composite

2.4 Plan of experiments (Taguchi's techniques)

For the elaboration of experiments plan, the method of Taguchi for five control factors at three levels is used, being understood by levels taken by the factors. Table 2 represents the factors to be studied and the assignment of the corresponding levels. The array chosen is the L_{27} (3^{13}) which has 27 rows corresponding to the number of tests (22 degrees of freedom) with 13 columns at three levels. The factors and the interactions are assigned to the columns. In practice, these factors can be assigned arbitrarily to any of the arrays columns, provided that all combinations must be included as per orthogonal array design. After assigning appropriate level settings, the S/N analysis (S/N: signal-to-noise ratio) is needed to evaluate the proposed experiment results. In S/N analysis, the greater the S/N, the better the experimental results:

$$\eta = -10 \log (\text{M.S.D.}) \quad (1)$$

where M.S.D. is the mean-square deviation for the output characteristic (erosive wear rate). As mentioned earlier, there are three categories of quality characteristics, i.e. lower-the-better, higher-the-better, and nominal-the-better. To obtain optimal performance, lower-the-better characteristic for erosion rate must be taken. The mean-square deviation (M.S.D.) for the lower-the-better characteristic can be expressed as (Glen, 1993):

$$\text{M.S.D} = \frac{1}{m} \sum_{i=1}^{m} T_i^2 \quad (2)$$

where m is the number of tests and T_i is the value of experimental result of the ith test.

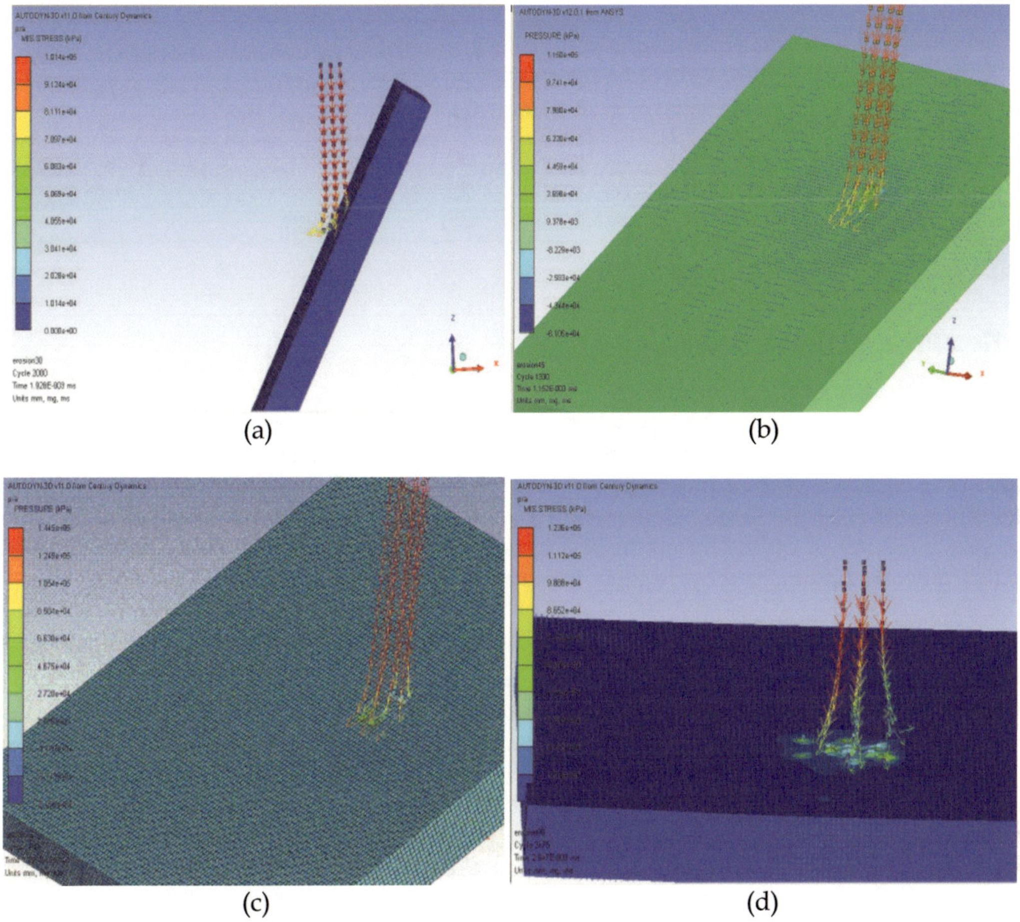

Fig. 2. Schematic diagram of target composite material and nozzle (a: 30° impingement angle, b: 45° impingement angle, c: 60° impingement angle and d: 90° impingement angle)

The plan of the experiments is as follows: the first column is assigned to impact velocity (A), the second column to CBPD content (B), the fifth column to impingement angle (C), the ninth column to stand-off distance (D) and the tenth column to erodent size (E), the third and fourth column are assigned to $(A\times B)_1$ and $(A\times B)_2$ respectively to estimate interaction between impact velocity (A) and CBPD percentage (B), the sixth and seventh column are assigned to $(B\times C)_1$ and $(B\times C)_2$ respectively to estimate interaction between the CBPD content (B) and impingement angle (C), the eight and eleventh column are assigned to $(A\times C)_1$ and $(A\times C)_2$ respectively to estimate interaction between the impact velocity (A) and impingement angle (C) and the remaining columns are used to estimate experimental errors. The output to be studied is erosion rate (E_r) and the tests are repeated twice corresponding to 54 tests. Furthermore, a statistical analysis of variance (ANOVA) is performed to identify the process parameters that are statistically significant. With the S/N and ANOVA analyses, the optimal combination of the process parameters can be predicted to a useful level of accuracy. Finally, a confirmation experiment is conducted to verify the optimal process parameters obtained from the parameter design.

Control factor	Level I	II	III	Units
A:Velocity of impact	43	54	65	m/sec
B:CBPD content	0	10	20	%
C:Impingement angle	30	60	90	degree
D:Stand-off distance	65	75	85	mm
E:Erodent size	250	350	450	μm

Table 2. Levels for various control factors

3. Results and discussion

3.1 Erosive wear of the composites

The erosive wear output characteristics are systematically determined by the Taguchi's technique as a multi-parametric optimization model, since the erosive wear may be a functional variable of five independent control factors and their interactions. Using this technique (a) firstly the steady state erosion characteristics of the composites are determined for selected level of optimally controlled operating variables and (b) subsequently control factors have been identified by Taguchi experimental techniques.

3.1.1 Effect of impingement angle on erosion rate

Solid-particle erosion is a complex wear mechanism, influenced by a number of factors such as impact velocity, impingement angle, erodent size, erodent temperature, stand-off distance, materials properties, particles geometry, environment temperature etc. Among these, angle of impingement is the one of the most important factor and widely studied parameter in the erosion study of materials (Hutchings, 1992; Tsuda et al., 2006). When the erosion rate is measured as a function of impingement angle, two types of material behavior generally observed in the target material i.e. ductile and brittle nature. The ductile behavior of materials is characterized by maximum erosion at acute angle (15-30°) and for brittle behavior of materials, the maximum erosion is observed at normal impingement angle (90°). But as far as polymer composites are concerned the composite materials show versatile in nature depending upon the fabrication techniques and the erosion rate show slightly different behavior. The reinforced composites show a semi-ductile behavior having the maximum erosion rate in the range of 45-60° (Hutchings, 1992), unlike the above two categories. This classification, however, is not absolute as the erosion of material has a strong dependence on erosion conditions such as the properties of target material. In the present study, Figure 3 shows that maximum erosion rate appears at 60°, indicating semi-ductile erosion behavior irrespective of fiber loading.

3.1.2 Surface morphology

Generally surface morphology of eroded surfaces indicates whether solid particle erosive wear has been occurred either by a ductile or brittle mechanism. Hence, scanning electron microscopy (SEM) studies have been done to ascertain the wear mechanism at 30°, 60° and 90° impingement angles. Figure 4a show micrographs of eroded surfaces of 10wt% short glass fiber reinforced polyester composites at 30° and 90° impingement angles at constant impact velocity of 43m/s, stand-off distance 75mm and erodent size 450μm. It is evident from the micrograph (Figure 4a) that the material removal in the composite is dominated by micro-ploughing, micro-cutting and plastic deformation. The plastically deformed material subsequently removed from the surface by micro-cutting leads to maximum erosion rate at 60° impingement angle; most material is lost when a maximum strain in the target is exceeded and formation of micro-cracks and embedment of fragments of sand particles is evident from the micrograph (Figure 4b).

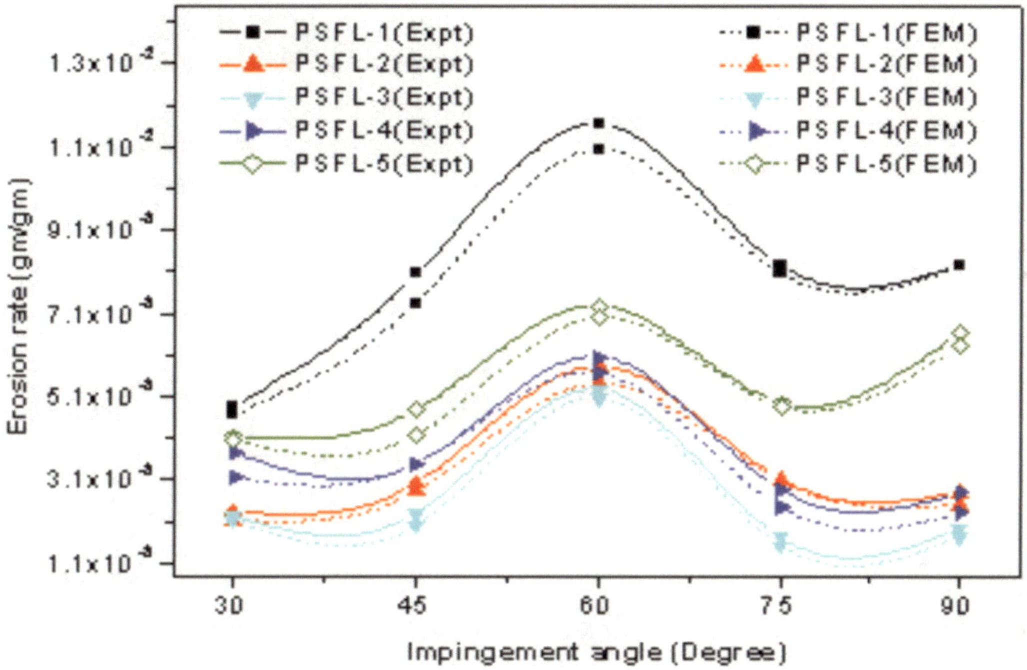

Fig. 3. Influence of impingement angle on erosion rates of unfilled glass fiber-polyester composites (Impact velocity: 43m/sec, stand-off distance: 75mm and erodent size: 450μm).

Micrograph (Figure 4c) of eroded surface of 20wt% glass fiber based composite shows deeper micro-cracks on the composite surface due to brittle nature of polyester resin. However, the normal impact did not result in higher erosion rate like brittle materials. During normal impact the largest part of the initial energy is converted into heat and hence matrix is softened which resulted in embedment of sand particles (Figure 4d). The embedded sand particles control the further erosion of the target surface. Figures 4e and 4f shows micrographs of eroded surfaces of 30wt% and 40 wt% glass fiber reinforced polyester composite respectively. At oblique impingement angle (60°), micrographs (Figures 4e and 4f) show matrix is plastically deformed and amount of deformation is proportional to impact velocity of particles. At 43 m/sec impact velocity removal of matrix along the length of the fiber and subsequently getting removed can be seen from the micrograph (Figure 4e). At same impact velocity for 40wt% glass fiber based composite little high degree of plastic deformation and removal of matrix and hence fibers protruded out from the matrix can be

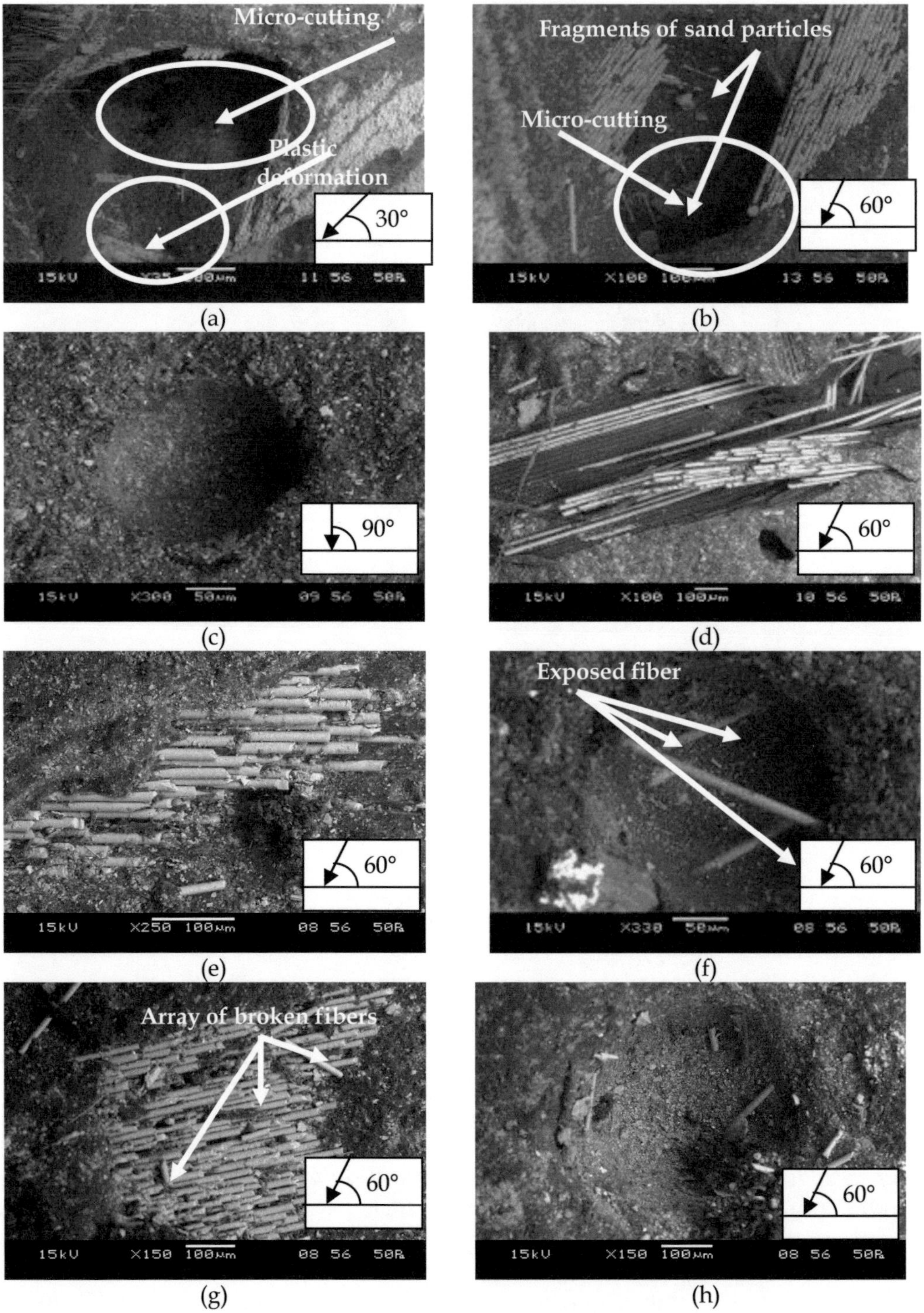

Fig. 4. SEM micrographs of the eroded surfaces of the unfilled glass fiber-polyester composites

seen in the micrograph (Figure 4f). Similarly, for 50wt% glass fiber based composites (Figures 4g, h) show similar behaviour as shown in Figures 4e and 4f but the target surface is completely eroded out and exposure of glass fibers at 60° impingement angle. The exposed fibers are broken into small fragments resulting in excessive wear.

3.1.3 Influence of impact velocity

In order to study the effect of impact velocity on erosion rate, erosion tests are performed on the short glass fiber reinforced polyester composites by varying the impact velocity from 43 to 65 m/s. As seen in Figure 5, erosive wear rates of the samples are remarkably higher with the increase in impact velocity conforming to the theoretical expectations. Particles have a higher kinetic energy at higher impact velocity, which results in greater impingement effect and results in increased erosion rate.

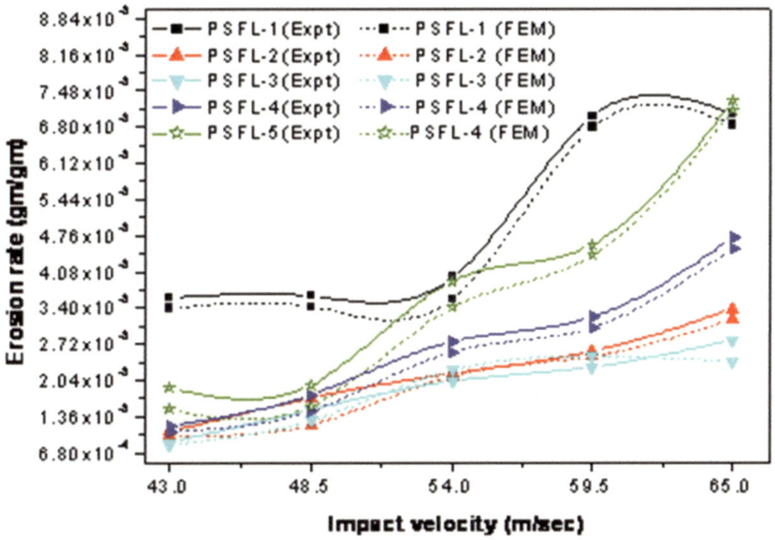

Fig. 5. Influence of impact velocity on erosion rates of unfilled glass fiber-polyester composites (Impingement angle: 60^0, stand-off distance: 75mm and erodent size: 450μm).

It is clear from Figure 5 that steady-state erosion rate of all the composites increases with increase in impact velocity. Similarly, in computationally the finite element simulated results seems to be parallel to the experimental results as shown in Figure 5. It was observed that the composite with 30 wt% of short glass fiber reinforced composite is most resistant to erosion rate followed by the composites with 20, 50 and 40 wt% fiber contents in a decreasing order. Thus it is clearly from Figure that velocity of the erosive particles has a very strong significant effect on erosion rate however, their real mechanistic aspects may need further investigation.

3.2 Steady state erosion rate (for particulate filled composites)

3.2.1 Influence of impingement angle

In case of CBPD filled short glass fiber-polyester composites, the erosion rate increases monotonically with the increase in impingement angle and reaches maximum at 60°

impingement angle for both particulate filled composites and unfilled composite. However, particulate filled composites shows lesser erosion rate as compared to unfilled composites as shown in Figure 6.

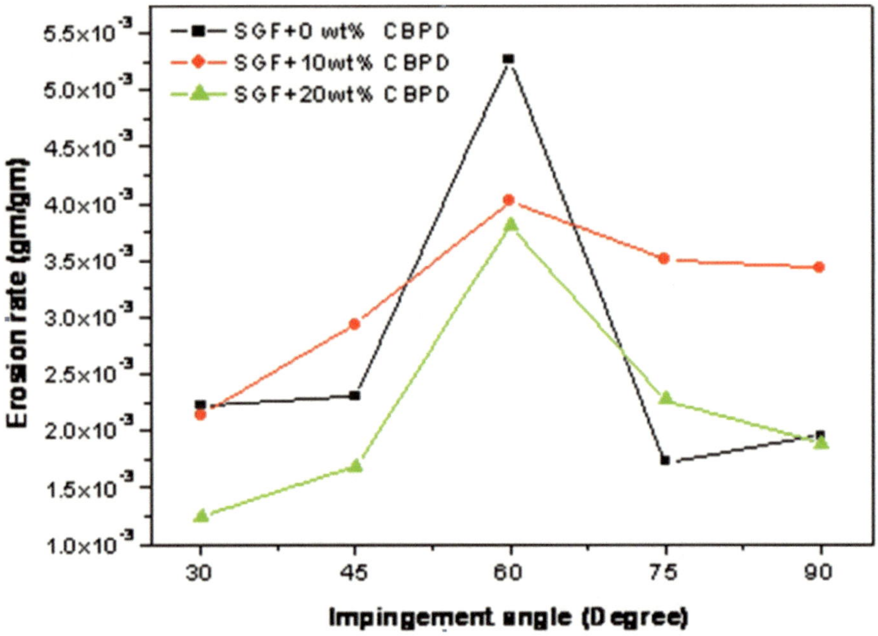

Fig. 6. Influence of impingement angle on erosion rates of particulate filled composites (Impact velocity: 43 m/sec, stand-off distance: 75mm and erodent size: 450μm).

At low impingement angle the surface damage is seemed to be less and the erosion rate is also very low. Earlier studies have also reported that the crack length decreases with decreasing angle, i.e. at lower impingement angle it is small and attain maximum towards 90° (Suresh & Harsa, 2006). Longer cracks imply more rapid erosion rate taking place and the volume of material removal is approximately proportional to square of the surface crack length. This has only happened in case of normal impact velocity and hence the erosion rate shows stronger dependence on impact velocity and the angle of impingement. The maximum erosion rate for unfilled and particulate filled composites show at an angle of 60° as shown in Figure 6 under similar condition as studied in case of unfilled composites (Figure 4). But earlier studies have proved that two different processes contribute erosion on brittle materials, i.e. when particles impact at near normal incidence, lateral cracks induce mass loss, but the radial cracks do not contribute significantly to materials removal. For impact at oblique incidence, however, radial cracks can lead to materials removal (John Rajesh et al., 2001). In the present study, all the composites neither show brittle response nor does it behave ductile. It shows semi-brittle-ductile nature as shown in Figure 6.

3.2.2 Influence of impact velocity

Similarly, the variation of erosion rate of unfilled and CBPD filled composites with impact velocity is shown in Figure 7. Erosion trials are conducted at five different impact velocities. It is seen, in this figure that for all the composite samples, the erosion rates gradually increases by the increase in impact velocity. The increase in erosion rate with impact velocity can be attributed to increased penetration of particles on impact as a result of dissipation of greater amount of particle thermal energy to the target surface. This leads to more surface damage, enhanced sub-critical crack growth etc. and consequently to the reduction in erosion resistance.

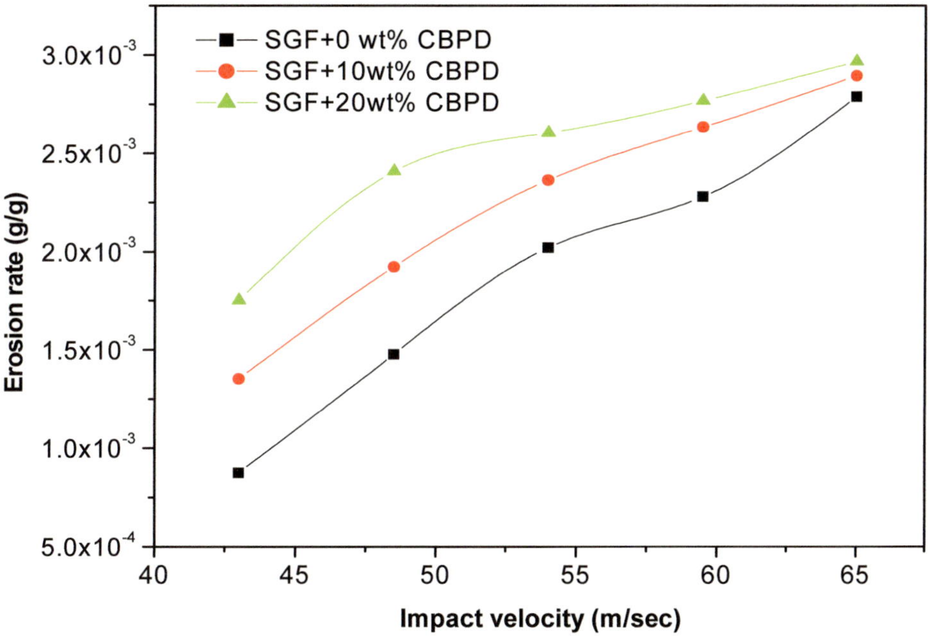

Fig. 7. Influence of impact velocity on erosion rates of particulate filled composites (Impingement angle: 60°, stand-off distance: 75mm and erodent size: 450μm).

3.2.3 Taguchi analysis and response optimization

As an evaluation tool for determining robustness, the signal-to-noise (S/N) ratio is the most important component of a parameter design. In the Taguchi method, the term 'signal' represents the desirable target i.e. erosion rate and the term 'noise' represents the undesirable value. The analysis is made using the computational software used for design of experiment applications (MINITAB 15). The response table (S/N ratio) for erosion rate is mentioned in Table 3. The overall mean for the S/N ratio of erosion rate is found to be 61.76db. Effect of control factors on erosion rate is shown in Figure 8. It is clear from Figure that factors A_1, B_3, C_1, D_3 and E_3 have been observed to be the most significant level which experimentally influences the erosion rate on the basis of smaller-the-better characteristics. The respective interaction plots are observed in the Figures 9a-c.

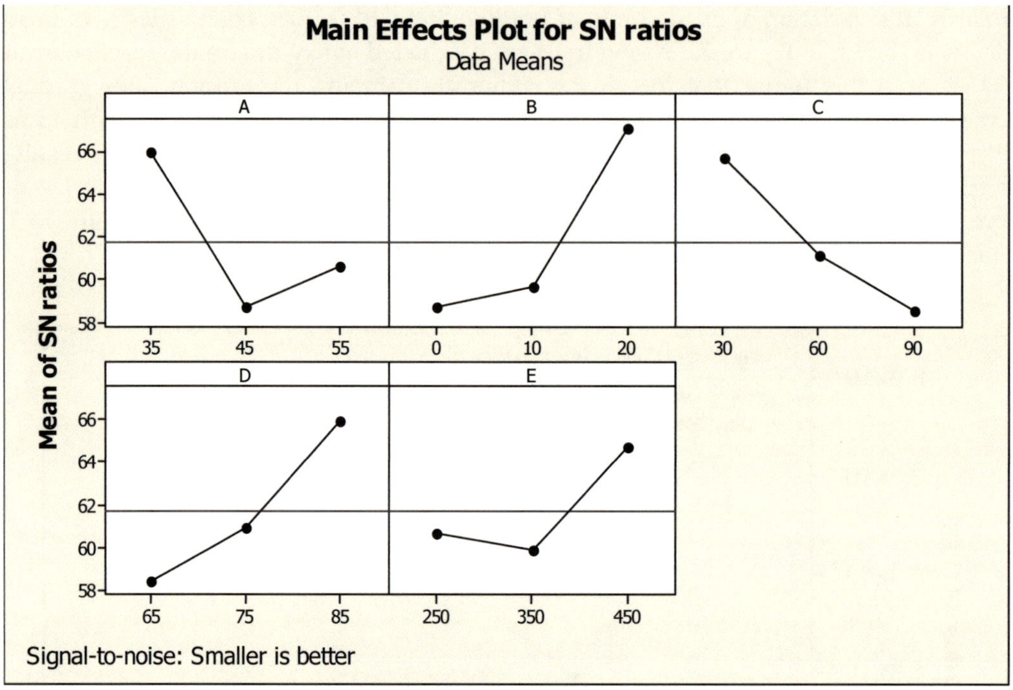

Fig. 8. Effect of control factors on erosion rate

3.2.4 Surface morphology of Taguchi experimental results

To verify the erosion mechanisms of glass fiber reinforced polyester composites, eroded surfaces of the composites are observed as per Taguchi experimental orthogonal array design (see Table 3) by scanning electron microscopy (SEM). Figure 10a, 10b shows the SEM micrographs of eroded composite sample studied at lower impingement angle (see Table 3, Experiment 1) shows the random distribution of short glass fiber and removable of matrix material on the composite surface. Figures 10c and 10d show for 10wt% of CBPD filler composite that the erosion rate decreases (see Table 3, Experiment 6) with lower particle size (250μm). When the erodent particle size is more (450μm) (see Table 3, Experiment 5), it has more energy to chip-off the target material as shown in Figure 10d. But in the case of a small particle size, it has insufficient energy to create the crack in the target material and thus lead to low erosion rate. In other hand, when the erodent particle strikes the composite surface, at 90° impingement angle the fiber damage resulting from recurring micro-ploughing in combination with the fibers getting slowly worn out from the matrix surfaces are observed. But it is not the case as tested same composite with 350μm particle size with impingement angle 60° (Figure 10e, f) shows more erosive surface with higher erosion rate (see Table 3, Experiment 4).

It has been discussed that the impact on brittle materials at an oblique angle produced radial cracks at an angle to the surface and they can contribute to only matrix material loss (Scattergood et al., 1981; Lawn, 1993). Radial cracks can also contribute to material loss when they drive through a relatively thin wall. In such a case, the material loss will occur without the formation of a lateral crack. Larger erodents produce deeper radial cracks. The tendency for material loss to occur from radial cracking should increase with erodent diameter (Lee etal., 2005; Milman et al., 1999). With the increasing of the CBPD particles in the composites from 10wt% to 20wt%, the wear rate of the composites increased gradually, reached a maximum and then declined gradually. The peak wear resistance for 20wt% CBPD particulate filled glass-polyester composites shows 60° impingement angle (see Figure 6). As shown from Figure 10f, the worn surface of particulate filled glass fiber-polyester composites when 350µm erodent particle stricks at an impact velocity 54m/sec the composite surface at lower impingement angle (30°) (see Table 3, Experiment 16) shows only removal of matrix material on the composite surface. Figure 10g shows a hole formed after

Expt. No.	Impact Velociy (A)(m/s)	CBPD content (B) (%)	Impingement angle (C) (Degree)	Stand-off Distance (D)(mm)	Erodent size (E) (µm)	Erosion rate (Er) (gm/gm)	S/N Ratio (db)
1	43	0	30	65	250	0.0003303	69.6224
2	43	0	60	75	350	0.0002466	72.1588
3	43	0	90	85	450	0.0001246	78.0908
4	43	10	30	75	350	0.0008917	60.9959
5	43	10	60	85	450	0.0005550	65.1141
6	43	10	90	65	250	0.0047442	46.4768
7	43	20	30	85	450	0.0001267	77.9468
8	43	20	60	65	250	0.0005667	64.9334
9	43	20	90	75	350	0.0012350	58.1667
10	54	0	30	75	450	0.0014625	56.6981
11	54	0	60	85	250	0.0028121	51.0194
12	54	0	90	65	350	0.0027000	51.3727
13	54	10	30	85	250	0.0001375	77.2339
14	54	10	60	65	350	0.0033938	49.3864
15	54	10	90	75	450	0.0041087	47.7258
16	54	20	30	65	350	0.0008175	61.7502
17	54	20	60	75	450	0.0001844	74.6860
18	54	20	90	85	250	0.0011706	58.6316
19	65	0	30	85	350	0.0024783	52.1171
20	65	0	60	65	450	0.0045143	46.9082
21	65	0	90	75	250	0.0031857	49.9359
22	65	10	30	65	450	0.0005443	65.2835
23	65	10	60	75	250	0.0012014	58.4060
24	65	10	90	85	350	0.0005114	65.8243
25	65	20	30	75	250	0.0003300	69.6297
26	65	20	60	85	350	0.0004295	67.3403
27	65	20	90	65	450	0.0003138	70.0667

Table 3. Experimental design using L_{27} orthogonal array

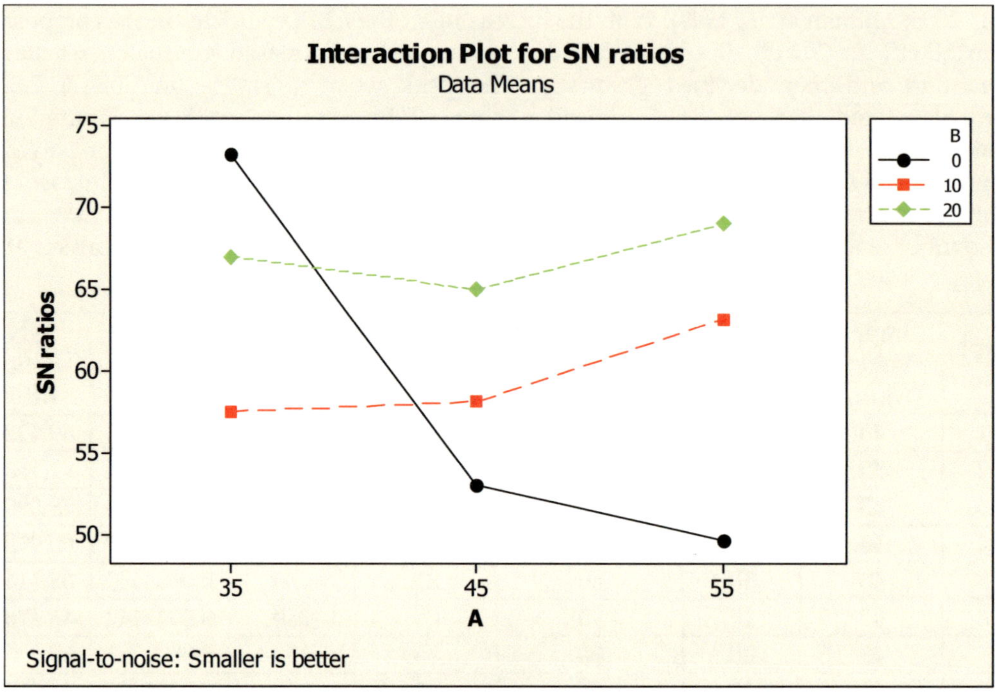

Fig. 9a. Interaction graph between factor A and factor B (A×B) for erosion rate

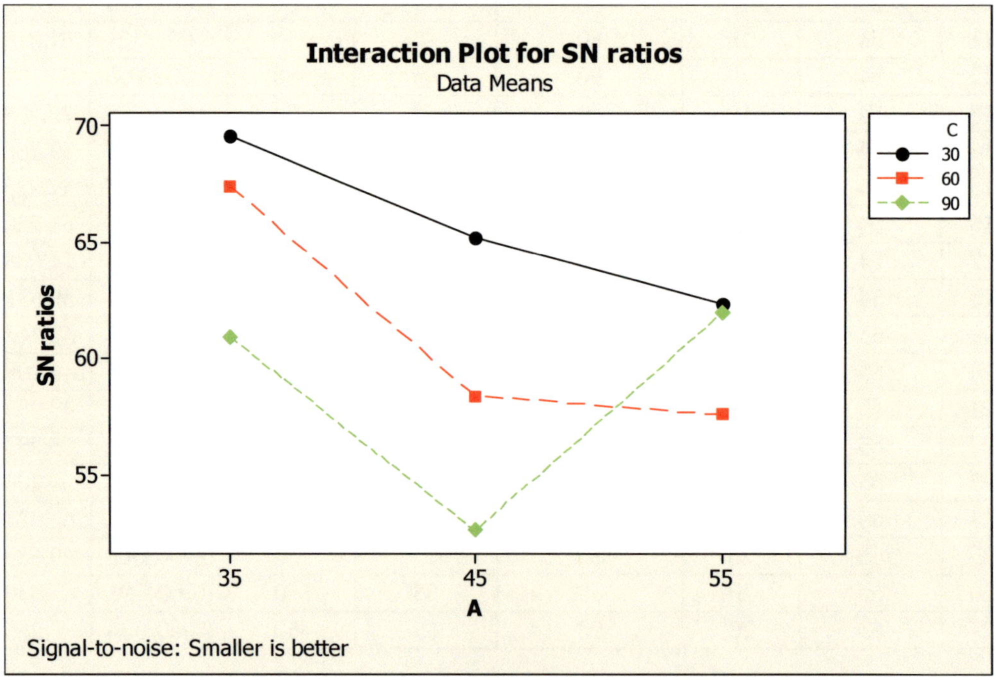

Fig. 9b. Interaction graph between factor A and factor C (A×C) for erosion rate

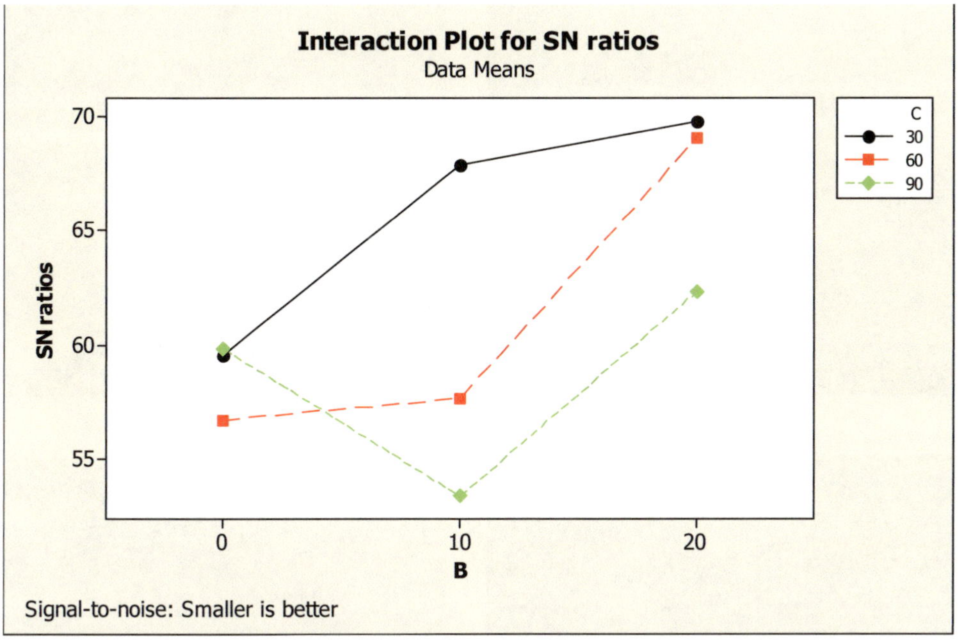

Fig. 9c. Interaction graph between factor B and factor C (B×C) for erosion rate

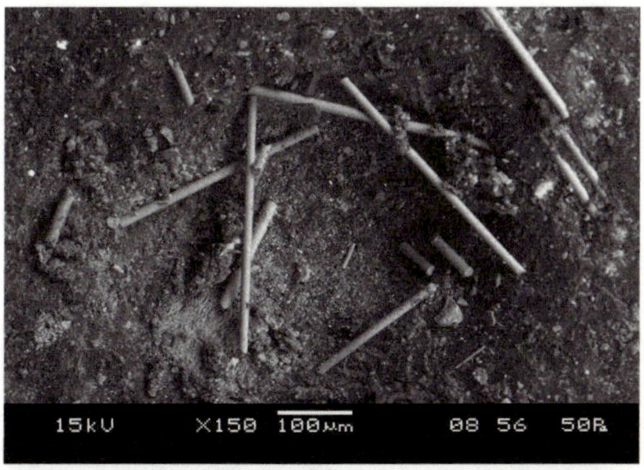

(a)

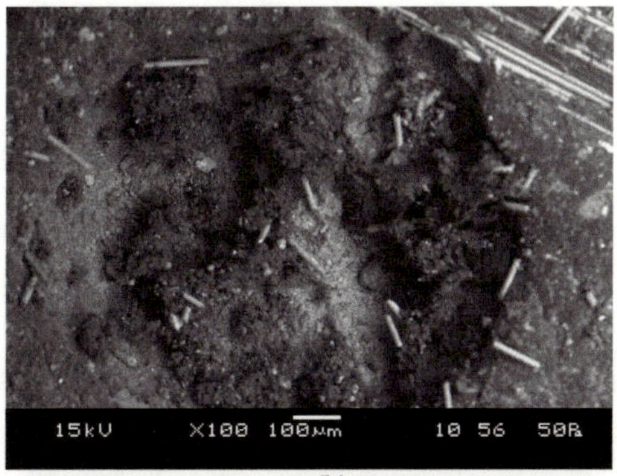

(b)

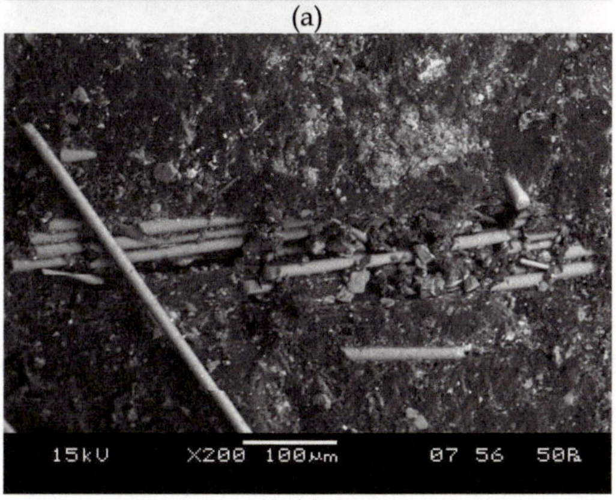

(c)

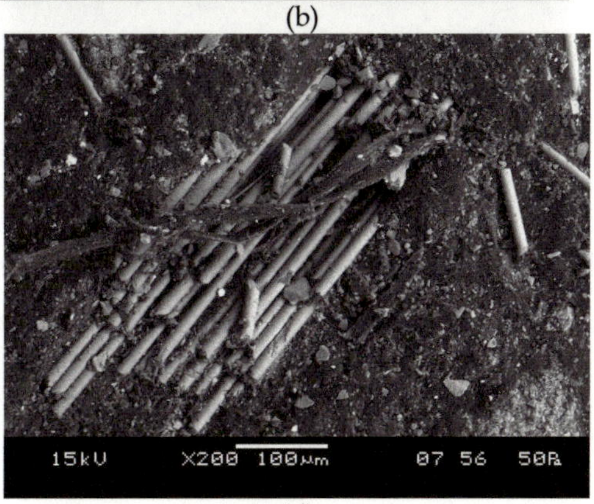

(d)

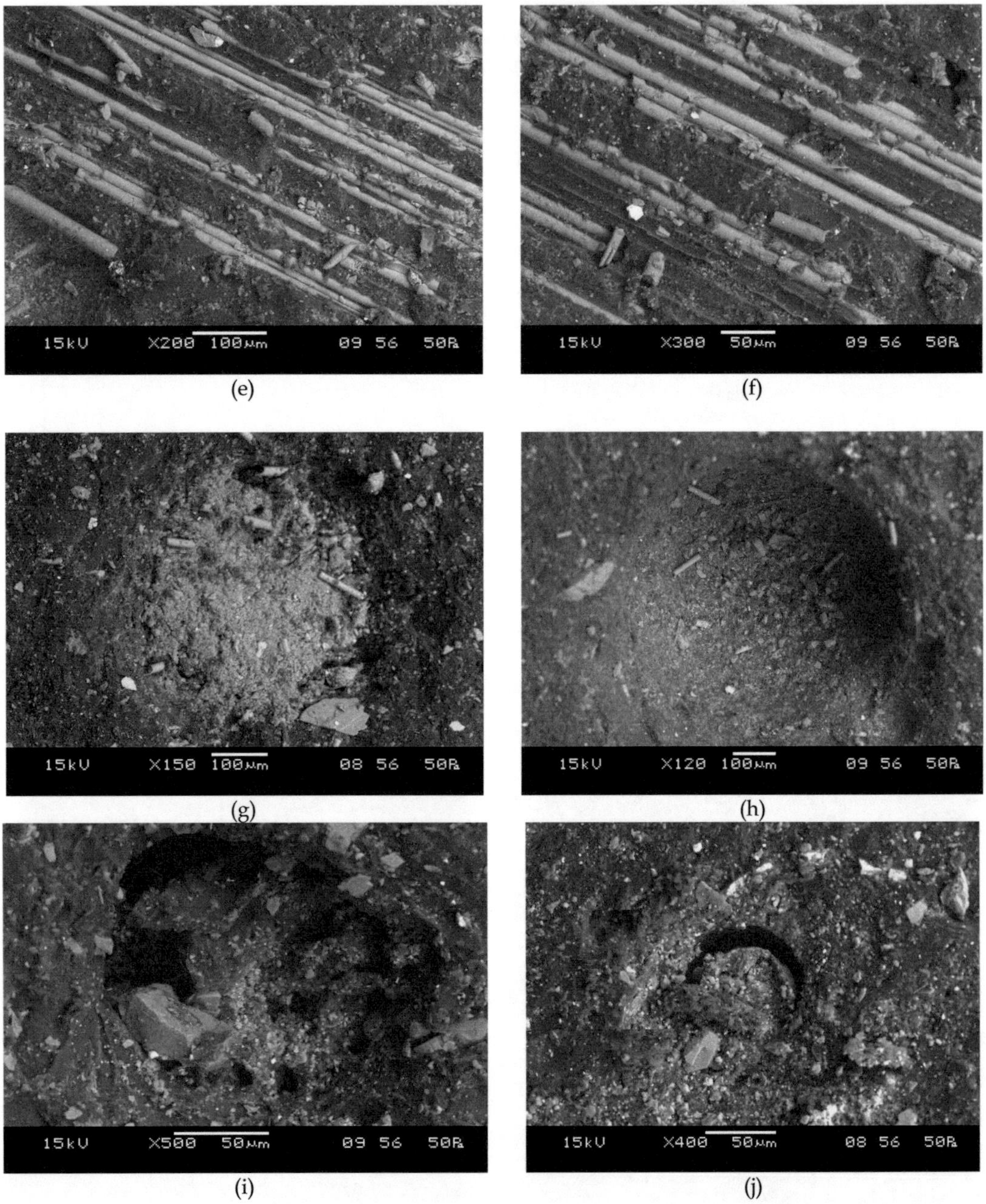

Fig. 10. SEM micrographs of the eroded glass fiber-polyester composites filled with CBPD

an CBPD particle is removed from the surface. The inside surface of the hole seemed very smooth and clear which indicated that one CBPD particle wholly debonded from the matrix with the propagation of interfacial cracks (Figure 10h) due in part to the poor interfacial bond strength. However, on further increase in erodent particle from 250µm to 350µm size and impingement angle from 30° to 60° (see Table 3, Experiment 26) the matrix materials removed from the surface along with exposure of glass fibers as shown in Figure 10i. The excessive amount of material loss in the composite is due to, too much CBPD particles were added (20 wt%), the interface area may have been too large and particles may have touched directly, so interfacial defects and cracks could largely emerge. Thus, CBPD particles could be more easily removed and could not improve the wear resistances. So the wear resistance of the composites declined with the increased content of the CBPD particles. In Figure 10j, an CBPD particle with a flat top surface protruding slightly over the worn surface, around whose back half interface distributed a clear gap (see Table 3, Experiment 25). The morphology showed that the top surface of the CBPD particle was heavily polished and became flat by the impact of the erodent particles with high velocity (65m/sec) at an angle of 60°, and the back part of the CBPD particle debonded and slipped from the matrix along the interface.

3.3 ANOVA and the effects of factors

Table 4 shows the results of the ANOVA with the erosion rate (E_r) for CBPD filled short glass fiber reinforced polyester composites. The objective of ANOVA is to analyze the influence of impact velocity (A), CBPD content (B), impingement angle (C), stand-off distance (D) and erodent size (E) on the total variance of the results. This analysis was undertaken for a level of significance of 5% that is for a level of confidence of 95%.

Source	DF	Seq SS	Adj SS	Adj MS	F	P
A	2	252.51	252.51	126.26	2.53	0.195
B	2	377.09	377.09	188.54	3.77	0.120
C	2	240.40	240.40	120.20	2.41	0.206
D	2	262.48	262.48	131.24	2.63	0.187
E	2	121.18	121.18	60.59	1.21	0.388
A×B	4	809.70	809.70	202.42	4.05	0.102
A×C	4	163.40	163.40	40.85	0.82	0.575
B×C	4	211.27	211.27	52.82	1.06	0.479
Error	4	199.89	199.89	49.97		
Total	26	2637.93				

Table 4. ANOVA table for erosion rate

From Table 4 it is observed that CBPD content (p = 0.120), stand-off distance (p = 0.187), impact velocity (p = 0.195) and impingement angle (p = 0.206) have great influence on the out put performance (erosion rate). The interactions of impact velocity and CBPD content (p = 0.102) has greater significant influence on erosion rate but the factor erodent size (p = 0.388), interaction between impact velocity and impingement angle (p = 0.575) and CBPD content and impingement angle (p = 0.479) have less significant contribution on erosion rate.

3.4 Confirmation experiment

The confirmation experiment is the final step in the first iteration of the design of experiment process. The purpose of the confirmation experiment is to validate the conclusions drawn during the Taguchi experimental design process. The confirmation experiment is performed by conducting a new test condition in combination of the significant factors and their interaction levels on erosion rate as reported in Table 4. The final step is to predict and verify the improvement of the erosion resistance. The predictive value $\overline{\eta}_1$ using the optimal levels of the input parameters can be calculated as:

$$\overline{\eta}_1 = \overline{T} + (\overline{A}_2 - \overline{T}) + (\overline{B}_2 - \overline{T}) + \left[(\overline{A}_2\overline{B}_2 - \overline{T}) - (\overline{A}_2 - \overline{T}) - (\overline{B}_2 - \overline{T}) \right] + (\overline{C}_3 - \overline{T}) + (\overline{D}_2 - \overline{T}) \quad (3)$$

$\overline{\eta}_1$ Predicted average

$\overline{T}$ Overall experimental average

$\overline{A}_2$, $\overline{B}_2$, $\overline{C}_3$ and $\overline{D}_2$ Mean response for factors and interactions at designated levels.
By combining like terms, the equation reduces to

$$\overline{\eta}_1 = \overline{A}_2\overline{B}_2 + \overline{C}_3 + \overline{D}_2 - 2\overline{T} \quad (4)$$

After solving the above predictive equation the erosion rate is found to be $\overline{\eta}_1 = 51.03\,\text{dB}$ and the experimental result is 51.54 dB. The resulting model seems to be capable of predicting wear rate to a reasonable level of accuracy. An error of 1.78% for the S/N ratio of wear rate is observed. This validates the development of the mathematical model for predicting the measures of performance based on knowledge of the input parameters.

4. Conclusions

1. The particle reinforced composites show good tribological properties. The erosion rate for unfilled composites 30wt% glass fiber polyester composites show better erosion resistance whereas, for particulate filled composites 20wt% filled content shows superior erosion resistance as compared with the rest of the filled and unfilled composites.
2. The variation of erosion rate with impingement angles, the material loss is dictated mainly more at 60^0 impingement angles, in the steady state erosion rate with other control factors are remain constant. However, compared with the experimental results, the FE model (ANSYS/AUTO-DYN) is much closer to the experimental results. The major advantages of simulated results are during experimental study it is very difficult to analysis the flow direction and particularly at low impingement angle most of the erodent particles are sliding on the target material instead of reback of erodent particles. However, in finite element simulated model the above conditions can be measured easily including the residual stress and the depth of penetration which is difficult to determine by experimental method.
3. In eroded samples observed in scanning electron microscope shows mostly two types of wear mechanisms i.e. micro-cutting and micro-ploughing actions. It shows formation of lips and their flow along the direction of the erodent. At a medium angle of impingement (60°), the material loss is governs mostly by cutting action. It also depicts formation of coarser and larger lips. At a higher angle of impingement (90°), the material loss is dictated by micro-fracturing mechanism. However, CBPD particulates provide the protection against erosive wear in the composite.

5. Acknowledgement

The authors are grateful to the financial supports of the research project Ref. No. SR/FTP/ETA-49/08 by Department of Science and Technology, India.

6. References

Adler, W.F. (1995). Waterdrop Impact Modeling, *Wear*, Vol. 186–187, pp. 341–351.

Al-Jabri, K. S.; Taha, R. A.; Al-Hashmi, A. & Al-Harthy A. S. (2006). Effect of Copper Slag and Cement By-pass Dust addition on Mechanical Properties of Concrete, *Construction and Building Materials*. Vol. 20, pp. 322-331.

Batis, G.; Rakanta, E.; Sideri, E.; Chaniotakis, E. & Papageorgiou, A. (2002). Advantages of Simultaneous use of Cement Kiln Dust and Blast Furnace Slag, *In: Proceedings of the International Conference on Challenges of Concrete Construction*, University of Dundee, Dundee, UK.

Biswas, S. & Satapathy, A. (2009). Tribo-Performance Analysis of Red Mud Filled Glass-Epoxy Composites Using Taguchi Experimental Design, *Materials & Design*, Vo. 30, No. 8, pp. 2841-2853.

Biswas, S. & Satapathy, A. (2010). A Comparative Study on Erosion Characteristics of Red Mud Filled Bamboo-Epoxy and Glass-Epoxy Composites, *Materials & Design*, Vol. 31, No. 4, pp.1752-1767.

El-Sayed, H. A.; Gabr, N. A.; Hanafi, S. & Mohran, M. A. (1991). Re-utilization of By-pass Kiln Dust in Cement Manufacture, *In: Proceedings of the International Conference on Blended Cement in Construction*, Sheffield, UK.

ElTobgy, M. S.; Ng, E. & Elbestawi, M. A. (2005). Finite Element Modeling of Erosive Wear, *International Journal of Machine Tools and Manufacture*, Vol. 45, pp. 1337-1346.

Glen, S.P. (1993). *Taguchi Methods: A Hands-on Approach*, Addison-Wesley, New York.

Hsu, S. M.; Shen, M. C. & Ruff, A. W. (1997). Wear Prediction for Metals, *Tribology International*, Vol. 30, No. 5, pp. 377-383.

Hutchings, I. M. (1992). Ductile brittle transitions and wear maps for the erosion and abrasion of brittle materials, *Journal of Physics D: Applied Physics*, Vol. 25, pp. A212-A221.

John Rajesh, J.; Bijwe, J.; Tewari, U. S. & Venkataraman, B. (2001). Erosive Wear Behaviour of Various Polyamides, *Wear*, Vol. 249, pp. 702-714.

Komvopoulos, K. & Choi, D. H. (1992). Elastic Finite Element Analysis of Multi-Asperity Contacts, *ASME Journal of Tribology*, Vol.114, pp. 823-831.

Krazowski, L. & Emery, J J. (1981). Use of Cement Kiln Dust as a filler in Asphalt Mixes, *In: Proceedings, ORF/CANMET Symposium on Mineral Fllers, Ontario Research Foundation*, Toronto.

Lawn, B. (1993). *Fracture in Brittle Solids*, 2nd ed., Cambridge University Press, Cam-bridge, UK.

Lee, S. H., Lee, Y. I.; Kim, Y.W.; Xie, R. J.; Mitomo, M. & Zhan, G. D. (2005). Mechanical Properties of Hot-Forged Silicon Carbide Ceramics, Scripta Materialia, Vol. 52, pp. 153-156.

Ling, F. F. & Pan C. H. T. (1986). Approaches to Modeling of Friction and Wear, *Proceedings of the Workshop on the Use of Surface Deformation Models to Predict Tribology Behavior*, Springer–Verlag, New York.

Ludema, K. C. & Bayer, R. G. (1991). Tribological Modeling for Mechanical Designers, ASTM STP1105, *American Society for Testing and Materials*, Philadelphia.

Milman, Y.V.; Chugunova, S. I.; Goncharova, I. V.; Chudoba, T.; Lojkowski, W. & Gooch, W. (1999). Temperature Dependence of Hardness in Silicon-Carbide Ceramics with Different Porosity, *International Journal of Refractory Metals and Hard Materials*, Vol. 17, No. 5, pp. 361-368.

Patnaik, A.; Kaundal, R.; Satapathy, A.; Biswas, S. & Pradeep K. (2010a). Solid Particle Erosion of Particulate Filled Short Glass Fiber Reinforced Polyester Resin Composites, *Advanced Materials Research*, Vol.123-125, pp. 213-216.

Patnaik, A.; Abdulla, Md., Satapathy, A.; Biswas, S. & Satapathy, B. K. (2010b). A Study on Possible Correlation between Thermal Conductivity and Wear Resistance of Particulate Filled Polymer Composites, *Materials and Design*, Vol. 31, pp. 837-849.

Patnaik, A.; Satatpathy, A.; Chand, N.; Barkoula, N. M. & Biswas, S. (2010c). Solid Particle Erosion Wear Characteristics of Fiber and Particulate Filled Polymer Composites: A Review, *Wear*, Vol. 68, No. (1-2), pp. 249-263.

Patnaik, A.; Satatpathy, A.; Dwivedy, M. & Biswas, S. (2010d). Wear behaviour of Plant-fiber (Pine-Bark) and Cement-kiln-dust Reinforced Polyester Composites using Taguchi Experimental Model, *Journal of Composite Materials*, Vol. 44, No. 5, pp. 559-574.

Patnaik, A.; Tejyan, S. & Rawal, A. (2010e). Solid Particle Erosion Behavior of Needlepunched Nonwoven Reinforced Composites, *Research Journal of Textile and Apparel*, Vol. 14, No. 3, pp. 12-22.

Patnaik, A.; Satatpathy, A., Mahapatra, S. S. & Dash. R. R. (2010f). Modified Erosion Wear Characteristics of Glass-Polyester Composites by Silicon Carbide Filling: A Parametric Study using Taguchi Technique, *International Journal of Materials and Product Technology (Materials Processing Technology)*, Vol. 38, No. 2-3, pp.131-152.

Patnaik, A.; Mahapatra, S. S. & Dash, R. R. (2010g). Flyash Filling for Improved Wear Resistance of Glass-Polyester Composites : An Experimental Study, *The Journal of Solid Waste Technology and Management*, Vol. 36, No. 1, pp.26-37.

Rothon, R. N. (2002). Particlulate Fillers for Polymers, *RAPRA Review Reports*, Vol. 12, No. 9, pp. 164.

Routbort, J. L.; Gulden, M. E. & Marshall, E. (1981). Particle Size Distribution Effects on the Solid Particle Erosion of Brittle Materials, *Wear*, Vol. 71, pp. 363-373.

Scattergood, R. O.; Routbort, J. L. & Turner, A. P. L. (1981). Velocity and Size Dependence of the Erosion Rates in Silicon, *Wear*, Vol. 67, pp. 227-232.

Sri Ravindrarajah, R. (1982). Usage of Cement Kiln Dust in Concrete, *International Journal of Cement Composites and Lightweight Concrete*, Vol. 4, No. 2, pp. 95-102.

Strzepa, P.; Zamirowski, E. J.; Kupperman, J. B.; Goretta, K. C. & Routbort, J. L. (1993). Indentation, Erosion and Strength Degradation of Silicon-Alloyed Pyrolytic Carbon, *Journal of Materials Science*, Vol. 28, pp. 5917–5921.

Suresh, A. & Harsha, A. P. (2006), Study of Erosion Efficiency of Polymers and Polymer Composites, *Polymer Testing*, Vol. 25, pp. 188-196.

Tsuda, K.; Kubouchi, M.; Sakai, T.; Saputra, A. H. & Mitomo, N. (2006) General Method for Predicting the Sand Erosion Rate of GFRP, *Wear*, Vol. 260, pp. 1045-1052.

Woytowitz, P. J. & Richman, R. H. (1999). Modeling of Damage from Multiple Impacts by Spherical Particles, *Wear*, Vol. 233-235, pp. 120-133.

7

Review Fabrication of Functionally Graded Materials under a Centrifugal Force

Yoshimi Watanabe and Hisashi Sato
Nagoya Institute of Technology
Japan

1. Introduction

The formation of gradients of chemical composition, phase distribution or microstructure represents a now fervently pursued concept in the design of advanced engineering components. In light of this concept, a significant progress has been made in the area of functionally graded materials (FGMs). FGMs are of practical interest because a wide gradation of physical and/or chemical properties can be achieved across a given material depending on the material design (Suresh & Mortensen, 1998, Miyamoto et al., 1999, Uemura et al., 2010). Thus, FGM holds continuous changes of the microstructure, the composition and the properties in some specific directions, *i.e.*, inhomogeneous on both macroscopic and microscopic scales. On the other hand, particle-, fibre- or platelet-dispersed composites are regarded as macroscopically homogeneous and microscopically inhomogeneous.
There are two types of graded structures of the FGMs, namely continuous structure shown in Fig. 1 (a), and stepwise structure shown in Fig. 1 (b). In the first type, the change in composition and/or microstructure occurs continuously with position. On the other hand in the second type, the microstructure feature changes in a stepwise manner, giving rise to a multilayered structure with interfaces existing between discrete layers (Miyamoto et al., 1999). As will be described in more detail below, the continuous graded structure can be created by a centrifugal force.

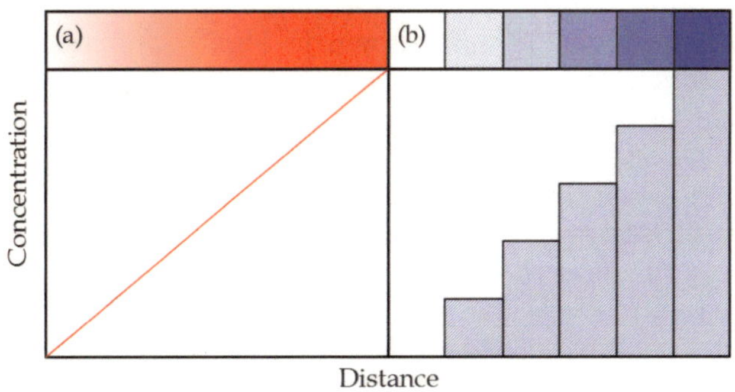

Fig. 1. Two types of graded structures. (a) Continuous structure, and (b) stepwise structure.

In the past, many kinds of processing methods for FGM have been proposed (Suresh & Mortensen, 1998, Miyamoto et al., 1999, Uemura et al., 2010). Powder metallurgy is one of the most important methods of producing FGMs. An example of a typical fabrication process by the powder metallurgy is schematically illustrated in Fig. 2. At first, material A and material B are weighed and mixed, as shown in Fig. 2 (a). The each mixed-powder is mixed uniformly by a V-shape mill, as shown in Fig. 2 (b). Next step is stepwise staking of premixed powder according to a predesigned spatial distribution of the composition (Fig. 2 (c)). Last step is a sintering. Spark plasma sintering (SPS), shown in Fig. 2 (d), is one of the more advanced sintering methods, and it makes possible sintering high quality materials in short periods by charging the intervals between powder particles with electrical energy and high sintering pressure. However, usually the FGM fabricated by this method should have the stepwise structure, and it is difficult to produce the FGMs with continuous gradients.

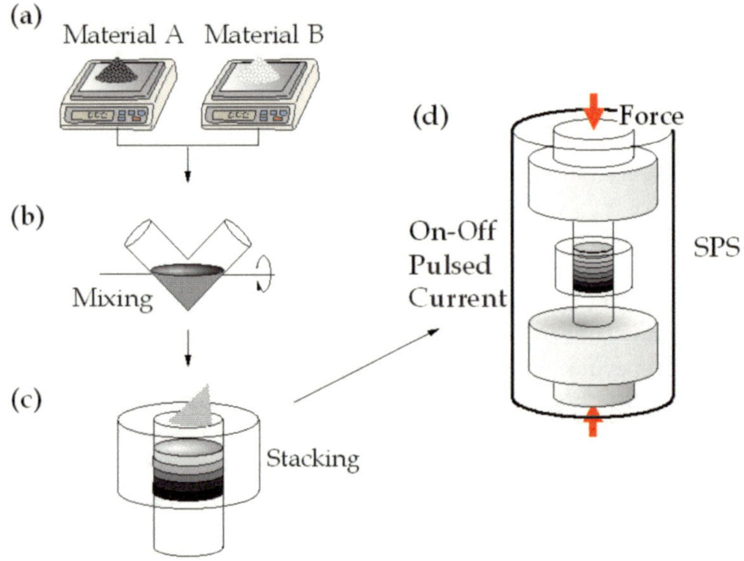

Fig. 2. Example of typical fabrication process of the FGMs by powder metallurgy method.

FGMs can be also fabricated under a centrifugal force, by which it is possible to produce the FGMs with continuous gradients. Fabrication methods of FGMs under the centrifugal force are classified into three categories, as shown in Fig. 3, namely centrifugal method (application of centrifugal casting, and shown in Fig. 3 (a)), centrifugal slurry method (centrifugal sedimentation, and shown in Fig. 3 (b)) and centrifugal pressurization method (simple pressurization by the centrifugal force, and one example is shown in Fig. 3 (c)).

In case of centrifugal method shown in Fig. 3 (a), a centrifugal force applied to a homogeneous molten metal, dispersed with ceramics particles or intermetallic compound particles, drives the formation of the desired gradation. The composition gradient is then achieved primarily by the difference in the centrifugal force produced by the difference in density between the molten metal and solid particles (Fukui, 1991, Watanabe et al., 1998). It is known that the motion of particles in a viscous liquid under a centrifugal force obeys the Stokes' law (Kang & Rohatgi, 1996, Watanabe et al., 1998, Ogawa et al., 2006)

$$\frac{dx}{dt} = \frac{|\rho_{\mathrm{p}} - \rho_{\mathrm{m}}| G g D_{\mathrm{p}}^2}{18\eta} \tag{1}$$

where dx/dt, ρ, G, g, D and η are velocity, density, G number (the ratio of the centrifugal force to the gravity), gravitational acceleration, particle diameter and viscosity of the molten metal, respectively. The subscripts 'p' and 'm' denote particle and matrix, respectively. When the desired gradation is achieved, motion of solid particles will be stopped by solidification of molten metal, and the solidified metal becomes a matrix of the FGM.

In contrast, slurry with two types of solid particles, high-velocity particle and low-velocity particle, is subjected to the centrifugal force during the fabrication of FGMs by the centrifugal slurry method (Watanabe et al., 2010), as shown in Fig. 3 (b). After complete sedimentation occurs, liquid part of the slurry will be removed, and therefore, it does not become a part of FGM.

Third one is the centrifugal pressurization method, by which the centrifugal force is only used for simple pressurization. In this method, compositional gradation should be formed prior the application of centrifugal force. Watanabe et al. are developing a centrifugal mixed-powder method, shown in Fig. 3 (c) (Watanabe et al., 2009), and a reactive centrifugal casting method (Watanabe et al., 2004a), which belong this category. In this chapter, some results from the centrifugal method, centrifugal slurry method, centrifugal mixed-powder method and reactive centrifugal casting method will be described.

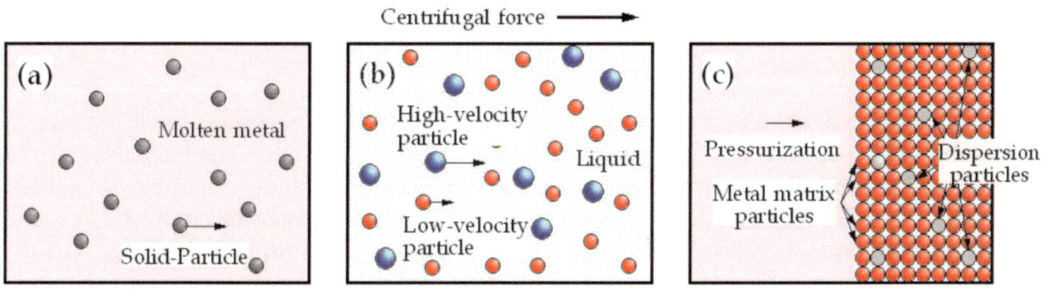

Fig. 3. Three types of fabrication methods of FGMs under the centrifugal force. (a) centrifugal method, (b) centrifugal slurry method, and (c) centrifugal pressurization method (Watanabe et al., 2010).

2. Centrifugal method

Centrifugal casting is a pressure casting, in which the force of gravity is enhanced by spinning the mould. Generally, the segregation, caused by the density-difference between particles and molten metal, must be avoided. However, it is possible to create a compositional gradient due to the difference in material density (Fukui, 1991, Watanabe et al., 1998).

Figure 4 shows the apparatus used for the centrifugal method. The ingot is melted and then the plug is pulled out to cast the molten metal directly into the spinning mould through the inlet. The spinning mould is preheated before the casting. The magnitude of the centrifugal force is expressed in G number, which is the ratio of the centrifugal force to the gravity, g, as given by the following equation:

$$G = 2DN^2 \quad (2)$$

where D is the diameter of the cast ring (m) and N is the rotation speed of the mould (s^{-1}). After casting, the mould-preheating furnace is removed and the mould is cooled until complete solidification occurs.

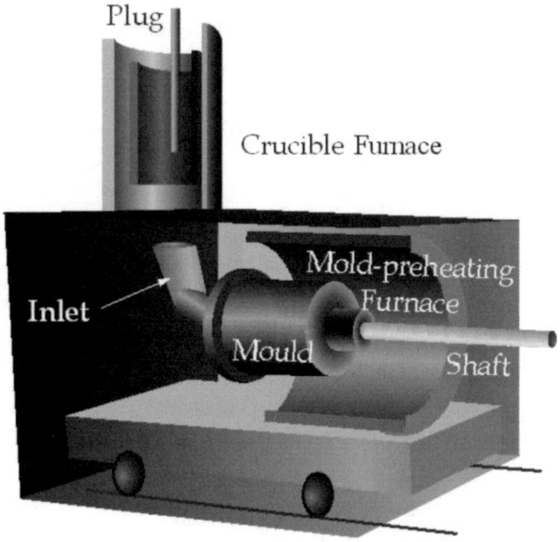

Fig. 4. The apparatus for the centrifugal method.

The composition gradient formed by the centrifugal method is affected by the difference in density between particles and a molten metal, the applied *G* number, the particle size, the viscosity of the melt, the mean volume fraction of particles, the thickness of manufactured ring and the solidification time (Watanabe et al., 1998). It is obvious that both density and viscosity are the materials' constants and both the volume fraction and thickness are the constants of products. Moreover, applied *G* number and solidification time show a mutual relation. One of the easily changeable parameters for the control of graded composition is the particle size.

2.1 Ceramic-particle dispersed FGMs fabricated by the centrifugal method

The typical microstructure of the Al/SiC FGM fabricated by the centrifugal method is shown in Fig. 5 (Fukui & Watanabe, 1996), where the applied *G* number is 129. Figs. 5 (a) and (b) are taken at different positions, namely they are 4.1 mm and 0.5 mm from outer periphery of the ring, respectively. As can be seen, the amount of SiC particles changes from place to place. The moving direction of the particle under the centrifugal force is determined by the relative magnitudes of densities. Since the densities of SiC and molten Al at 700 ^{o}C are 3.15 Mg/m^3 and 2.37 Mg/m^3, respectively, the SiC particles move toward the outer periphery of the ring under the centrifugal force.
Figure 6 shows the histogram of the ceramic-particle volume fraction of the Al/SiC FGM fabricated by the centrifugal method (Fukui & Watanabe, 1996). In this figure, the abscissa represents the position in the thickness direction of the ring, normalized by the thickness, *i.e.*, 0.0 and 1.0 correspond to the inner and outer peripheries, respectively. It is clear that the distributions of the particles in both specimens are graded, suggesting that the ceramic-dispersed FGMs can successfully be fabricated by the centrifugal method.

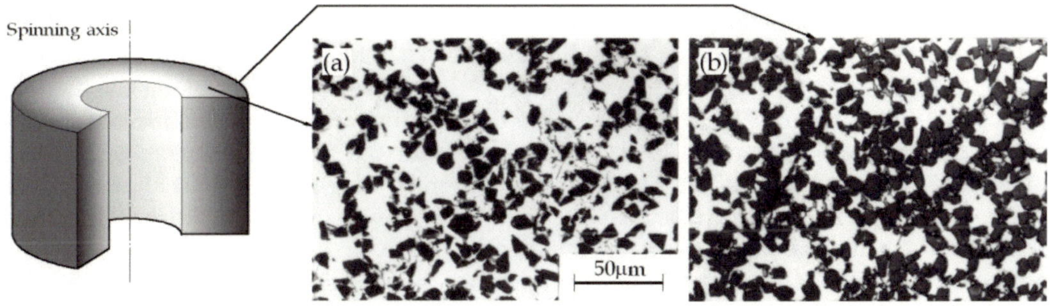

Fig. 5. Typical microstructure of the Al/SiC FGM fabricated by the centrifugal method (Fukui & Watanabe, 1996). Black and white are SiC and Al matrix, respectively.

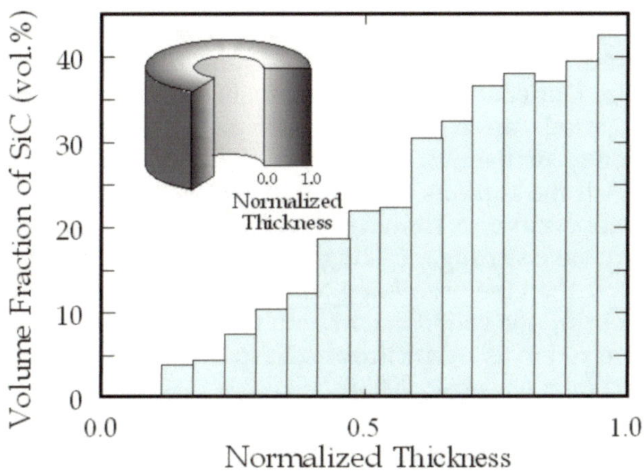

Fig. 6. Distribution of the volume fraction of SiC particles in the FGM fabricated by the centrifugal method (Fukui & Watanabe, 1996).

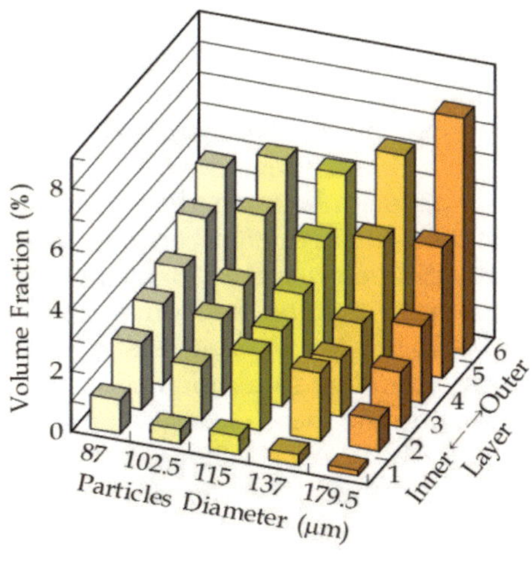

Fig. 7. The particle distribution in the FGM fabricated by the centrifugal solid-particle method under G=15 (Watanabe et al., 2002).

As described in introduction, the motion of ceramic particles in a viscous liquid under a centrifugal force can be estimated using Stokes' law. Therefore, in the FGMs fabricated by the centrifugal method, the mean particle size is distributed gradually along the applied centrifugal force. Figure 7 shows the particle distributions in the FGM fabricated by the centrifugal method (Watanabe et al., 2002). Note that the volume fraction of particles gradually increases towards the outer region. The most remarkable result shown in this figure is that the average particle size at the outer region is larger than that at the inner region, and the average particle size is gradually distributed in the FGMs. Although the data are not presented here, it has also been found that the particle size gradient in the FGM becomes steeper with increasing the *G* number or with decreasing the mean volume fraction of particles (Watanabe et al., 2002). These results are in agreement with the Stokes' law; the migration distance is greater for larger particles.

2.2 Intermetallic-particle dispersed FGMs fabricated by the centrifugal method

Intermetallic compound particles could be also applicable for the centrifugal method as dispersed particles. The fabrication of the intermetallic compound particles dispersed FGMs by the centrifugal method can be classified into two categories based on the relation between the processing temperature and the liquidus temperature of master alloy (Watanabe et al., 2005). If the liquidus temperature of the master alloy is significantly higher than the processing temperature, as shown in Fig. 8 (a), the dispersed phase remains solid in a liquid matrix during the centrifugal casting. Alternatively, if the liquidus temperature of the master alloy is lower than the processing temperature, as shown in Fig. 8 (b), centrifugal force can be applied during the solidification both to the dispersed phase and to the matrix. These methods are referred to as a centrifugal solid-particle method and a centrifugal *in-situ* method, respectively (Watanabe et al., 2005).

2.2.1 Centrifugal solid-particle method

Since the liquidus temperature of Al-5mass%Ti alloy, containing 11vol% Al_3Ti platelets in the Al matrix, is about 1160 °C and the processing temperature is 840 °C of a liquid-solid coexisting temperature, the Al_3Ti platelets remain solid in the liquid Al matrix and a centrifugal force is worked directly to platelets during the casting process.

Figure 9 shows typical microstructures of an Al/Al_3Ti FGM fabricated under G = 30 (Watanabe et al., 2001). It is found that the volume fraction of the Al_3Ti platelets increases towards the outer region of the ring. The steeper distribution profile of the Al_3Ti platelets is formed in case of larger applied *G* specimen. It is worthwhile to notice that the Al_3Ti particles are oriented with their platelet planes nearly perpendicular to the radial direction of the ring. The graded distribution of orientation becomes steeper as the platelet size and mean volume fraction increases (Sequeira et al., 2007). In this way, the orientation of the Al_3Ti platelets, as well as the mean volume fraction of the particles, was found to be gradually distributed in the Al/Al_3Ti FGMs (Watanabe et al., 2001).

It is known that volume fraction, size, shape and orientation of reinforcements in composite play an important role in improving the mechanical properties of the materials. Therefore, the Al/Al_3Ti FGMs with oriented Al_3Ti platelets should have an anisotropic wear resistance (Watanabe et al., 1999). The anisotropic wear resistances in the Al/Al_3Ti FGMs were measured in three directions, namely along the longitudinal direction on outer surface of the ring (A), along the radial direction on the radial plane (B), and along the hoop direction on the radial plane (C), and results are shown in Fig. 10 (Watanabe et al., 2008). The result for a pure Al specimen made by the same process is also shown in the figure for comparison. The

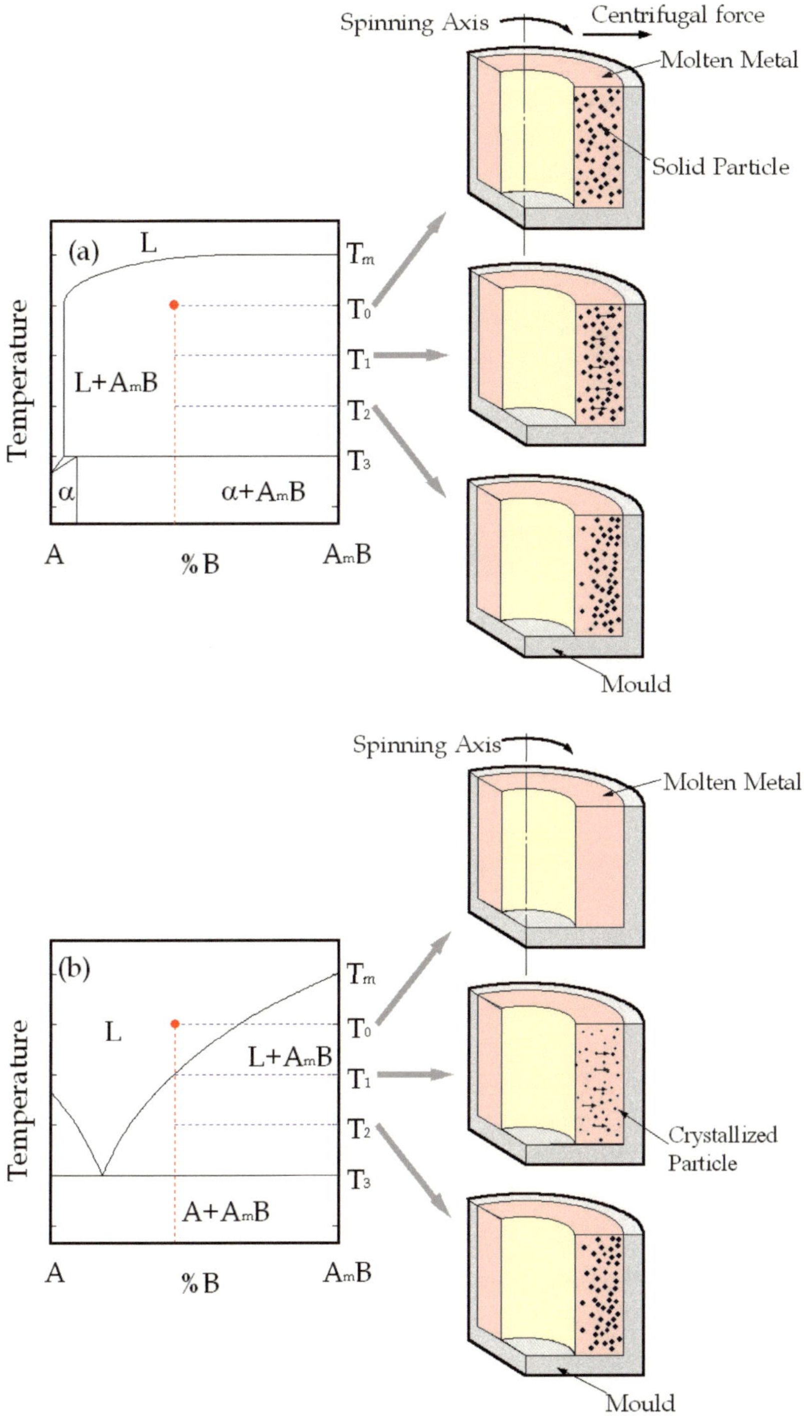

Fig. 8. (a) Centrifugal solid-particle method and (b) a centrifugal *in-situ* method (Watanabe et al., 2005) .

wear volumes in the Al/Al_3Ti FGM are much smaller than that of pure Al. Anisotropic wear resistance was found to be dependent on the direction of the test wear relative to the Al_3Ti platelet orientation. Specimen tested along the Al_3Ti platelets thickness direction shows the smallest wear resistance among the three orientations due to the ease with which the Al_3Ti platelets broke. Although the data is not presented here, a greater anisotropy in wear resistance was found for specimens with larger orientation parameters (Watanabe et al., 2008).

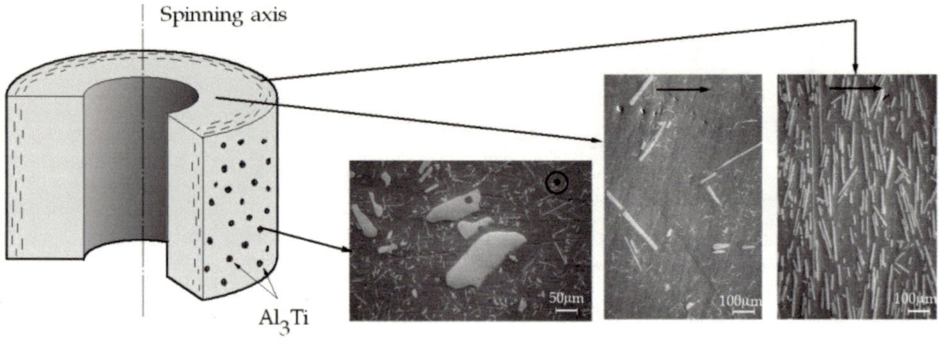

Fig. 9. Typical microstructures of an Al/Al_3Ti FGM fabricated under G = 30 (Watanabe et al., 2001). The arrows and ◎ mark within the photographs indicate the direction of the centrifugal force.

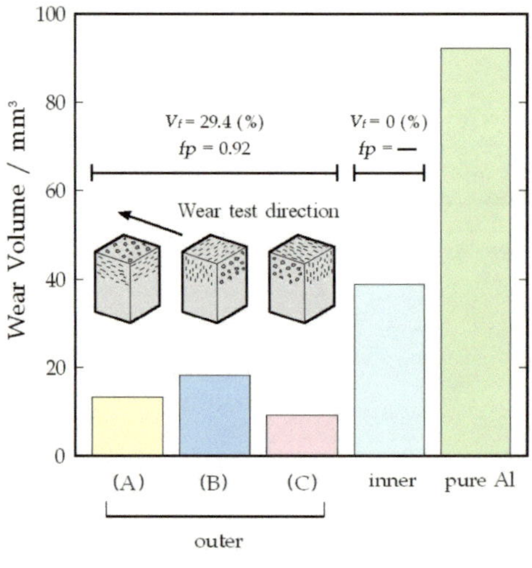

Fig. 10. Wear volumes of the FGM fabricated under G = 50. The result for a pure Al specimen made by the same process is also shown for comparison (Watanabe et al., 2008).

2.3 Centrifugal *in-situ* method

The master alloy ingot used for the fabrication of the Al/Al_3Ni FGMs contains 20mass%Ni. The liquidus temperature of Al-20mass%Ni is about 780 °C, whereas the processing temperature is 900 °C. Therefore, the centrifugal force is applied directly to a liquid phase. Figures 11 (a), (b) and (c) show typical microstructures of an Al/Al_3Ni FGM taken at the inner, the interior, and the outer regions of the ring, respectively (Watanabe et al., 2004b). As can be seen, the Al_3Ni primary crystals are distributed graded manner in the specimen. Moreover, there is remarkable position dependence in the particle size, namely the smaller particles are found at the outer region of the ring, and *vice versa,* as shown in Fig. 12. Since the motion of solid particles in a viscous liquid obeys Stokes' law, the particle size at the outer region is larger than that at the inner region by the centrifugal solid-particle method, as shown in Fig. 7. Such results are contradictory to the present observation. Therefore, the mechanisms to form the graded composition by the centrifugal solid-particle method and the centrifugal *in-situ* method are different each other.

It is also found that as the *G* number becomes larger, the particle size at the ring's outer region becomes smaller (Watanabe et al., 2004b). It has been accepted in general that the crystallized particle size is varied depending on the solidification process. In case of the centrifugal method, it is reported that the cooling rate at the outer region of the ring is larger than that at the inner region (Kang & Rohatgi, 1996, Hattori et al., 2010). Moreover, the larger cooling rate for larger *G*-number specimens is found. Therefore, it is concluded that the difference in the particle size distributions of the Al/Al_3Ni FGM fabricated by the centrifugal *in-situ* method should be mainly originated from a gradation of the cooling rate (Watanabe et al., 2004b).

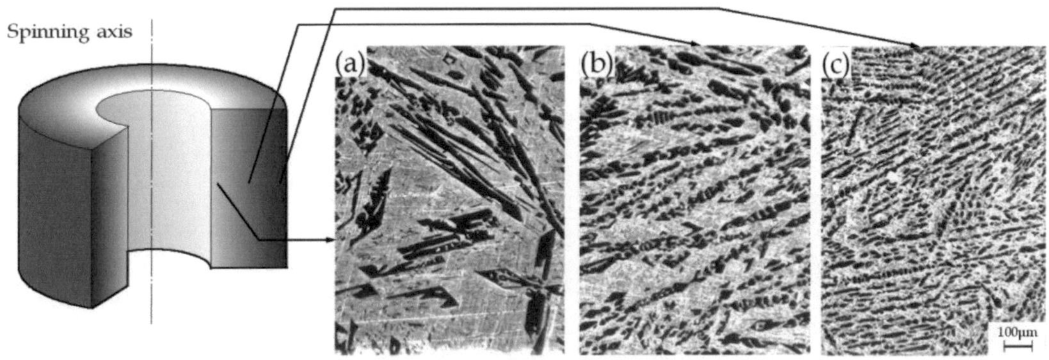

Fig. 11. Typical microstructures of an Al/Al_3Ni FGM fabricated under $G = 50$ (Watanabe et al., 2004b).

To discuss the formation mechanism of the graded composition during the centrifugal *in-situ* method, this method is employed for Al-33mass%Cu eutectic alloy, which do not have any primary crystals. The results are shown in Fig. 13 (Watanabe and Oike. 2005). Noteworthy, it has been observed that the graded composition is appeared in the Al-33mass%Cu eutectic alloy sample. The origin of the graded structure could, therefore, not be explained by migration of the primary crystals under the centrifugal force.

The formation mechanism of the graded composition in the A–B alloy by the centrifugal in-site method could be summarized as follows (Watanabe and Oike. 2005). First, due to the density difference, partial separation of A and B elements in the liquid state occurs. Then, a chemical composition gradient is formed before the crystallization of the primary crystal. The primary crystal in the matrix appears to depend on local chemical composition. The primary crystal migrates according to density difference, and a further compositional gradient is formed.

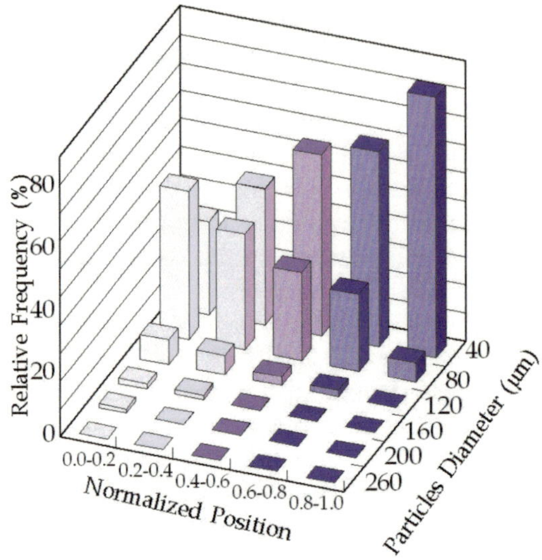

Fig. 12. The particle distribution in the Al/Al_3Ni FGM fabricated by the centrifugal *in-situ* method under G=50 (Watanabe et al. 2004b).

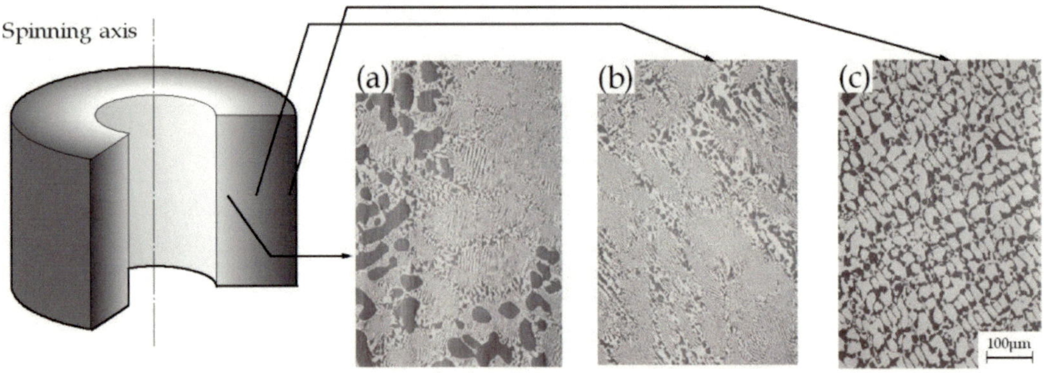

Fig. 13. Microstructures of a centrifugal cast Al/Al_2Cu sample (Watanabe and Oike. 2005). White parts are Al_2Cu phase.

In the case of FGMs fabricated by the centrifugal *in-situ* method, the size of the primary crystal particles was smaller at the outer ring region, where the volume fraction of the particles is increased. Therefore, the particle size gradient emphasizes the gradients in mechanical properties. This may be one of the advantages of the FGMs fabricated by the centrifugal *in-situ* method.

3. Centrifugal slurry method

Although the powder metallurgy has many advantages to fabricate the FGMs, it is difficult to produce the FGMs with continuous gradients. By combination of the powder metallurgy and a centrifugal slurry method, this shortcoming can be overcome. For the centrifugal slurry method, slurry with two types of solid particles will be used, namely high-velocity

particle with larger density and/or larger particle size and low-velocity particle with smaller density and/or smaller particle size, judging by eq. (1). The particles gradients can be controlled by the difference of migration rate between the two kinds of particles. After complete sedimentation occurs, liquid part of the slurry will be removed, and a green-body with continuous gradient can be obtained. The green-body is, then, subjected to sintering by SPS or other sintering methods, and finally an FGM with continuous gradient can be fabricated.
Figure 14 shows migration velocities of Ti and ZrO_2 particles under the centrifugal force (Watanabe et al., 2010), where the densities of Ti particle and ZrO_2 particle are 4.5 Mg/m^3 and 5.95Mg/m^3, respectively. It is obvious that the velocity of ZrO_2 particle is higher than that of Ti particle, when the particle size is same, due to its larger density. In this condition, Ti particle and ZrO_2 particle become low-velocity particle and high-velocity particle, respectively. On the other hand, if the slurry contains the smaller ZrO_2 particles and larger Ti particles, the Ti particles can have high velocity in a specific condition.

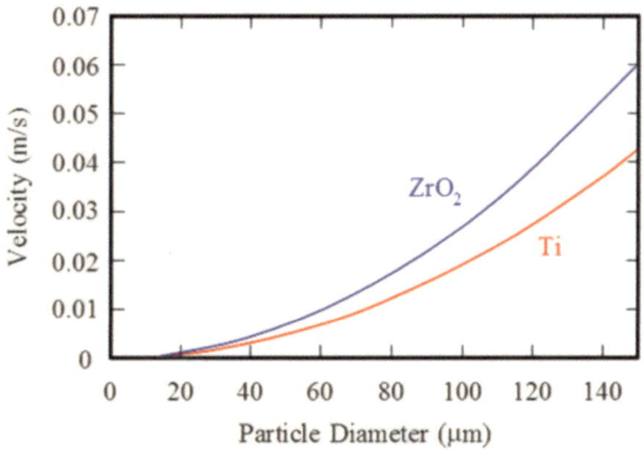

Fig. 14. Velocities of Ti and ZrO_2 particles under the centrifugal force (Watanabe et al., 2010).

Compositional gradient of the FGM fabricated by the centrifugal slurry method is calculated by simulation for the system with 90~150 μm particle sized Ti and 38~75 μm particle sized ZrO_2 particle. In this case, Ti particle and ZrO_2 particle becomes high-velocity particle and low-velocity particle, respectively. Results are shown in Fig. 15 (a) (Watanabe et al., 2010). Here, the horizontal axis is the normalized position of the green-body, and 0.0 and 1.0 correspond to the top and bottom surfaces of the settled green-body, respectively. It is clear from this figure that continuous gradient can be obtained by the centrifugal slurry method. However, large compositional gradient, from 0 vol% at one end to 100 vol% at another end of the FGM for specific component, cannot be achieved. This is because the low-velocity particles (ZrO_2 particle) placed at the bottom region before the sedimentation still may remain around same region after the complete sedimentation.
To overcome the above shortcoming, a slurry pouring method has been proposed to fabricate the FGM with large compositional gradient (Watanabe et al., 2010). First, solvent of the slurry is inserted into a spinning mould, as shown in Fig. 16 (a). Next, the slurry with two types of solid particles is poured into the spinning mould with solvent zone, as shown in Fig. 16 (b). Then, two types of solid particles migrate toward the centrifugal force

direction, as shown in Fig. 16 (c). The existence of solvent zone increases the sedimentation period. As a result, large compositional gradient, for example from 0vol% at one end to 100vol% at another end, can be fabricated.

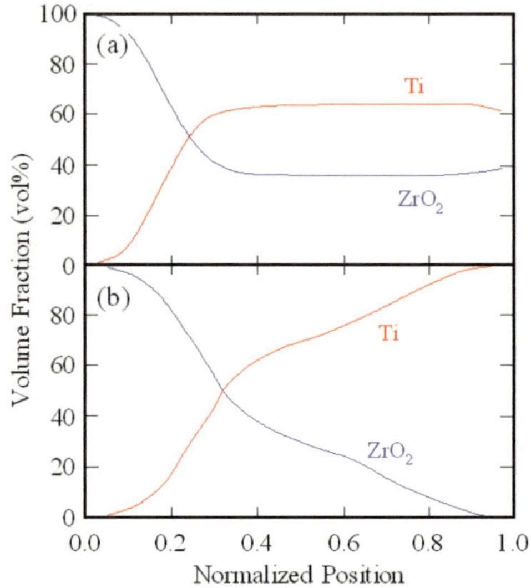

Fig. 15. Volume fraction distributions of Ti and ZrO_2 particles within the FGMs by the centrifugal slurry method (a) and by the centrifugal slurry-pouring method (b) obtained by the computer simulation (Watanabe et al., 2010). Width of solvent zone is 100mm.

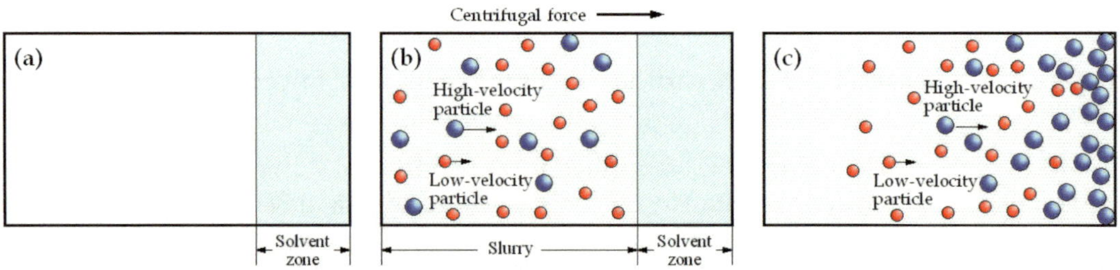

Fig. 16. Schematic illustrations of the centrifugal slurry-pouring method.

The computer simulation is conducted for the centrifugal slurry-pouring method, and results are shown in Fig. 15 (b), here the width of solvent zone is 100mm (Watanabe et al., 2010). It is clear from this figure that the FGM has a continuous gradient. It should be pointed out that the volume fraction of Ti at the normalized position of 0.0 is 0%, while 100% at 1.0 position. Thus, a large compositional gradient can be achieved by the centrifugal slurry-pouring method.

In order to verify and confirm the above simulation results, experiments were also conducted without and with the solvent zone. For simplicity, centrifugal force is not applied and particles were allowed to settle by gravity. After complete settlement of the particles, liquid is removed and the green-body is dried. The green-body was sintered by SPS method at 1300 °C for 5 minutes under applied stress of 30 MPa. FGMs obtained by SPS method have cylindrical shape with 20 mm in diameter.

Figures 17 (a) and (b) show experimental results of volume fractional gradients within the FGMs fabricated by the centrifugal slurry method (width of solvent zone is 0mm) and centrifugal slurry-pouring method (width of solvent zone: 100mm), respectively (Watanabe et al., 2010). Without the solvent zone, compositional gradient is limited as shown in Fig. 17 (a), whereas large compositional gradient, the range is between 0vol% and 100vol%, is achieved by the centrifugal slurry method with the solvent zone (the centrifugal slurry-pouring method). Good agreement is found between the experimental and calculated profile.

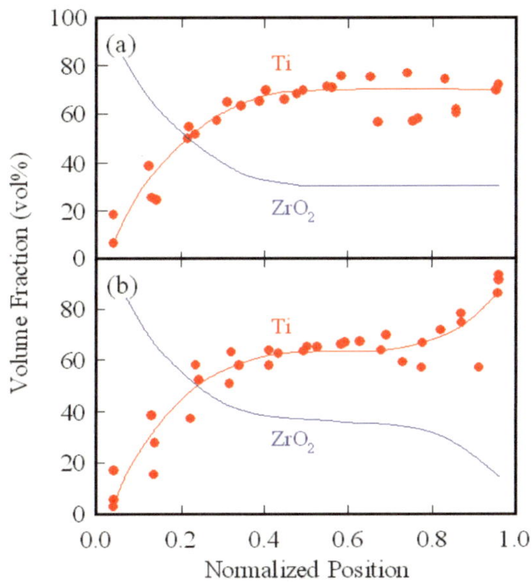

Fig. 17. Experimental results of volume fractional gradients within the FGMs fabricated by (a) the centrifugal slurry method (Width of solvent zone: 0mm) and (b) centrifugal slurry-pouring method (Width of solvent zone: 100mm), (Watanabe et al., 2010).

4. Centrifugal pressurization method

4.1 Centrifugal mixed-powder method

As a more developed fabrication method of FGMs under the centrifugal force, the centrifugal mixed-powder method was recently proposed by Watanabe et al. (Watanabe et al., 2009). As a first step of the process, a powder mixture of matrix metal particles, *A*, and dispersion-particles, *B*, is inserted into a spinning mould, as shown in Fig. 18 (a). Then, matrix metal ingot, *A*, is melted and poured into the spinning mould with powder mixture *A* + *B*, as shown in Fig. 18 (b). As a result, the molten matrix metal, *A*, penetrates into the space between the particles by the centrifugal force pressure, as shown in Fig. 18 (c). At the same time, powder of matrix metal, *A*, is melted by the heat from the molten matrix poured from a crucible, as shown in Fig. 18 (d). Finally, an FGM ring with dispersion-particles, B, distributed on its surface, can be obtained, as shown in Fig. 18 (e).

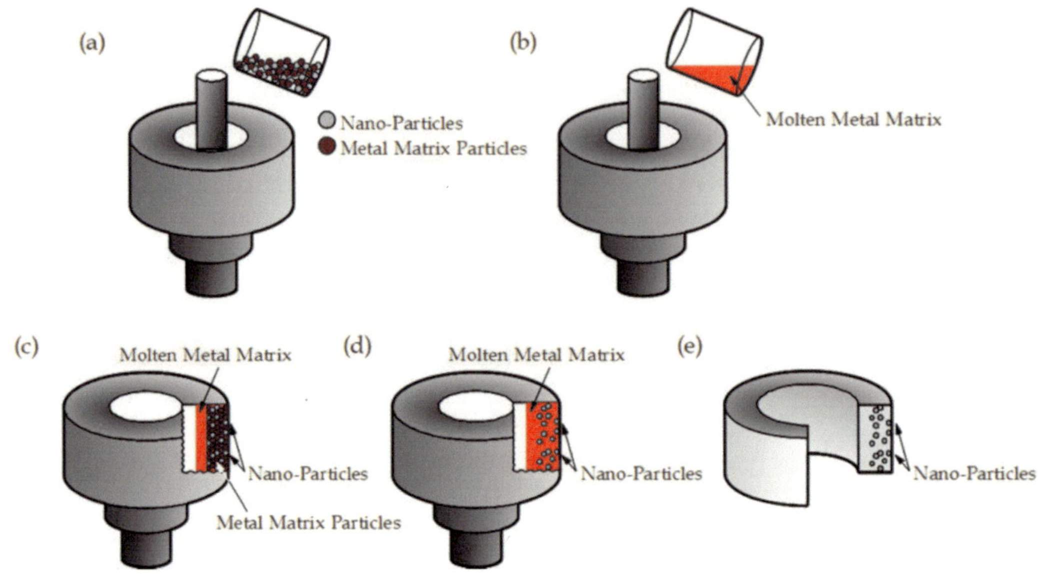

Fig. 18. The schematic description of the centrifugal mixed-powder method (Watanabe et al., 2010).

Cu-30vol%SiC mixed-powder was fabricated using pure Cu particles (99.9%, 1 mm and < 45μm in diameter) and SiC particles (150 μm). Using this mixed-powder, Cu/SiC FGM was fabricated by the centrifugal mixed-powder method using vertical-type centrifugal casting machine. The applied centrifugal force was $G = 100$, and the spinning mould containing the powder mixture was heated up to 800 °C. Then, molten Cu with purity of 99.9% was poured into the spinning mould (Watanabe et al., 2009).

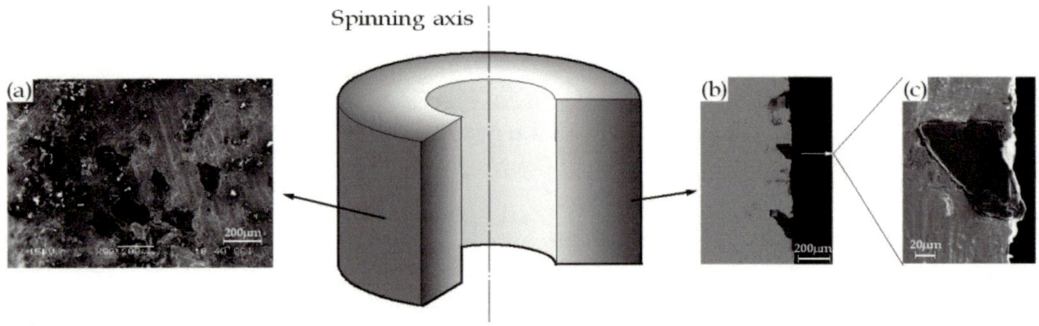

Fig. 19. Microstructures of a Cu/SiC FGM fabricated by the centrifugal mixed-powder method (Watanabe et al., 2009).

When the Cu/SiC FGM was removed from the mould, no powders dropped out. Therefore, it is considered that all of the SiC particles in the powder mixture remained in the Cu/SiC FGM. SiC particles are observed on the outer surface of the specimen, as shown in Fig. 19 (a) (Watanabe et al., 2009). Moreover, the SiC particles are embedded in Cu matrix, as shown in Figs. 19 (b) and (c). It is found that SiC particles are successfully distributed on the surface of the FGMs by the centrifugal mixed-powder method, and the SiC particles are homogeneously distributed on the surface

4.2 Reactive centrifugal casting method

Matsuura et al. developed a new technique named reactive casting, which involves an exothermic reaction between elemental liquids, and enables one to produce the liquid of a high melting point intermetallic compound without the need for external heating (Matsuura et al., 2000). A combination of the reactive casting and centrifugal casting can be applied to the fabrication of a Ni-aluminide / steel clad pipe having an excellent resistance to corrosion and oxidation as well as a considerable level of toughness. This novel method is named as a reactive centrifugal casting method (Watanabe et al., 2004a, Watanabe et al., 2011). Ni powder was placed on a spinning steel pipe, as shown in Fig. 19 (a), and molten Al was poured into the steel pipe, as shown in Fig. 19 (b). The molten Al and Ni powder exothermically reacted and produced a composite layer consisting of Ni-aluminides on the inner surface of the steel pipe, as shown in Fig. 19 (c). The heat generated by the exothermic reaction melted the inner surface of the steel pipe and bonded the composite layer to the steel. Since inexperience Al ingots are used instead of the experienced Al powder, this process will reduce the production cost.

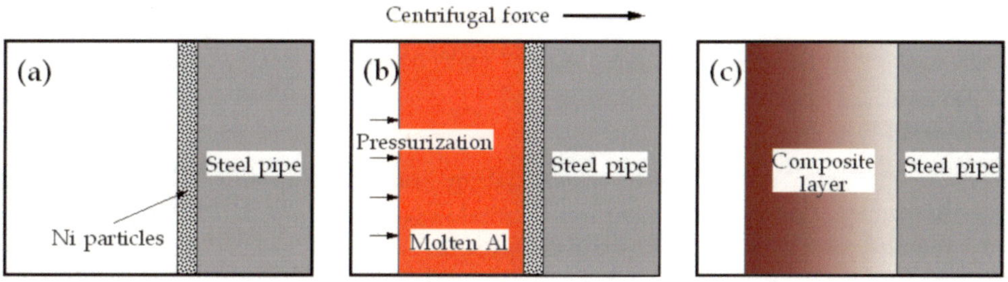

Fig. 20. Schematic illustrations of the reactive centrifugal casting method.

Figure 20 shows the SEM photograph of specimen fabricated under $G = 80$. The pouring temperature of the Al liquid and the preheating temperature of nickel powder were 1200 °C and 700 °C, respectively. A wide region having homogeneous microstructures can be observed in the composite layer of the sample. In a region away from the joint interface, however, the graded microstructure was formed. It was shown that the reaction is remarkably promoted by increasing the pouring temperature of Al, the preheating temperature of the nickel and the centrifugal force. It is also found that the amounts of initial Al and Ni play an important role in the control of the microstructure.

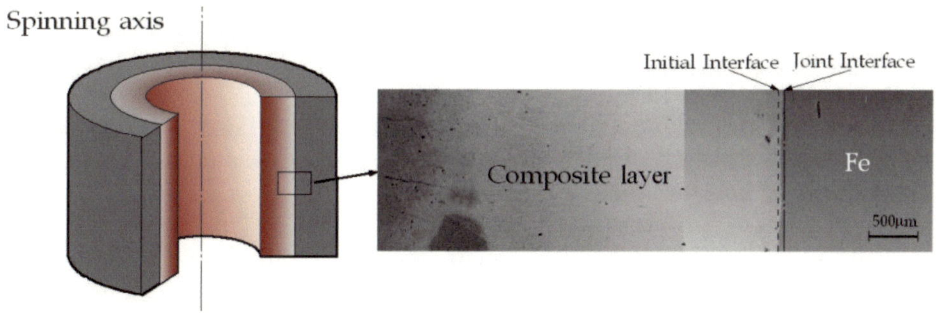

Fig. 21. Typical SEM photograph of the Ni-aluminide/steel clad pipe fabricated by the reactive centrifugal casting method (Watanabe et al., 2004a).

5. Conclusion

Functionally graded materials (FGMs) can be fabricated under a centrifugal force, by which it is possible to produce the FGMs with continuous gradients. Fabrication methods of FGMs under the centrifugal force are classified into three categories, namely centrifugal method, centrifugal slurry method and centrifugal pressurization method. We have emphasized the use of the FGM fabrication methods under the centrifugal force as ones of the practical methods, since it has the feasibility of scaling up to mass production at a low cost. Although this chapter is mainly based on the studies by the authors, some other investigators have also fabricated FGMs using a similar method. The research activity in this field is continually increasing.

6. Acknowledgments

This work is supported by "Tokai Region Nanotechnology Manufacturing Cluster in KNOWLEDGE CLUSTER INITIATIVE" by the Ministry of Education, Culture, Sports, Science and Technology of Japan. One of the authors (YW) gratefully acknowledges the financial support from "The Light Metal Educational Foundation Inc. of Japan".

7. References

Fukui, Y. (1991). Fundamental Investigation of Functionally Gradient Material Manufacturing System using Centrifugal Force. *JSME Int. J. Series III*, Vol.34, No. 1, (March, 1991), pp. 144-148, ISSN 0914-8825.

Fukui, Y. & Watanabe, Y. (1996). Analysis of Thermal Residual Stress in a Thick-Walled Ring of Duralcan Base Al-SiC Functionally Graded Material, *Metal. Mater. Trans. A*, Vol. 27A, No. 12, (December, 1996), pp. 4145-4151, ISSN 1073-5623 (Print), 1543-1940 (Online).

Hattori, Y.; Sato, H.; Miura-Fujiwara, E. & Watanabe, Y. (2010). Estimation of the Cooling Rate Distribution by means of Lamellar Spacing of Al-Al_2Cu Eutectic Structure during Fabrication of FGM under Centrifugal Force, *Functionally Graded Materials*, Vol. 24, (2010), pp. 17-22, ISBN 978-4-9901902-6-2.

Kang, C. G. & Rohatgi, P. K. (1996). Transient Thermal Analysis of Solidification in a Centrifugal Casting for Composite Materials Containing Particle Segregation. *Metall Mater Trans B*, Vol. 27B, No. 2 (April, 1996), pp. 277-285, ISSN 1073-5615 (Print) 1543-1916 (Online).

Matsuura, K.; Jinmon H.; Hirashima, Y.; Khan, T. I. & Kudoh, M. (2000). Reactive Casting of Ni-Al-Fe Ternary Intermetallic Alloys. *ISIJ Int.*, Vol. 40, No. 2 (February, 2000), pp. 161-166, ISSN 0915-1559

Miyamoto, Y.; Kaysser, W. A.; Rabin, B. H.; Kawasaki, A. & Ford, R. G. (Edts.), (1999). *Functionally Graded Materials: Design, Processing and Applications*, Kluwer Academic Publishers, ISBN 0-412-60760-3, Boston, USA.

Ogawa, T.; Watanabe, Y.; Sato, H., Kim, I-S & Fukui, Y. (2006). Theoretical Study on Fabrication of Functionally Graded Material with Density Gradient by a Centrifugal Solid-Particle Method, *Composites Part A*, Vol. 37, No. 12, (December, 2006), pp. 2194-2200, ISSN 1359-835X.

Sequeira, P. D.; Watanabe, Y.; Eryu, H.; Yamamoto, T. & Matsuura, K. (2007). Effects of Platelet Size and Mean Volume Fraction on Platelet Orientation and Volume Fraction Distributions in Functionally Graded Material Fabricated by a Centrifugal Solid-Particle Method. *Trans. ASME, Journal of Engineering Materials and Technology,* Vol. 129, No. 2, (April, 2007), pp. 304-312, ISSN 0094-4289.

Suresh, S. & Mortensen, A. (1998). *Fundamentals of Functionally Graded Materials, Processing and Thermomechanical Behaviour of Graded Metals and Metal-Ceramic Composites,* Communications Ltd., ISBN 1-86125-063-0, London, UK.

Uemura, S.; Noda, Y.; Shinohara, Y. & Watanabe, Y. (ed.) (2010). *Development and Technology of Functionally Graded Materials* (in Japanese), CMC Publishing Co.,Ltd., ISBN 978-4-7813-0123-5, Tokyo, Japan.

Watanabe, Y.; Yamanaka, N. & Fukui, Y. (1998). Control of Composition Gradient in a Metal-Ceramic Functionally Graded Material Manufactured by the Centrifugal Method. *Composites Part A,* Vol. 29A, No. 5-6, (1998), pp. 595-601, ISSN 1359-835X.

Watanabe, Y.; Yamanaka, N. & Fukui, Y. (1999). Wear Behavior of Al-Al_3Ti Composite Manufactured by Centrifugal Method. *Metall. Mater. Trans. A,* Vol. 30A, No. 12, (December, 1999), pp. 3253-3261, ISSN 1073-5623.

Watanabe, Y.; Eryu, H. & Matsuura, K. (2001). Evaluation of Three-Dimensional Orientation of Al_3Ti Platelet in Al based FGMs Fabricated by a Centrifugal Casting Technique. *Acta Mater.,* Vol. 49, No. 5, (March, 2001), pp. 775-783, ISSN 1359-6454.

Watanabe, Y.; Kawamoto, A. & Matsuda, K. (2002). Particle Size Distributions of Functionally Graded Materials Fabricated by Centrifugal Solid-Particle Method. *Comp. Sci. Tech.,* Vol. 62, No. 6, (May, 2002), pp. 881-888, ISSN 0266-3538.

Watanabe, Y.; Watanabe, S. & Matsuura, K. (2004a). Nickel-Aluminides/Steel Clad Pipe Fabricated by Reactive Centrifugal Casting Method from Liquid Aluminum and Solid Nickel. *Metall. Mater. Trans. A,* Vol. 35A, No. 5, (May, 2004), pp. 1517-1524, ISSN 1073-5623.

Watanabe, Y.; Sato, R.; Matsuda, K. & Fukui, Y. (2004b). Evaluation of Particle Size and Particle Shape Distributions in Al-Al_3Ni FGMs Fabricated by a Centrifugal *in-situ* Method. *Science and Engineering of Composite Materials,* Vol. 11, No. 2-3, (2004), pp. 185-199, ISSN 0334-181X.

Watanabe, Y.; Kim, I-S. & Fukui, Y. (2005). Microstructures of Functionally Graded Materials Fabricated by Centrifugal Solid-Particle and *in-situ* Methods. *Metals and Materials International,* Vol. 11, No. 5, (October, 2005), pp. 391-399, ISSN 1598-9623.

Watanabe, Y. & Oike, S. (2005). Formation Mechanism of Graded Composition in Al-Al_2Cu Functionally Graded Materials Fabricated by a Centrifugal in situ Method. *Acta Mater.,* Vol. 53, No. 6, (April, 2005), pp. 1631-1641, ISSN 1359-6454.

Watanabe, Y.; Sato, H. & Fukui, Y. (2008). Wear Properties of Intermetallic Compound Reinforced Functionally Graded Materials Fabricated by Centrifugal Solid-particle and *In-Situ* Methods. *Journal of Solid Mechanics and Materials Engineering,* Vol. 2, No. 7, (July, 2008), pp. 842-853, EISSN 1880-9871.

Watanabe, Y.; Inaguma, Y.; Sato, H. & Miura-Fujiwara, E. (2009). A Novel Fabrication Method for Functionally Graded Materials under Centrifugal Force: The Centrifugal Mixed-Powder Method. *Materials,* Vol. 2, No. 4, (December, 2009), pp. 2510-2525, EISSN 1996-1944.

Watanabe, Y.; Miura-Fujiwara, E. & Sato, H. (2010). Fabrication of Functionally Graded Materials by Centrifugal Slurry-Pouring Method and Centrifugal Mixed-Powder Method. *J. Jpn. Soc. Powder Powder Metallurgy*, Vol. 57, No. 5, (May, 2010), pp. 321-326, ISSN 0532-8799.

Watanabe, Y.; Inaguma, Y. & Sato, H. (2011). Cold Model for Process of a Ni-aluminide/Steel Clad Pipe by a Reactive Centrifugal Casting Method. *Mater. Lett.*, Vol. 65, No. 3, (February 2011), pp. 467–470, ISSN 0167-577X.

8

Synthesis and Properties of Discontinouosly Reinforced Aluminum Matrix Composites

Dusan Bozic and Biljana Dimcic
Institute of Nuclear sciences "Vinca" University of Belgrade
Serbia

1. Introduction

Due to the very high performance requests, a new type of, so called, "high tech" materials which are used in aircraft, rocket or automobile industry, need to have a good combination of properties under static and dynamic loading. One group of materials that can meet such requirements are metallic composites which consist of the metallic matrix (metal or alloy) reinforced whit whiskers or directionally oriented fibers of secondary phase.

Currently, there are many different composite materials available, enabling one to easily select the most suitable combination of metallic matrix and reinforcing phase, depending on the uniquely defined application requests. By selecting proper type of matrix and suitable chemical composition, shape and the amount of reinforcing phase, many composites with different mechanical properties can be produced. Potential application of such materials is very wide, but the limiting factor is still their quite high price due to the complexity of their production.

In the last two decades there was an intense development of one class of composites with aluminum alloy matrix best known as the discontinuously reinforced aluminum materials (DRA). These materials consist of aluminum alloy matrix reinforced by ceramic particles. Although, DRA composites can be produced in many ways, powder metallurgy technique offers the best results.

This article presents results of both fundamental and development studies of sintered composite materials with the Al-Zn-Mg-Cu matrix and the SiC reinforcing phase, conducted in last couple of years. As this is the composite which is already commercially applied or on the verge of being applied, all gathered results are thoroughly analyzed from the aspects of its synthesis, microstructure and several mechanical and fracture properties.

2. Matrix properties of sintered DRA materials

Mostly used alloys in the aircraft industry belong to the Al-Zn-Mg-Cu system. Usually in these alloys, traces of transient elements (below 1 mass %) such as: Fe, Si, Co and Ni can also be found. Commercially these alloys are known as 7000 or 7xxx class. One that has been used for this study is best known as CW67 alloy. Compared to the Aluminium alloys of the same composition produced by different techniques (such as melting and casting), sintered alloys have much better properties such as strength, fracture toughness and stress corrosion resistance (Tietz & Palmour, 1986). Mentioned advantages of these alloys can be related to the high amount of fine secondary precipitates and undiluted dispersed particles.

It is known that the phase particles which consist of elements such as Co or Fe+Ni are less prone to coarsening than the particles consisting of Cr, Ti or Mo (Tietz & Palmour, 1986). Precipitating phases Co_2Al_9 and $FeNiAl_9$, which are formed in the Al-Zn-Mg-Cu alloy, are very fine, ranging from 0.05-2 μm in size.Even finer particles can be produced by forming the dispersed oxide phase due to presence of oxygen (0.05-0.08 mass %) in this type of alloy.

Increased stress corrosion resistance of sintered DRA is favored by adequate grain morphology and presence of the higher amount of Co_2Al_9 precipitates in the structure. Presence of higher amount of Co has a very big influence on the higher fatigue resistance values of sintered alloys compared to casted ones (Tietz & Palmour, 1986). In order to obtain finer and more stable structure, a small amount of elements, such as Zr (0.2-0.8 mass %) can be added. Within casted alloys, amount of Zr cannot be higher than 0.12 mass %.

For further improvement of mechanical properties of sintered aluminum alloys, mechanical alloying can be applied (Tietz & Palmour, 1986). By performing mechanical alloying, multiple strengthening effects can be achieved: strengthening of the solid solution (commonly with Mg as an alloying element), precipitation strengthening and dispersed phase strengthening. It has been found that higher stress corrosion resistance of mechanically alloyed alloys of 550 MPa and high value of the Young's modulus of 76.5 GPa compared to other sintered alloys with values of 310 MPa for the stress corrosion resistance and 73GPa for the Young's modulus, make these alloys very interesting candidates for the further development and characterization.

Improvement of many properties (strength, Young's modulus, toughness etc.) of sintered Al-Zn-Mg-Cu alloys compared to the casted ones is achieved by using the contemporary powder metallurgy techniques. Production of alloy powders by rapid cooling process (rate of cooling in the range from 10^4 to 10^9 K/s), vacuum degassing of powders or compacts and powder compacting at elevated temperatures (hot pressing) are just a few of mentioned powder metallurgy techniques which can improve alloy properties (Figure 1).

An initial powder which is used for the production of sintered aluminium alloys can be made by several techniques, but mostly used one is pulverization. This technique is equally used in industry and smaller laboratories and there are several different types of it (Figure 2): gas atomization, ultrasound gas atomization, splat quenching, melt spinning *etc.* (Tietz & Palmour, 1986).

As it can be seen from the Fig.1, there are two mostly used compaction processes. Both processes consist of several steps. The first one starts with encapsulation, followed by powder degassing ending with compaction at higher temperatures (Tietz & Palmour, 1986). In the second process, powder is iso-staticaly pressed up to 70% of the theoretical density, degassed and only afterwards compacted at high temperatures up to theoretical density (Bozic et al., 1997). Both of these production processes are widely used, but the first one is more complicated (due to encapsulation/decapsulation), while the second one is less effective in removal of hydrogen and oxygen originating from the absorbed moisture in hydrated aluminum oxides. Degassing temperatures are in both cases in the range from 400 to 500 °C. Powder compaction at elevated temperatures (up to 500 °C) is done by hot pressing, hot iso-static pressing or hot extrusion which is mainly used for the production of the final products. By hot pressing at 450 °C during 2h and applied pressure of 35 MPa, a pore free compact can be produced (Fig. 3).

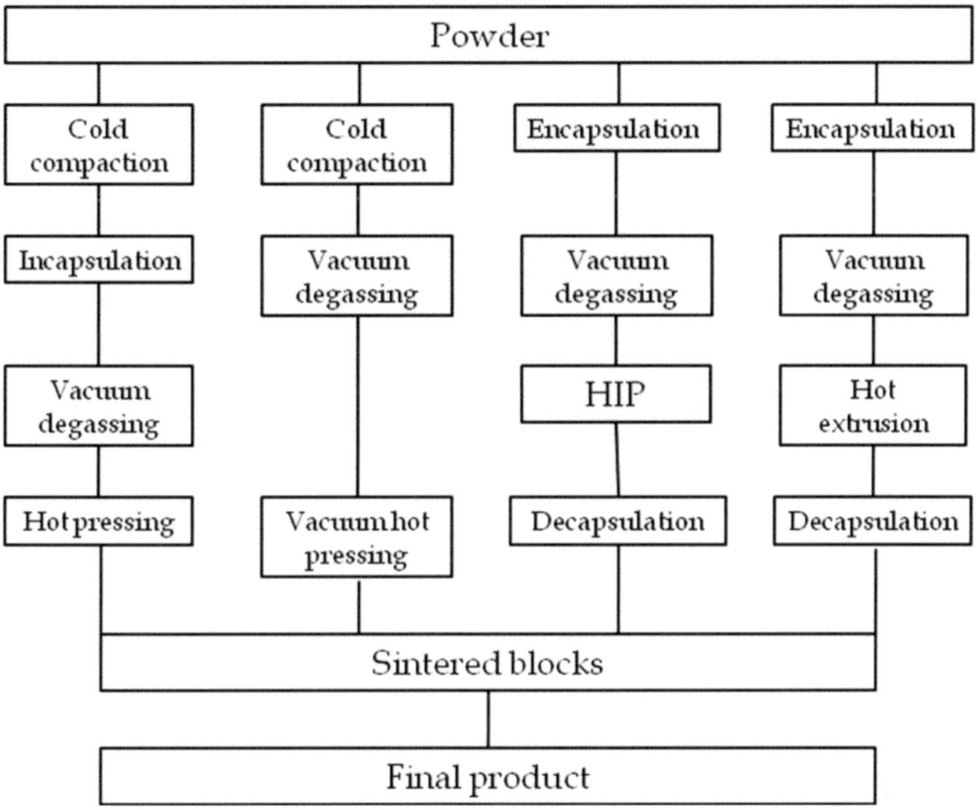

Fig. 1. Schematic presentation of high strength aluminum alloys production techniques (Tietz & Palmour, 1986).

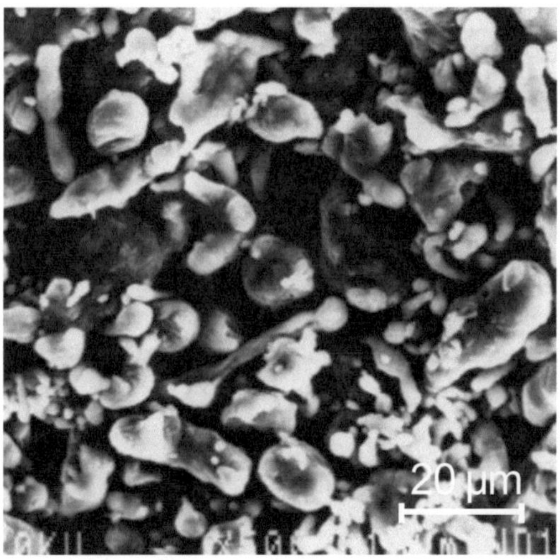

Fig. 2. SEM. Aluminum alloy powder (CW67) produced by gas atomization.

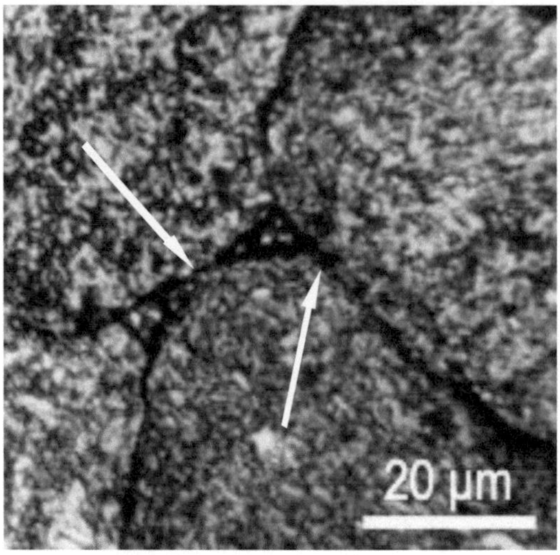

Fig. 3. SEM. Microstructure of hot-pressed Al-Zn-Mg-Cu alloy. Arrows denote a few prior particle boundaries.

Primary particle boundaries can be seen in the microstructure of the hot pressed compacts (arrows in Fig. 3), and most of the unwanted oxides which are deteriorating mechanical properties of the materials are located at these places (Gnjidic et al., 2001). Therefore, it is necessary to deform these materials by hot extrusion or rolling before use for the high risk applications (Bozic et al., 1997).

Since the Al-Zn-Mg-Cu alloy matrix is the precipitation strengthened alloy, it is important to perform adequate thermal treatment of the alloy after compaction. This treatment consists of the solution annealing at 475°c for 1h followed by water quenching and further ageing at 120°C for 24h or 160°C for 16h.

Mechanism of the thermally induced precipitation in Al-Zn-Mg-Cu alloys with the high Zn : Mg ratio can be presented in this way:

$$\text{Saturated solid solution - Gunier-Preston (GP) zones - } \mu' (MgZn_2) \text{ - } \mu(MgZn_2)$$

Gunier-Preston (GP) zones are precipitating from the saturated solid solution, and they have coherent interfaces with the parent phase. Semi coherent μ' ($MgZn_2$) phase has the monoclinic type of the crystal lattice, while the stable μ($MgZn_2$) phase has the hexagonal crystal structure. These phases are precipitating at different temperatures in the different time interval. When GP zones and the small amount of μ' ($MgZn_2$) are present in the structure, alloy exhibits the highest strength values. Addition of Cu above 1 mass % can also increase the strength (strengthening due to alloying) but the precipitating mechanism still remains the same. When there is more Cu in the chemical composition of the alloy, it can be involved in the precipitation by replacing the Zn atoms.

3. Reinforcing phase properties of sintered DRA materials

Selection of an appropriate composition, morphology and volume fraction of reinforcing phase is a matter of trade off between the requested mechanical properties and its economic value. Typical properties which have the biggest effect on the selection of the particular reinforcing phase are: Young's modulus value, wear resistance, fracture toughness, fatigue resistance and thermal expansion coefficient. A proper choice of the type and geometry of the reinforcing phase is very important for achieving the best combination of required properties and costs.Mostly used reinforcing phases for the production of discontinuously reinforced composites are: SiC in the shape of whiskers, particles of SiC or Al_2O_3, short Al_2O_3 or graphite fibers. Compared to fibers, particles and whiskers are easier for production; they are less expensive and have stable properties.

Whiskers are monocrystalline materials with a high length/width ratio (50 to 100). This shape of reinforcing phase is mostly used for composites that will be submitted to extrusion, rolling or forging. Unlike the particle reinforced composites, materials reinforced with whiskers are slightly less isotopic and their price is much higher.

Reinforcing phase in particle shape is not very expensive. Maximal/minimal particle dimension ratio is quite low (1 to 5). They have more isotropic properties compared to whiskers, resulting in better characteristics of composite materials. This property enables obtaining better characteristics of composite materials. It is important to know that this type of reinforcing phase is most commonly made by conventional methods used for the production of the metallic materials as well.

Technical SiC, which is mostly used as a particle reinforcing phase is often produced by reaction between SiO_2 and coke (in excess amount). SiC powder obtained in this way may still contain a small amount of un-reacted SiO_2 or free graphite (Fig. 4). Free graphite in a SiC powder can reduce the mechanical properties of the composites and therefore it has to be removed. Removal of this graphite can be done by thermal treatment of the powder at 900 °C for 2h (Gnjidic & Bozic, 1999).

Unlike composites produced by melting and casting that consist of particles with dimensions of hundreds of microns, composites produced by powder metallurgy techniques can use only much finer particles. Usual size of the SiC particles lies in the range from submicron size up to around 20 μm (Figs. 5-7).

As it can be seen from the images, SiC powder characteristics can differ depending on its particle shape, which can be: spherical (Figs. 5 and 6), polygonal (Fig. 6) or irregular with sharp edges (Fig. 7). In submicron powders, appearance of the agglomerates is common (Figs. 5).

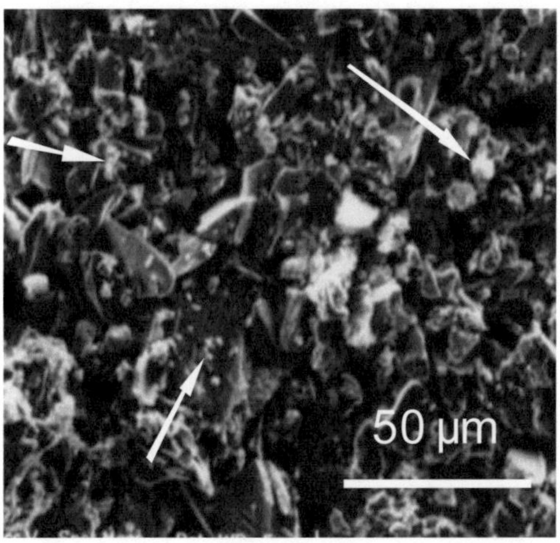

Fig. 4. SEM. SiC powder particles before heat treatment. Arrows appointing free graphite.

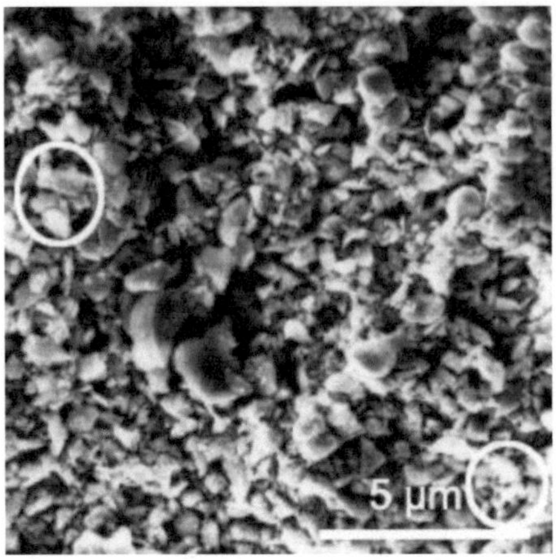

Fig. 5. SEM. Size and shape of SiC powder particles (average size 0.7 μm). Circles denote agglomerates.

Fig. 6. SEM. Size and shape of SiC particles (average size 10 μm).

Fig. 7. SEM. Different shapes of SiC particles (average size 15 μm).

4. Mixing process

Mixing or homogenization of powders is one of the most critical processes in production of DRA materials by powder metallurgy techniques. If it is done properly, particles of the reinforcing phase can be uniformly distributed in the composite metallic matrix which improves the structural and mechanical properties of these materials. The most important parameters that influence effectiveness of the mixing process are: mixing dish filling rate, rotation velocity and mixing time (German, 1994). Optimal values of these parameters are mostly confidential or patent protected. This means that for every combination of powders (shape, size, amount of reinforcing phase) and mixing dish type it is necessary to experimentally find optimal mixing conditions, *i.e.* optimal values of the previously mentioned parameters.

One method for quantitative analysis of the SiC particle distribution homogeneity is based on the concept of homogeneity index, Q, which is a quantitative measure of the one component powder particle distribution homogeneity in the mixture of a few powders (Gray, 1973). It should be noted that this concept can only be valid under assumption that the volume fraction of the observed particles is equal to their surface fraction.

Standard deviation (α) of particle surface measurements is defined by Eq. (1):

$$\alpha = \frac{1}{N}\sum\left|A_i - A_f\right| \tag{1}$$

Where N is the number of repeated measurements, A_i is the individual area fraction and A_f is the mean value of all measured particle surfaces.

Equation (2) describes the case in which completely separated system appears (complete segregation) or, in other words, system in which the whole measuring surface belongs to the particles of just one powder component.

$$\alpha = 2A_f(1 - A_f) \tag{2}$$

By combining Eqs. (1) and (2), homogeneity index, Q, can be calculated as:

$$Q = \alpha / \alpha_{seg} \tag{3}$$

Values of homogeneity index vary from 0 (perfect uniform distribution of particles in the mixture) to 1 (complete non-uniformity of the powder particles).

Using this method, a homogeneity index for all mixtures can be determined by analyzing the eight randomly chosen measuring fields split into 16 measuring units. The size of the measuring unit can be calculated from the Eq. (4):

$$A = 1/N_A \tag{4}$$

Where N_A is the number of SiC particles in the measuring unit.

By using any type of software for the quantitative analysis of the microstructure, homogeneity of SiC particle distribution in metal matrix can be calculated (Mc Kimpson et al., 1999).

In our study, which is based on the data found in the literature (German, 1994), mixing dish rotation velocity was taken to be constant and equal 70 rpm for all regimes. Beside the rotation velocity, mixing time of 30 min was taken as a constant for the first five regimes as well. The amount of powders was different and it was increased by 5 vol. % for each following regime. In the second five regimes, mixing dish filling rate was kept constant while the time of mixing was changed. The selected mixing dish filling rate was the optimal value from the first five experimental regimes (Table 1).

Mixing of powders occurred due to 'diffusion" in the cylindrical dish without any mixing accelerators. Results of quantitative microstructural analysis showing the effect of the mixing dish filling rate on the homogeneity index values are presented in Fig. 8.

All mixtures were well mixed and differences between their homogeneity indexes were in the range of 15%. Dependence of homogeneity index on the mixing dish filling rate for three different volume fractions of SiC particles in the mixture exhibits the same trend. The optimal value was achieved for the mixture with the lowest amount of the reinforcing phase.

The resulting force that affects particles in powder mixture during mixing is a result of unified influences of centrifugal, gravitational and frictional forces.The latter one exhibits the predominant effect on the mixing (homogenization) results. It was found in this study that optimal mixing dish filling rate was the lowest one (20 vol. %). The reason for this might be the fact that inter-particle contact is minimal during mixing since the available length for the free fall of particles is maximal, and therefore, agglomerates are hard to form and the ones already formed can easily be destructed (Fig. 8). With increase of the mixing dish filling rate, amount of SiC powder also increases forming more agglomerates that need to be destroyed. Available length for the free fall of particles also decreases resulting in overall decrease in powder mixture homogeneity. Negative effect of agglomerate presence in the microstructure can also be increased with the increase of SiC particle amount. It is known that a very large number of parameters influence the efficiency of particle stacking in the volume of the powder (Gray, 1973). The most important parameters are shape, size, physical and chemical characteristics of the powder particles as well as the shape, size and material of the mixing dish and many more. The real influence of only one parameter is very hard to predict since it often overlaps with the influence of another one. In the case of heterogeneous system there is one additional parameter and that is the volume fraction of the second powder. Higher amount of additional powder decreases the efficiency of particle stacking but not very dramatically since homogeneity indexes do not differ much between the different analyzed mixtures (Fig. 8).

Regime	Mixing dish filling rate (vol. %)	Mixing time (min)
I	20	30
II	25	30
III	30	30
IV	35	30
V	40	30
VI	30	60
VII	30	90
VIII	30	120
IX	30	150

Table 1. Different mixing regimes for metallic-ceramic composites

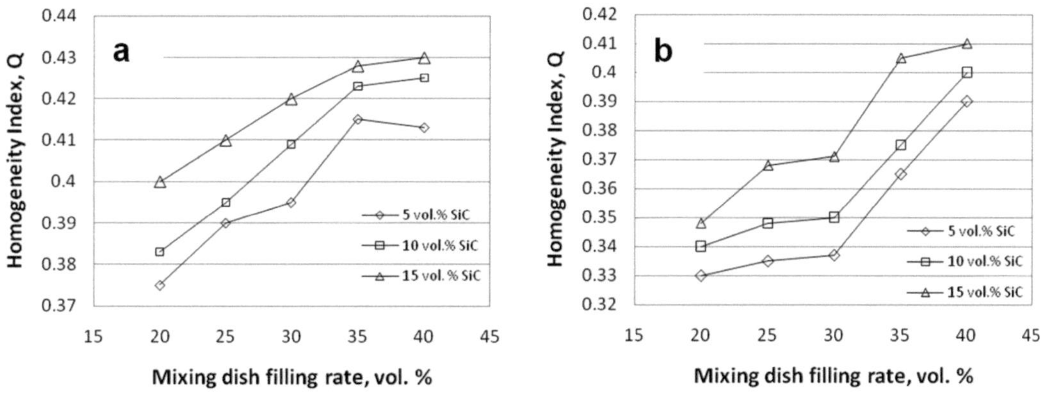

Fig. 8. Plot of homogeneity index values *vs.* amount of the mixing dish filling rate for alloys with a) 0.7 μm and b) 15 μm sized reinforcing powder particles.

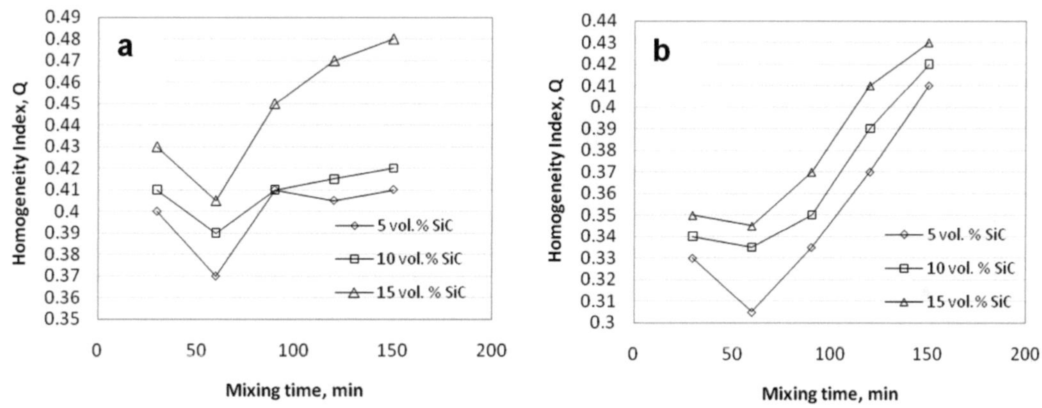

Fig. 9. Plot of homogeneity index values *vs.* mixing time for alloys with a) 0.7 μm and b) 15 μm sized reinforcing powder particles.

It can be noticed from Fig. 9 that influence of the mixing time on the homogeneity index values is much higher than the influence of the mixing dish filling rate (Fig. 8). Values of homogeneity index increase up to 40% with the increase of mixing time compared to the optimal one.

Plots of homogeneity index values *vs.* mixing time, for the optimal value of mixing dish filling rate, exhibit the same trend for all mixtures. The best homogeneity is, like in the previous case (Fig. 8), obtained for the mixtures with 5 vol. % SiC. Mixing time of 60 min was optimal from the aspect of secondary (reinforcing) phase distribution. It is known that during longer mixing time the effect of particle "diffusion" is more pronounced (German, 1994). All plots display the increase of homogeneity index with increase of the amount of reinforcing phase in the mixture during constant mixing time. That kind of behavior is somewhat expected because of increase of the number of reinforcing phase particles that need to be mixed. Figs. 10 a-c and 11 a-c display the microstructures of samples with the optimal homogeneity which are produced under the optimal mixing conditions. Figs. 10 d and 11 d illustrate so called diffusion mixing as a result of prolonged mixing time. Scanning electron images clearly show clusters of SiC as well as areas without SiC presence.

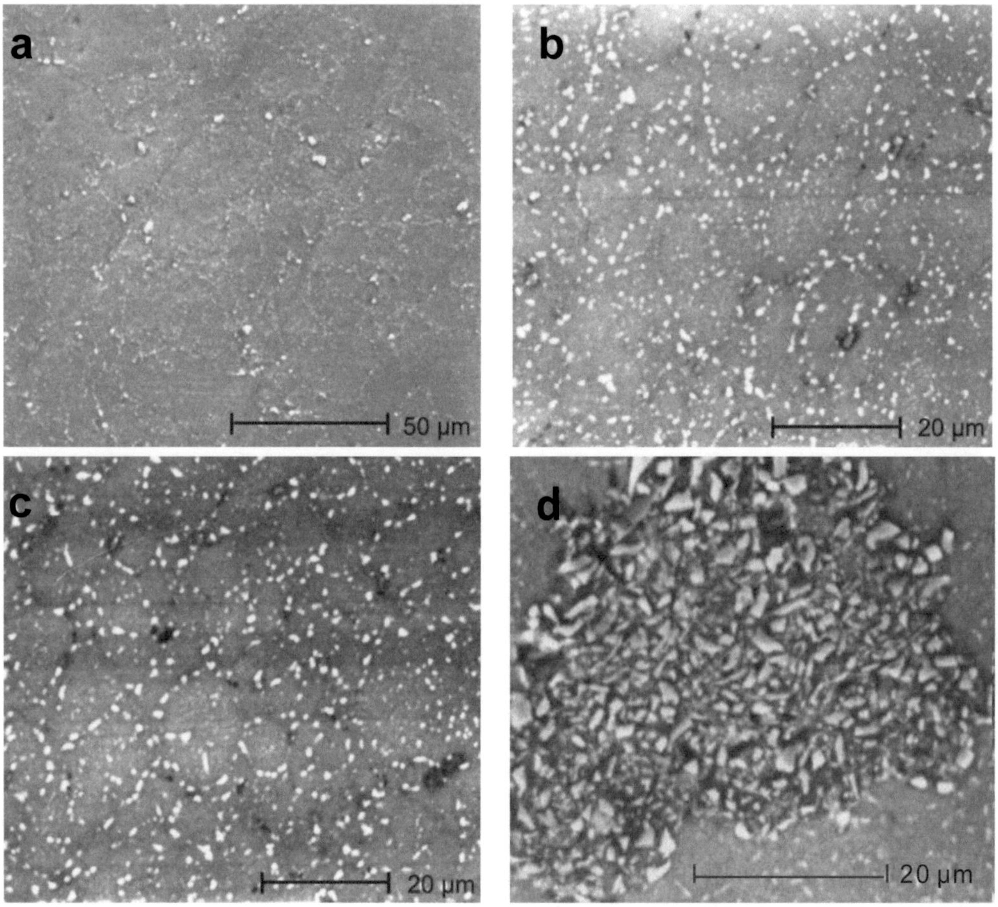

Fig. 10. SEM. Composite microstructure (d_{SiC}=0.7 μm)-optimal homogeneity. a) 5 vol. % SiC, b) 10 vol. % SiC, c) 15 vol. % SiC, and example of prolonged mixing time effect, d) 15 vol. % SiC, 150 min mixing time.

SiC powder with medium size particles of 15 μm, used in this study is characterized by wider particle size distribution. Around 20 mass.% of this granulation contained particles larger than 20 μm, and a certain percent of 1-2 μm sized particles. Several authors (Flom and Arsenault, 1989) found that when a SiC particle size is 20 μm or larger, fracture process changes from mainly matrix controlled failure to particle cracking controlled. On the other hand, particles of around 1 μm are often prone to agglomeration during mixing. Therefore, even under optimal mixing conditions, segregation of particles, presence of surface defects and cracking of large particles can be expected with this kind of reinforcing material.

By sieving this fraction, a powder with the medium size SiC particles of 10 μm, characterized by the considerably narrower particle size distribution than the previous one can be extracted, Fig. 6.

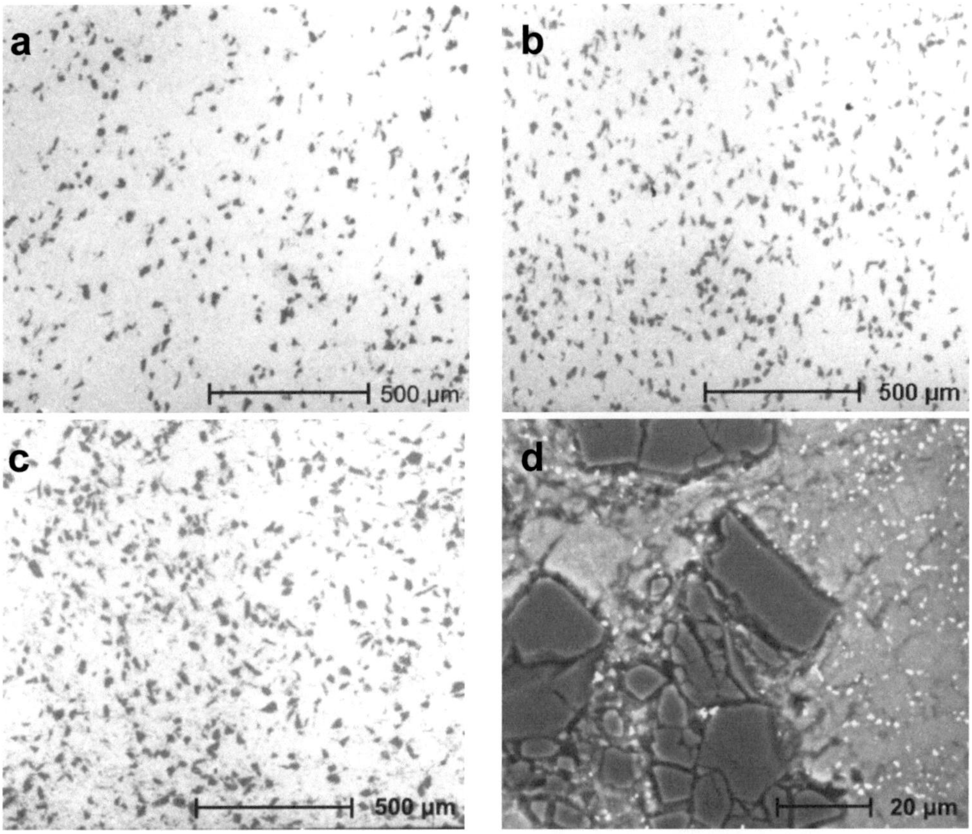

Fig. 11. LM. Composite microstructure (d_{SiC}=15 μm). a) 5 vol. % SiC, b) 10 vol. % SiC, c) 15 vol. % SiC, and example of prolonged mixing time effect, d) SEM. 15 vol. % SiC, 150 min mixing time.

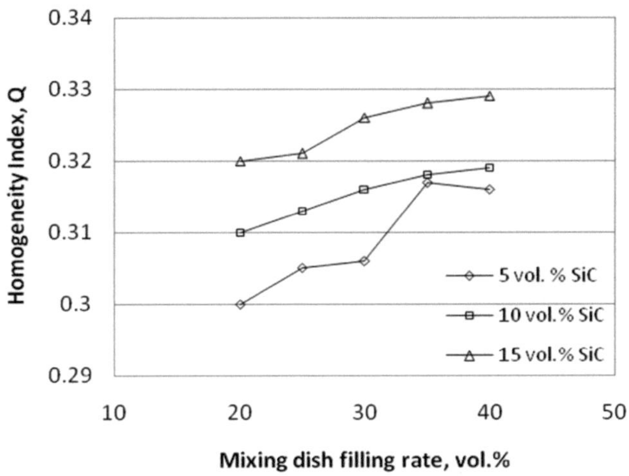

Fig. 12. Plot of homogeneity index values *vs.* mixing dish filling rate for the alloys with 10 μm size reinforcing phase particles.

Mixing of aluminum alloy and SiC powders was conducted under the same conditions as in the previous cases in this study. Dependence of the homogeneity index on the mixing dish filling rate and mixing time is shown in Figs. 12 and 13. Although the character of the plots is similar to the previous ones in Figs. 8 and 9, we can see some differences. In this case, higher degree of homogenization is achieved and better distribution of reinforcing particles is observed even for longer mixing times. Microstructures of such mixtures are presented in Fig. 14 a and b.

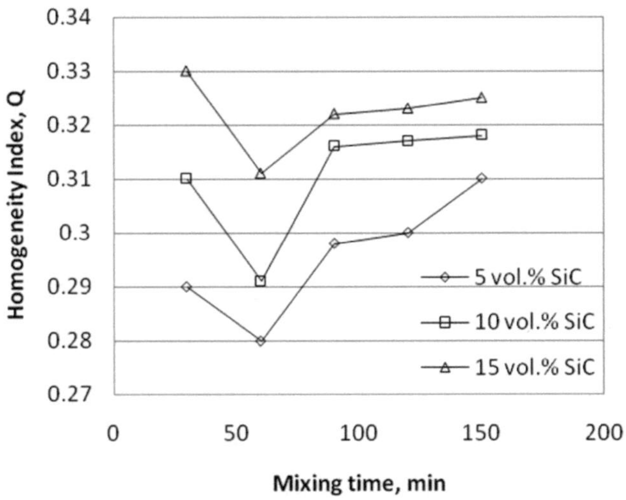

Fig. 13. Plot of homogeneity index values *vs.* mixing time for the alloys with 10 μm size reinforcing phase particles.

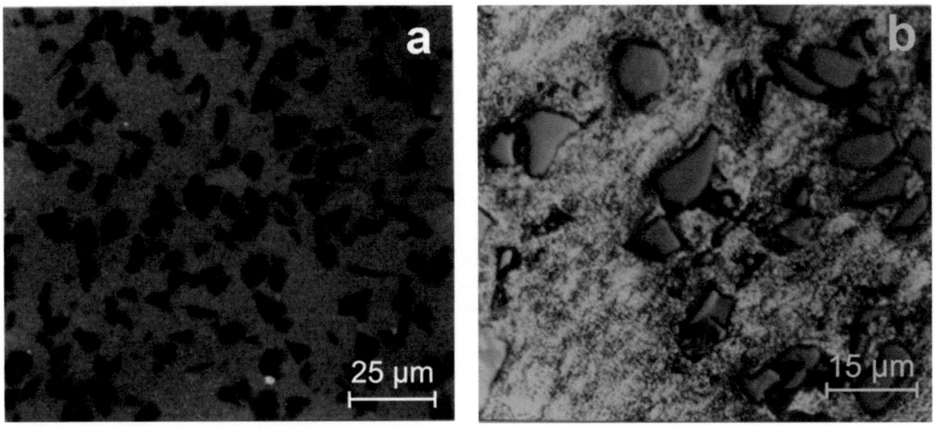

Fig. 14. Composite microstructure (d_{SiC}=10 μm). a) LM. Optimal homogeneity reached-15 vol.% SiC and b) SEM. Effect of prolonged mixing time -10 vol.% SiC.

Different values of mixing dish filling rates and mixing time have great influence on the SiC particle distribution in aluminum alloy and hence on the mechanical properties an mechanical behavior of the produced DRA material.

5. Methods of powder mixture consolidation

In order to make a fully dense product, powder mixture needs to be consolidated. Powder metallurgy consolidation techniques consist of three basic steps:

1. Cold isostatic pressing (CIP) - enables formation of compacts with density of 65-75% of the theoretical density;
2. Compacts degassing at high temperatures-removal of absorbed gases and moisture;
3. Hot or hot vacuum pressing, hot iso-static pressing or hot extrusion -production of compacts with theoretical density.

Existence of the open porosity in compacts after cold iso-static pressing enables further vacuum degassing. Degassing temperature is usually in the range from 400 to 500°C. Incomplete degassing results in detaining a large amount of oxides on the particle surface which cause weaker particle bonding in the following stages of consolidation process. In this kind of material, formation of cracks and fracture along the primary particle boundaries is likely to occur. On the other hand, too high degassing temperatures or very long degassing times are not very favorable since they can lead to decrease of the Zn amount in the composite matrix (in 7000 alloys).

Mostly used consolidation techniques for production of DRA materials are hot pressing and hot vacuum pressing. Difference between these two is that in the first case mixture of powders or porous compact is encapsulated, degassed and then hot pressed in air, while in the second case degassing is done without encapsulation of the material just before the hot pressing under vacuum. It is accepted that the latter one offers better results.

To control a large number of parameters and to understand certain mechanisms of above mentioned processes is very important for the production of composites. By presenting the densification process with densification maps it is much easier to comprehend these issues (German, 1996). Practical value of these maps is that they link the influence of the crystal structure and atomic bonds on the plastic flow of the materials. This further enables more simple analysis of mechanisms which are very important for the experiment design.

Densification maps presented in Figure 15 (Bozic et al., 2009) are determined for the constant pressure of 35 MPa and for the constant size of matrix powder particles (2R=125 μm). A model that is used is universal and it allows the comparison of densification mechanism for different applied pressures and different particle sizes. Other densification process parameters can easily be changed by changing boundaries between zones of dominating densification mechanisms. Due to the modular model structure, complete modulus can be corrected or changed. This fact is particularly useful since it enables potential corrections of the model.

A model used in this study was made for analyzing different densification mechanisms during the hot pressing, such as plastic flow, power-law creep and diffusion. From the chosen examples it can be noted that the process of densification is slowing down as the fraction of the reinforcing phase increases (slope of the dashed lines) which is in agreement with the experimental studies. A change in densification mechanism during hot pressing induced by the change of fraction of reinforcing phase can also be observed from this maps (different size of the zones for certain densification mechanism).

Densification maps provide a possibility of predicting dominant deformation mechanism, densification rate and time of pressing for the given conditions of hot pressing process.

Densification parameters are slightly changed compared to the optimal hot pressing parameters for the DRA materials production. Poreless structure was obtained at higher temperature (470 °C) and during longer time (5h). Only under these conditions densification can be fully performed when low pressure (35 MPa) is applied.

In the presented case, complete densification was achieved for aluminum alloy and DRA composite with 15 μm SiC, but not for the composite containing the smallest (0.7 μm diameter) SiC particles.

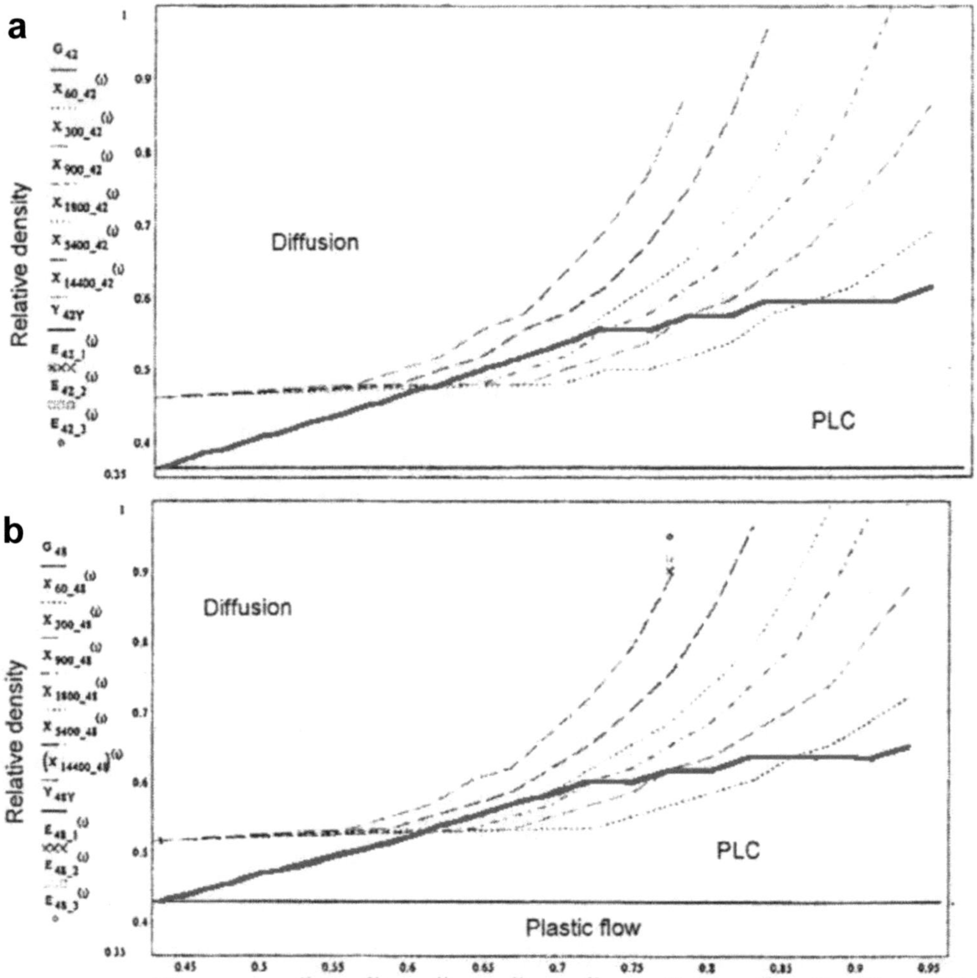

Fig. 15. Deformation mechanism maps (X_{t_D0} is a time curve, where *t* is the time of densification and D_0 is the starting density of the powder, and E_{D0_n} is the group of experimental values; thick line represents the boundary between power-law creep and diffusion zones. a) Pure Al powder; b) Al-10 vol.% SiC composite. Al particle radius in the composite was *2R=125 μm*, and SiC particle radius was *2R=33 μm*. Applied pressure was *P=35 MPa (Bozic et al.,* 2009)

Microstructures of the hot-pressed composite materials show that mixing parameters used in the present work enabled formation of relatively uniformly distributed SiC particles in the aluminum matrix (Fig. 16 a and b). Porosity existing in the composites reinforced with 0.7 µm SiC, which has been detected after density measurements (Table 2) is illustrated in Fig. 17. Formation of agglomerates during mixing was the primary reason for such behaviour.

Sample	Theoretical density, ρ_t (g/cm^3)	Density, ρ (g/cm^3)	Percentage of theoretical density
CW67	2.88	2.87	100
CW67-0.7 µm SiC	2.90	2.80	97.5
CW67-15 µm SiC	2.93	2.93	100

Table 2. Density values of the hot pressed aluminum alloy (CW67) and two selected composites.

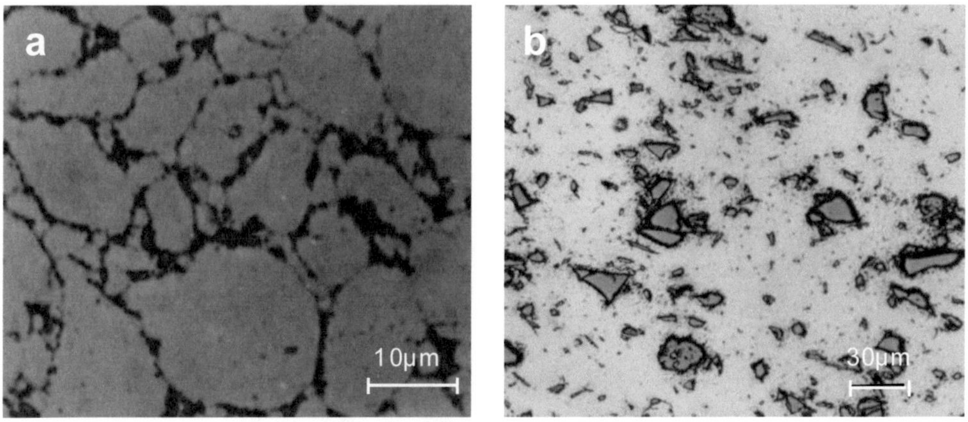

Fig. 16. Microstructures of hot-pressed samples. a) LM. Composite containing 10 vol.% - 0.7 µm SiC, b) SEM. Composite containing 10 vol.% - 15 µm SiC.

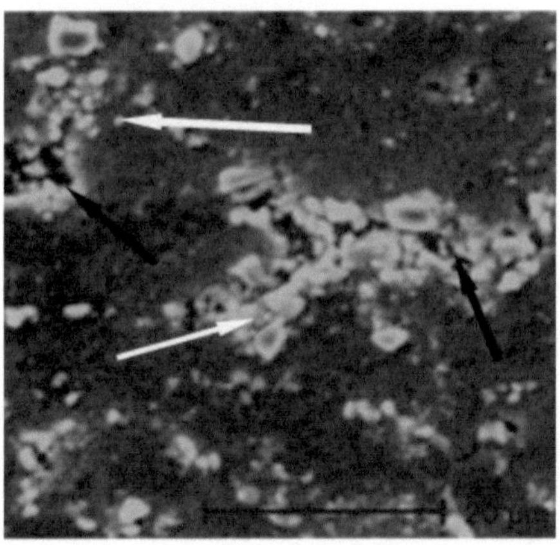

Fig. 17. SEM. Microstructure of hot-pressed composite containing 10 vol.% - 0.7 µ SiC. White arrows -particle agglomeration; Black arrows-pores.

6. Influence of the reinforcing phase on mechanical and fracture characteristics of DRA materials

Prior to compressive testing, all hot-pressed materials were subjected to two types of thermal treatments: under-ageing and peak-aging. Initially, all samples have been homogenized by solution thermal treatment (1h at 475 ˚C) followed by water quenching to room temperature. Under-ageing was performed at 120 ˚C for 1h, while the peak-ageing was done at the same temperature of 120 ˚C for 24h for the CW67 alloy and 22h for the composites. Presence of SiC particles significantly influences the heterogeneous nucleation of strengthening phases. Capability for heterogeneous nucleation can be partially attributed to high dislocation density in the composite matrix, which is present due to the large mismatch between the thermal expansion coefficients of the matrix and the reinforcing phase. It is well known that dislocations act as nucleating sites for precipitation during aging, but they also promote accelerated ageing in the matrix. For that reason, peak-aging conditions for composites are reached in shorter times than for the monolithic alloy (Fig. 18).

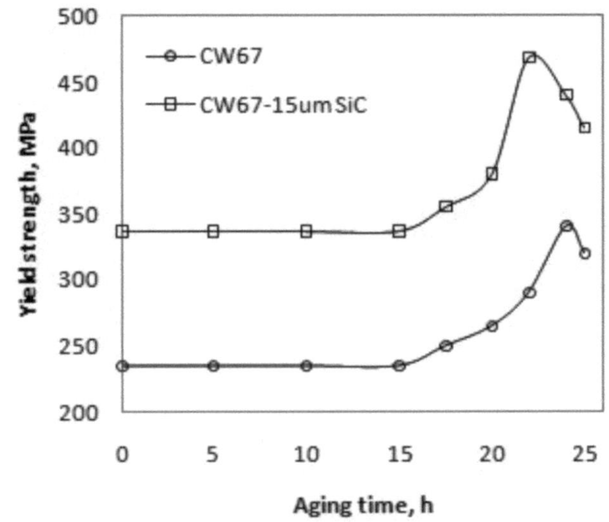

Fig. 18. Yield strength of the CW67 alloy and its corresponding composite *vs.* aging time.

Influence of different SiC particle sizes on the yield strength (σ_{yc}), ultimate strength (σ_{uc}) and ductility (strain to failure) (ε) under compression at room temperature for both under-aged (UA) and peak-aged (PA) samples is listed in Table 3.

Values of yield strength (measured at 0.2 % offset strain) for all composites, irrespective to heat treatment, are higher than those of the monolithic alloy. On the other hand, the ultimate strength and ductility values of the composite samples are lower than those obtained for the unreinforced alloy. It can also be noted that 0.2 % offset strain occurred at much higher fraction of the total strain-to-failure in the composite than it did in the case of the CW67 alloy. Main reason for such behavior could be the influence of the SiC particle sizes, their nature and agglomerate formation. These results also imply that smaller number of particles is damaged when the applied stress reaches yield stress in the composites. Stress will continue to accumulate after reaching the yield strength, leading to rapid failure and therefore low values of the ultimate strength and ductility.

Analysis of results indicates that the samples submitted to peak-aging generally exhibited higher yield strength and ultimate strength values but also lower ductility when compared to under-aged samples.
The highest compressive property values are obtained for the composite reinforced with 10 μm SiC particles. It is worth noting that the lowest values are obtained for the composite material with 0.7 μm SiC particles. This result is unexpected since it is known that particle size decreasing is associated with improved mechanical characteristics of the particle reinforced composites.

Material	σ_{YC}(MPa)		σ_{UC}(MPa)		ε (%)	
	UA	PA	UA	PA	UA	PA
CW67	235	340	860	950	38	35
CW67-0.7 μm SiC	312	400	720	740	22	19
CW67-10 μm SiC	400	490	850	930	27	23
CW67-15 μm SiC	331	437	743	830	24	20

Table 3. Compressive properties of pure aluminum alloy and reinforced materials in under-aged (UA) and peak-aged (PA) conditions.

Generally, mechanical behavior of composites is affected by the presence of reinforcing particles dispersed in the metal matrix. When analyzing the yield behaviour of these materials, it is important to recognize that the yielding strength of the composite is determined by the yield strength of the matrix. Most probably yield strength may not be the same as yield strength of the monolithic alloy processed and heat-treated in a similar way (Doel & Bowen, 1996). The local yield strength of the matrix may, in fact, be higher than that of the monolithic material due to the increased dislocation density and reduced grain size. Namely, matrix grain size is reduced by particles acting as nucleation sites during solidification. Also, grain size is smaller due to formation of subgrains after dislocation rearrangement into boundaries within the grain (Rees, 1998). These subgrains are formed in grains with high dislocation density surrounding a SiC particle. This rearrangement is a recovery process driven by the energy stored within the distorted matrix at the interface of two phases. Also, SiC particles act as obstacles for dislocation motion.
There are several factors affecting the local stress in the matrix. First of all, there is a tensile residual stress field in the matrix due to the difference in coefficient of thermal expansion (CTE) between aluminum alloy and SiC. Partial relief of this stress can result in an increase of dislocation density (Miller & Humphreys, 1991). Further, any applied load cannot be evenly distributed between the matrix and the SiC particles because of difference in stiffness. Several authors have previously reported that the reinforcing phase can support more applied load than the matrix (Doel & Bowen, 1996). Thus, local stress in the matrix may be lower than the nominal applied stress. In addition, SiC particles can withstand this stress until it becomes sufficiently high to cause particle fracture or failure either at or near the interface. It is believed that damage will first occur by particle fracture because the interface between SiC particles and aluminum is very strong. Hence, as soon as the particle breaks the stress in the undamaged material increases. In time, micro voids are beginning to

form in the matrix as a result of dislocation pile-up, due to the applied load. These voids continue to grow causing failure of composite materials.

In order to comprehend the exact influence of SiC particle size on the compressive properties of the composite, a comparison was made between the properties of three composites in both under-aged and peak-aged condition.

Composite sample with 0.7 μm SiC particle size had the smallest increase of yield strength, presumably due to the presence of the agglomerates in the structure (Fig. 17), originating from the starting powder (Fig. 6). In this case, matrix around the agglomerates is subjected to higher stress because they cannot support the same amount of stress as non agglomerated particles. Therefore, matrix is forced to yield at lower applied stress and, therefore, yield strength of the composites is also lower. Fracture surface of this composite show evidence of both trans-crystalline and inter-crystalline fracture mechanisms (Fig. 19). A factor that might have a detrimental effect on the compressive properties of this particular composite material is the homogeneity of the particle distribution.

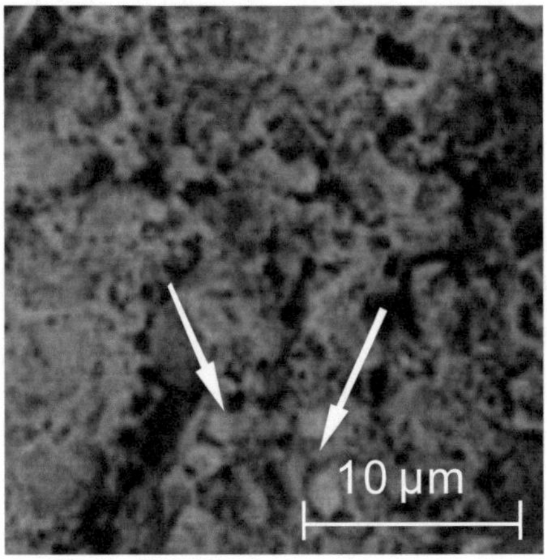

Fig. 19. SEM. Fracture surface of composite with 0.7μm SiC particles in peak-aged condition. White arrows appointing agglomerates.

Composites reinforced with 10 and 15 μm SiC particles are generally both stronger and more ductile than the material reinforced with 0.7 μm particles. Of the two analyzed composites, one containing 10 μm SiC particles has the highest increase in yield strength and the lowest decrease in ultimate compressive strength and ductility for both ageing conditions.

Main reason for the superior mechanical characteristics of the medium sized particles composites is uniformity of reinforcing phase particle dimensions (Fig. 20) and structure without agglomerates.

By making a mixture with narrower particle size distribution, the number of fracture initiation sites in DRA material is reduced making the matrix/SiC decohesion a dominant fracture mechanism (Fig. 20).

Formation of the crack and the crack path depend on the microstructure homogeneity of the alloys. Uniform path of the existing crack is typical for the less homogenous alloys, while the zig-zag path characterizes homogenous alloys (Fig. 21). The latter type of crack propagation is a consequence of homogenous distribution of secondary (SiC) particles

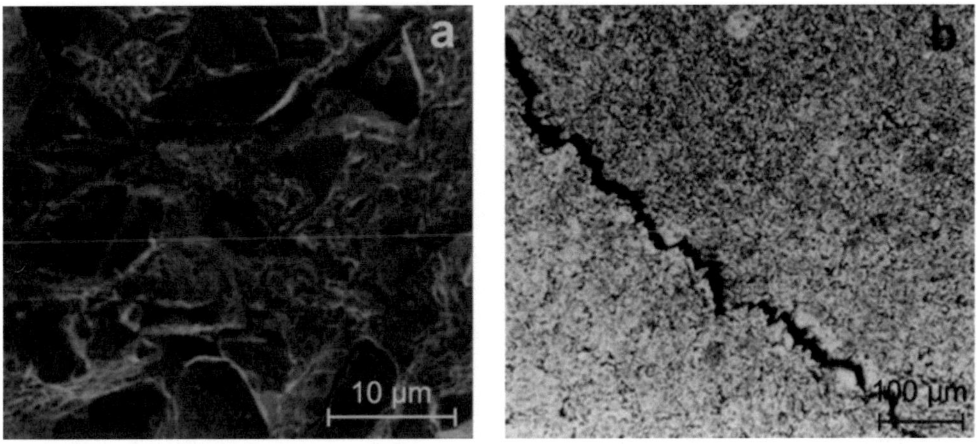

Fig. 20. Fracture surface of DRA materials after 60 min mixing. a) SEM. CW67- 10 vol.% SiC, d_{SiC}=10µm; b) LM. CW67- 10 vol.% SiC, d_{SiC}=10µm.

which block the propagating crack and deflect it. By deflecting the crack, the energy level at the crack tip which is necessary for its propagation decreases disabling its further propagation. Therefore, more energy (larger applied force) has to be introduced to the system for the continuation of crack movement. This fact explains why these, more homogenous alloys, have better mechanical properties (Fig. 22).

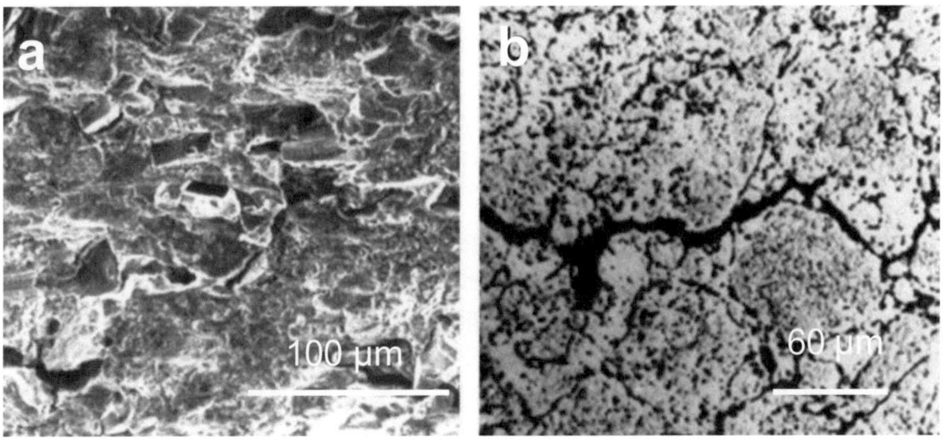

Fig. 21. Fracture surface of DRA materials after 60 min mixing. a) SEM. CW67- 10 vol.% SiC, d_{SiC}=15µm; b) LM. CW67- 5 vol.% SiC, d_{SiC}=0.7 µm.

In the case when the reinforcing particles are not evenly distributed in the alloy matrix, crack propagates easily through the matrix (Fig. 23 a and b). This behaviour corresponds to poor mechanical properties of the DRA material with such microstructure. When mixing time is too long a formation of SiC agglomerates is noticeable. Due to the very weak bonding between the SiC particles in the agglomerates, a particle decohesion will easily occur under applied stress and the crack will appear (Fig.24).

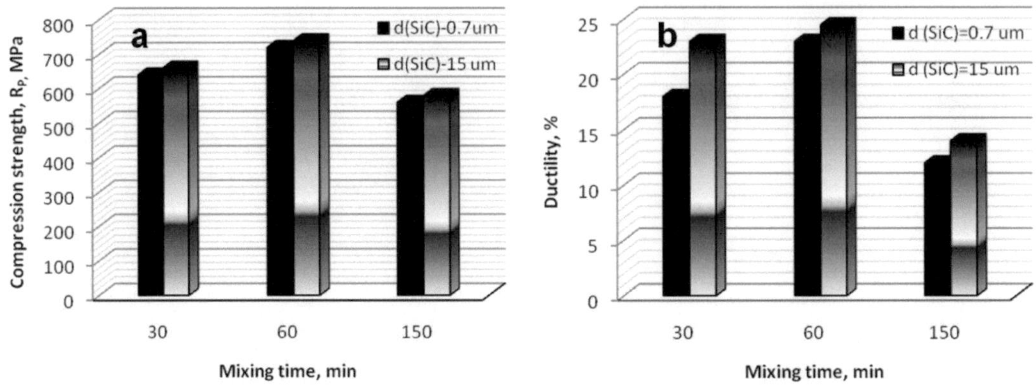

Fig. 22. CW67- 5 vol. % SiC composite compression properties dependence on time of mixing. a) compression strength; b) ductility.

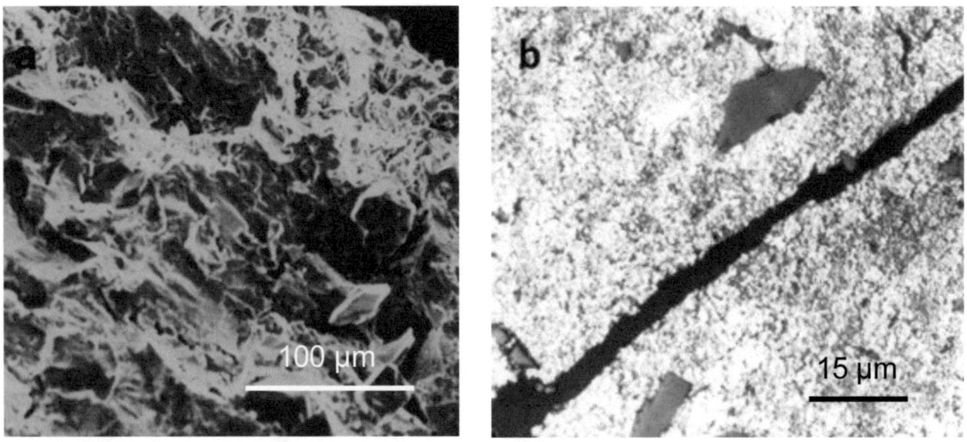

Fig. 23. Fracture surface of DRA material after 30 min mixing. a) SEM. CW67-10 vol.% SiC, d_{SiC}=15 μm and b) LM. CW67-10 vol.% SiC, d_{SiC}=15 μm.

As observed by Humphreys (1988), an addition of a brittle reinforcement phase in high strength aluminum alloys may even decrease the ultimate tensile strength. Indeed, in the present composites this was found to be the case. Indeed, in the present composites this was found to be the case. The values of the ultimate strength were lower than the values for the monolithic material (Table 3). The presence of SiC particles can be detrimental to the ultimate compressive strength of the composite materials because of more additional cracking mechanisms compared to monolithic alloy. These mechanisms are: particle cracking (Fig. 25), particle matrix debonding (Fig. 20 a) and particle agglomerate decohesion (Fig. 19). The latter two mechanisms are of a secondary importance when particles are well distributed and strongly bonded. Thus, it is particle cracking that has a major influence on the ultimate compressive strength of SiC/Al composite materials.

Extensive studies of the room temperature mechanical behavior of these composites have been conducted, but the information about mechanical properties of DRA materials at elevated temperatures is still limited. Therefore, some results of the temperature effect on the compressive properties and fracture behaviour of a SiC particle reinforced CW67 aluminium alloy will be presented.

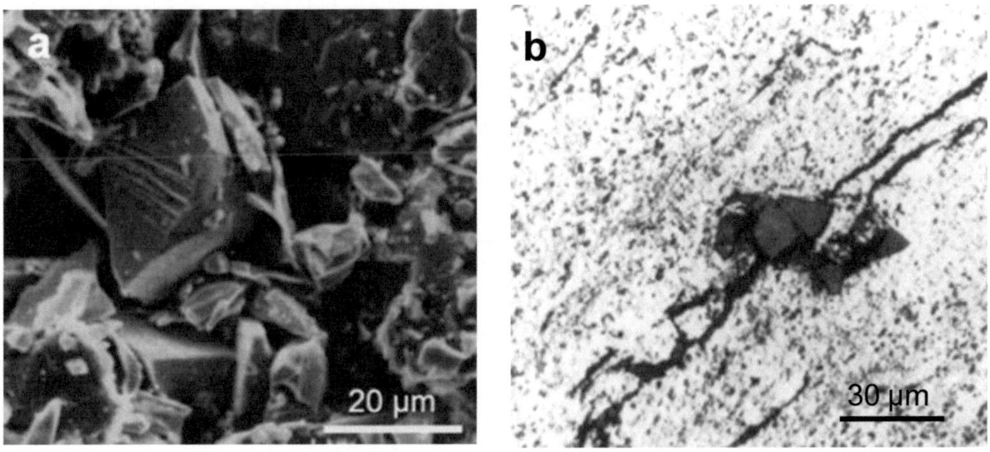

Fig. 24. Fracture surface of DRA material after 150 min mixing. a) SEM. CW67-5 vol.% SiC, d_{SiC}=15 µm; b) LM. CW67-5 vol.% SiC, d_{SiC}=15 µm.

Fig. 25. SEM. Fracture surface of the composite with 15 vol.% SiC.

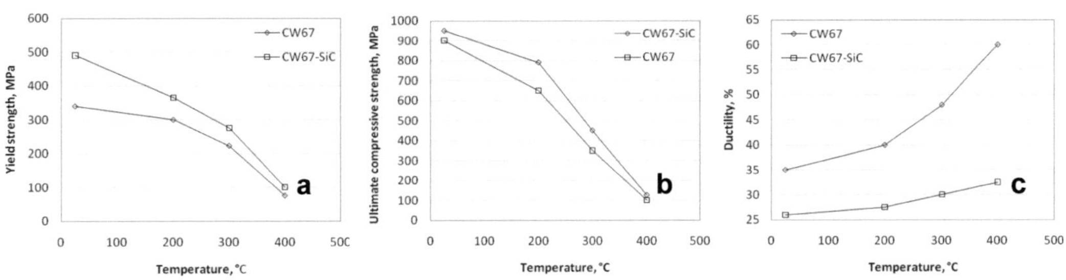

Fig. 26. a) to c) Compressive properties of CW67 and CW67- 15 vol. % SiC composite as a function of temperature.

With increasing temperature matrix becomes softer and stress accommodation around the particles caused by difference in CTE becomes easier compared to the room temperature. The constraint around the particles in the matrix can be relaxed by a dynamic recovery process. It is therefore expected that any misfit strain gradient can be relaxed by the recovery process which could lead to a decrease in work hardening and strength, resulting in a small difference in strength between the composite and the monolithic alloy at higher temperatures (Fig. 26 a, b). Increase of composite ductility with temperature (Fig. 26 c) can be a consequence of the improved ductility of the matrix. However, ductility of the composite is lower than that of the monolithic alloy, which is caused by the particle cracking or particle matrix debonding effects.

Microstructures of composites strained at room temperature show that SiC particles may fracture more frequently in the regions of agglomerates or by particle cracking of large particles, rather than by debonding between the matrix and SiC particles.

Presence of cavities is observed in microstructures of samples tested at elevated temperatures at the matrix/reinforcing phase interface (A in Figs. 27 and 28 b), particle boundaries (B in Fig. 27 b), and in the regions where the SiC particles are agglomerated (C in Fig. 27).

At 200˚C, small voids were found at the particle boundaries with tiny linkage between voids. At 300˚C and 400˚C these voids become larger. It has been observed (Fig. 27 a-c) that debonding between particles and the matrix accelerates above 200˚C.

Fig. 27. LM. Longitudal sections of the composites tested at: a) 200°C, b) 300°C and c) 400°C.

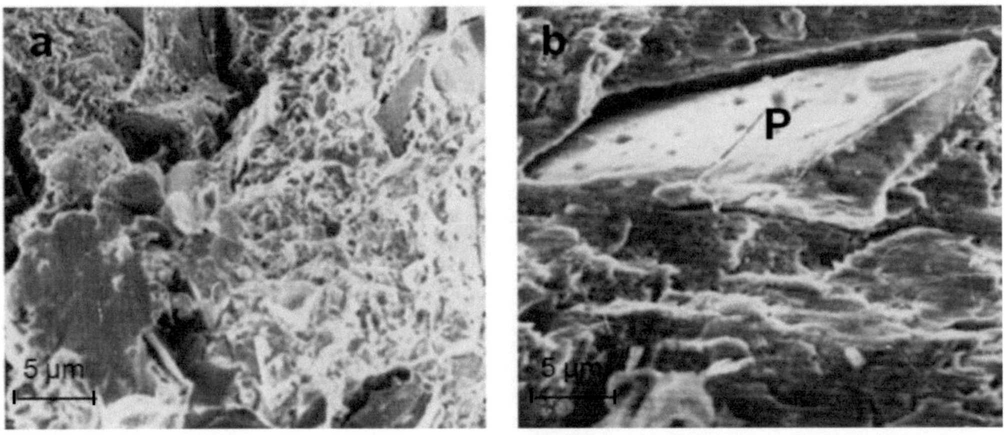

Fig. 28. SEM. a) and b) Fracture surface of the composite CW67-SiC tested at 400°C. (Mg, Cu)Zn_2 precipitates on SiC (indicated with P).

7. Conclusions

The basic conclusion is that much more research is needed for further enhancement of the DRA materials properties. Computer simulation and mathematical modeling of composite fabrication would enable faster achieving of results and therefore these methods need to be addressed. In this area some scattered work has been done but it still needs further development. By finding and establishing the proper models actual experimentation and prediction of properties, shape, size, volume fraction of the participating phases and mixing parameters would largely be cut down.
As mentioned before, dispersed phase size, shape, volume fraction, wettability and distribution play the most important role in the properties attained in metal-matrix composites. A lot of research is focused on optimizing these parameters and based on the results, property designing of composites is possible. However, all the experimentation has stopped at decreasing the dispersed phase level below 10μm, possibly due to the difficulty in dispersing uniformly and without coagulation of finer particles in the matrix. This is especially pronounced when the liquid metallurgy route is adopted for making the composites. If the coagulation problem cannot be solved when the particle size is decreased below 10 μm, alternate methods of fabrication have to be attempted. Most promising method so far, seems to be in-situ production of composites, which is recently a very modern trend in science. In the present age of nanomaterials, methods of dispersion or making composites with nano sized dispersed phase holds a lot of potential for commercial exploitation in the future.

8. Acknowledgements

Authors are in debt to the Ministry of Science of the Republic of Serbia for the financial support realized through the Project No. 172005.

9. References

Adler P.N., Delasi R., Geschwind G., Influence of microstructure on mechanical properties and stress-corrosion susceptibility of 7075 aluminum-alloy, *Metallurgical Transaction*, vol.3 12 (1972), pp. 3191-3200.

Bozic D., Pavlovic M., Mitkov M. and Nikacevic M., Mechanical and structural properties of CW67 alloy produced by powder metallurgy techniques (in Serbian), *Metalurgija* 3 (1997), p. 201.

Bozic D., Dimcic O., Dimcic B., Vilotijevic M., Riznic-Dimitrijevic S., Modeling of densification process for particle reinforced composites, *Journal of Alloys and Compounds*, vol. 487 1-2 (2009), pp. 511-516.

Doel T.J.A., Bowen P., Tensile properties of particulate-reinforced metal matrix composites, *Composites Part A*, vol. 27 8 (1996), pp.655-665.

Flom Y.and Arsenault R.J., Effect of particle size on fracture toughness of SiC/Al composite material, *Acta Metall* 37 (1989), pp. 2413–2423.

German R.M., *Powder Metallurgy Science,* Second Edition, 1994, Metal Powder Industries Federation, Princeton, NJ.

German R. M., *Sintering Theory and Practice*, John Wiley and Sons Inc., New York,NY; 1996

Gnjidić Ž., Božić D., *Tehnika,* vol. 8, 6 (1999), pp.5.

Gnjidić Ž., Božić D., and Mitkov M., The influence of SiC particles on the compressive properties of metal matrix composites, *Mater Charact* 47 (2001), pp. 129–138.

Gray W.A., *The packing of solid particles*, Champan and Hall Ltd., London (1973).

Humphreys F. J., In: *Proceedings in Mechanical and Physical behavior of metallic and ceramic composites*, ed. Anderson S et al. 51 (1988).

Mc Kimpson M.G., Heimi A.N., Hung X. In: *Proceedings of structural applications of mechanical alloying*, vol. 203. Materials Park, Ohio: ASM; 1999.

Miller W. S., Humphreys F. J., Strengthening mechanism in metal matrix composites, *Scripta Metall. Mater.* 25 (1991), pp.

Rees D.W.A., deformation and fracture of metal matrix particulate composites under combined loading, Composites, Part A, 1998, 29A: 171-82.

Tietz T.E. and Palmour G., *Advances in powder technology* vol. 9, ASM, Metals Park, Ohio (1986) p. 189.

9

Modelling Reaction-to-fire of Polymer-based Composite Laminate

Damien M. Marquis and Éric Guillaume
Laboratoire national de métrologie et d'essais (LNE)
France

1. Introduction

The use of polymer based laminate composite has greatly increased in the last 20 years. These materials offer many advantages in engineering applications. However, unlike other materials such as steel and aluminium, current composites materials commonly used in structure are reactive. They decompose and release heat and smoke when exposed to high temperature fires. The diverse range of uses for composite materials shows that they can be exposed to a variety of fire threats. In some cases, this use can increase the risk or the likelihood of severity. Several case studies of fire in aircrafts and ships have demonstrated the importance of understanding the fire reaction of composite.

The knowledge of the fire behaviour of polymer composite material is of great importance from a practical point of view because of their widespread use in several domains (aircraft, shipbuilding, railways,etc.). The fire behaviour of a diverse variety of composite material has been analyzed since the 1970s. Usually, it is defined by their *fire reaction* and *fire resistance* [ISO13943 (2001)]. Fire reaction is described as "*the response of a material in contributing by its own decomposition to a fire to which it is exposed under specified conditions*". Fire resistance is defined as "*the ability of an item to fulfil for a stated period of time the required stability and/or integrity and/or thermal insulation, and/or other expected duty specified in a standard fire-resistance test*" .

The aim of this chapter is to give a global point of view on the modelling reaction-to-fire of polymer-based composite. It defines the elements of knowledge and the parameters for the characterization of these materials and the modelling of various processes related to combustion. This chapter will describe how heating up a composite material leads to flaming ignition up to surface fire growth. The discussion will be focused on composite material with fuel polymer matrix (thermoplastic and thermoset resin) reinforced with non-combustible fibres (carbon or glass). Special attention is given by Mouritz & Gibson (2006) on fire behaviour of thermoplastic and thermoset matrix, consequently this part is not treated in this chapter. The reaction-to-fire of polymer matrix can be synthetically divided into five (major) parameters, as shown on Fig 1 that are heating, decomposition, ignition, combustion, and surface flame spread. The combustion mechanisms, thermochemistry thermodynamics and dynamics of fire are more complex, and are beyond the scope of this chapter. The topic of fire and flame is reviewed in several extensive overviews [Drysdale (1998), Quintiere (2006)]. The main focus of the present chapter is to help the reader to understand phenomena assumptions and simplification embedded in different models and tests that attempt to predict thermal

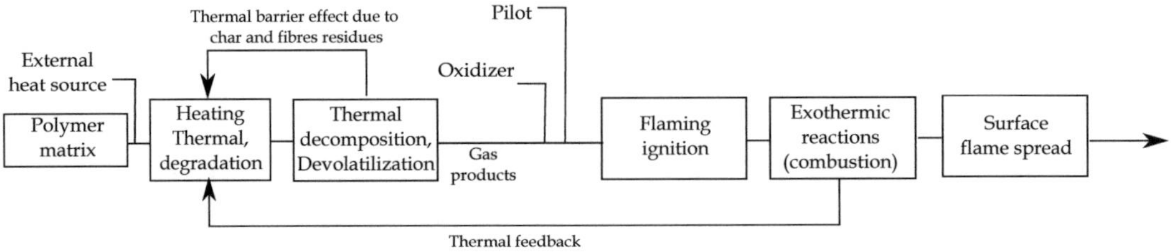

Fig. 1. Reaction-to-fire processes of polymer matrix.

decomposition, ignition, combustion and surface flame spread. All processes involved will be divided in two parts: those associated with the solid phase and those associated with the gas phase.

2. Thermal, physical and chemical processes in solid phase

The devolatilization process depending on thermal decomposition results from exposure to heat and depends strongly on the oxygen content. In an oxidizing flow environment, for example, it is instead the first step of the flaming combustion process, which is defined by a complex physico-chemical process of oxidation, which is accompanied by a release of energy and gas. The devolatilization process of the solid phase of a polymeric material in general and of composite material in particular is a complex and heterogeneous process consisting of several partial reactions. It involves reaction pathway to extensive and irreversible chemical and physical changes caused by heat. This process has been studied since the end of the 19th century by chemists with works on heterogeneous reactions in applications other than fire. The interest of scientists and fire researchers in the problem of how a solid potentially becomes a fuel gas is particularly motivated by the need to quantify the heat and gases source term in numerical simulations and by the increase of calculation capacities. The accurate prediction of the source term can only be attained by the very precise knowledge of the physics and chemistry of the decomposition process. It other words, the pyrolysis products are combustible compounds with high chemical energy that are converted into heat in the flame region. Thus, the prediction of the fire growth requires the quantification of the dynamics of the solid fuel.

2.1 Terminology

Phenomena occurring during "*devolatilization*" are of primary interest in fire safety engineering since the rate of thermal decomposition controls fire growth, spread velocity, release of toxic gases, dripping, production of liquid by products, fire propagation, etc. Thermal degradation and thermal decomposition are different concepts, although these two terms are often considered as synonyms in the literature. Thermal decomposition concerns "*the changes in the chemical structure caused by heat*". Thermal degradation is "*a process whereby the action of heat or elevated temperature on a material, product, or assembly causes a loss of physical, mechanical, or electrical properties*" [ASTM (E176)]. Thermal degradation is mostly related to materials applications. Changes in thermal properties with temperature highly influences the heat transfer into the solid and the heat and mass transfer towards environment.

The behaviour of a previously defined composite material strongly depends on the environment and more particularly on the oxygen content. So, according to the O_2 concentration, it is possible to separate thermal decomposition by *thermolysis* and those by *thermo-oxidative decomposition*. Etymologically, thermolysis is composed of Greek roots "*thermo*", denoting the temperature, and "*lysis*", derivative of "*lusis*" which defines the loss or disintegration. Therefore, thermolysis defines the thermal decomposition of a polymer under the effect of heat only. The scientific community used the thermolysis decomposition when the process occurs in the absence of oxygen. It uses the term *thermo-oxidative decomposition* when the devolatilization process results from the combined effects of heat and oxygen. In a more global way, the term "*pyrolysis*" translates the decomposition under the action of a fire. During combustion, oxygen is lacking and it is often difficult to separate the processes of thermolysis from those related to oxidation because of their overlapping nature. In the remainder of this chapter, the term pyrolysis is used whatever the environment (oxidizer or inert) to describe thermal decomposition in a fire is.

2.2 Heating up

Consider a "semi-infinite" composite material initially at temperature T_0. It is exposed to an external radiative heat flux $\dot{q}_i''|_{x=0}$ on one face at $x = 0$ as shown on Fig. 2. This moment is defined as the onset of the decomposition process $t = 0$. Heating of the composite material is caused either by an external heat source, or when combustion is initiated by the exothermicity of reaction (thermal feedback). In the solid phase, a heat wave δ_t move up in-depth mainly by thermal conduction phenomena $\dot{q}_c''$ from the sample surface where the heat is absorbed. Heat conduction is a transport phenomenon of internal energy due to heterogeneity of molecular agitation. The highest temperatures are reached close to the surface, and in-depth energy transfer by conduction and radiation will result in an increase in temperature of the deeper parts of the slab. The temperature $T(x,t)$ vary in depth and in time. Thermal degradation leads to a significant variation of thermophysical properties in-depth and in time. The radiative absorption $\kappa(x,t)$ in-depth depends on the intrinsic characteristics of composite material (opacity, etc.). Progressively, a series of physical and chemical processes are initiated as the energy reaches the composite surface.

2.3 Thermal decomposition

In their reference paper, Hirschler & Morgan (2008) explain these phenomena. The authors explain that the nature of the volatile products of thermal decomposition is dictated by the chemical and physical processes and their interactions.

2.3.1 Chemical processes

In the absence of oxygen, a solid polymer experiences a physical degradation and a chemical decomposition when it is subjected to a heat source. At a low temperature ($t \sim 100°C$), it undergoes no chemical decomposition but a thermal degradation. Furthermore, it can release residual monomers depending of the initial product and or the additives. The chemical decomposition and devolatilization appears beyond a certain temperature, called *temperature of decomposition*. In a general way, the literature shows that the thermal decomposition of a polymer is an endothermic phenomenon, which occurs before the flaming ignition, in an interval of temperature bordered between 200 and 400°C. Beyond the decomposition temperature, i.e. when the energy contribution is increased to bond dissociation energy of macromolecules, an irreversible change begins. The thermal decomposition of a polymer is often initiated by the dissociation of chemical bond (covalent bond) to form radicals. Bond dissociation energy depends on the nature of the atom making up the bond and also of the

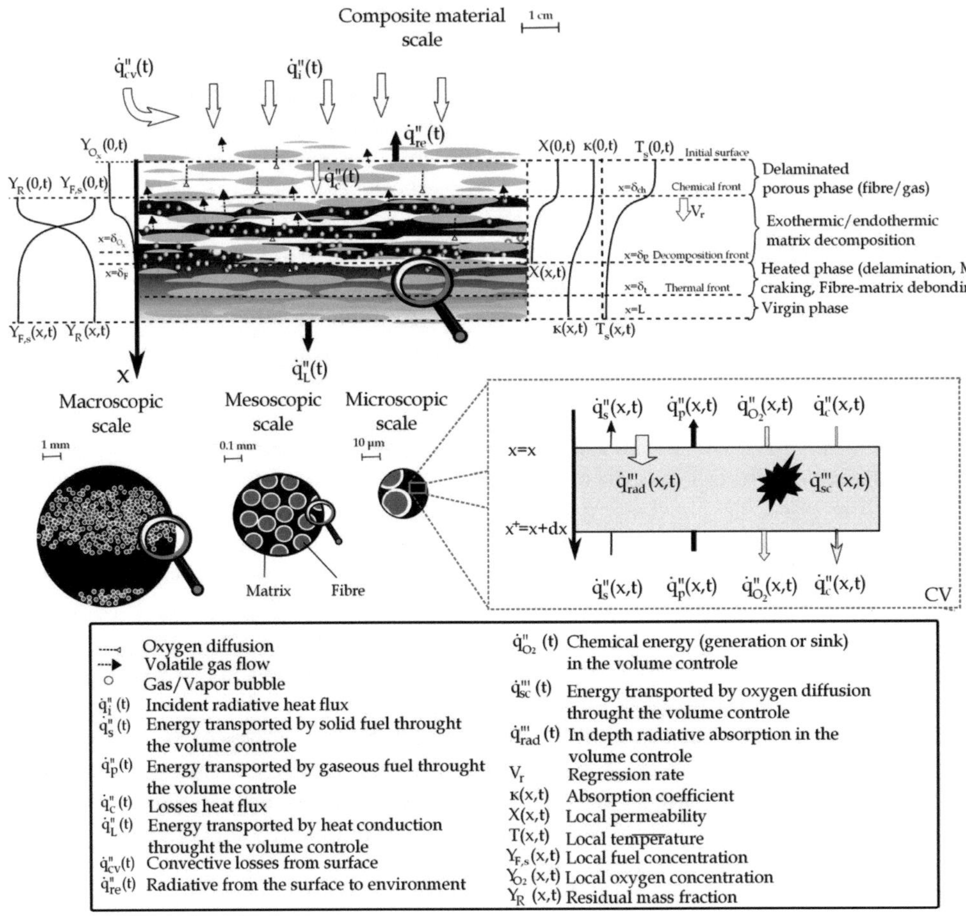

Fig. 2. Schematic of the reaction processes in the through thickness direction of a decomposing polymer composite prior to ignition induced by an external radiant heat source. Several scales of observation of a polymer reinforced fibres composites laminate. Volume control (VC) showing the main heat transfer mechanism in the composite material.

structure environment in which the bond occurs. This endothermic process, more collectively called "*thermal cracking*", results from the thermal excitement of atoms and depends naturally on the temperature and on the presence or not of oxygen. This phenomenon is random but depends on a macro-molecular chain. Four general mechanisms are important for the thermal decomposition of polymer matrix [Troitzsch (2004)] : (1) random chain-scission, in which the polymer backbone is randomly split into smaller fragments; (2) chain-end scission, called *unzipping*, in which the polymer depolymerises from the chain-ends; (3) elimination of pendant groups without breaking of the backbone; and (4) cross linking. In fact, only a few polymers decompose predominantly through one mechanism. Indeed, in many cases a combination of two or more mechanisms can coexist even if generally one of them prevails [Vovelle & Delfau (2000)]. The dominant reaction is generally the random chain scission. Matrix polymers that decompose by random scission and depolymerization are usually more flammable than polymers that cross-link or remove than groups.

As opposed to the thermal decomposition by anaerobic thermolysis, the thermo-oxidative decomposition occurs in the presence of oxygen. Bolland & Gee (1946) have tried to understand the phenomenon by studying the thermal-oxidation of polyolefin and rubbers. The mechanism they have developed could be applied successfully to a large number of polymers such as a polymer matrix of composite materials. The decomposition process is accelerated by an oxidant (such as air or oxygen). In a composite, the oxidative process occurs generally at the surface region. The low permeability of the composite and devolatilization from the decomposition zone impedes the ability of oxygen to diffuse into the condensed phase. Atmospheric oxygen does not have of a major influence on the process of in-depth decomposition because the front moves deep into the condensed phase. Therefore, decomposition tends to be driven in-depth only by the heat. The oxidation process leads on surface to the formation of a continuous "crust", whose thickness increases temporally to an asymptotic value. The position of this asymptote does not vary much with the temperature or the nature of the polymer matrix. In this layer called "*char*", the concentration of the products of oxidation is maximal on the surface and decreases gradually within the polymer. The formation of this layer can be accompanied by the appearance of superficial cracks, which is then deported to the decomposition process into a solid.
Devolatilization processes lead to the formation of gas fuel and toxic and corrosive volatile species (generally with aliphatic and aromatic hydrocarbons light as well as oxygenated compounds). The gas mixture with an oxidizer can be at the origin of the flaming ignition. This phenomenon will be detailed in this chapter.

2.3.2 Physical processes

2.3.2.1 Polymer matrix decomposition

The physical processes of laminate and sandwich composite in fire has been a topic of intensive investigation in recent years. A complete review is given by Mouritz & Gibson (2006) and Hirschler & Morgan (2008). The heating of a laminate composite is led by the surface irradiance level and the thermal diffusivity of the virgin material. The thermal diffusivity of the laminate is particularly low in the thickness direction because of the thermophysical properties of fibres. Thermal diffusion causes a degree of expansion within the matrix due to the thermo-mechanical strain [Grenier (1998)]. When the surface of a composite material reaches a sufficiently high temperature, the decomposition and devolatilization of the polymer matrix start. Polymer matrix degrades in the sequence of endothermic reaction that usually occurs by chain-scission depending on composition. Volatiles resulting from this scission have a cooling effect on the laminate as a whole and it results in a contraction of the composite. The pressure exerted by the trapped gases onto the soft complaint matrix can lead to matrix crack delamination, i.e. the separation of fibres plies which depend on the architecture of fibre. It was marked by an audible tearing sound. The delamination causes the formation of an unbounded interface between fibre plies, which produced a significant rise of the thermal resistance (with the presence of a gaseous barrier). The shielding effects of emerging voids can slow down the decomposition rate of the virgin condensed phase.
Before flaming ignition can occur, fuel polymer matrix needs to be transformed in the gas phase. Reaction rates associated with thermal decomposition can be considered negligible up to the decomposition temperature and therefore, the composite material will not undergo any chemical transformation but can have some physical changes. As the temperature increases,

reaction rates increase and the solid fuel starts changing. Given the temperature distribution within the material, the rates of decomposition are a function of the depth x. Fig. 2 shows a through-thickness view of a composite exposed to one side heating-up in a fire, and the damage is seen to occur in a different zone through the material:

1. the *decomposed porous phase* ($0 < x < \delta_{ch}$), which characterizes the region where the polymer matrix is fully decomposed. This phase composed of fibres and gas, acts as an insulator, inhibiting the heat conduction to the chemical phase ;
2. the *chemical phase* ($\delta_{ch} < x < \delta_p$), which defines the part of the polymer which produces the vast majority of irreversible chemical reaction ;
3. the *heated phase* ($\delta_p < x < \delta_t$), which is the depth to which a solid is heated enough to change its physical properties without deteriorating its chemical properties ;
4. the *condensed phase* ($\delta_t < x < L$), which designates a solid in its virgin state (non-thermally and physically altered).

The mass fraction of flammable gases present in the local productions of decomposition is described with the means of a single variable $Y_{F,s}(x,t)$. This function, which represents a global contribution of all products that can be further oxidized, varies in-depth and in time. It is minimal at the surface $Y_{F,s}(\delta_{ch},t)$. Some composites are highly permeable and allow to transport species in and out the solid. If the pyrolysis rate is small compared to the diffusion of oxygen, oxygen can diffuse in-depth up to $x = \delta_{O_x}$. In-depth oxygen distribution ($Y_{O_x}(x,t)$) reaches ambient values at the surface ($Y_{O_x}(\delta_{ch},t)$). For other composites, oxidation occurs on a small thickness close to the surface. The oxygen and fuel gas concentration are controlled by the permeability of the solid $\chi(x,t)$, the production/comsummation rate and indirectly from temperature distribution $T(x,t)$. Here, the permeability function is a combination of permeability, porosity and any damages that can occur within the composite material.
During the decomposition, the frame of reference $x = \delta_{ch}$ moves with a velocity V_r towards the unexposed surface as the fuel matrix is consumed and the surface regresses. The polymer content is negligible at the surface because the matrix has completely degraded and any residual organic component has condensed into char. The endothermic decomposition of the polymer matrix continues until the reaction zone reaches the rear-face of the laminate. The energy generated by the reaction is now fed back to continue the decomposition process. The gas fuel encounters the ambient oxidizer close to the solid surface and if the condition is sufficient, these gas mixtures can ignite.
All the polymer composites do not break down in a similar way [Drysdale (1998)]. As previously defined, the decomposition process depends on the chemical structure of the specified polymer, in particular on its polymerization degree.
For example, the progressive heating up of certain polymer matrix such as thermoplastic resin or thermoset resin (epoxy, vinylester, polyester) promotes the softening of polymer matrix without any irreversible change, and then the onset of nucleation sites resulting from the break down of molecular chains from the decomposition zone. The devolatilization promotes bubbling that congregate in cavities through this region and diffuse to the surface under the effect of pressure gradients and buoyancy. The presence of gas-phase oxygen reduces the viscosity of the molten polymer matrix, leading to a higher bubbling frequency and a less-violent bursting process. These molecules can escape and undergo further reactions to yield the actual species, usually hydrocarbons and hydroxy radicals, which can lead to a flammable mixture with the oxidizer. The in-depth degradation is going to depend partially,

but not only, on radiative properties of the polymer and more particularly on its emissivity and on its transmissivity. According to the faculty of the radiative property of composite, the reaction can occur in various depths. The holes formed by the bubbles at the surface allow oxidant to penetrate in a small thickness ($\sim 1\ mm$), thereby increasing the depth of the oxygen affect region beyond that, which is possible by diffusion. On the surface, The oxygen diffusion $Y_{O_x}(x,t)$ produces a thermo-oxydative reaction which leads to the formation of a viscous carbon residue [Pal & Mackásy (1991)], commonly called *tar*. This thin oxidized layer δ_{O_x}, strongly affects the process and kinetics of thermal decomposition [Brausman (1988), Kashiwagi (1994)].

Other polymers matrix such as phenolic resin lead to the formation of a solid carbonaceous residue known as *char*. Char is a highly porous material that consists in graphitic and amorphous region. It should be noted here that the proportion of fuel gases and char products depends on the matrix composite and the environment in which the devolatilization takes place (pressure, temperature, oxygen concentration). This char results in random chain scission reactions and cross-linking of the condensed phase. It reduces the heat transfer between the heat source and the virgin condensed phase. This phase limits the reaction rate [Kimoto & Tanaka (2004)]. The amount of char formed in a composite material is dependent on the chemical nature of the polymer matrix and, if present, organic fibres. During the carbonization of the solid, the pressure gradients generate the formation of cracks in the carbon layer. These allow repulsing the gases produced by the decomposition of the solid. Parker (1992), Levchik & Wilkie (2000) specify that these gases are going to cool down by convection of the residues whose temperature is much higher than gas.

The fire condition also has a large influence on the fire behaviour of composite material. Most experimental studies were performed under well-ventilated condition. However, the atmosphere of some fires can be substantially different for example in vitiated condition. In this cases, the lack of oxygen leads to a modification of decomposition by physical (e.g. flame displacement) and chemical (e.g. O_2 diffusion) ways. The temporal evolution of gaseous compounds and decomposition rate of the solid phase are altered by oxygen depletion, such as gaseous species. Hshieh & Beeson (1997) investigated the effect of oxygen concentration in a fire atmosphere on epoxy and phenolic material. It shows that the ignition time decreases slowly with increasing oxygen content and this is because more oxygen is available to react in the flame with the devolatilization products of thermoset matrix. Marquis et al. (2011) have shown for a thermoset-based sandwich composite that when the inlet oxygen level is close to 10 %, the decomposition rate of solid phase is halved compared to the ambient level. The slowest of the mass loss rate with O_2 depletion is the result of the reduction of oxygen diffusion processes in the solid phase. Below 10%, the influences on the decomposition and smoke production are negligible. So, devolatilization is purely a thermolysis process and the solid phase oxidation can then be neglected.

2.3.2.2 Fibre reinforcement

Various types of fibres can be used to reinforce polymer composite, most notably glass, carbon, aramid or less commonly extended chain polyethylene. Other type of reinforcement (bore, polyester, basalt, silica...) can be found but are not considered in this section. In a global way, fibres reinforcements are chemically inert in fire and retain chemical and physical stability at high temperatures. Fibre reinforcement and architecture do not contribute directly to the heat release but they can considerably modify the reaction-to-fire of polymer matrix. Le Bras et al. (1998) have studied the influence of fibre reinforcement on the fire reaction of composite. In fact, the increase of fibre content reduces the available quantity of fuel matrix in the composite material. Furthermore, it modifies the thermophysical properties of laminate and therefore the heating rate and decomposition rate. The architecture of fibre has also some influence on the thermal behaviour of composite.

The organic agents used as surface treatment of fibres (e.g. to improve adhesion with polymer matrix) are not inert with fire. Fibres are usually covered with a thin layer of organic sizing agent to provide chemical adhesion and specific properties (anti-static, abrasion resistance...). The quantities are very small and the heat release can be negligible compared to the combustion of polymer matrix.

2.3.3 Decomposition rate and reaction pathway of condensed phase.

The decomposition process depends on fuel and heating characteristics, which can distinctively lead to different reaction pathways. The mechanism can be a compendium of reaction that may be sequential or concomitant, and depend on the oxygen concentration. In fact, the thermal decomposition and devolatilization of a heated polymer matrix solid is the net result of hundred of reactions occurring simultaneously [Troitzsch (2004), Mouritz & Gibson (2006)]. Furthermore, exothermic and endothermic reactions can occur simultaneously, which complicates the analysis of thermal behaviour.

Various experimental techniques can be used to study and provide information on chemical reactivity and thermodynamics of a solid polymers matrix in temperature and chemical species. Experimental investigations are generally conducted through thermogravimetry (TG), differential scanning calorimetry (DSC), or differential thermal analysis (DTA). In each case, very small samples are used to reduce heat and mass transfer effects. These analysis of a testing procedure in which mass loss, heat absorption or evolution of the chemical changes of polymer matrix, are recorded as it is dynamically heated inside and over under controlled atmosphere and controlled temperature. It provides qualitative and quantitative information regarding the reaction taking place in a heated solid. Nevertheless, it is often necessary to use a combination of methods to understand the occurring processes. Some simplifications can be defined to hypothesize that the behaviour observed with measurement instruments is caused mainly by the mechanism of decomposition and is not very much influenced by external noise [Bustamante-Valencia (2009), Marquis et al. (2011)].

Generally, these researches are carried out for various heating rates β and two atmospheres (pure nitrogen and air). The basic nature of TG analysis requires an heating rate of 0.1 to 50 $K.min^{-1}$, which is generally slower than typical heating rates in fires (100-150 $K.min^{-1}$). Although these values are much lower than those observed in real fires, they were adopted to clearly separate the reactions. One difficulty associated with using the higher heating rate $\beta \geqslant 50 K.min^{-1}$ in TG is that the thermal lag between the sample temperature and the atmosphere increases with the heating rate, especially if the decomposition process is endothermic. Additional observations performed on the released gaseous compounds can be used to define decomposition mechanisms in accordance with the chemical process [Bustamante-Valencia (2009), Marquis (2010a)]. The measurement of gas release can be performed using a Fourier Transform Infra Red spectroscopy gas analyzer (FTIR) or mass spectroscopy (MS). It allowed the identification of chemical products released by each reaction in the solid phase.

Comprehensive and quantitative mechanisms can be defined using the experimental results coming from the data of the mass loss rate of the condensed phase, the kinetic of release of main gases and the measurement of thermal properties. The approach most commonly used to describe the decomposition of a solid is not to distinguish the various products of decomposition, but to gather them in "*pseudo-species*" and then group reactions according to their simultaneity. The thermal decomposition of polymer matrix can be implemented in the form of a comprehensible multiple-step reaction pathway with anaerobic thermolysis reactions and oxidation reactions. They are generated based on the hypothesis that each peak of the mass loss rate curves in thermogravimetry represents a reacting species in the solid phase [Ohlemiller (1985), Rein et al. (2006)]. As an example, a reduced comprehensible kinetic mechanisms for polyester resin was found by Marquis et al. (2011), as shown in Table 1. It was defined in accordance with the chemistry of the thermal process. The first reaction

Reactions	Reagents	Condensed products+gas products
Thermolysis	resin	$\rightarrow v_p$.char+$(1-v_p).[Y_k+Y_j+HCOH+CH_4]$
Oxidation	resin + O_2	$\rightarrow v_o$.tar+$(1-v_o).[Y_k+Y_j+HCOH+CH_4]$
Oxidation	tar + O_2	$\rightarrow v_{o,t}$.char+$(1-v_{o,t}).[Y_k+Y_j+HCOH+CH_4+CO+NO+HCl]$
Oxidation	char + O_2	$\rightarrow v_{o,c}$.ash+$(1-v_{o,c}).[Y_j+CO+NO+HCl]$

$Y_j = CO_2 + H_2O$; $Y_k = CH_3CH_2CHO + C_8H_8$; CO_2 : carbon dioxide; CO : carbon monoxide; H_2O Water; NO: Nitric oxide; HCl: hydrochloric acid; $HCOH$: Formaldehyde; CH_4: methane; CH_3CH_2CHO : Propionaldehyde; C_8H_8: styrene

Table 1. Comprehensible reaction pathway of polyester resin considering species of the solid and gas phases according to Marquis (2010a).

purely encompasses the thermolysis reaction, while the last three reactions include oxidation. Two intermediate products are formed from the initial decomposition of resin polyester, char and tar. The term "*Reagents*" represents the substance that is transformed during the reaction γ. The term "*condensed products*" corresponds to the condensed phase (liquid or solid) that remains in the sample holder at the end of the reaction. Gas effluents, listed in the column "*gas products*", are associated to each reaction γ.

2.3.4 Modelling the thermal decomposition rate

One of the most important factors contributing to the fire hazard of fuel solid composite is its thermal decomposition kinetics. Since the ignition of composite material is usually kinetically controlled, devolatilization is strongly influenced by thermal stability. Although techniques exist for measuring the rate constants of the elementary gas phase, analogous techniques have not been developed yet for kinetics in solid. Nonetheless, some fairly advanced diagnostic tools are used in the field of thermal analysis to investigate the kinetics and thermodynamics of decomposing solids. Since the middle of the 20th century, two main methods have been used to analyze and to model the thermal decomposition of solids in fire applications: the *model-fitting* (modelistic) method and the *isoconvertional* method. The former consists in selecting from a long list of models the one that best fits TG non-isothermal experimental curves. However, this method is applied to single-step reactions and usually various models may lead to indistinguishable fits of experimental data with crucially different Arrhenius parameters [Vyazovkin (2000)]. The latter method was introduced by Kissinger (1957) and improved by Friedman (1963). This method allowed the detection of multi-step decomposition processes, but numerical problems with the temperature integral term were often found. With the improvement of calculation power, a new multi-step decomposition method was proposed by Ohlemiller (1985). Basically, this method consists in predicting the change of mass in thermogravimetry while the kinetic parameters are calculated by numerical iteration. It is used by several authors [Rein et al. (2006), Matala (2008), Bustamante-Valencia (2009), Lautenberger (2009)] to create a numerical simulation of the decomposition of various materials, and by Marquis (2010a) on composite material.

Traditional methods of parameter estimations from TG data become inefficient or impossible to apply for a multiple-step reaction pathway, such as the one proposed here. The method of parameter estimation employed here is estimated from a mathematical model of the mass fraction equations observed in TG (MLR). A lumped model of the mass fraction processes is developed to simulate transient decomposition during TG experiment according to the above mechanism.

The kinetic model (finished kinetics) is based on the description of thermal decomposition in various stages. Each one of these reactions γ of the comprehensible reaction pathway has an Arrhenius reaction rate $\dot{\omega}_\gamma$ (s^{-1}) defined in Eq. 1.

$$\dot{\omega} = A\, Y_{Ox}^{m}\, Y_{F,s}^{n}\, e^{-E/RT} \tag{1}$$

where Y_F, s is the solid fuel mass fraction, Y_{Ox} is the oxidant mass fraction and the superscript m and n are constants. E is the activation energy ($J.mol^{-1}$), A is the frequency factor or pre-exponential factor (s^{-1}), T is the temperature (K) and R is the universal constant of gases (8,314 $J.K^{-1}.mol^{-1}$). Eq. 1 represents at the moment the most successful model. One shortcoming of using this equation to model the decomposition of thick solids is that the reaction rate within the solid should depend on the local oxygen concentration within the decomposing solid, not only at the surface. However, modelling the penetration of oxygen into a solid to determine the local oxygen concentration is a difficult task. Furthermore, it should be highlighted that there is some debate regarding the interpretation of the kinetic parameter (A E, n and v) as well as the physical trueness of modelling the rate constant using an Arrhenius form. The validity of using the law to treat a heterogeneous reaction has been widely questioned in recent years. In spite of a lot of reactions taking place, these parameters do not have much physical significance. Though the Arrhenius function is often used because it describes the processes remarkably well, the kinetic parameter is nowadays seen as "statistical values". They are dependent of the model and experimental conditions. So, results from two different models are rarely comparable between them (e.g. from the literature). Therefore, it is not illogical to obtain several sets of data for the same reaction.

The mass loss rate (or the time derivatives of mass fraction) of each condensed products j, is expressed as the product of the reaction rate $\dot{\omega}_\gamma$ and the residual mass fraction Y_{R_γ} (where $0 \leq Y_{R_\gamma} \leq 1$).

$$\frac{d}{dt} Y_j = Y_{R_\gamma} \dot{\omega}_\gamma \tag{2}$$

The total mass remaining in the TG sample holder at a time t, [See Eq. 3] can be obtained by summation of the individual condensed products j (solid or liquid) mass losses following Eq. 2.

$$\frac{d}{dt} Y = \sum_{j=1}^{M} \frac{d}{dt} Y_j = \sum_{j=1}^{M} \sum_{\gamma=1}^{N} (Y_{R_{j,\gamma}} - 1) \dot{\omega}_{j,\gamma} \tag{3}$$

The solution to this system of ordinary differential equations requires a stiff numerical solver. The unknowns (kinetic parameters) in the mass balance presented in Eq. 1 are: A_γ, E_γ, n_γ and Y_{R_γ} ($v = 1...N$). The optimum kinetic parameters are the ones that best fit the experimental curves (mass and mass loss rate from TG).

In fact, the current trend for the optimization of kinetic parameters from mass loss rate dY/dt is addressed using an inverse approach, called "*genetic algorithm*"(GA). This indirect method is based on the principles of Darwinian evolution. It is used for the minimization of the differential and integral form of the objective function simultaneously taking into account the difference between the mass-loss model prediction and mass loss measured for all heating rates data. The GA method has lots of advantages such as the ability to treat highly non-linear problems and search spaces that have a high dimensionality [Rein et al. (2006), Houck & Joines (1995)]. However, a major disadvantage of the inverse problems is that the uniqueness of the solution (the kinetic-parameter set) cannot be guaranteed because very complex physical processes are being simulated with a quite simple mass-loss model. For these reasons, an efficient heuristic method, such as genetic algorithms, is therefore necessary, from predefined ranges. The best fitness is obtained in a particular predefined range of input parameters. The uniqueness of the solution is a concern in inverse problem. To resolve the issue, the approach used is to verify that the parameters are applicable at several heating rates. For example, Fig. 3 provides a comparison between the experimental and calculated measurement in nitrogen and air for polyester resin at several heating rates (10, 20 and 50 $K.min^{-1}$).

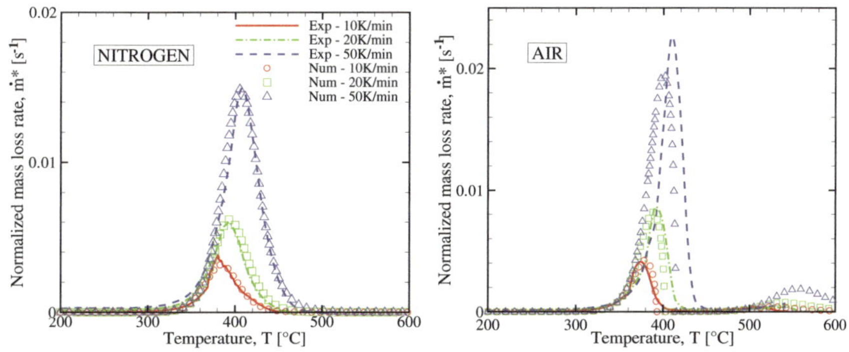

Fig. 3. Solid phase processes of polyester matrix during thermal decomposition for various heating rates β (10, 20 and 50 $K.min^{-1}$) and two atmospheres (Nitrogen and air). a) Plots of mass loss. b) Plots of mass loss rate. [Marquis et al. (2010b)]

An interesting discussion should take place among the scientific community of fire in order answer to the following questions: Is it useful to keep on devoting time and efforts to fit numerically curves by using parameters that do not have a physical meaning and that have a very uncertain application domain? How could the models be improved in order be able to use the experimental results? What is the relevance of these kinetics parameters at another scale?

2.4 Modelling physical transport processes in solid phase during thermal decomposition

Modelling the chemical reaction and transport processes taking place during the thermal decomposition of the polymer solid phase is very difficult because of the lack of knowledge on physical and chemical processes occurring. Most of the pyrolysis models developed for the polymer and composite material are extensions or modifications of those established for heterogeneous polymers like cellulose, most notably by Kung (1972) and Kansa & Perlee (1977). Many of the processes that occur in burning wood are similar to those described for composite material. Since then, these models have been adapted for modelling the heat and mass transfer through fibre-polymer composite material in fire. Several researchers [Tant et al. (1985), Henderson & Doherty (1987a), Henderson & Wiebelt (1987b), Florio et al. (1989), Sullivan & Salamon (1992a), Sullivan & Salamon (1992b), Perring et al. (1980), Mc Manus & Springer (1992a), Mc Manus & Springer (1992b), Dimitrienko (1995), Davies et al. (1995a)Dimitrienko (1997), Milke & Vizzini (1991), Gibson et al. (1995),Davies & Wang (n.d.), Looyed et al. (1997), Gibson et al. (2004), Lattimer & Ouelette (2006), Trelles & Lattimer (2007)] have developed different approaches but the most influential work in this area came from Henderson et al. (1985). Research on the effects of fire on sandwich composite has used the same principle as single monolithic composite in terms of modelling [Davies et al. (1995a), Davies et al. (1995b), Looyed et al. (2001), Krysl et al. (2004), Galbano et al. (n.d.), Marquis (2010a)]. In all of these cases, some assumptions are made. The main hypothesis is that the thermal decomposition was modelled using a Arrhenius law.

The simplest of these models considers that the pyrolysis occurs at the surface (as it is the case for some thermoplastic polymers) whereas the most detailed considers that it is about a volume, which degrades, taking into account the heat diffusion within the solid matrix. It is particularly the case when we study materials, which generate chars, where gas diffusion processes are important. It is then advisable to characterize the thermal transfers, the oxygen diffusion as well as the degradation kinetics and char formation, which are all relevant parameters. The main difference between the models is that the kinetics of thermal decomposition are considered as infinitely fast. In addition, for reasons of time saving and simplification of calculation, a large majority of these models is based on a kinetic description of pyrolysis, using a single global reaction, but not taking into account all stages of decomposition. The strong nuances between the various mechanisms are the number of reactions proposed, the order of reaction, the number of species within the condensed phase, the description of phenomena and diffusive transport as well as taking into account or not the effective thermal properties of materials. Another limit of these models is not taking into account the influence of oxygen concentration, yet very influential kinetic parameter [Kashiwagi & Ohlemiller (1982), Kashiwagi et al. (1987), Moghtaderi et al. (1996), Dakka et al. (2002)].

Despite these assumptions, models are able to predict a lot a phenomena. Nevertheless, models with a comprehensible treatment of the dynamics of both heat and mass transfer should be developed together with experimental measurements to improve the understanding of the decomposition behaviour of polymer based composite. The next section provide a comprehensible mathematical pyrolysis model for composite materials.

2.4.1 Global comprehensible pyrolysis model

In order to determine the fuel production, it is necessary to define the evolution of the temperature inside the solid fuel. This can be achieved by defining a comprehensive energy equation. The key challenge to the modelling of composite in fire is the complexity of the processes which controls the devolatilization:

- thermal processes (heat conduction, gas diffusion, exo/endothermic reaction, radiation, convection [see Fig. 2]);
- chemical processes (softening, melting, pyrolysis, char formation, oxidation).
- physical processes (thermal expansion, internal pressure build up, vaporization of moisture, delamination damage, matrix cracking, surface ablation and fibres reaction).

Modelling underlies understanding and analysing all these processes as accurately as possible. It is further complicated because many processes are intertwined and do not occur independently of each other. Understanding the way these processes interact is essential to analyse and model the fire behaviour of composite materials.

The study of these phenomena requires knowledge of transfer properties (mass, momentum, energy). As for all the heterogeneous multi-phase systems, these properties obviously depend on the morphology of the matrix and phenomena that develop and interact in the various phases. The approach that is usually used to simulate thermal decomposition is inspired by the works of Whitaker (1977). This one is directly inspired by methods that are traditionally used in continuum mechanics to search for expressions of local conservation laws. The formalism of Whitaker (1977) is based on a "*multiscale homogenization approach*" of the physical equations relative to the microscopic level of each phase (solid matrix, fibre, gas/vapor, moisture) present in the composite material. This approach puts into practice a change of scale, as shown in Fig 2. To simplify the problem, the decomposition process is usually described in a one-dimensional semi-infinite slab with coordinate x for the depth. To determine the pyrolyzate production, it is necessary to solve the evolution of the temperature inside the solid fuel. External radiant heat flux $\dot{q}_i''|_{x=0}$ is applied to the top surface at $x = 0$. The temperature of the solid, initially at temperature T_0, increases as the heat reaches the surface of the slab. In the model formulation that follows, a decomposing material is modelled as consisting of both a condensed phase (resin matrix, char, ash, etc.) and a gas phase (pyrolyzate, oxidant, nitrogen, etc.). Processes in the gas phase, the mixing with the oxidizer and the formation of a flame are not considered. Given the complicated character of the thermal response of a composite material, the mathematical description also uses further assumptions: (1) Pyrolysis kinetics is modeled by a single-step first-order reaction; (2) the composite material forming the skins consists of the volumetric fractions X_f (glass fibres), X_p (polymer), X_m (moisture), and X_g (gas and/or vapours); (3) the effective density of each condensed-phase component of the panel is constant whereas the corresponding volumetric fraction varies to take into account chemical and physical transformations occurring during the exposition to fire; (4) the total volume occupied by each skin does not change as a consequence of thermal decomposition and combustion of the polymeric resin; (5) the solid, liquid and gas/vapour phases are in a local thermal equilibrium; (6) the transport by diffusion of volatile species is small; (7) the transport by convection and diffusion of the liquid-phase moisture is negligible; (8) the gas/vapour mixture obeys to the ideal gas law.

By applying these assumptions and the formalism of Whitaker (1977), the local energy balance on Figure 2 can be reduced to a single equation following the approaches of Kung (1972) and Kansa & Perlee (1977) [Perring et al. (1980), Henderson et al. (1983), Florio et al. (1989), Gibson et al. (2004) Mouritz & Gibson (2006), and Galbano et al. (2009). Estimation of the net heat transfer into a composite material leads to a change in energy accumulation E_{CV} within a control volume CV. [Torero (2008)]. The global comprehensible pyrolysis model can be summarized by Eq. 4:

$$\begin{aligned}\frac{\partial E_{CV}}{\partial t} &= \left[\dot{q}_s''(x^+,t) + \dot{q}_p''(x^+,t) + \dot{q}_{O_2}''(x,t) + \dot{q}_c''(x,t)\right] \\ &- \left[\dot{q}_s''(x,t) + \dot{q}_p''(x,t) + \dot{q}_{O_2}''(x^+,t) + \dot{q}_c''(x^+,t)\right] \\ &+ \dot{q}_{rad}'''(x,t)dx + \dot{q}_{sc}'''(x,t)dx\end{aligned} \tag{4}$$

where $E_{CV} = h_{CV}(x,t)\rho_{CV}(x,t)dx$. h and ρ denote respectively the variation of enthalpy and the density within the control volume. The gaseous species of devolatilization ($\dot{q}_p''$) and oxygen diffusion into the solid ($\dot{q}_{O_2}''$) carry energy in and out the control volume. The mass flow of these gases ($\dot{m}_p''$ and $\dot{m}_{O_2}''$) incorporates the regressions rate. The conductive heat flux is defined in the control volume by ($\dot{q}_c''$). The terms $\dot{q}_{rad}'''$ and $\dot{q}_{sc}'''$ are respectively the in-depth radiative absorption and the chemical energy (generation or sink).
Usually, the considered system consists of a gas phase, a moisture phase, a solid phase (matrix + fibres) and a char. The global comprehensible pyrolysis model takes into account the highly unsteady character of the process, solid and gaseous rise of enthalpy, devolatilization and combustion of the polymeric component of the laminate composite both including finite-rate kinetics, heat transfer by convection and conduction, convective mass transfer, variation of component volumetric fractions, moisture evaporation, surface ablation of the solid residue left after the conversion of the polymeric resin as soon as the temperature reaches a critical value, heat transfer (convection and radiation), and variable properties. Sub-models are developed for the chemical and physical processes with intrinsic values for all the properties. Detailed information about the mathematical description of the various processes and a list of property values are provided below. The unsteady enthalpy conservation equation of the system is written as:

$$\begin{aligned}&\frac{\partial}{\partial t}\left[X_p\rho_p h_p + X_c\rho_c h_c + X_f\rho_f h_f + X_m\rho_m h_m + X_g\rho_g h_g\right] \\ &= \frac{\partial}{\partial x}\left(k_{s,net}\frac{\partial T}{\partial x}\right) - \dot{\omega}_m'''\Delta h_v + \sum_{\gamma=1}^{N}\dot{\omega}_\gamma'''\Delta h_{r_\gamma} \\ &+ \frac{\partial}{\partial x}\left[\sum_{j=1}^{M}\dot{m}_g''h_g\right] - \frac{\partial}{\partial x}\left[\dot{m}_{O_x}''h_{O_x}\right] \\ &+ \frac{\partial(\rho_s V_r C_{p_s} T)}{\partial x} + \dot{q}_{rad}'''\end{aligned} \tag{5}$$

This equation is coupled with the mass conservation equations and the Darcy equation, becoming a stiff system of differential equations. $X_i(x,t)$ denotes the volume fraction of element i, $k_{i,net}(x,t)$ the thermal conductivity of element i, $C_{p_i}(x,t)$ the specific heat of element i ($W.m^{-1}.K^{-1}$), $\rho_i(x,t)$ the density of the element i ($kg.m^{-3}$), $V_r(t)$ the regression rate of polymer matrix, $\dot{m}''$ the mass flow ($kg.s^{-1}.m^{-2}$), ε the radiative properties of material, h the specific enthalpy ($J.kg^{-1}$), Δh_v the latent heat of water vaporization ($J.kg^{-1}$) and Δh_{r_γ} the heat of reaction γ ($J.kg^{-1}$). Phase changes are incorporated into the heat sinks where some rate function is defined to describe the conversion from one phase to another. The subscripts m, c, f, g, O_x, p and s are used to denote the moisture, the char, the fibres, the gas and vapour, the oxidizer, the polymer matrix, and the solid respectively.

The boundary condition to the top surface ($x = 0$) is a balance between incoming thermal radiation $\dot{q}_i''$ and heat losses by radiant emission $\dot{q}_{re}''$ and Newtonian cooling $\dot{q}_{cv}''$. It is possible to specify (through the value of the radiative absorption κ for each condensed phase species) whether radiation is absorbed at the surface or in-depth. If $\bar{\kappa}|_{x=0} \neq \infty$, radiation is absorbed volumetrically in-depth. The boundary conditions on the condensed phase energy equation are:
for $x = 0$

$$-k_s \frac{\partial T}{\partial x}\bigg|_{x=0^+} = \varepsilon_s \dot{q}_i'' - \varepsilon_s \sigma \left(T^4(0,t) - T_0^4\right) - h_{cv}(t)\left(T(0,t) - T_\infty\right) \qquad (if\ \bar{\kappa}|_{x=0} = \infty) \quad (6)$$

$$-k_s \frac{\partial T}{\partial x}\bigg|_{x=0^+} = -\varepsilon_s \sigma \left(T^4(0,t) - T_0^4\right) - h_{cv}(t)\left(T(0,t) - T_\infty\right) \qquad (if\ \bar{\kappa}|_{x=0} \neq \infty) \quad (7)$$

for $x \rightarrow L$:

$$0 = -k_s \frac{\partial T}{\partial x}\bigg|_{x=L^-} \quad (8)$$

The calculation of the total mass loss rate (sum of the reactions γ) can be evaluated from a local point of view, by the expression [Torero (2008)]:

$$\dot{m}_p'''(x,t) = Y_R(x,t) \sum_{j=1}^{M} \sum_{j\gamma=1}^{N} \dot{\omega}_{j\gamma}''' = Y_R(x,t) \sum_{j=1}^{M} \sum_{j\gamma=1}^{N} \left[A_{j\gamma} Y_{0_x}^{m_{j\gamma}}(x,t) Y_{F,s}^{n_{j\gamma}}(x,t) e^{-E_{j\gamma}/RT(x,t)} \right] \quad (9)$$

It is important to note that fuel produced in-depth does not necessarily reach the surface and, in many cases, that pressure increases within the fuel structure can be observed. The effects of permeability and pressure are combined in a complex manner to define the flow within the composite material. That actually remains an unresolved problem. Furthermore, the use of a simple variable such as χ can be justified to take into account these phenomena. The total fuel production at the surface per unit area can be defined by integrating equation 9 across the thickness L including the permeability function described above:

$$\dot{m}_p''(x,t) = \int_0^L \chi(x,t) \left[Y_R(x,t) \sum_{j=1}^{M} \sum_{\gamma=1}^{N} \left(A_{j,\gamma} Y_{0_x}^{m_{j\gamma}}(x,t) Y_{F,s}^{n_{j\gamma}}(x,t) e^{-E_{j\gamma}/RT(x,t)} \right) \right] dx \quad (10)$$

where $Y_{F,s}(x,t)$ denotes the local fuel mass fraction, $Y_{Ox}(x,t)$ the local oxidant concentration, $Y_R(x,t)$ the residual fuel fraction, $\chi(x,t)$ is the local permeability, $T(x,t)$ the local temperature and $A_{j\gamma}$, $E_{j\gamma}$, $n_{j\gamma}$ and $m_{j\gamma}$ are the constants kinetics. Throughout this work, the index j is used to denote the condensed product j and the index γ is used to denote the chemical reaction γ.

2.4.2 Thermophysical properties of composites under decomposition

An accurate prediction of thermal decomposition of composite in fire needs to register a large amount of data on the thermo-physical properties of each element constituting the material on the temperature range interest. These properties need to be quantified over the temperatures expected range, which is not easy. For the current simulations, this entails determining the density $\rho_j(T)$, the thermal conductivity $k_j(T)$ and the specific heat capacity $C_{p,j}(T)$ and the gas

permeability χ_j of both virgin condensed-phase species and fully decomposed material, the heats of reaction Δh_{r_γ}, the effective heat of combustion(Δh_c), radiative properties (emissivity ε and absorption coefficient κ) etc. The properties change with temperature and it is necessary to consider the temperature dependence in the analysis.
These properties depend on the weaving style, the characteristic and the density of the different components of other factor introduced during weaving, such as crimp angle, make the prediction very fastidious. In order to achieve the prediction, the conventional method to determine the thermophysical properties of composite material is based on the *multiscale homogenization approach* [Fig 2]. Most methods for modelling the thermophysical properties of composite materials use the concept of homogenization whose purpose will be to link the local thermophysical properties to the global effective thermophysical properties. The large numbers of models vary in mathematic complexity depending on the type of laminate being analyzed. More sophisticated models consider the influence of fibre architecture [Ott (1981)] whereas simple model determine the equivalent thermophysical properties in two direction [Henderson et al. (1985)] : one in the fiber direction called *longitudinal* or *harmonic* and the other perpendicular to the fiber called *transverse* of *Arithmetic*.
The change of thermophysical properties (thermal conductivity and specific heat) in temperature for the laminate and the char is often expressed using empirical curve fit polynomial equations [Mouritz & Gibson (2006)].

$$\phi_v = \phi_0 + \phi_1.T \qquad (11)$$

where T is the temperature and ϕ is used to denote the specified property, such as thermal conductivity or specific heat. A complete review is given by Ott (1981), Henderson et al. (1985), Gibson et al. (2004) and Mouritz & Gibson (2006). The temperature dependence of thermal conductivity can also be expressed as a linear combination of the virgin material and char conductivities, weighed by the ratio of the solid material mass to the initial virgin mass or final char density [Henderson et al. (1985)]. These models are not suited for all types of composite material and more complex models are needed to consider the influence of fibre architecture. Furthermore, the prediction capabilities of these models have not been rigorously evaluated, particularly due to the difficulties associated with accurately measuring the temperature dependent thermal conductivity of reagents such as char. It is important to note that the thermal properties of composite may change with the evolution of fire induced damage. For example, the delamination of the virgin phase can reduce the thermal conductivity, slowing the heat transport by creating an air gap between debonded ply layers. Most thermal models ignore this effect of delamination, which leads to an over-estimation of the temperature in the virgin material shielded by the cracks.
In principle, the required properties could be measured with multiple specialized small-scale tests but this approach is not realistic for most practitioners and researchers, due to the expense associated with running multiple specialized tests. For these reasons, Lautenberger (2009) used another approach and estimated the physical properties of condensed phase species from an optimization routine. Nevertheless, some preliminary tests of this code have shown that it costs a lot of time in terms of calculation and generates results that are difficult to interpret. Furthermore, it increases the number of optimized input parameters, which can affect the quality of numerical results, leading to non-physical but mathematical solutions.

2.5 Flaming ignition of laminate composite

The sequence of events leading to the ignition of a gas phase flame is described in this section. It is assumed that gaseous fuel emerges from the heating and the decomposition of the polymer matrix following the description provided in the previous section. The terms of flaming ignition, describe the flame appearance in the immediate neighbourhood vicinity of a polymer. In fact, polymer matrix does not ignite. Only gas burn. To ignite the polymer-based composite, it is necessary to be able to raise the temperature of a composite surface until it exists a volatile flammable mixture of species and oxygen on surface. In other words, the gas pressure must correspond to the lower flammable limit (LFL). If the fuel gas concentration from the surface of the material is insufficient, this ignition results only from a flash point. When the mass flow of product fuel gas is sufficient, a diffusion flame is then established persistently above the material. This condition is not sufficient and flaming requires local contributions of energy. This amount of energy required to start the flaming is called *activation energy*. It depends on the composition and architecture of the composite, pressure and temperature. Since it is made by an external source (flame or electric spark) the ignition is *called piloted*. The ignition can be *spontaneous*, also called *self-ignition*, when energy intake does not result from the exothermic oxidation reactions in the gas phase.

2.5.1 Inert solid assumption

The flaming ignition of a solid fuel in general and of a composite material in particular depends on the ability of heat to diffuse into the solid. To predict flaming ignition, it is necessary to solve the heat equation when it is exposed to a radiant heat flux on the surface. The ignition of polymer and composite material has been successfully modelled using semi-infinite solid heat transfer models, integral heat transfer models and finite difference models. Extensive reviews of these modeling efforts can be found in Kashiwagi (1994), Di Blasi (2000), Moghtaderi (2006), Lautenberger & Fernandez-Pello (2004) and Torero (2008). Consider the composite as a "*homogeneous semi-infinite solid*" initially at temperature T_0. It is exposed to a constant external radiative heat flux $\dot{q}''_i|_{x=0}$ in $x = 0$. For some composite, the velocity of the reference frame V_R is very small and can be assumed to be negligible ($V_r \simeq 0$). Assume that equivalent thermo-physical properties of composite material are considered constants ($\rho_s(x,t) \simeq \overline{\rho}_s$, $k_s(x,t) \simeq \overline{k}_s$, $C_{p,s}(x,t) \simeq \overline{C}_{p,s}$) up to the decomposition temperature and the local permeability is negligible ($\chi \simeq 0$ so $\dot{m}''_O \sim 0$). However, this assumption is questionable between the degradation temperature ($T \sim 100°C$) and the temperature of thermal decomposition ($T \sim 200°C$). In reality, thermophysical properties vary with temperature but a global set of equivalent properties can be established to give a good estimation of ignition. To simplify the problem, the solid is considered as inert until ignition ($\dot{m}''_p = 0$). As a result of the assumption, the energy balance is dramatically simplified. Despite the fact that this limitation is not justified, it still is the backbone of all standard test method analyses for ignition. These assumptions allow to reduce the study to a simple problem of one dimension conductive heat flux into the solid. Eq. 5 can be reduced to:

$$\rho_s C_{p,s} \frac{\partial T_s}{\partial t} = \frac{\partial}{\partial x}\left(k_s \frac{\partial T}{\partial x}\right) + \dot{q}'''_r \qquad (12)$$

This last term is justified because composite materials are semi-transparent and in-depth radiative absorption is observed. The subscript s define the solid composite. The boundary conditions are given in $x = 0$ by Eq. 6 and 7 in neglecting the term $\dot{q}''_i$, while in $x \to L$, it is defined by:

$$0 = -k_s \frac{\partial T}{\partial x}\bigg|_{x=L^-} + k_L \frac{\partial T_L}{\partial x}\bigg|_{x=L^+} \qquad (13)$$

where the subscripts L denoted the back face.
If the composite is considered as a solid opaque to radiation whatever the wavelength is, the most of incident irradiance level is absorbed at the surface ($a = \varepsilon = 1$ following the Kirchhoff law where a is the absportivity and ε, the thermal emissivity). Therefore, the in-depth radiative absorption $\dot{q}_r'''$ is negligible in Eq. 12 and the boundary conditions are given by Eq. 6 and 8. This hypothesis is more critical. Small justifications exist in the literature to support these assumptions for composite material. In the case of composite material, the absorptivity (or emissivity) can approach a unit in the visible spectrum.

2.5.2 Thermal behaviour of composite material

One of the critical factors influencing the flaming ignition of combustible materials such as polymer-based composite is the thermal thickness. Actually, the heat wave velocity into the material depends on the thermal effusivity and thickness of the solid. A thin solid can ignite within a very short time when it is exposed to a hot fire because its temperature quickly tends to become uniform due to his weak thermal effusivity. On the other hand, a thick solid is going to tend to heat diffusion by conduction within the condensed phase, which limits the rise of its surface temperature, and consequently its ignition. Consequently, the ignition of a solid is strongly conditioned by its thermal behaviour.
A solid can be defined as *thermally thin* when the thermal wave is absorbed so rapidly that there is no significant temperature gradient (quasi-homogeneous temperature) through it. In contrast, in a *thermally thick* material, a significant thermal gradient exist through the solid. For a homogeneous material exposed to a radiant heat flux, a simple criterion based on the radiant modified Biot number (Bi_r) can be used to establish if the material is thermally thin or thick. $Bi_r = h_r L.k_s^{-1}$ where h_r is the radiative exchange coefficient and k_s the thermal conductivity of the specimen. This approach can however not be used for inhomogeneous material such as the composite material. Another approach is therefore to assess the characteristic depth of thermal diffusion $\delta_t(t)$. It is defined by $\delta_t(t) > A\sqrt{\alpha_s t}$ where α denotes the thermal diffusivity ($m.s^{-2}$). A is a constant (between 1.13 and 4) which evolves from one author to another [Dusimberre (1998), Dembsey & Jacoby (2000)]. This equation is only valid for simple radiative thermal ignition and does not consider heat loss by emission $\dot{q}_{re}''$ which depends on the surface emissivity. This approach considers that a solid is thermally thick if $\delta_t(t)$ does not exceed its sample thickness L (m). In practice this means that most composite material are thermally thin when less than $\sim 1 - 2\,mm$ thick.

2.5.3 Ignition properties

The ignitability of polymer material can be characterized by an empirical parameter, often called improperly "*ignition temperature*" [Simms (1963) Martin (1965)]. This temperature is the lowest temperature at which the surface must be carried to ignite. This parameter is determined through standardized tests or when not using an external energy source. These temperatures depend largely on the size and shape of the specimen, the ambient environment (pressure, concentration of oxygen, ventilation), the heat source, etc. For this reason, the published data relative to the controlled temperature or spontaneous temperature can vary significantly from one document to another. If the sample is suddenly exposed to an external radiant heat flux, then the time delay between exposure and ignition is named the "*ignition delay time*", t_{ig}. The heating rate directly depends on the "*thermal effusivity*" b of the solid, i.e. its ability to exchange heat energy with its environment. A final link can be made to establish a critical ignition condition. If the ignition delay time is infinitely long, there will be no temperature gradients within the composite material. Surface heat losses will then be equivalent to the external incident heat flux. This represents the minimum heat flux required to achieve T_{ig}, and thus flaming ignition of the solid fuel. This heat flux is named the "*minimum heat flux for ignition*", $\dot{q}''_{min}$. These extrinsic parameters of composite material represent then the entire process of ignition. They can be experimentally measured or defined by theoretical mathematics from the simplifications made in this section. Among the work done on the subject, those of Quintiere & Harkleroad (1985), Delichatsios et al. (1991), Janssens (1993), Hopkins & Quintiere (1996), Moghtaderi et al. (1997), and Tewarson (1995), Spearpoint & Quintiere (2001) and Babrauskas (2003) should be highlighted.

2.5.3.1 Surface temperature

Consider a semi-infinite solid, initially at T_0 and exposed to a radiant heat flux $\dot{q}''_i|_{x=0}$. The surface temperature for a constant property inert solid that is opaque to thermal radiation takes different forms as shown in Table 2 [Lautenberger & Fernandez-Pello (2004)]. These solutions are valid only if the thermophysical properties do not vary with temperature. Although this approximation is usually made, the properties of real composite material generally vary with temperature. The ignition temperature can be evaluated by fixing $x = 0$ and $T(0,t) = T_s = T_{ig}$ when $t = t_{ig}$.

	BOUNDARY CONDITIONS	SOLUTIONS
A	$-k\frac{\partial T}{\partial x}\Big|_{x=0} = \dot{q}''_i$ $T\Big|_{x\to\infty} \to T_0$	$T(x,t) = T_0 + \dot{q}''_i.\frac{\delta}{k}\left[\frac{1}{\sqrt{\pi}}.e^{(-x/u)^2} - \frac{x}{u}\text{erfc}\left(\frac{x}{u}\right)\right]$
B	$-k\frac{\partial T}{\partial x}\Big|_{x=0} = \dot{q}''_i - h_{ig}(T-T_0)$ $T\Big|_{x\to\infty} \to T_0$	$T(x,t) = T_0 + \frac{\dot{q}''_i}{h_{ig}}\left[\text{erfc}\left(\frac{x}{u}\right) - \text{e}^{[\frac{x}{k/h_{ig}} + \frac{t}{t_c}]}\text{erfc}\left(\frac{x}{u} + \sqrt{\frac{t}{t_c}}\right)\right]$
C	$-k\frac{\partial T}{\partial x}\Big|_{x=0} = \dot{q}''_i$ $\frac{\partial T}{\partial x}\Big|_{x=L} = 0$	$T(x,t) = T_0 + \frac{\dot{q}''_i}{u/k}\sum_{n=0}^{\infty}\left[\text{ierfc}\left(\frac{2n+x/L}{u/L}\right) + \text{ierfc}\left(\frac{2(n+1)-x/L}{u/L}\right)\right]$
D	$-k\frac{\partial T}{\partial x}\Big|_{x=0} = \dot{q}''_i - h_{ig}(T-T_0)$ $\frac{\partial T}{\partial x}\Big|_{x=L} = 0$	$T(x,t) = T_0 + \frac{\dot{q}''_i}{h_{ig}}\left[1 - \sum_{n=1}^{\infty}\left(\frac{4\sin(\lambda_n L)}{2\lambda_n L + sin(2\lambda_n L)}\cos(\lambda_n(L-x))e^{(-\frac{k}{\rho c}\lambda_n^2 t)}\right)\right]$

where $u = \sqrt{4\alpha t}$, $t_c = \frac{k\rho c}{h_{ig}^2}$ and $\text{ierfc}(x) = \frac{e^{-x^2}}{\sqrt{\pi}} - x.\text{erfc}(x)$

A : Semi-infinite solid exposed to a net radiant heat flux (no surface heat losses).

B : Semi-infinite solid exposed to a net radiant heat flux with surface heat losses by convection on the surface (the radiative heat losses are linearized into the convective term h_{ig}).

C : Solid of thickness L exposed to a net radiant heat flux (no surface losses) and perfectly insulated at back face.

D : Solid of thickness L exposed to a net radiant heat flux with surface heat losses by convection (the radiative heat losses are linearized into the convective term h_{ig}).

Table 2. Solutions of the heat equation

2.5.3.2 Ignition delay time

Ignition delay times are generally estimated experimentally because of the difficulty in modelling the ignition mechanism of composite material. Although all these theories have been proposed for calculating the time-to-ignition for combustibles materials, they have never been validated for polymer-based composite. It is difficult to theoretical model the flaming ignition of composite material due to the complex physical and chemical processes that occur during decomposition. When composite is exposed to a heat source, it can crack and delaminate. The complexity of these processes does not allow easy analytical predictions. Nevertheless ignition models proposed in this section have been found to give good estimation of the ignition delay time of some composite.

Usually, the ignition delay time t_{ig} can be defined as [Quintiere (2006)] :

$$t_{ig} = t_{ch} + t_r + t_d \tag{14}$$

where t_{ch} is the denoted time (s) needed for the flammable mixture to proceed to combustion once at the pilot; t_r is the diffusion time (s) needed for the flammable fuel concentration and oxidizer to reach the pilot; and t_d is the condition heating time (s) for the composite to achieve the decomposition temperature. Therefore, the for piloted ignition, it appears that $t_{ig} \approx t_d$ [Torero (2008)].

For high incident heat fluxes where the ignition temperature is attained very fast, thus t_{ig} is much less than the characteristic time $t_c = (k\rho C_p)/h_{ig}^2$, Carslaw & Jaeger (1959) defined analytically the ignition delay time of a thermally thick solid with the expression:

$$t_{ig} \approx \frac{\pi}{4} \frac{(k_s \rho_s C_{p,s})(T_{ig} - T_0)^2}{(\dot{q}_i'')^2} \tag{15}$$

This term reflects that the ignition delay time is independent of the total heat transfer term h_{ig} but depend on the incident heat flux $\dot{q}_i''$, the global thermophysical properties and the ignition temperature.

For low incident irradiance level, where $t_{ig} \geq t_d$, the ignition delay time is defined by:

$$t_{ig} \approx \frac{h_{ig}^2}{\pi k_s \rho_s C_{p,s}} \left[1 - \frac{(\dot{q}_i'')^2}{h_{ig}^2 (T_{ig} - T_\infty)^2} \right] \tag{16}$$

A very similar analysis can be conducted for thermally thin composite material:

$$t_{ig} \approx \frac{L \rho_s C_{p,s} (T_{ig} - T_0)}{\dot{q}_i''} \tag{17}$$

where L is the thickness. A comprehensible data review of this expression is provided by Babrauskas (2003). In this expression it is assumed that the heat losses are negligible and the subtract is an inert thermally thick and opaque solid. These equations have been proven to be accurate in the theoretical determination of the ignition delay time for thermal thick specimen such as wood and plastics but not polymer composite.

Various expressions derived from Carlaw et Jaeger, have been performed to predict the time-to-ignition [Bushman (n.d.), Abu Zaid (1988), Smith & Satija (1981), Grenier (1998), Mikkola & Wichman (1989), Shields et al. (1994), Mikkola (1992) and Babrauskas (2003)] indicate however that there is much confusion about how to present and interpret experimental data (see correlations Hallman (1971), Quintiere & Harkleroad (1985), Tewarson (1993)). It also states that the correlations derived from experimental data to determine the time-to-ignition must have a theoretical basis. In fact, the author advocates the use of correlations proposed by Abu-Zaid, Mikkola and Wichman or those proposed by Janssens (1993).

2.5.3.3 Ignition temperature and thermal effusivity

The minimum incident heat flux can be evaluated using the approach of Janssens (1993). It suggests to plot the transformed experimental ignition time ($t_{ig}^{0.55}$) versus the incident radiant heat flux $\dot{q}_i''$. This approach leads to a hyperbolic curve with the asymptote given the value of the minimum ignition energy of $\dot{q}_{min}''$. Below this value, the material does not ignite because the energy intake is too low. The minimum value of heat flux is an extrinsic property that depends on the material boundary conditions associated with heat transfer equations. Once $\dot{q}_{min}''$ is found, the ignition temperature T_{ig} can be obtained by an iterative approach in solving Eq. (9).

$$a_s\dot{q}''_{min} = h_{cv}(T_{ig} - T_0) + \varepsilon\sigma\left(T_{ig}^4(t) - T_0^4\right) \tag{18}$$

When the thermophysical properties are not known, it is possible to estimate the thermal effusivity b $(J.K{-}1.m^{-2}.s^{-1/2})$ for a thermally thick composite material from the slope a_{slope} of the line:

$$b = \sqrt{(a_{slope}\, 0.73\, \dot{q}''_{min}\, h_{ig}^{-1.1})} \tag{19}$$

and for a thermally thin material:

$$\rho c = \frac{1.03 h_{ig}}{L\, \dot{q}''_{min} a_{slope}} \tag{20}$$

From these observations, Tewarson (1993) introduces the thermal response parameter (TRP) concept $[kW.s^{0,5}.m^{-2}]$ which characterize the heating up of material.

$$TRP = (T_{ig} - T_0)\sqrt{b} \tag{21}$$

3. Chemical processes in gas phase

When the onset of thermal decomposition occurs, the matrix polymer of composite begins to devolatilize in very small quantities in-depth at $x = \delta_p$ on Fig. 2. Heating up the composite material leads to increase the fuel mass flux [Eq. 10]. The gas fuel encounters the ambient oxidizer close to the solid surface and if the condition is sufficient, this gas mixture can ignite. However, gas mixtures consisting of combustible, oxidizer, and inert gases are only flammable under certain conditions. Ignition of gas fuel is possible if and only if the fuel concentration are comprised between the lower flammable limit (LFL), i.e. the leanest mixture that still sustains a flame, and the upper flammability limit (UFL) which gives the richest flammable mixture.

As discussed in the previous section, this condition is not sufficient and flaming requires local contributions of energy. When the process includes an outside source (pilot flame or electric spark) the gas phase processes are strongly simplified and the influence of environmental variables are reduced. In these cases, ignition can be assumed at the moment where a LFL is attained at the local of the pilot. The process of auto-ignition is more complex and it is not practical to rely the auto-ignition to describe the susceptibility of composite material to ignite. It must be highlighted that in reality, most ignition scenarios can occur following the pilot flame process. To attain the LFL at the pilot location, it is necessary to resolve the momentum and mass transport equations simultaneously. This can be done using the surface boundary conditions explained above.

Once ignition has been achieved, a flame can propagate through the regions where a flammable mixture is present, consuming the reactants. Independent of the flow field, it is most likely that a flammable mixture is established close to the solid composite surface. Flaming combustion is a phenomenon of oxidation defined by a large number of irreversible chemical reactions in gaseous phase. The result is an exothermic reaction in the gas phase that leads to a self-sustained burning of the fuel. The exothermicity of reactions leads to a heat release resulting from the rupture of bonds between atoms of fuel and the creation of new molecules more stable chemically. This exothermic reaction can be represented by expressions similar to those in Eq 1 that could include several hundred different reaction steps in the gas phase. These reagents are disappearing to form products at a rate that depends on the concentrations of gaseous species, temperature and pressure according to very complicated mechanisms. The products of intermediate reactions are often unstable and prone to react with radicals, especially to oxidize. Under ideal conditions, the oxygen in the air combines with radical carbon and hydrogen to produce carbon monoxide and water. However, some radicals can react together to form polycyclic aromatic hydrocarbons. In flames, these compounds react with oxygen molecules so that small particles condense and agglomerate to form the *soot*.

When the flame is well established, the devolatilization rate determines the combustion condition: if a flame can continue to exist or if the combustion reaction ceases after the gas phase mixture is consumed. As discussed previously, the combustion and pyrolysis rate constitute an auto-catalytic process. The flame heat feedback enhances the devolatilization. If pyrolysis rates are not sufficient, the flame is extinguished but continuous pyrolysis leads once again to the formation of a flammable mixture and subsequent re-ignition. This manifests itself as a sequence of flashes that precede the establishment of a persistent flame over the combustible solid. This process is generally associated to "flash point" of the liquid fuels and can be used for solid fuels [Atreya (1998)]. The flow field and the supply of fuel define the characteristics of the diffusion flame established on a composite surface. The rate at which both reactants reach the flame zone defines the flame temperature and thus the characteristic chemical time. Furthermore, the burning of polymer matrix plays an important part in the fire growth and the mechanisms of fire. It determines the fuel consumption (or burning rate), the flame height and the maximum heat and smoke release. These ones influence some characteristics of fire such as the reduction of oxygen, the strapline of the flame, the solid temperature and the radiation on adjacent surfaces. The fundamental understanding of the mechanisms controlling the combustion of a combustible surface seems necessary to characterize the material as it provides reliable information on the behaviour of the solid in a real fire.

3.1 Heat released

The essential property characterizing quantitatively the intensity of fire is the heat release rate $\dot{Q}''$ or HRR ($kW.m^{-2}$). It is the rate at which the combustion reaction produces heat. It is a function of the decomposition rate of polymer matrix $\dot{m}''_p(x,t)$ ($kg.s^{-1}$). The heat release rate is dependent on material (chemical composition, fibre content and architecture) geometry, environment condition and external irradiance levels as shown by Gibson et al. (2004). The relationship of these two quantities can be expressed as:

$$\dot{Q}'' = \Delta h_c \, \dot{m}''_p \tag{22}$$

where Δh_c is the effective heat of combustion ($kJ.kg^1$). Prediction of the heat release rate is a complex phenomena that involves calculating the pyrolyzate rate [See Eq. 9] based mainly on two factors, the heat transfer into the solid and a pyrolysis kinetic mechanism.

3.2 Gaseous species released

During the thermal decomposition and combustion of material, the generation of gaseous products are issued either from the solid phase or from reactions in gas phase, as shown in 4. Gases molecules (as free radical) and some of toxic gases come from the breakdown of the chemical bond in the solid phase. For example, the stable molecules such as hydrogen chloride HCl or hydrogen bromide HBr are slightly affected by the flame when material ignites. Others gas such as aldehydes and hydrogen cyanide HCN are dependent on the time residence in the flame and can be strongly modified by the flame. The radical chain reactions in the gas phase lead to the end products of combustion, typically carbon dioxide (CO_2), sulfur dioxide (SO_2) or nitrogen oxide (NOx) etc. Species such as carbon monoxide (CO) can be generated during the devolatilization of the solid phase or produced in gas phase by oxidation reaction of hydrocarbon. Velocity of CO formation depends of the type of hydrocarbon, the temperature, the local concentration in O2 and the residences times. Velocity of the CO formation depend of the type of hydrocarbon, the temperature, the local concentration in O_2 and the residence times. A variety of other gases can be produced during the combustion of composite material [Hume (1992), Sorathia (1993), Tewarson (1993), Sastri et al. (1999), Marquis et al. (2011)].

The amount of gaseous products and the diversity of the species can be influenced by: (1) the composition of the matrix; (2) the reinforced fibre and architecture; (3) the geometry and orientation of the composite; (4) the decomposition processes (char formation, delamination, crack...); and (5) the fire conditions (oxygen content, pressure and temperature).

For a given composite, the analysis of Marquis et al. (2011) shows that the kinetics of mass loss and production of species depend on the oxygen concentration and the incident radiant heat flux on the material (or ambient temperature). The depletion of oxygen concentration leads to an increase of unburnt, namely: CH_4, C_2H_4 and $HCOH$. These ones depend on the composition of the matrix. For a given O_2 concentration, it can be seen that the mass loss rate and the gas production rate regress linearly with the irradiance level. This behaviour is the result of both of the reduction of oxygen diffusion processes in the solid phase and the thermal balance at the surface of the specimen. This strong dependence of thermal degradation and gases release in bi-variables makes it difficult to model the combustion of a composite material in the gas phase.

To understand the mechanisms operating in the gas phase, it is often necessary to accurately study the oxidation kinetics of gaseous fuels. Depending on the reaction mixture composition and temperature of the medium, there are many reaction pathways to describe the transition

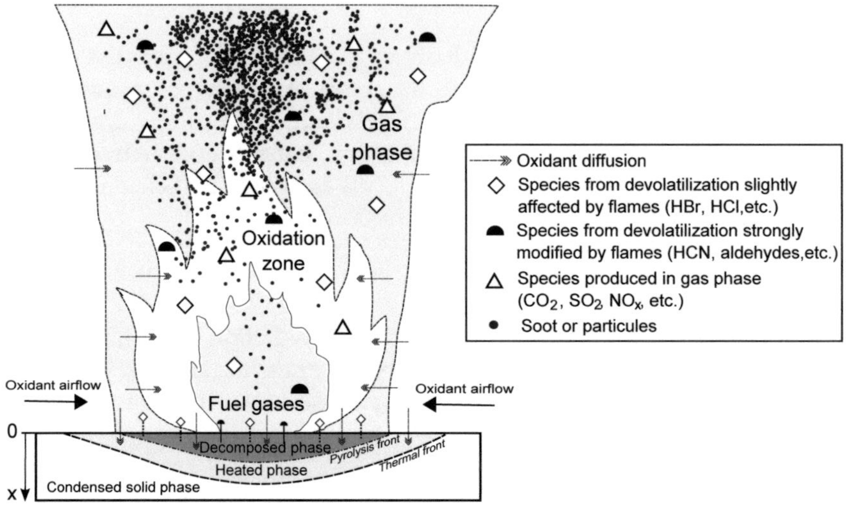

Fig. 4. Production of gaseous species from the devolatilization process of solid phase.

from a reactive to a product, usually irreversibly. This reaction mechanism may be represented at different levels of complexity, ranging from a simple overall reaction (less than 10 reactions and about 4 or 5 species) to a detailed mechanism including sometimes more than hundreds of species. Generally, these detailed mechanisms are established to model the combustion of a single gas species. Establishing a model reflecting the oxidation of gaseous fractions from a polymer material is tricky because this material produces a gas mixture rich in CO_2, CO and light hydrocarbons. In fact, the size of a detailed reaction scheme for describing the oxidation of a gas mixture representative of the gases emitted by the polymers can quickly become a serious handicap. Actually, using detailed combustion chemistry is impossible within fire simulation models. This constraint implies reducing the size of the reaction mechanism while trying to conserve its predictive potential. Because of the simplicity of the models used to represent the chemistry of combustion, it is futile to try predicting precisely the definition of chemical species of a multi-fuel sandwich composite from a global reaction [Marquis (2010a)].

4. Surface flame spread and fire growth

The previous section has given the reader a feel for the various approaches taken for pyrolysis modelling, identified some of the strengths and weaknesses and examined the fundamental transport occurring within composite material. Although microscale transport phenomena were emphasized, the ultimate goal of modelling is the prediction of large-scale fire behaviour. For this reason, the prediction of surface flame spread and fire growth are fundamental to model the fire contribution of composite material.

4.1 Surface flame spread

The rate at which a fire develops depends on how rapidly a flame can spread from point to ignition to involve an increasingly larger area of composite material. So the surface flame spread on composite material is a subject of interest on fire safety because it influences the fire growth and the heat release rate. Surface flame spread is the result of interactions between the gaseous phase and the condensed phase. It can be represented by a moving combustion wave in the vicinity of a pyrolyzing region on the surface which acts as a fuel source: the flame front acts as an ignition pilot that initiates the ignition of pyrolyzed fuel leaving the surface. Enough heat must be transferred from the flame to the unburnt material ahead of the flame to decompose the polymer matrix and allow the surface flame spread. During the devolatilization, the gaseous species is then diffused through the decomposed phase (fibre/gas) and convected away from the surface, mixing with the oxidizer and generating a flammable mixture ahead of the flame leading edge, which is then ignited by the flame. The flame spread rate is therefore determined by the ability of the flame to transfer the necessary heat to devolatilize the polymer matrix and to ignite the fuel mixture ahead of it.

Flame spread characteristics are affected not only by the mechanism in solid decomposition (thickness, polymer composition, geometry, fibre content, etc.) but also by other factors such as flow configuration, the possible presence of an external source, environmental condition (oxygen level, temperature, pressure, gravity) and the orientation of solid composite. Flame spread over solid fuel includes two characteristic flame configurations. Either (1) *flow-aided flame spread* where the flame spread in the same direction as the oxidizing flow due to wind or buoyancy or (2) *opposed flow flame spread* which refers the flame spread in the opposite direction as the oxidizing gas flow. Furthermore, the surface flame spread process on composite material depends on mechanisms occurring both in the gas phase and the solid phase. In the gas phase, the dominant phenomena are the flow configuration, the geometry of the flame, chemical reactions and mechanisms of mass transport and heat transfer (radiation, convection and conduction) from the hot mixture of reacting gases and the combustion products above the pyrolyzed region to the unburned fuel surface. In the condensed solid phase, the flame spread is determined by the composition and geometry of composite, the heat flow on the surface, the heat diffusion unto the composite (thermal thickness) and the kinetics of pyrolysis. Progresses were made in the understanding of phenomena through theoretical sophistication and laboratory experiments. An extensive research literature exists for these class of flame spread problem, see the comprehensible reviews of [Drysdale (1998), Quintiere (2006), Fernandez-Pello (1984), Hasemi (2008) Hirano (1991) Williams (1976)].

Some noteworthy basic studies of surface flame spread allow to model surface flame spread. A great amount of theoretical and experimental work of various degrees of complexity relative to the flame spread problem has been carried out. As long as surface flame spread is discussed for the assessment of fire safety of composites materials, these phenomena are generally modelled as a thermal process. Generally these formulae are able to describe qualitatively the dependence on environmental conditions and property values of the solid fuel, when phenomena are controlled by heat transfer mechanisms. Given the strong interactions between the solid and gas phases, it is necessary to make some simplifications to the basic model. More comprehensible mathematical model includes balance equation for both gas and solid phases and need a numerical treatment.

4.2 Fire growth on composite structure

Until recently, models able to analyze the fire behaviour of composite material were not available. Thus, the conventional approach to evaluate the fire response of composite material consisted in performing fire tests at real scale on selected components of the considered whole structure. These tests can offer a lot of information about the structure integrity of a given design for specific test conditions. However, the results of such tests are usually difficult to extrapolate to other scenarios, making very costly and complex performance demonstration protocols relying only on experiments tests.

From the 2000s, several approaches of fire growth models have been developed. Despite the simplified nature of this type of fire growth modelling and its inherent uncertainties, good agreement between calculated and prediction has been demonstrated for complex material such as fire retarded composite [Lattimer et al. (2003)]. The major progress has been made towards the coupling of first principles based condensed phase fuel general models to computational fluid dynamics (CFD) that simulate the gas-phase fluids dynamics, combustion processes, and heat transfer aspects of a fire [Mc Grattan et al. (2010)]. Numerical simulation may be a powerful tool for the prediction of reaction to fire of composite material, provided that all the main chemical and physical processes are taken into account even with some simplifications. The primary advantage of this approach is the flexibility and the possibility of a rapid low-cost assessment of the fire performance compared to an analytical approach. Nevertheless, a CFD code based fire model requires an understanding of the processes occurring both in the solid phase and the gas phase. A coupled pyrolysis/CFD fire model offer the possibility to consider complex geometries and ignition scenarios, and to evaluate the impact design change. Solid phase pyrolysis models have been coupled to CFD for simulating large scale fire growth [Lautenberger (2009), Marquis et al. (2009), and (2010b)] and real scale fire [Gutierrez et al. (2008), Marquis (2010a)].

A major obstacle impeding heal world application of fire growth modelling is the difficulty associated with determining thermophysical properties and understanding thermal, chemical, physical and failure processes which control the degradation (e.g. species formation, devolatilization, chemical kinetic, structural change, evolution of thermophysical properties, fibre presence, fibre interactions, the link between in-depth decomposition and its effects on the structural behaviour). The need to understand the mechanisms in gas phase (formation and oxidation of soot, transport processes) is therefore added, to ensure a correct understanding of heat transfer into the solid phase. This strong dependence between theses two phases leads to a physical system whose analysis is not easy.

The phenomenon analysis is even more complicated because many processes are interdependent and do not occur when isolated from each other. Understanding these processes and the way they interact is essential to analyze and to predict the fire behaviour and surface flame spread. The task is more complicated for composite material, because the polymer composite used in everyday life represents such a diversity of compositions (copolymers, presence of additives ...), physical or geometrical (architecture of fibre...), that work is still needed before assuming a prediction of certain conditions of ignition and flame spread.

Due to a lack of knowledge on the subject and the complexity of the physical and chemical processes, it is necessary to have a methodology that includes successive validation steps at increasing scales. This approach allows better assessment of the final quality of numerical simulations to predict all the phenomena of fire growth on a composite material (decomposition process, ignition, combustion and surface fire spreads) whatever the geometry and the orientation is. This complex approach was successfully applied by Marquis (2010a) and (2010b), to simulate with a CFD code, fire behaviour and fire growth on thermoset-based sandwich composite structure in well ventilated fire condition. To achieve this, the author uses a combined test-simulation approach, with increasing scale and complexity. The analysis of the fire behaviour of the material was carried out from matter scale, up to an end use representative scale, through five successive scale levels. This work shows that the multi-scale methodology development seems essential to simulate the fire growth on composite structures in a reliable way. Nevertheless, the approach showed its limits with the thermoplastic materials (Bustamante-Valencia (2009)), mainly because of the lack of capacity of the model to treat melting, bubbling and related phenomena of thermoplastic materials during their combustion.

5. Perspective

The last decade has seen a surge of modelling being applied to fire growth composite. But the robustness of the state-of-the-art of fire growth modelling is currently stalled by a poor capability to model the burning and decomposition processes of fuel solid polymer. Current fire modelling tools fail to predict properly the fire development (e.g. flame spread and fire growth) on composite material because of the complex phenomena and hypotheses used. The consequence is that currently fire modelling cannot predict the transient evaluation of the pyrolyzate rate and mainly the heat release rate in a non-trivial scenario, particularly in an under-ventilated scenario. Mains key areas in which progress is needed concern the behaviour of the condensed phase and the heat and mass transport into the composite: (1) the consideration of the oxygen diffusion on the surface, then within the material, to develop a system of devolatilization equations including the local concentration in oxygen; (2) The consideration of the movements of the condensed phase, condition needed to model the thermoplastic matrix. Furthermore, some improvements in the flame-spread model would deal with a better understanding and description of the chemical processes both in the solid and the gas phase and radiative heat transfer. Several studies are developing methodologies to overcome this limitation and the few applications available until now offer promising results but require further investigations. This topic will be one of the most important research objectives in fire engineering of the incoming decade.

6. References

Abu Zaid, M. (1988). Ph.D. thesis, University of Michigan State (USA).

ASTM (E176). Standard terminology of fire standards, *Annual book of ASTM standards ASTM West Conshohoken PA.* 4.07.

Atreya, A. (1998). *Philos. T. Roy. Soc. A.* 356: 2787–2813.

Babrauskas, V. (2003). *Ignition handbook*, Fire science.

Bolland, J. & Gee, G. (1946). *Trans. Far. Soc.* 42: 244.

Brausman, S. (1988). *Journal of polymer science* 26: 95–105.

Bushman, A. (n.d.). *NBS Report 7306* .

Bustamante-Valencia, L. (2009). Ph.D. thesis, University of Poitiers (France).

Carslaw, H. & Jaeger, J. (1959). *Conduction of heat solids*, 2th edition, Oxford university Press.

Dakka, S., Jackson, G. & Torero, J. (2002). *29th Int. Symp. of Combustion, the Combustion Institute*, pp. 281–287.

Davies, J., Dewhurst, D. & J., M. (1995a). *Offshore application Sandwich construction 3*, Southanpton UK Emas.

Davies, J., Dewhurst, D. & J., M. (1995b). *Offshore application Sandwich construction 3*, Southanpton UK Emas.

Davies, J. & Wang, H. (n.d.). *7th int. conf. on fire science and engineering Interflame.*

Delichatsios, M., Panagiotou, T. & Kiley, F. (1991). *Combustion and flames* 84: 323–332.

Dembsey, N. & Jacoby, D. (2000). Fire and materials, 24: 91–100.

Di Blasi, C. (2000). *Polymer international* 49: 1133–1146.

Dimitrienko, Y. (1995). *International Journal of Heat and Mass Transfer* 38: 139–146.
Dimitrienko, Y. (1997). *International Journal of Heat and Mass Transfer* 28A: 453–461.
Drysdale, D. (1998). *An introduction to fire dynamics*, 2nd Edt J Wiley and Sons.
Dusimberre, G. (1998). *Heat transfer by finite differences methods*, Edt Wiley.
Fernandez-Pello, C. (1984). *Combustion science and technology* 39: 121–134.
Florio, J., Henderson, J. & Test, F. (1989). *High temperature-high pressure* 21: 157–165.
Friedman, H. (1963). *Journal of Polymer Science Part A: Polymer Chemistry* C6: 183–195.
Galbano, A., DiBlasi, C., Branca, C. & Millela, E. (n.d.). *Polymer Degradation and Stability* pp. 1267–1280.
Gibson, A., Wright, P., Wu, Y. & Mouritz, A. (2004). *Journal of composition material* 38: 1283–1307.
Gibson, A., Wu, Y., Chandler, H. & Wilcox, J. (1995). *Revue de l'institut franï£¡ais du pï£¡trole* 50(1): 69–74.
Grenier, A. (1998). *Fire safety journal* 30: 137–159.
Gutierrez, J., Breuillard, A., Corrignan, P., Marquis, D., Chivas, C., Guillaume, E. & Pavageau, M. (2008). *5th int. conf. of Composite In Fire*, Newcastle (UK), pp. 11–15.
Hallman, J. (1971). Ph.D. thesis, University of Oklahoma.(USA).
Hasemi, Y. (2008). *in* 4th (ed.), *SFPE Handbook of fire protection Engineering*, National fire protection association, Quincy (USA), pp. 2.278–2.290.
Henderson, J. & Doherty, M. (1987a). *High Temperatures High Pressures* 19: 95–121.
Henderson, J., Wiebeld, J., Tant, M. & Moore, G. (1983). *Thermochim Acta* 57: 161–171.
Henderson, J. & Wiebelt, J. (1987b). *Journal of Composite Materials* 21: 373–393.
Henderson, J., Wiebelt, J. & Tant, M. (1985). *Journal of Composite Materials* 19: 579–595.
Hirano, T. (1991). *3rd international symposium of fire safety science*, Boston (USA), pp. 27–44.
Hirschler, J. & Morgan, A. (2008). *in* 4th (ed.), *SFPE Handbook of fire protection Engineering*, NFPA edt, Quincy (USA), pp. 1.113–1.143.
Hopkins, D. & Quintiere, J. (1996). *Fire safety journal* 29: 241–268.
Houck, C. & Joines, J. (1995). *Report NCSU IE TR* 95(09).
URL: *www.ise.ncsu.edu*
Hshieh, F. & Beeson, H. (1997). *Fire and Materials* 21: 41–49.
Hume, J. (1992). *Int. conf. on Material and Design Against Fire*, Londons, pp. 11–15.
ISO13943, E. (2001). *Sï£¡curite au feu, vocabulaire*, AFNOR norme franï£¡aise.
Janssens, M. (1993). *2nd Int. Conf. Fire and Material*, Interscience communication, Londons, pp. 37–46.
Kansa, E. & Perlee, H. (1977). *Combustion and flame* 29: 311–324.
Kashiwagi, T. (1994). *25th Int. Symp. on combustion, The combustion institute*, pp. 1423–1437.
Kashiwagi, T. & Ohlemiller, T. (1982). *19th Int. Symp. on Combustion, the Combustion Institute.*, pp. 815–523.
Kashiwagi, T., Ohlemiller, T. & Werner, K. (1987). *Combustion and Flame* 69: 331–345.
Kimoto, K. & Tanaka, T. (2004). *Fire science and technology* 23: 170–190.
Kissinger, H. (1957). *Analytical Chemistry* 29(11): 1702–1706.
Krysl, P., Ramroth, W., Stewart, L. & Asaro, R. (2004). *Int. Journal for Numerical Methods in Engineering* 61: 49–68.
Kung, H. (1972). *Combustion and flame* 18: 185–195.
Lattimer, B., Hunt, S. & Sorathia, U. (2003). *Fire safety journal* 38: 771–796.
Lattimer, B. & Ouelette, J. (2006). *Composites : Part A.* 37: 1068–1081.
Lautenberger, C. (2009).
URL: *http://code.google.com/p/gpyro/*

Lautenberger, C. & Fernandez-Pello, A. (2004). *in* B. Sundén & M. Faghri (eds), *Transport phenomena in fire*, WIT Press, Southampton (UK), pp. 209–259.

Le Bras, M., Bourbigot, S., Mortaigne, B. & Cordellier, G. (1998). *Polymer and polymer composite* 6: 535–539.

Levchik, C. & Wilkie, C. (2000). Char formation, *in* W. Grand & Dekker (eds), *Fire retardancy of polymeric materials*.

Looyed, M., Bettess, P. & Gibson, A. (1997). *Int. Journal of Numerical Methods for Heat & Fluid Flow* 7(6): 609–625.

Looyed, M., Rados, K. & Bettess, P. (2001). *Finite Elements in Analysis and Design* 37(11): 913–927.

Marquis, D. (2010a). Ph.D. thesis, Ecole des Mines de Nantes (France).

Marquis, D., Guillaume, E. A., Pavageau, M. & Chivas-Joly (2009). *11th Int. Conf. on Fire and Material*, San Francisco (USA), pp. 121–134.

Marquis, D., Guillaume, E. A., Pavageau, M., Chivas-Joly, C., Gutierrez, J. & Cueff, C. (2010b). *12th Int. Conf. on Fire Science and Engineering Inferflam*, Nottimgham (UK), pp. 407–419.

Marquis, D., Guillaume, E., Camillo, A., Pavageau, M. & Rogaume, T. (2011). *12th Int. Conf. on Fire and Material*, San Francisco (USA), pp. 121–134.

Martin, S. (1965). *10th Int. Symp. on combustion/ The combustion institute.*

Matala, A. (2008). MSc Thesis,University of Technology, Helsinki. (Finland).

Mc Grattan, K., McDermott, R., Hostikka, S. & J., F. (2010). *User's guide, Nist publication* 1019-5: 210.

Mc Manus, H. & Springer, G. (1992a). *Journal of Composite Materials* 26: 206–229.

Mc Manus, H. & Springer, G. (1992b). *Journal of Composite Materials* 26: 206–229.

Mikkola, E. (1992). *in* Babrauskaes & Grayon (eds), *Heat release rate in fires*, Elsevier applied science, Barking (UK), pp. 225–232–356.

Mikkola, E. & Wichman, I. (1989). s, *Fire and material* 14: 87–96.

Milke, J. & Vizzini, A. (1991). *J. of Comp Technol Res.* 13: 145–151.

Moghtaderi, B. (2006). *Fire and Material* 30: 1–34.

Moghtaderi, B., Novozhilov, N., Fletcher, D. & Kent, J. (1996). *Journal of Applied Fire Science* 6: 189–201.

Moghtaderi, B., Novozhilov, N., Fletcher, D. & Kent, J. (1997). *Fire safety journal* 29: 41–59.

Mouritz, A. & Gibson, A. (2006). Edt Springer, Solid mechanics and its applications.

Ohlemiller, T. (1985). *Progress in Energy and Combustion Science* 11: 277–310.

Ott, H. (1981). *Plastics, rubber rocessing and application* 1: 9–24.

Pal, G. & Mackásy, H. (1991). *Plastics their behavior in fire, Studies in polymer science*, Elsevier.

Parker, W. (1992). Wood material, part a, prediction of the heat release rate from basic measurement, *in* B. Grayon (ed.), *Heat release rate in fires*, Elsevier applied science, Barking (UK), pp. 333–356.

Perring, G., Farell, P. & Springer, G. (1980). *Journal of composite materials* 14: 54–66.

Quintiere, J. (2006). *Fundamental of fire phenomena*, Wiley and sons.

Quintiere, J. & Harkleroad, M. (1985). *T.Z. Harmathy Fire safety : Science and Engineering ASTM. SP 882*, pp. 239–268.

Rein, G., Lautenberger, C., Fernandez-Pello, C., Torero, J. & Urban, D. (2006). *Combustion and flame* 146: 95–108.

Sastri, S., Armistead, J., Keller, T. & Sorathia, U. (1999). *42th Int. SAMPE Symp.*, pp. 1032–1038.

Shields, T., Silcok, G. & Murray, J. (1994). *Fire and Material* 18: 243–254.

Simms, D. (1963). *Combustion and flames* 7: 253–261.

Smith, E. & Satija, S. (1981). *20th joint ASME/AICHE national heat transfer conf.*, Conference location.

Sorathia, U. (1993). *1st Int. workshop on composite material for offshore operation*, Houston (USA), pp. 309–317.

Spearpoint, M. & Quintiere, J. (2001). *Fire safety journal* 36: 391–415.

Sullivan, R. & Salamon, N. (1992a). *Int. Journal of Engineering Science* 30: 431–441.

Sullivan, R. & Salamon, N. (1992b). *Int. Journal of Engineering Science* 30: 939–951.

Tant, M., Henderson, J. & Boyer, C. (1985). *Composites* 16: 121–126.

Tewarson, A. (1993). *Journal of Fire Science* 11: 421–441.

Tewarson, A. (1995). *in* 2nd (ed.), *SFPE Handbook of fire protection Engineering*, NFPA edt, Quincy (USA), pp. 3.53–3.124.

Torero, J. (2008). *in* 4th (ed.), *SFPE Handbook of fire protection Engineering*, NFPA edt, Quincy (USA), pp. 2.260–2.276.

Trelles, J. & Lattimer, B. (2007). *Fire and Materials* 31: 147–171.

Troitzsch, J. (2004). *Plastics flammability handbook ;: Principles, regulations, testing and approval*, Hanser.

Vovelle, C. & Delfau, J. (2000). *Technique de l'ingénieur* AM3 170: 18.

Vyazovkin, S. (2000). *Int. Review of Physical Chemistry* 16(1): 45–60.

Whitaker, S. (1977). *International Journal of Heat and Mass Transfer* 13: 119–203.

Williams, F. (1976). 16th int. symp. on combustion - the combustion institute, Pittsburg (USA), pp. 1281–1294.

10

Production, Characterization, and Mechanical Evaluation of Dissimilar Metal/Ceramic Joints

José Lemus-Ruiz, Leonel Ceja-Cárdenas,
Egberto Bedolla-Becerril and Víctor H. López-Morelos
Instituto de Investigaciones Metalúrgicas,
Universidad Michoacana de San Nicolás de Hidalgo
Morelia, Michoacán
México

1. Introduction

Ceramics and metals are two of the oldest established classes of technologically useful materials. While metals dominate engineering applications, ceramics are used extensively to provide thermal and electrical insulation and are emerging as important structural materials. Properties of individual ceramics and metals can vary widely, but the characteristics of most materials in the two classes diffier significantly (Richerson, 1992). The properties reflect the different lattice binding characteristics of metals and ceramics. Ceramics are neither purely ionic nor purely covalent, and the relative degree of ionicity or covalency of ceramics can be related to the electronegativity of its components. Modern ceramics are gradually becoming important in structural applications because of their high strength to weight ratio, high modulus, high corrosion resistance, excellent high temperature properties, and abundance. The manufacture of component shapes from monolithic engineering ceramics such as Si_3N_4 is difficult, and this has generated a continued interest in the use of joining technologies to produce complex configurations from assembles of simples shapes.
The use of ceramics and ceramics based composites in industrial applications mainly for high temperature has received extensive attention recently. However, the utilization of the full capabilities of ceramics demands an ability to join ceramics to themselves or metals. Significant differences in the mechanical properties of ceramics and metals make it extremely difficult to obtain ceramic/metal joints with adequate mechanical integrity. Silicon nitride (Si_3N_4) and tungsten carbide (WC) are two of the most attractive advanced ceramics materials for specific applications since the resulting microstructure of these ceramics after sintering is similar to that of whisker or particulate-reinforced composites, respectively. Joining ceramics to metals is a key of technology in the use of advanced ceramics in complex structures. Moreover, the use of advanced ceramics depends on the reliability of ceramic/metal joining processes and the properties of the resulting interfaces. Several methods for joining ceramics to metals have been developed, some need an intermediate liquid phase, brazing, thermal spraying, and others are produced by solid state bonding and co-sintering (Tinsley et al., 1998; Tomsia, 1993). In all cases some problems

have to be overcome: (i) morphological adaptation of the surfaces in contact; (ii) formation of bonds: van der Waals bonds, chemical bonds by simple recovery of the molecular orbital or by chemical reaction, where new phases are grown at the interface; (iii) strict control over the residual stresses developed in both materials; (iv) changes in mechanical properties of the alloy and ceramic close to the interface. Mechanical aspects are essential for the last two points and for the first one in the case of the solid state bonding (Treheux et al., 1994).

In the absence of a liquid phase, ceramic/metal joining can only be accomplished with temperature, pressure, and time sufficient to promote mobility and interdiffusion of reacting species, producing adequate interfacial contact between the ceramic and the metal. Initially, the two components contact only at their asperities. In the first stage of the bonding process, more intimate contact is achieved by plastic deformation, which results in two partly joined surfaces with residual voids in between. When ceramic/metal bonds are being formed, and deformation of the metal is permissible, plastic flow generally occurs within the metal, and in the second stage of the bonding process, creep and diffusion eliminate the voids. The key step in the formation of a ceramic/metal interface is to achieve adequate contact between the two materials at the interface. The driving force for the formation of a ceramic/metal interface is the reduction in free energy when intimate contact is established between the ceramic and metal surfaces. In order for this to occur, both surfaces have to be brought into intimate contact, which is usually accomplished by controlling the temperature and atmosphere during interface formation. The application of pressure results in displacement of impurities and adjustment of irregularities due to localized deformation.

Metal ceramic interlayers play an important role in the fabrication of both metal-ceramic joints and in the manufacture of metal-ceramic or metal-matrix composites. In the case of joining interlayers are used to alleviate thermal expansion mismatch or create bonding between the ceramic and metal. Similarly, they may be employed as coatings to ceramic particles to activated their normally inert surface or act as protection from reaction with liquid metal. Therefore, in order to understand the mechanical performance of joints it is important to understand the mechanisms of interface formation between the metal and ceramic.

2. Solid-state diffusion bonding

Diffusion bonding is a solid-state process for the fabrication of metal-metal, ceramic-ceramic and ceramic-metal joints that is conceptually simple. The process requires no localized melting of components or introduction of foreign bonding materials, but merely that mating surfaces are brought into intimate, atomic scale contact so that an interface can be formed by interdiffusion to create a structural continuum. Such interfaces, whether between metals, ceramics or between a metal and a ceramic, can have good mechanical integrity even at high temperatures. The simplest form of diffusion bonding involves the application of a low pressure at a high temperature to achieve bonding of the components. It is important they have smooth and well-matched mating surfaces. Understanding the relationship between processing, structure and properties of metal/ceramic interfaces is becoming increasingly important as performance requirements demand a combination of these different materials in applications ranging from electronic packaging to high temperature aircraft structures (Howe, 1993).

There are two possible mechanisms of the solid phase bonding of ceramics to metals without the intermediary of a liquid filler metal: when the solid-phase ceramic and metal are brought into contact, the metal may undergo plastic deformation, enter the surface irregularities of the ceramic, adhere and bond to the ceramic. In another possible mechanism, the metal may diffuse through the interface, react with the ceramic and form a continuous layer (Okamoto, 1990). Diffusion bonding has attracted interest as a means of bonding Si_3N_4 and successes have been achieved by controlling the microchemistry and microstructure of the interfaces formed. The first requirement for solid-state diffusion bonding is the creation of intimate contact between the surfaces to be bonded in order that the atomic species come into intimate contact. In addition to a good contact, there should be enough diffusion between the materials in a reasonable time. Pressure can be applied uniaxially (hot-press) or isostatically (hot-isostatic press) on a diffusion couple. Bonding of ceramics by uniaxial pressing has been achieved, but it is not yet an important and widely used industrial process. Figure 1 shows an illustrations of events during metal/ceramic diffusion bonding in solid-state (Nicholas, 1990).

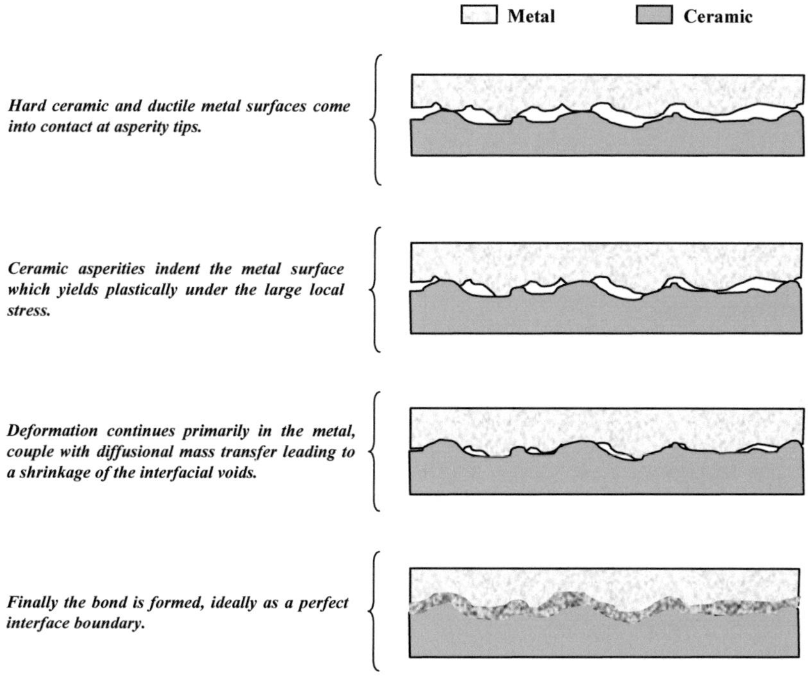

Fig. 1. Sequence of events during metal-ceramic diffusion bonding (Nicholas, 1990)

There are several advantages of solid-state joining. When ceramic have to be bonded to metals, it is a common practice to introduce a metal interlayer between the components. The interlayer should be ductile so that it can deform readily to achieve intimate contact with both mating surfaces at various pressures and temperatures, that it should act as a stress relieving buffer layer if the thermal expansivities of the metal and ceramic components differ significantly, and of course that it should adhere strongly to both the metal and ceramic components. The technique has been applied, so far, mainly for the joining of Al_2O_3 and SiO_2, but its utility has also been demonstrated for joining of Si_3N_4 and other new engineering ceramics like WC. The technological advantages of diffusion bonding are low deformation which enables parts to be joined with little distortion, the ability to join large areas, the applicability of diffusion bonded joints at high service temperatures, and the possibilities for joining materials in a none conventional way (Dunford & Wisbey, 1993). However, the major disadvantages of diffusion bonding are high capital costs, and the restricted joint geometries that can be produced (Peteves et al., 1996). In addition, specific problems exist when applying diffusion bonding to Si_3N_4. The misfit in the coefficient of thermal expansion *(CTE)* of the joining materials can result in areas of high residual stresses at the interface during the cooling process (Locatelli et al., 1997).

2.1 Bonding mechanisms

Diffusion bonding can be defined as the creation of an intimate bond or joint between two materials by thermally assisted processes occurring in the solid state. In order to understand the mechanisms and driving forces of diffusion bonding, the evolution of the bond microstructure must be appreciated. The bonding process can be viewed as two steps operating in parallel. The first is the transition from two surfaces contacting at their asperities to an intimate interfacial conformity. This must involve the elimination of a large volume of interfacial voids accommodated by mass transfer mechanisms, plastic flow and diffusion. In parallel with this, but sequential to each individual contact, there must be an adhesion process giving the interphase boundary strength. A third step, with possible destructive consequences, is a subsequent chemical reaction between the metal and ceramic in contact to form a third phase at the boundary. The driving force for the formation of an interface between materials is the energy decrease of the system resulting from its establishment. The interfacial energy should reach the lowest achievable value as the bond is formed, otherwise further changes that could degrade the stability of the bond may occur under operating conditions.

The mechanisms of diffusion bonding two identical materials and similar surfaces have been studied since the 1960s, and it is now generally accepted that joint formation occurs by collapse of interface voids produced by a number of diffusion and creep mechanisms. The collapse of interfacial voids can be brought about by a number of mechanisms analogous to those occurring in pressure sintering, and these are best grouped in terms of sources and sink for matter and are (Chen & Argon, 1981; Almond et al., 1983).

A. *Surface diffusion from a surface source to a neck.*
B. *Volume diffusion from a surface source to a neck.*
B'. *Evaporation from a surface source to condensation at a neck.*
C. *Grain boundary diffusion from an interfacial source to a neck.*
D. *Volume diffusion from an interfacial source to a neck.*
E. *Plastic yielding resulting in deformation of original surface asperities*
F. *Power law creep.*

An illustration of the various routes of material transfer is contained in Figure 2. These mechanisms are normally separated in two main stages. ***Stage 1****: Plastic deformation.* The contact area of asperities, though initially small, will rapidly grow until the application load can be supported, which means that the local stress falls below the yield strength of the material. ***Stage 2:*** *Diffusion and power law creep.* The driving force for mechanisms A, B, and B' is the difference in surface curvature. Matter is transferred from the point of least curvature (sharp neck of the void at the bond interface) to the point of greatest curvature. Thus, as the voids change from an elliptical to a circular cross section, the rates of these mechanisms will approach zero because the aspect ratio of the voids tends to unity. In addition to these stages, recrystallization and grain growth may occur during bonding. Interface formation must be accompanied by the collapse and annihilation of voids created at first contact. The driving force for this collapse is identical to what drives diffusion bonding in metals. Void closure results in a net approach of the two surfaces being joined. This allows mechanical work to be done by the bonding pressure. The reduction in void volume is accompanied by a reduction of void surface energy, which is a further driving force. A number of additional competing mechanisms can occur during the bonding of dissimilar materials such as metals and ceramics.

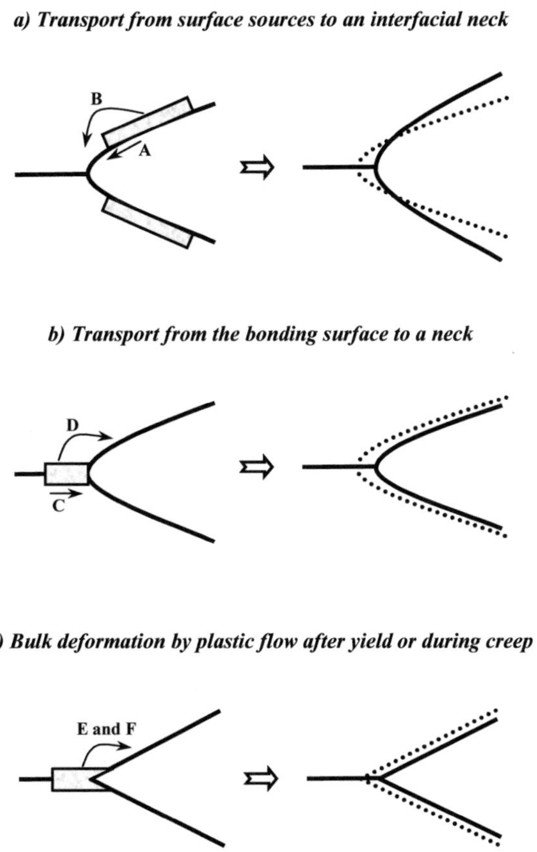

Fig. 2. Schematic illustration of material transfers for various mechanisms involved in diffusion bonding (Chen & Argon, 1981; Almond et al., 1983)

2.2 Metal/ceramic interfaces

Having selected suitable joining materials and decided upon a joint design, consideration can be given to the choosing of equipment and processing parameters that will ensure effective manufacture of the joints. While choices about materials and joint designs usually have to be made based on the nature of the components and the proposed service performance of the bonded assembly, those concerned with process parameters are numerous not only because of their variety but also because of the possibility of making incremental adjustments during trial runs to optimize joint quality by fine tuning. The main variables that affect solid-state bonding are the pressure, temperature, time, surface roughness, and, in some cases environment (Akselsen, 1992).

2.2.1 Effect of bonding temperature

Temperature is the most important parameter in the bonding process due to the fact that: *(i)* in thermally activated processes, a small change in temperature will result in the greatest change in process kinetics, diffusion, creep, compared with other parameters; and *(ii)* virtually all mechanisms in diffusion bonding are sensitive to temperature, plastic deformation, diffusion, creep. Temperature increases interaction across a metal/ceramic interface by increasing the mobility of atoms and also the mobility of dislocations in the metal during bonding. Since the mobility of dislocations increases with temperature and the flow stress correspondingly decreases, the pressure required for bonding decreases with increasing temperature. In general, the temperature required to obtain sufficient joint strength is typically within the range *0.6 and 0.95* T_m, where T_m is the absolute melting point of the base material. The effect of temperature is shown for Si_3N/Nb joints in Figure 3 (Lemus-Ruiz el al., 2006). It is seen that the interface in samples hot-pressed at 1400°C consists of a layer with one phase and thickness average of 2 μm, however the interface in samples hot-pressed at 1600°C consist of one layer with two intermixed phases and thickness average of 18 μm.

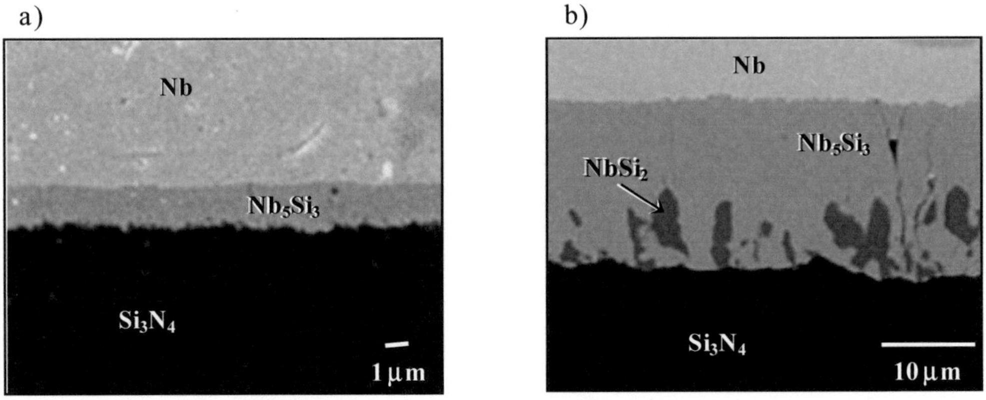

Fig. 3. Cross-section of the Si_3N_4/Nb interface for a sample hot-pressed at a) 1400°C for 120 minutes and b) 1600°C for 60 minutes in vacuum

On joining of metals to silicon nitride (Si_3N_4), an additional difficulty is introduced by the presence of nitrogen gas at the interface. Nitrogen gas is formed upon the dissociation of Si_3N_4, and it may or may not diffuse into the metal, depending on its solubility at the bonding temperature. In general the chemical reactions occurring when silicon nitride is in contact with a metal (Me) follows one of the following three routes (Richerson, 1992; Baker, 1992):

$$\text{Me} + Si_3N_4 = \text{Me silicide} + \text{Me nitride} \quad (1)$$

$$\text{Me} + Si_3N_4 = \text{Me silicide} + N_{2(g)} \quad (2)$$

$$\text{Me} + Si_3N_4 = \text{Me silicide} + \text{Si} \quad (3)$$

Comparing the thermodynamic stability of Si_3N_4 with those of metal silicides and nitrides, it is possible to predict the likely products of diffusion bonding. When the bonding temperature is above the minimum temperature required for reaction (2), it proceeds and forms intermetallic compounds at the interface. The free nitrogen produced by this reaction diffuses into the metal when it has a high solubility for nitrogen. However, strong nitride formers, such titanium (Ti) and niobium (Nb), promote reaction (1) above the minimum temperature of reaction (2) and below the minimum temperature in the case of reaction (3). This effect can be observed for Si_3N_4/Ti and Si_3N_4/Nb joints in Figure 4 (Lemus & Drew, 2003; Lemus-Ruiz et al., 2006).

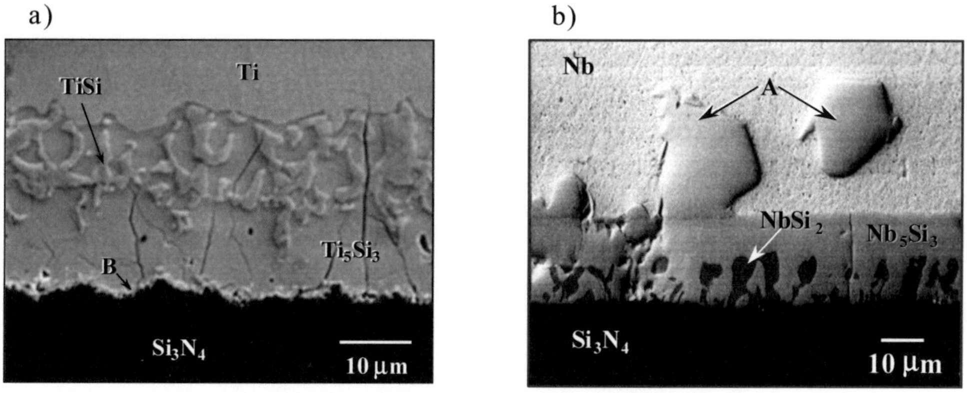

Fig. 4. Cross-section of samples hot-pressed for a) Si_3N_4/Ti at 1400°C for 180 minutes and b) Si_3N_4/Nb at 1600°C for 90 minutes in vacuum

In bonding of Si_3N_4/Ti at 1400°C for 180 minutes (Figure 4a) the interface consists of one layer with three intermixed phases. EPMA performed on this sample indicated that these phases were Ti_5Si_3, TiSi and the zone (B), which is a mixture of TiN containing Y, Al, O and Si. This is associated with segregation of the sintering aids, Y_2O_3 and Al_2O_3, during the decomposition of the ceramic to form an amorphous alumino-silicate. On the other hand, when Si_3N_4 is joined to Nb at 1600°C for 90 minutes (Figure 4b), the interface consists of one layer with Nb_5Si_3 and $NbSi_2$ intermixed phases and a zone (A) inside the metal, which are a mixture of niobium nitride and a trace of silicon. This is associated with the saturation of nitrogen in solution in the interface to form a niobium nitride phase. Because diffusion is the dominating reaction mechanism in solid state joining, silicide transformations may or may not be observed depending on the joining parameters, especially bonding temperature and time. These parameters affect the concentration of diffusing Si at the interface, and therefore, the nature of the resulting silicide. The optimum bonding temperature occurs at a point where the strength reduction due to residual stresses starts to balance the strength enhancement as a result of void elimination.

2.2.2 Effect of bonding time

The pressure applied in diffusion bonding is typically some small fraction of the room temperature yield stress of the base metal to avoid macroscopic deformation of the materials. In addition to establishing contact between the metal and ceramic, an important role of the pressure is to destroy the stable oxide film, through plastic deformation at the bonding temperature, present on the surface of most metals and this has a large effect on the integrity of the metal/ceramic bond. Process parameters as bonding temperature, time and pressure are strongly inter-related. Bonding times may vary from one second to several hours depending on the metal/ceramic combination and the joining temperature (Schwartz, 1990). A relatively short bonding time is usually required to form a strong bond between a metal and ceramic under temperature and pressure conditions which are sufficiently high to cause rapid deformation of the metal. When prolonged times or higher temperatures are used, chemical reactions can progress at the interfaces of some systems and significantly influence bond quality. These reactions can initially enhance bonding, forming "chemical bridges", but often ultimately cause degradation because their growth progressively generates volume mismatch strains and stresses within the reaction product or at the product/workpiece interfaces.

Figure 5 shows results of bonding cermet of WC-Co to Ni samples produced at 980°C for 25 minutes (Lemus-Ruiz et al., 2009). It can be seen a continuous and homogeneous bonding line interface can be seen. The bonding process is governed by diffusion of Ni through the Co of the WC and diffusion of Co through Ni, forming a thin inter-diffusion zone. Line analysis using electron probe micro-analysis shows an overview of the Co, W, and Ni components through the interface. It can be clearly observed from the micro-analysis profiles of Ni and Co that the accumulation of these components in the bonding line with inter-diffusion of Ni to Co and Co to Ni.

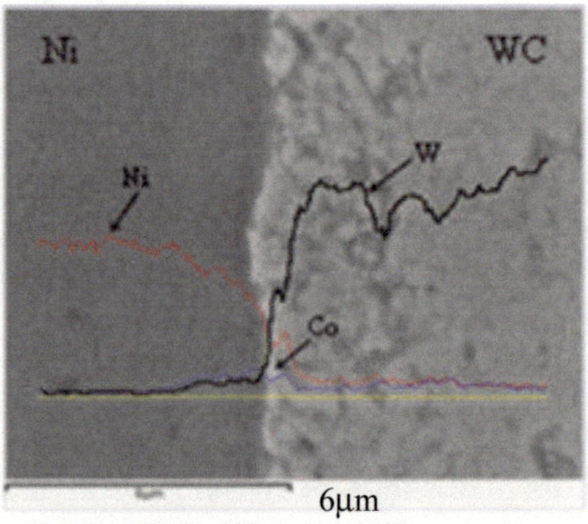

Fig. 5. Cross-section of the interface obtained in a WC/Ni sample bonded at 980°C for 25 minutes

2.2.3 Effect of surface roughness

Surface preparation before bonding is usually necessary for joint components and inserts. There are three main objectives: the first is the production of surfaces that are free of mobile contaminants such as oil films and surface layers of fragile or thick corrosion products, and any other protective layer that might inhibit bonding. Having produced clean surfaces, it is then necessary to ensure that there is macroscopic conformity between the mating surfaces of the components to be bonded. Without this, a lack of surface conformity of components to be diffusion bonded may limit good intimate contact when joining either directly or via thin interlayers. Finally, it is important to create microscopic topographic features on each material surface in order to encourage bonding. The presence of asperities on the bonding surfaces prevents large-scale plastic deformation at the surface because the metal is affixed between the asperities and limits the total area of contact between the metal and oxide-free ceramic, and, in addition, large voids are rarely closed (Treheux et al., 1994). Therefore, it is necessary to polish the bonding surfaces prior to joining, improving the initial area of contact between the metal and ceramic and prevent the formation of large voids. The removal of rust and other thick corrosion products and the creation of desired macroscopic and microscopic topographies can often be achieved by mechanical treatment of the surfaces. Grinding can be used to prepare both metal and ceramic surfaces. In practice it can be useful also to avoid producing very smooth surface finishes because microscopic asperities can assist the disruption of surface oxides.

In order to establish the effect of the surface roughness of the materials to be joined on the interface growth, a series of unpolished $Si_3N_4/Ti/Si_3N_4$ samples were hot-pressed. These samples were as-ground, without final polishing. The average surface roughness of polished Si_3N_4 and Ti surfaces was *63.4 nm* and *53.4 nm*, respectively. By contrast the values for unpolished samples were *584 nm* and *1140 nm* for Si_3N_4 and Ti, respectively. Figure 6 shows a plot of the thickness of the interfaces as a) function of the square root of holding time for the samples hot-pressed at joining temperatures of 1400°C and 1500°C and b) as a function of temperature. The error bars correspond to the standard deviation for the average thickness of at least twenty measurements for each sample. Figure 6 shows an excellent parabolic fit of the growth of the reaction zone, which is characteristic of processes of reactive diffusion through an interface that is growing. It can be observed in Figure 6b that the interface obtained in samples hot-pressed with unpolished materials is significantly lower than that obtained with polished samples, however only a slight change is observed when the temperature is increased.

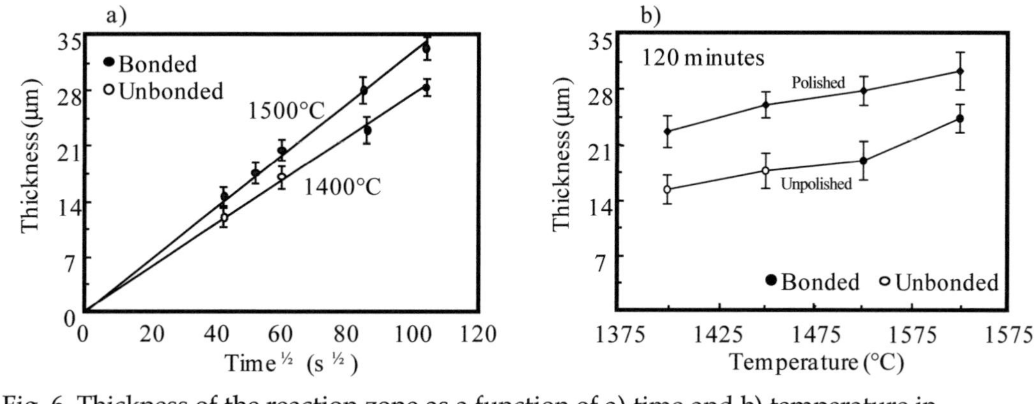

Fig. 6. Thickness of the reaction zone as a function of a) time and b) temperature in $Si_3N_4/Ti/Si_3N_4$ joining samples

3. Liquid-state diffusion bonding

The use of a liquid or a solid foil that flows readily under low applied stress to join materials can have advantages. Flow of a wetting liquid or of the ductile solid can fill irregularities in the surface and therefore imposes less stringent demands on surface preparation and the degree or extent of surface mating required. Indirect joining is the most common method of achieving high integrity joints using a wide range of intermediate bonding materials. The major categories of joining using an intermediate layer include joining with adhesives, cements, glasses or glass-ceramics, and brazes. The indirect bonding of ceramics includes those techniques in which a liquid medium is responsible for bonding. Liquid phase bonding using molten metals has been widely used to create ceramic/ceramic and ceramic/metal joints. In cases where ceramics are being introduced into predominantly metal structures, brazes provide a more familiar joining material. Brazing is commonly used in the manufacture of high-integrity joints with good mechanical properties. However, brazing requires wetting of the ceramic into contact angles less than 90°C, which is often difficult to achieve because of the high stability of ceramic compounds. One way to promote wetting is to metallize the ceramic surface. Another alternative is to add small percentages of a reactive metal, e.g. Ti, which reacts with the ceramic and generally facilitates wetting via the formation of a more metallically bonded reaction product at the interface (Loehman et al., 1990), in order to promote wetting of the surface by the braze. A solid/liquid interface can be formed if the liquid wets or spreads thereby penetrating between irregularities at the solid surface. It has been widely considered that wetting is an essential prerequisite for the creation of a good bond. It is understood, however, that wetting alone is not sufficient to guarantee good adhesion since wetting can occur with either van der Waals or stronger 'chemical' bonding across the interface (Savov et al., 1997; Mortensen & Jin, 1992).

3.1 Wettability and contact angle

The wettability is the capability of solids to build interfaces with liquids, in other words, it is the ability of a liquid to spread over a solid surface. It describes the extent of intimate contact between a liquid and a solid; but it does not represent an index of the strength of the interface. The wettability of a given solid/liquid couple can be measured by considering the equilibrium forces in a system consisting of a drop resting on a flat solid surface in a given atmosphere, this method for evaluating wettability is called the sessile drop method. This technique enables measurement of both, θ and γ_{SG}, however, this method is an oversimplification of true and complex phenomena. The contact angle, θ, of a liquid on the solid surface is the parameter used to measured the degree of wetting; $\theta > 90°$ means non-wetting, $\theta = 0°$ means perfect wetting and $\theta < 90°$ indicates partial wetting. It is, however, ordinary practice to say that a liquid wets a solid when $\theta < 90°$. The shape of the drop results from the balance between the surface force and the interfacial forces that are trying to minimise the surface free energy of the system. Under thermodynamic equilibrium and steady state conditions, the contact angle is related to the three tensions γ_{SG}, γ_{SL} and γ_{LG} of the interfaces solid/gas, solid/liquid and liquid/gas, respectively, by the well-known Young equation;

$$\gamma_{LG} \cos \theta = \gamma_{SG} - \gamma_{SL} \quad (4)$$

High temperatures and extended contact times, usually promote chemical reactions enhancing or inducing wetting (Landry et al., 1996; Asthana, 1998). The Figure 7 shows the effect of time at temperature of 900°C on the contact angles of several Al-alloys with TiC ceramic in argon (Contreras et al., 2001). The criteria for selection of brazing alloys are that they must wet or coat the ceramic, form a chemical bond at the interfaces resulting in a strong joint, and they should cause minimal degradation of the base material or materials. Successful brazing alloys produce bonds that are, strong, reliable, and relatively inexpensive to manufacture. As is the case for any joining process, there are also some important constraints and concerns, many of which are the direct consequence of the presence and action of the reactive metal.

A popular technique adopted to enhance wetting is to add suitable alloying elements to the melt. Alloying additions can promote wetting with a solid surface by three mechanisms; 1) by decreasing the surface tension of the liquid due to adsorption onto` the surface of the liquid, 2) by decreasing the solid/liquid interfacial tension due to segregation of solute to the interface, and 3) by inducing a chemical reaction at the solid/liquid interface which in turn decreases the solid/liquid interfacial tension by forming a stable compound at the interface. Another method for improving wettability is the application of a metal coating to the surface of the ceramic which, in principle, increases the surface energy of the solid and acts as a diffusion barrier delaying the onset of reaction. Nickel and copper coatings are commonly used (Yokokawa et al., 1991).

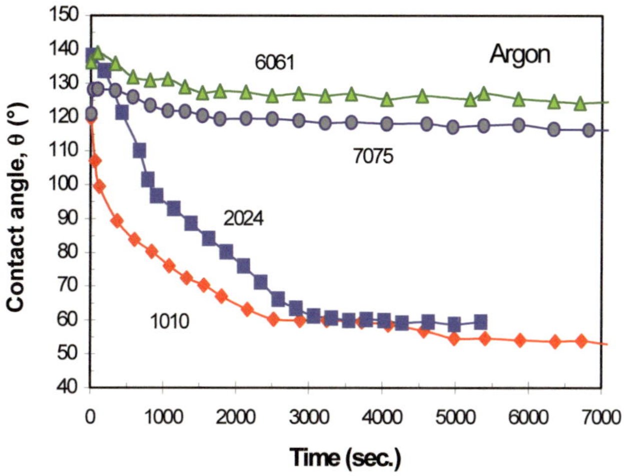

Fig. 7. Wetting behaviour of Al-alloys with TiC in Ar at 900°C (Contreras et al., 2001)

3.2 Brazing of silicon nitride

Joining ceramic/ceramic and ceramic/metal structures by brazing is attractive because of the flexibility of design that can be accommodated, the wide availability of the processing equipment, and the established use of brazing technique for high-integrity metal/metal structures such as aero-engine components. However, new and specially developed braze alloys are generally required to join ceramics because they are not wetted by conventional brazes based on Ag, Au, or Cu. These alloys, called "active brazes", contain components that react with the ceramics to change the chemistry of their surfaces to promote wetting and bonding. Many commercially available brazing alloys meet the main mechanical requirement of metal/ceramic joints, in which the more frequently used braze includes a Cu-Zn based alloy. Addition of small quantities of Ti increases the ceramic wettability and reactivity by diffusion of Ti across the interface (Zhang & He, 2010; Zou et al., 2009).

The Figure 8 shows a cross-section interface of diffusion bonding Si_3N_4/Cu/Ti combinations joined at temperature of 980°C. Liquid formation can occur by the interaction of Si with the Cu promoting bonding. Joining formed through the formation of a reactive interface on the metal side of the sample as a result of diffusion of N and Si and interaction of these species with Cu and Ti. Liquid formation can occur on the first step of the bonding, as a consequence of the interaction of Cu with Si and Ti.

The interface consisted of a continuous and homogeneous reaction layer free of porosity and containing several intermixed phases produced by chemical interaction, as well as the high affinity of Ti, Cu, Si and N. It can be observed that phase A correspond to reaction zone rich in Ti and Cu, however, in phase B increase the amount of Cu and Si, on the other hand, the phase C correspond to Cu interlayer remaining, and Ti and Si increase in phase D, which represent the diffusion-reaction produced during joining process.Liquid formation plays an important role in the joining process, because it increases the rate of the interface formation, improving the contact area between the bonding materials, and consequently the interaction is higher, promoting rapid diffusion of the material, since liquid diffusion is much faster than the diffusion in solid-state.

	Concentration (%at.)		
Phase	Ti	Cu	Si
A	52.34	33.54	14.12
B	41.30	40.62	18.07
C	11.20	80.95	7.85
D	57.51	14.97	27.53

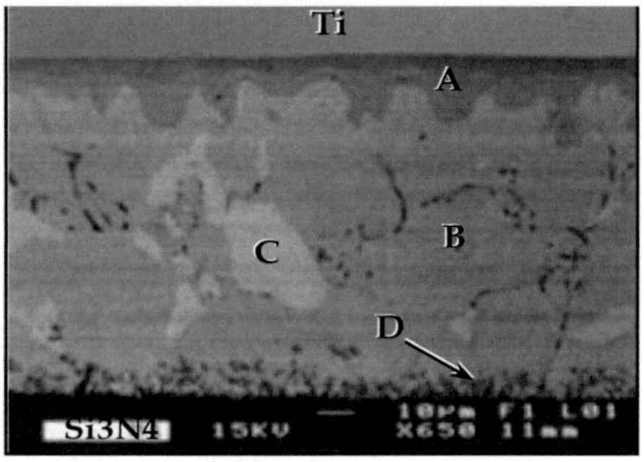

Fig. 8. Cross-section interface of a Si_3N_4/Cu/Ti sample bonded at 980°C for 5 minutes in Ar

3.3 Brazing of WC-Co cermets

On the other hand, during joining at 980°C of WC-Co cermets to Ni using Cu-Zn alloy as joining element, liquid formation occurs (melting point of 70Cu-30Zn (wt%)≅ 950°C) and joining takes place by liquid state diffusion of Co and Ni and interaction of these species with Cu and Zn of the liquid Cu-Zn alloy. Figure 9 shows a cross-section of the interface observed in brazing samples of WC/Cu-Zn/Ni produced at 980°C for 15 minutes (Lemus-Ruiz et al., 2008). It could be observed that increasing the bonding time, the diffusion increase forming a continuous bonding layer free of porosity. Electron probe micro-analysis performed on these samples indicated that Ni, Co, Cu, and Zn are in the diffusion interface, however no phases were detected and the components are in solution in the bonding interface. According to the thermodynamic, WC is stable at the bonding temperature; therefore W and C interactions with Ni or Cu-Zn are not expected in the joining zone. Diffusion is the dominating reaction mechanism; consequently the high affinity of Co and Ni for Cu and Zn resulted in immediate diffusion. Joining parameters, such as bonding temperature and time, affect the concentration of diffusion of the components at the interface, and therefore, the nature of the resulting interface.

An overview of the different components in the interface was obtained in a WC/Zn-Cu/Ni sample joined at 980°C for 15 minutes by line analysis using electron probe micro-analysis. The results are illustrated in Figure 10 where the Ni and WC are on the left and right, respectively. The scan line was chosen to start on the Ni side of the sample through the interface, Cu-Zn, finishing on the WC side.

Inter-diffusion of Cu-Ni and Cu-Co could be observed. The micro-analysis profile indicates the presence and even distribution of Zn concentration. In the region corresponding to diffusion zone high levels of Cu and Zn were observed, however evaporations of Zn during bonding could be occurred.

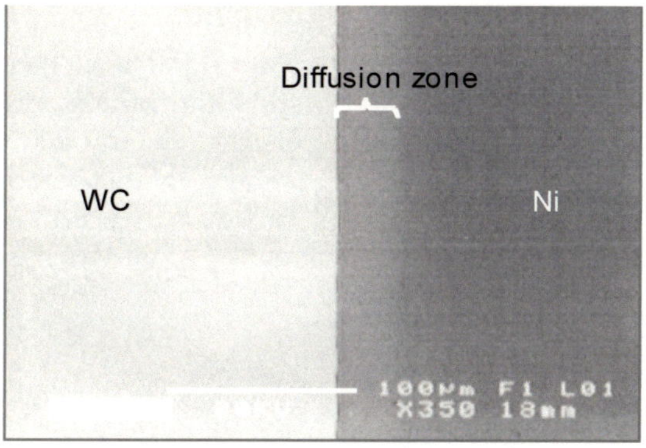

Fig. 9. Cross-section of the interface obtained in a WC/Cu-Zn/Ni sample bonded at 980°C for 15 minutes in Ar

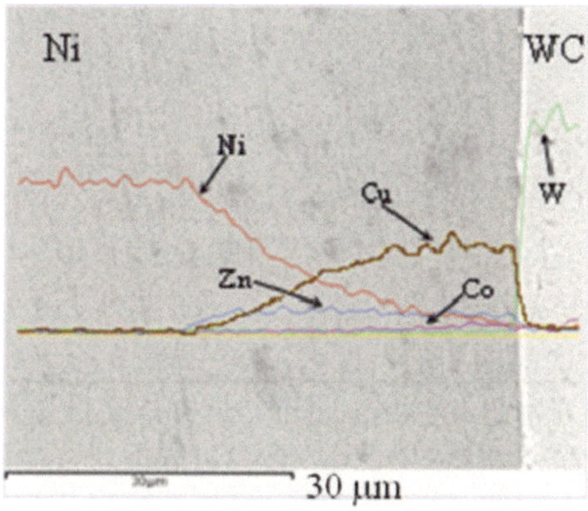

Fig. 10. Line analysis through the interface obtained in a WC/Cu-Zn/Ni sample joined at 980°C for 15 minutes in Ar

4. Mechanical evaluation

Several problems have been associated with the joint strength influencing the reliability of a joint. Figure 11 shows the schematic illustration of several important defect categories, which may cause scatter in strength directly (Suganuma, 1993). From the microscopic view, the reaction structure caused by wetting or by chemical and physical bond-ability between two faces may be of concern. These factors will reflect the distribution of un-joined or weakly bonded island-like defects on interfaces resulting in substantial reduction in joint strength. Un-joined areas reduce joint strength especially in solid-state joining. From the more macroscopic view, when a reaction layer grows, cracking occurs in the layer, which frequently influences joint strength. Thermal or residual stress in a joint becomes another important factor.

The development of residual stresses at the interface when the material is cooled down from the bonding temperature to room temperature is one of the major problems in ceramic/metal joining. These residual stresses reduce the strength of the bonded material and in some cases lead to joint failure during or after the joining process. The mechanical characterization of a metal/ceramic joint is a complex problem. There are a variety of different properties to be considered in ceramic/metal joints. Depending on the application of the joint, some properties are more important than others. However, the mechanical performance is one of the most important properties for any joint (Anderson, 1989). The strength of metal/ceramic joints materials is generally characterized by bend testing where the test specimen can have a circular, square, or rectangular cross section and is uniform along the complete length.

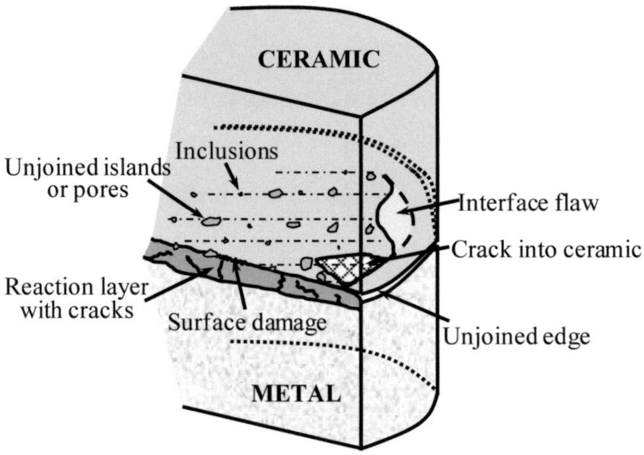

Fig. 11. Schematic illustration of various structural defects in ceramic/metal joint

The final goal for joining research will be in establishing a technique producing a tightly.
The interfacial strength of metal/ceramics joints is determined by four-point bending test using a universal testing system and a bending jig. The specimen is carefully placed in the bottom part of the jig, with the interface plane parallel to the plane of vertical displacement of the plunger. The load is applied at a low vertical speed of 0.5 mm/min until the applied load resulted in fracture of the specimen (Lemus-Ruiz & Aguilar-Reyes, 2004). The bend strength is defined as the maximum tensile stress at failure and is often referred to as the modulus of rupture *(MOR)*. The bend strength for a circular test specimen can be calculated using the general flexure stress formula described in equation (Richerson, 1992):

$$X = \sigma_{4\text{-pt}} = 16.F.d \,/\, \pi.D^3 \tag{5}$$

Where F is the load at fracture, d is the distance between the outer and inner span of the four-point bend jig and D is the diameter of the specimen. For each set of experimental conditions studied, temperature and time, an average of at least five samples must be used to determine the bending strength for each joining condition.

4.1 Bonding strength of Si_3N_4/Ti and WC/Ni

The strength of diffusion-bonded ceramics to metals depends on the nature and microstructure of the interface between the materials. In order to establish a mechanical evaluation of joints in hot-pressed Si_3N_4/Ti/Si_3N_4 samples, and WC/Cu-Zn/Ni and WC/Ni samples, the joint strength was measured using four-point bending (Lemus-Ruiz & Aguilar-Reyes, 2004; Lemus-Ruiz et al., 2009). The samples used for bend testing were of 50 mm in length and 7 mm in diameter. An example of a test sample is shown in Figure 12. The modulus of rupture (MOR) was calculated for Si_3N_4/Ti/Si_3N_4 samples hot-pressed at 1500°C and times varying from 30 to 180 minutes: the results obtained are shown in Figure 13. The error bars correspond to plus or minus the standard deviation for the average joint strength of at least five samples for each set of experimental conditions. It can be observed in Figure 13a that the strength of the joint increased from a value of 18 MPa and reached a maximum value of 147 MPa, when the time was increased from 30 to 120 minutes, respectively, and decreased beyond this time. The detrimental effect of the thickness of the reaction layer can be observed in the case of samples hot-pressed for three hours, where the strength of the joint decreased on compare with those samples hot-pressed for two hours. The effect of the growth of the reaction layer on the modulus of rupture of the Si_3N_4/Ti/Si_3N_4 samples can be more clearly observed in Figure 13b, which presents the results of the bending strength as a function of temperature for Si_3N_4/Ti/Si_3N_4 samples hot-pressed under vacuum for a constant time of 120 minutes. The effect of the temperature was more pronounced than the effect of the time. It can be observed that the strength of the joint increased from 64 to 147 MPa when the temperature was raised from 1400 to 1500°C. However, the joint strength decreased, above 1500°C.

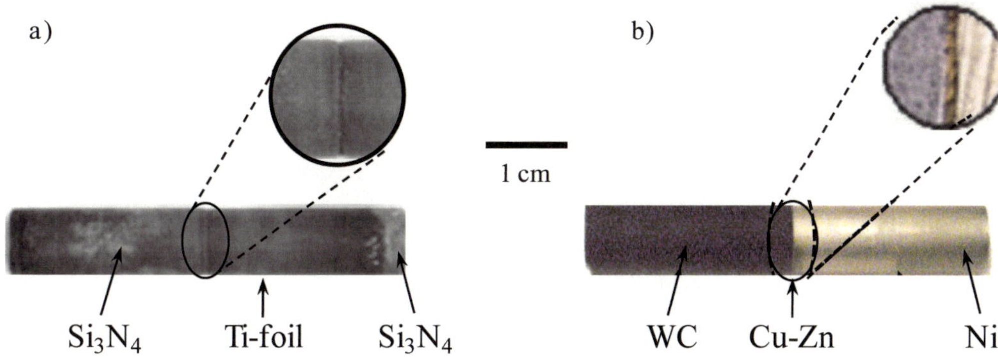

Fig. 12. Schematic representation of a) Si_3N_4/Ti/Si_3N_4 and b) WC/Cu-Zn/Ni samples

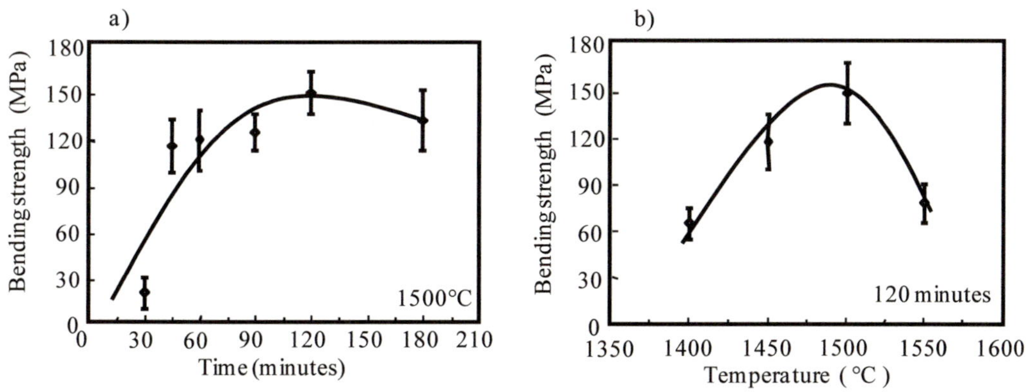

Fig. 13. Modulus of rupture in function of a) time and b) temperature for Si_3N_4/Ti/Si_3N_4 samples hot-pressed at 1500 C in vacuum

On the other hand, the results of modulus of rupture calculated for a) WC/Cu-Zn/Ni and b) WC/Ni samples diffusion (liquid and solid state, respectively) joining at 980°C and times varying from 5 to 60 minutes in argon atmosphere, are shown in Figure 14. In Figure 14a, it can be observed that the strength of the joint increased from a value of 233 MPa and reached a maximum value of 255 MPa, when the time was increased from 5 to 15 min, respectively, and decreased beyond this time. This improvement was attributed to the increase of interface reaction and formation of a strong chemical bridge between the two materials. On the other hand, the thickness of the reaction zone increases with time and may dominate in the final strength.

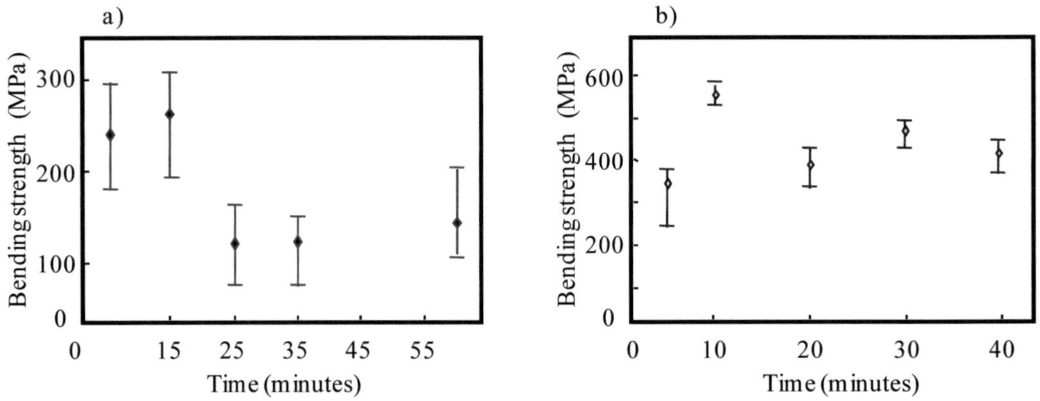

Fig. 14. Modulus of rupture for a) WC/Cu-Zn/Ni and b) WC/Ni samples joined at 980°C

Reaction products are generally brittle, and as the thickness of these phases increases, the joint strength, at first, rises due to the creation of a strong, integral bond and then reaches a maximum at a certain thickness and then decreases as the interface continues to grow. The strongest joint was obtained for joining conditions of 980°C and 15 minutes, with a resulting average bending strength of 255 MPa; however, joint strengths greater than 100 MPa in average were produced at 980°C and times vary from 5 to 60 minutes. In Figure 14b, four-point bending testing gave a maximum average joint strength higher than 300 MPa for WC/Ni samples produced by solid state diffusion bonding at 980°C. It can be observed a maximum average value joint strength of 576 MPa when the time was of 10 minutes, however, it value decreased when the time increase until 307 MPa at 5 minutes. The effect of a reaction layer on the interface strength depends on a number of factors such as the mechanical properties of the reaction layer, its thickness and morphology. The choice of suitable conditions to prepare ceramic/metal joints requires knowledge concerning the mechanism of reaction between the materials and the evolution of the interface. The thickness of the metal layer is very important, due that it affects directly the mechanical properties of a metal/ceramic joint. Therefore, the reaction layer thickness must be controlled in order to ensure good joint strength.

4.1.1 Interpretation of joint strength

The effect of a reaction layer on the interface strength depends on a number of factors such as the mechanical properties of the reaction layer, its thickness and morphology, the strength of the interfacial bond and the mode of loading at the interface. Reactions formed at metal/ceramic interfaces include solid solutions, amorphous and crystalline phases. Each of these reaction products forms a different type of interface between the metal and ceramic and the relative efficiency of these various types of reaction products on the strength of metal/ceramic interfaces is not fully understood. However, most reaction layers are brittle and therefore potentially detrimental to the interface properties. Although the concentration of residual stresses in joints is a function of the joining temperature, the thickness of the reaction zone may dominate in the final strength. Reaction products are generally brittle and as the thickness of these phases increases, the joint strength, at first rises due to the creation of a strong, integral bond and then reaches a maximum at a certain thickness and then decreases as the interface continues to grow. Therefore, the reaction layer thickness must be controlled ensure good joint strength. It is clear that the amount of interfacial reaction played a major role in determining the final mechanical properties of the joints. Furthermore, the nature of the reaction products may also have influenced the mechanical properties of the joints. The mechanical properties of a metal/ceramic interface depend upon many factors such as the elastic properties of the metal and ceramic, the thickness of the metal layer, the specimen geometry and the mode of loading.

In summary, the choice of suitable conditions to prepare ceramic/metal/ceramic joints requires knowledge about the mechanism of reaction between the materials and the evolution of the interface. The maximum value of 147 MPa was obtained for Si_3N_4/Ti/Si_3N_4 samples hot-pressed at 1500°C and 120 minutes. This strength is of the same order as the MOR reported for diffusion bonding of Si_3N_4/Si_3N_4 joints using a ZrO_2 interlayer made at 1550°C and 60 minutes (175 MPa) (Akselsen, 1992), and higher than that reported for brazing of Si_3N_4 to different metals, i.e. Mo/Si_3N_4 joints brazed with Cu-5%Cr alloy reports 120 MPa (Okamoto, 1990).

4.1.2 Fractography

All joints Si_3N_4/Ti/Si_3N_4 fracture samples showed the same type of fracture mode shown in Figure 15. The fracture originated and mainly propagated along the Si_3N_4/reaction zone interface and passed through the Ti metal to the other side of reaction zone/Si_3N_4 interface.

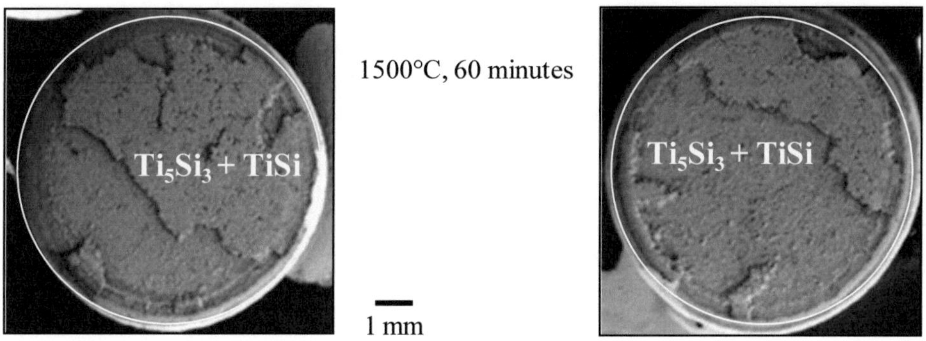

Fig. 15. Fractograph of Si_3N_4/Ti/Si_3N_4 samples after four-point bending test

The main part of the fracture surface was on the Ti-silicide of the Si_3N_4/reaction zone interface and probably initiated at the edge of the sample. Edges and corners are a major source of failure for metal/ceramic interfaces when there is a mismatch in either the thermal expansion or elastic modulus of the metal and ceramic. When a metal has either a larger thermal expansion coefficient or a lower modulus than the ceramic or both (which is often the case), the unconstrained metal develops a smaller lateral expansion at the interface than the ceramic. In order to maintain continuity at the interface, the metal must be uniformly extended by the application of edge forces. Surface forces which are equal in magnitude but opposite in sign must then be applied to the metal in the bonded state to achieve stress-free conditions at the surface and this introduces large normal and shear stresses near the edge of the metal/ceramic joint. A mismatch in modulus generates interfacial tensile stresses at the edge and thus always enhances the propensity to fracture. On the other hand, ceramics are more brittle, stiffer, and have lower thermal expansion coefficients than metals, thus, the interface between the ceramic and the metal will be in a state of stress when cooled from the joining temperature. If the metal, Ti, undergoes plastic deformation during the cooling cycle, some of the interfacial stresses could be relieved, resulting in an increased joint strength (El-Sayed et al. 1997). The amount of the plastic deformation depends on the yield strength of the metal. Generally the lower the yield strength, the more plastic deformation will occur and consequently this could lower the residual stresses. However the interface could have some defect or point of high stress concentration where fracture is initiated.

On the other hand, all WC/Cu-Zn/Ni joints fracture samples showed the same type of fracture mode where the fracture originated and mainly propagated along the WC/reaction zone interface and passed through the other side of reaction zone/Ni interface. However, in the case of WC/Ni samples the fracture did not occur in the interface of the joint, but the failure occurred in the metal Ni.

5. Conclusion

The amount of reaction between the ceramics and metal is an important factor in the mechanical reliability of the joints. Excessive growth of the reaction layer decreased joint strength.

Bending tests performed on $Si_3N_4/Ti/Si_3N_4$ joints hot-pressed at different joining temperatures confirmed the important relationship between the interface thickness and joint strength. Increasing the joining time further reduced the joint strength. Thickness of the reaction interface increases when bonding time increases. On the other hand, eutectic liquid formation during joining process plays an important role in promoting bond and interface formation. The surface roughness of the starting materials is an important factor affecting the extent of the interface growth. For similar joining conditions, hot-pressing of samples using unpolished materials produced thinner interfaces than polished samples.

6. Acknowledgment

The authors would like to thank to CONACYT-México and Universidad Michoacana a de San Nicolás de Hidalgo (UMSNH) for the financial support and facilities of this research.

7. References

Akselsen, O.M. (1992). Diffusion bonding of ceramics. *Journal of Materials Science,* Vol. 27, No. 3, pp. 569- 579

Almond, E.A., Cottenden, A.M. & Gee, M.G. (1983). Metallurgy of interfaces in hardmetal/metal diffusion bonds. *Metals Science,* Vol. 17, pp. 153-158

Anderson, R.M. (1989). Testing advanced ceramics. *Advanced Materials Procesing,* Vol. 3, pp-31-36

Asthana, R. (1998). Reinforced cast metals-part II evolution of the interface. *Journal of Materials Science,* Vol. 33, no. 8, pp. 1959-1980

Baker, H. (1992). *Alloy Phase Diagrams,* (Vol. 3), ASM handbook International, Materials Park, Ohio, USA

Chen, I.W. & Argon, A.S. (1981). Diffusive growth of grain-boundary cavities. *Acta Metallurgy,* Vol. 29, pp. 1759-1768

Contreras, A., Lopez, V.H., Leon, C.A., Drew, R.A.L & Bedolla, E. (2001). The relation between wetting and infiltration behavior in the Al-1010/TiC and Al-2024/TiC systems. *Advances in Technology of Materials and Materials Processing Journal,* Vol. 3, pp. 27-34

Dunford, D.v. & Wisbey, A. (1993). Diffusion bonding of advanced aerospace metallics. *Materials Research Society Symposium Proceeding,* Vol. 314, pp. 39-50

El-Sayed, M.H., Naka, M. & Schuster, J.C. (1997). Interfacial structure and reaction mechanism of AlN/Ti joints. *Journal of materials science,* Vol. 32, pp. 2715-2731

Howe, J.M. (1993). Bonding, structure and properties of metal/ceramic interfaces. *Materials Research Symposium Proceeding,* Vol. 314, pp. 27-37

Landry, K., Kalogeropuolou, S., Eustathopoulos, N., Haidich, Y. & Krasovsky, V. (1996). Characteristic contact angles in the aluminiumn/vitreous carbon system. *Scripta Materialia,* Vol. 34, No. 6, pp. 841-846

Lemus, J. & Drew, R.A.L. (2003). Joining of silicon nitride ceramics using a titanium foil Interlayer. *Materials Science and Engineering A,* Vol. 352, pp. 169-178, ISSN 0921-5093

Lemus-Ruiz, J. & Aguilar-Reyes, E.A. (2004). Mechanical properties of silicon nitride joints using a Ti-foil interlayer. *Materials Letters,* Vol. 58, No. 19, pp. 2340-2344, ISSN 0167-577X

Lemus-Ruiz, J., Ceja-Cárdenas, L., Salas-Villaseñor, A.L. & Bedolla-Becerril, E. (2009). Mechanical evaluation of tungsten carbide/nickel joints produced by direct diffusion bonding and using a Cu-Zn alloy, *Proceeding of the 4th International Brazing and Soldering Conference,* American Welding Society, ISBN 978-0-87171-751-1, Orlando, Florida, USA

Lemus-Ruiz, J., Ceja-Cardenas, L., Verduzco, J.A. & Flores, O. (2008). Joining of tungsten carbide to nickel by direct diffusion bonding and using a Cu-Zn alloy. *Journal of Materials Science,* Vol. 43, pp. 6296-6300, ISSN 0022-2461

Lemus-Ruiz, J., León-Patiño, C.A. & Aguilar-Reyes, E. (2006). Interface behaviour during the self-joining of si_3n_4 using nb-foil interlayer. *Scripta Materialia,* Vol. 54, No. 7, pp. 1339-1343, ISSN 1359-6462

Locatelli, M.R., Dalgleish, B.J., Nakashima, K., Tompia, A.P. & Glaeser, A.M. (1997). New approaches to joining ceramics for high-temperature applications. *Ceramics Internationals,* Vol. 23, pp. 313-322

Loehman, R.E., Tomsia, A.P., Pask, J.A. & Johnson, S.M. (1990). Bonding mechanisms in silicon nitride brazing. *Journal of the American Ceramic Society,* Vol. 73, No. 3, pp. 552-558

Mortensen, A. & Jin, I. (1992). Solidification processing of metal matrix composites. *International Materials Reviews,* Vol. 37, No. 3, pp. 101-128

Nicholas, M.G. (1990). *Joining of Ceramics; Advanced Ceramic Reviews-Institute of Ceramics,* Chapman and hall, New York, USA

Okamoto, T. (1990). Interfacial structure of metal-ceramic joints. *ISIJ International,* Vol. 30, No. 12, pp. 1033- 1034

Peteves, S.D., Ceccone, G., Paulasto, M., Stamos, V. & Yvon, P. (1996). Joining silicon nitride to itself and to metals. *JOM,* (January), pp-48-52

Richerson, D.W. (1992). *Modern Ceramic Engineering,* (2nd Edition), Marcel Dekker, New York, USA

Savov, L., Heller, H.P. & Janke, D. (1997). Wettability of solids by molten metals and alloys. *Metallurgy,* Vol. 51, No. 9, pp. 475-485

Schwartz, M.M. (1990). *Ceramic Joining,* ASM international, Materials Park, USA.

Suganuma, K. (1993). Reliability factors in ceramic/metal joining. *Research Society Symposium Proceeding,* Vol. 314, pp. 51-60

Treheux, D., Lourdin, P., Mbongo, B. & Juve, D. (1994). Metal/ceramic solid state bonding: mechanisms and mechanics. *Scripta Metallurgy et Materials,* Vol. 31, No. 8, pp. 1055-1060

Yokokawa, H., sakai, N., Kawada, T. & DokiYA, M. (1991). Chemical potential diagram of Al-Ti-C system: Al_4C_3 formation on TiC formed in Al-Ti liquids containing carbon. *Metallurgical and Materials Transactions A,* Vol. 22, pp. 3075-3076

Zhang, J. & He, Y. (2010). Effect of Ti contento n microstructute and mechanical properties of Si_3N_4 joints brazed with Ag-Cu-Ti+Mo composite filler. *Materials Science Forum,* Vol. 654-656, pp. 2018-2121 doi:10.4028/www.scientific.net.

Zou, J., Jiang, Z., Zhao, Q. & Chen, Z (2009). Brazing of Si_3N_4 with amorphous $Ti_{40}Zr_{25}Ni_{15}Cu_{20}$ filler. *Materials Science and Engineering A,* Vol. 507, pp. 155-160. www.elsevier.com/locate/msea.

11

Measurement of Strain Distribution of Composite Materials by Electron Moiré Method

Satoshi Kishimoto, Yoshihisa Tanaka,
Kimiyoshi Naito and Yutaka Kagawa
National Institute for Materials Science
Japan

1. Introduction

It is very important to measure the local strain and stress distributions for understanding the mechanical properties of structural materials. Therefore, there are many techniques to measure the strain or stress distribution such as the strain gage method, optical elasticity method, optical Moiré method, etc. In these methods, the optical Moiré method (Weller & Shepard, 1948; Morse et al., 1960; Sciammarela & Durelli, 1961; Durelli & Parks, 1970; Theocaris, 1969; Post et al, 1994; Chiang, 1982; Post, 1988) is one of the convenient methods to measure the deformation of the materials. However, these methods are difficult to apply for deformation measurements from a microscopic aspect.

To measure the micro-deformation in a very small area, the authors have developed an electron Moiré method (Kishimoto et al., 1991, 1993) , and J.W. Dally, D. T. Read (Read & Dally, 1994; Dally &.Read, 1993) and H. Xie (Xie et al., 2007) advocated it. This method keeps the main advantages of the moiré and laser moiré interferometry methods, and has the additional ability of measuring deformation in a micro-area with a high sensitivity. Besides, the electron moiré method also uses a wide range for measuring the deformation. The range of the measurable deformation is from 25microns to 0.1micron using a model grid with different pitches.

To measure the micro-deformation i.e. sliding and slip lines in a smaller area, micro-grid method is very useful. Compare these two methods, electron moiré method is easy to understand the strain distribution and the large sliding (Kishimoto et al., 1991, 1993).

In this study, In order to pursue the application of the electron moiré method, some typical experiments were performed. The strain distribution at the interface of the laminated steel, strain distribution of the fiber and the matrix in the fiber reinforced plastic, the thermal strain in or around the metallic fiber in Al alloy were observed.

2. Experimental procedure

2.1 The principle of the electron Moiré method

The principle of the optical moiré method is shown in Figure 1 (a). In this method, moiré fringes can be observed when two gird (model grid and master grid) are overlapped. From

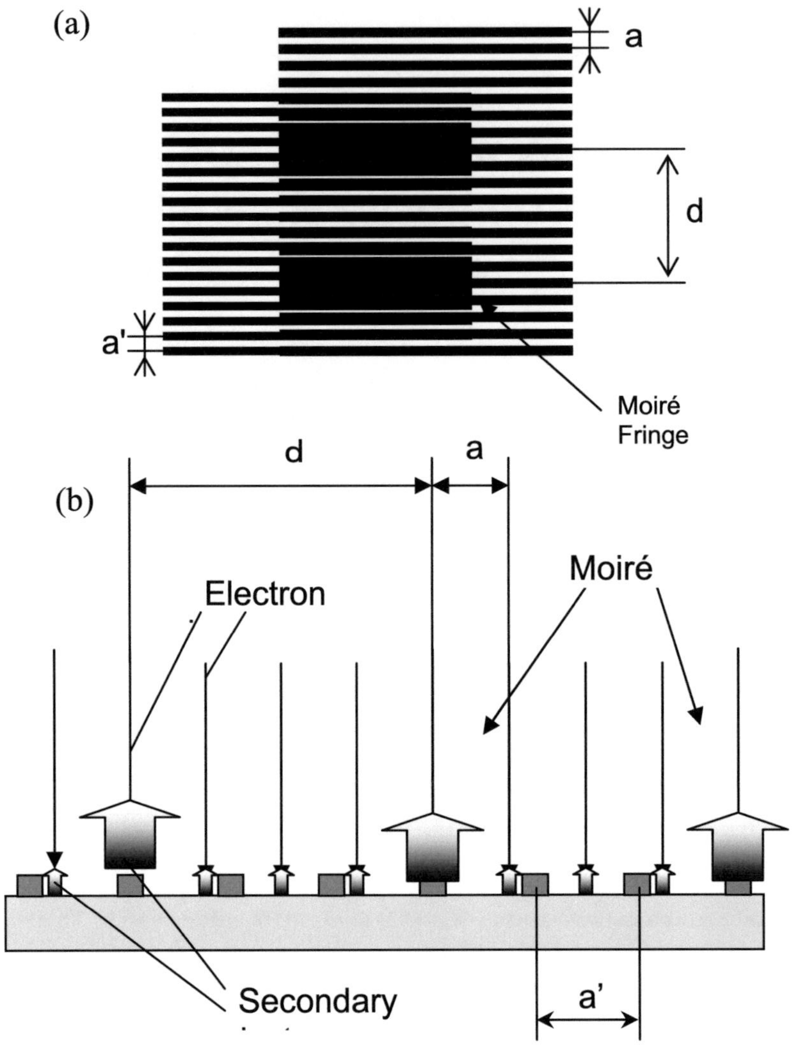

Fig. 1. Schematic formation of moiré fringe; (a) and the spacing of master grid ;(b) electrons after secondary.

the spacing of moiré fringes, and the spacing of master grid, the strain can be calculated. A principle of an electron moiré method is shown in Figure 1 (b) and as follows. A model grid is prepared on the surface of the specimen before deformation by using electron beam lithography or photo-lithography. Before and after deformation, the specimen is mounted on the specimen stage in a scanning electron microscope (SEM) and an electron beam scan having spaces almost same as that of the model grid can be used for the master-grid. The difference in the amount of the secondary electrons per a primary electron makes the Moiré fringes (electron Moiré fringes) that consists bright and dark parts.

2.2 Micro grid fabrication by electron beam lithograph

The procedure for producing a model by using the electron beam lithography (Kishimoto, et. al, 1993) is shown in Figure 2. The specimen must first be polished to a mirror-like finish (up to 0.05μm Al_2O_3 powder), then it is covered with an electron-sensitive layer (electron beam resist, Nippon Zeon ZEP-520-22 and Toray EBR-9), spinning at a speed of 2,500 RPM for 120 seconds, and then baked in an oven for 30 minutes at 453 K for ZEP-520 and spinning at a speed of 2,000 RPM for 120 seconds, and then baked in an oven for 30 minutes at 468 K for EBR-9.

The specimen was then mounted on the specimen stage in a TOPCON SX-40A SEM for the electron beam exposure. After electron beam exposure, the specimen coated by ZEP-520 was developed in a solution of ZED-N50 for 60 seconds, and then immediately rinsed in ZMD-B for 30 seconds. The specimen coated by EBR-9 was developed in a solution of Type 1 for EBR-9 for 60 seconds, and then immediately rinsed in 2-propanol for 30 seconds. The specimen was coated with a very thin layer (10-20 nm) of gold by plasma sputtering.

The difference in the emitted amount of the secondary electrons per a primary electron between the surface and the deposited layer must be large enough to get produce a contrast in the electron moiré fringe. After removing the resist using an organic solvent, a model grid is formed on the specimen surface.

2.3 Observation of electron moiré fringe and micro-grid

The specimen with micro-grid (model grid) was mounted on the specimen stage in a TOPCON SX-40A SEM for the observation of electron moiré fringe and TOPCON SM-510 for observation of micro-grid. The large difference in the emitted amount of the secondary electrons per a primary electron between the surface and the deposited layer makes it possible observes the micro-grids and the electron moiré fringes.

2.4 Three-point bending test of fiber reinforced plastic

Figure 3 (a) shows a schematic set-up of the three-point bending test specimens. A model grid was prepared near the bottom surface of the carbon fiber reinforced plastic (CFRP) as shown in Figure 2 (b) and a higher magnification SEM image is shown in Figure 3 (c). The strain of the bottom surface (bending strain) was measured by a strain gage. The specimen's surface was observed by an optical microscope (NIKON Opti-Photo) with a CCD camera, and the electron moiré fringes were observed by the scanning electron microscope (TOPCON SX-40A) equipped with a beam blanker and a pattern generator. The specimen was set on the stage of SEM (TOPCON SM-510) with bending devices. SEM images and electron moiré fringes were observed during bending.

2.5 Compression test of laminated steel

Eight sheets of austenite stainless steel (304 stainless steel sheets (SUS304)) of 10-mm thickness and 7 sheets of martensite steel (WT-780C) sheets of 12-mm thickness were alternately overlapped and rolled to 12.7mm at 1473 K after a 2h hold at the same temperature. The thickness of each layer was about from 0.7 mm to 0.9mm.

This laminated steel was cut and machined into specimens with a 10-mm width, 10-mm length and 12-mm height, and then the laminated sides were then polished. On the polished surface, some model grids with a cross grid of 2.9-μm spacing were prepared by electron beam lithography. The compressive tests were performed using a universal tensile test machine (Shimadzu AUTOGRAPH AG-100kND) at room temperature. The loading direction was perpendicular and parallel to the each layer. After the test, the nominal strain

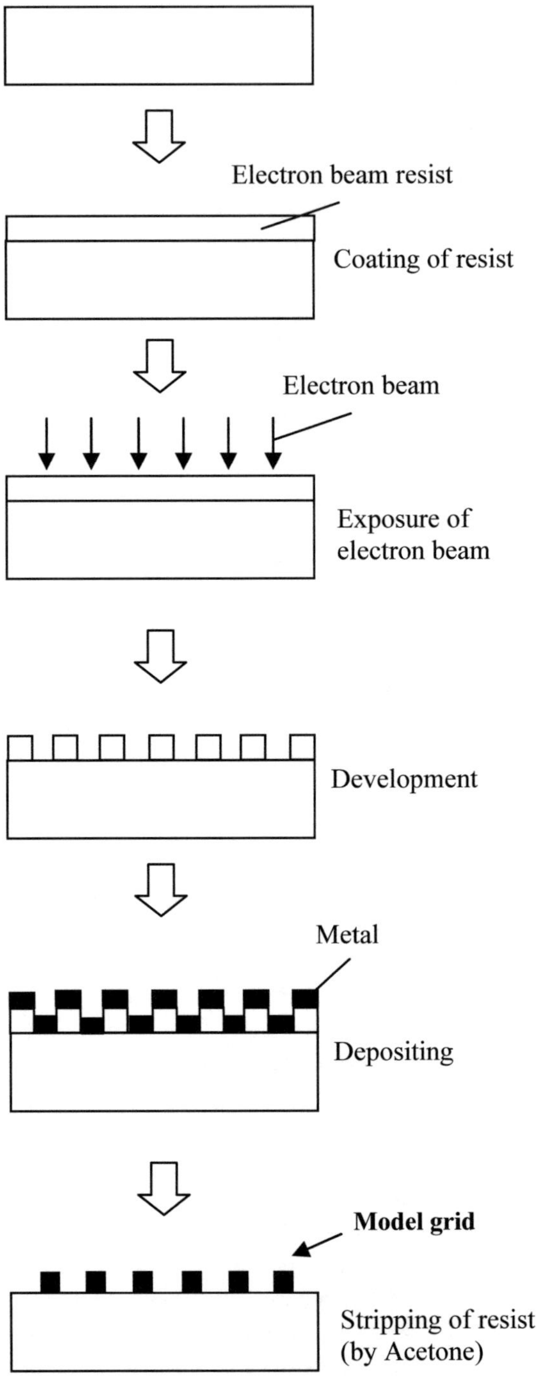

Fig. 2. Schematic image of model grid fabrication process by electron beam lithography.

was measured. The specimen was then placed on the stage of a scanning electron microscope (SEM), and the electron moiré fringe was observed by the electron moiré method. Figures 4 (a) and (b) show the schematic image of compression test specimen and model grid on the specimen's surface, respectively.

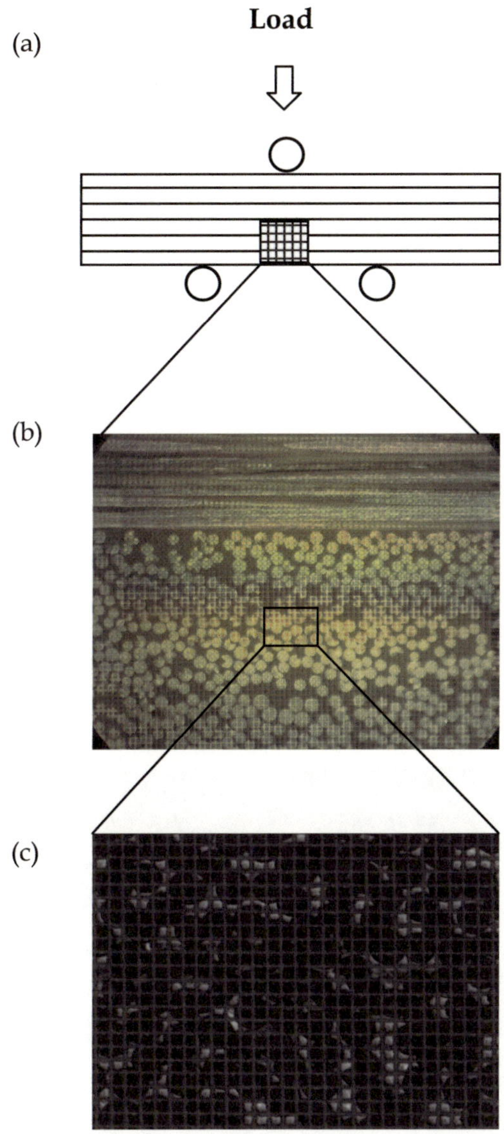

Fig. 3. Schematic image of setup of tree point bending test ;(a) , Optical microscope image on model grid; (b) and SEM image of model grid (spacing of the model grid is 5μm); (c).

2.6 Deformation by thermal expansion

The specimen was made of commercial aluminum alloy with Ti-Ni fibers. The diameter of the Ti-Ni fiber was about 600μm. A sample was cut from the bulk laminates perpendicular to one fiber axis using a diamond saw, and then polished into a film about 300µm thick. The specimen was put on the heating stage in SEM. The electron moiré fringes before and after heating (up to 353K) were observed. From these moiré fringes, strain change by thermal expansion was calculated and thermal expansion ratio was calculated.

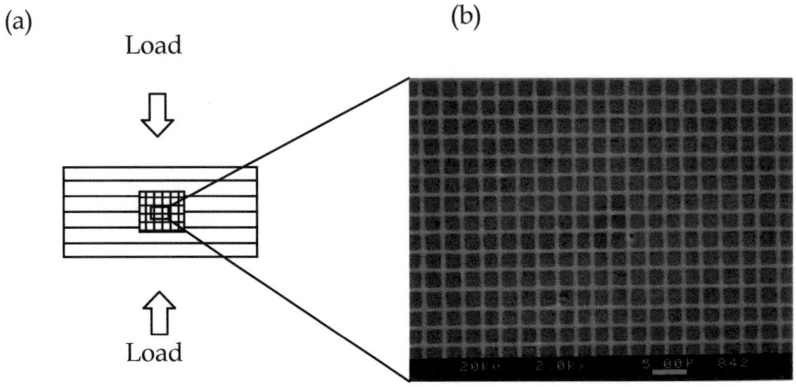

Fig. 4. Schematic image of setup of compression test ;(a) and SEM image of model grid; (b).

3. Result and discussion

3.1 Model grid and electron moiré fringe

Figure 3 (c) and Figure 4 (b) show the model grid fabricated by electron beam lithography. The specimen's surface was observed by a scanning electron microscope (SEM:TOPCON, SM-510), and the electron moiré fringes were observed by the SEM (TOPCON SX-40A) equipped with a beam blanker and a pattern generator. Figure 5 shows the electron moiré fringe created by the electron beam exposure with 3.3 µm spacing onto the model grid (2.9-µm spacing) of the specimen shown in Figure 4 (b).

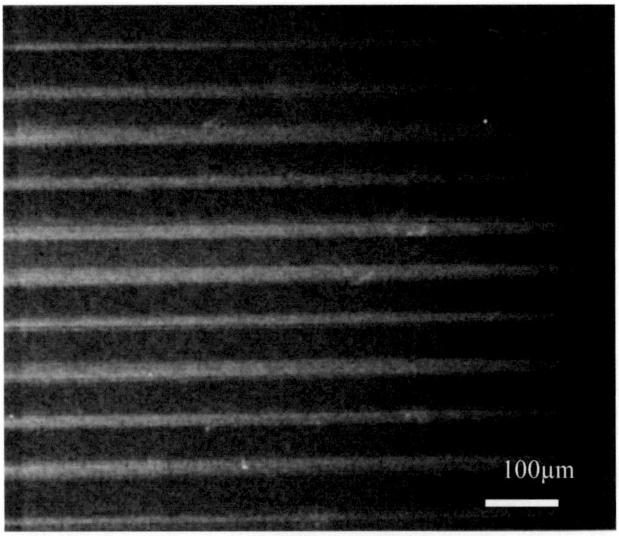

Fig. 5. Electron moiré fringe on the model grid of Figure 4 (b).

3.2 Strain distribution of fiber reinforced plastic under three-point bending test

The strain distribution of a carbon fiber reinforced plastic during a bending test and after initiation of cracks was measured. Figures 6(a) and (c) show SEM images of a same micro-grid on the FRP specimen during bending (maximum strain is 0.1%) and after initiation of cracks (see Figure 7), respectively. Figures 6 (b) and (d) show electron Moiré fringes of the same place during bending (maximum strain is 0.1%) and after initiation of cracks, respectively. The electron beam exposure with a 4.5μm spacing onto the model grid (5 μm spacing) of the specimen creates the electron moiré fringes. Figure 6 (b) and (d) show electron Moiré fringe fabricated by electron beam scan in the horizontal direction. Strain distribution in the horizontal direction, ε_x is also indicated in this figures. Strain ε_x can be calculated by using the images of the electron Moiré fringes and eq. (1). Using the moiré patterns, the strain ε can be calculated, which are expressed as follows, (Kishimoto et al., 1991, 1993)

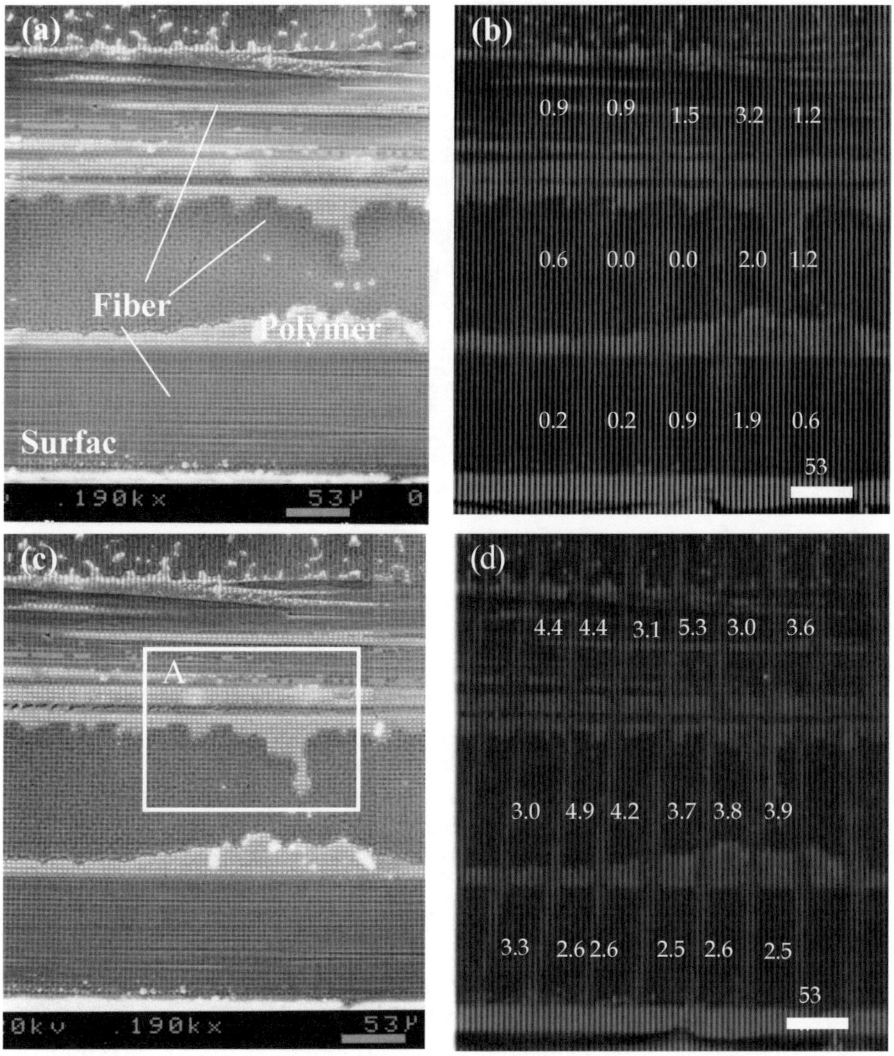

Fig. 6. SEM image; (a) and electron moiré fringe; (b) at 0.1% maximum strain and SEM image; (c) and electron moiré fringe; (d) after cracking of FRP specimen.

$$e = a/(d-a)-(a'-a)/a \quad (1)$$

where, **d, a** and **a'** are the spacing of the Moiré fringe, the spacing of the electron beam scan and the spacing of the model grid before deformation, respectively. The strain in a 100μm area can be measured using the electron Moiré method. Before initiation of the cracks, the strain ε_x is tensile and almost same. However, after initiation of the cracks, strain ε_x near the surface and inside is different. The strain ε_x near the surface is smaller than that of inside. It should be thought that stress near the surface of the specimen was released by the cracking.

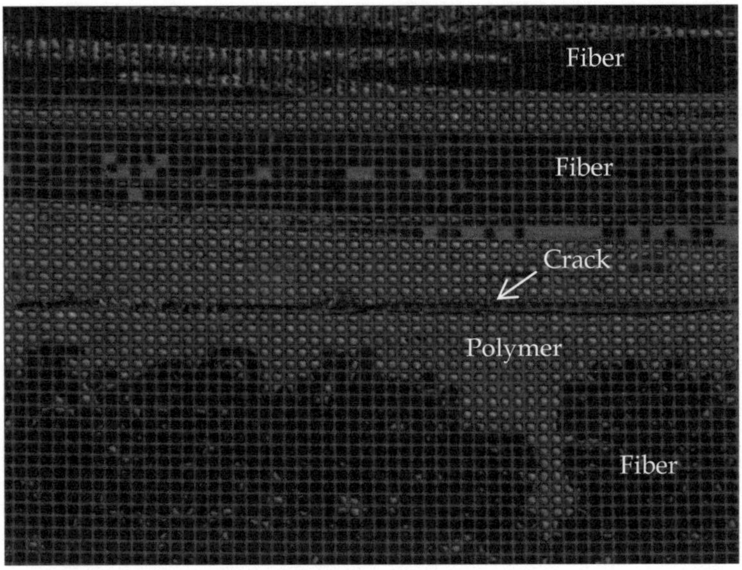

Fig. 7. Higher magnification SEM images of the area A in Figure 6 (c).

3.3 Compressive deformation of laminated steel

The strain distribution of laminated steel after a compression test (about -4.6%) was measured. igures 8 (a) and (d) show SEM images of the micro-grid area, Figures 8 (b), (c), (e) and (f) show an example of the electron Moiré fringe after the compression test. For Figure 8 (a), (b) and (c), the loading direction is perpendicular to the laminate sheets and for Figure 8 (d), (e) and (f), the loading direction is parallel to the laminate sheets. These electron moiré fringes are created by electron beam exposure with 3.3-μm spacing onto the model grid (2.9-μm spacing) of the specimen (Figures 8 (b) and (e)) and electron beam exposure with 2.5-μm spacing onto the model grid (3.2-μm spacing) of the specimen (Figures 8 (c) and (f)). Also, the electron moiré fringes in Figures 8 (b) and (e) is formed by the electron beam scan in the direction perpendicular to the loading direction and the electron moiré fringes in (Figures 8 (c) and (f) is formed by the electron beam scan in the loading direction. The boundaries between SUS304 and WT-780, the parts of SUS304 and WT-780 and the bottom surface are also shown in the same figures.

In Figure 8 (b), (c), (e) and (f), the electron beam moiré fringes are complicated. In Figure 8 (b), (c), (e) and (f), the shape of the electron moiré fringe is wavy and the spacing of moiré fringe is different in places. This means that the strain in the horizontal direction, ε_x and the strain in the vertical direction, εy is different at each point and distributed largely. Near the boundary between WT-780C and SUS304, the strain ε_y is larger than that of other places.

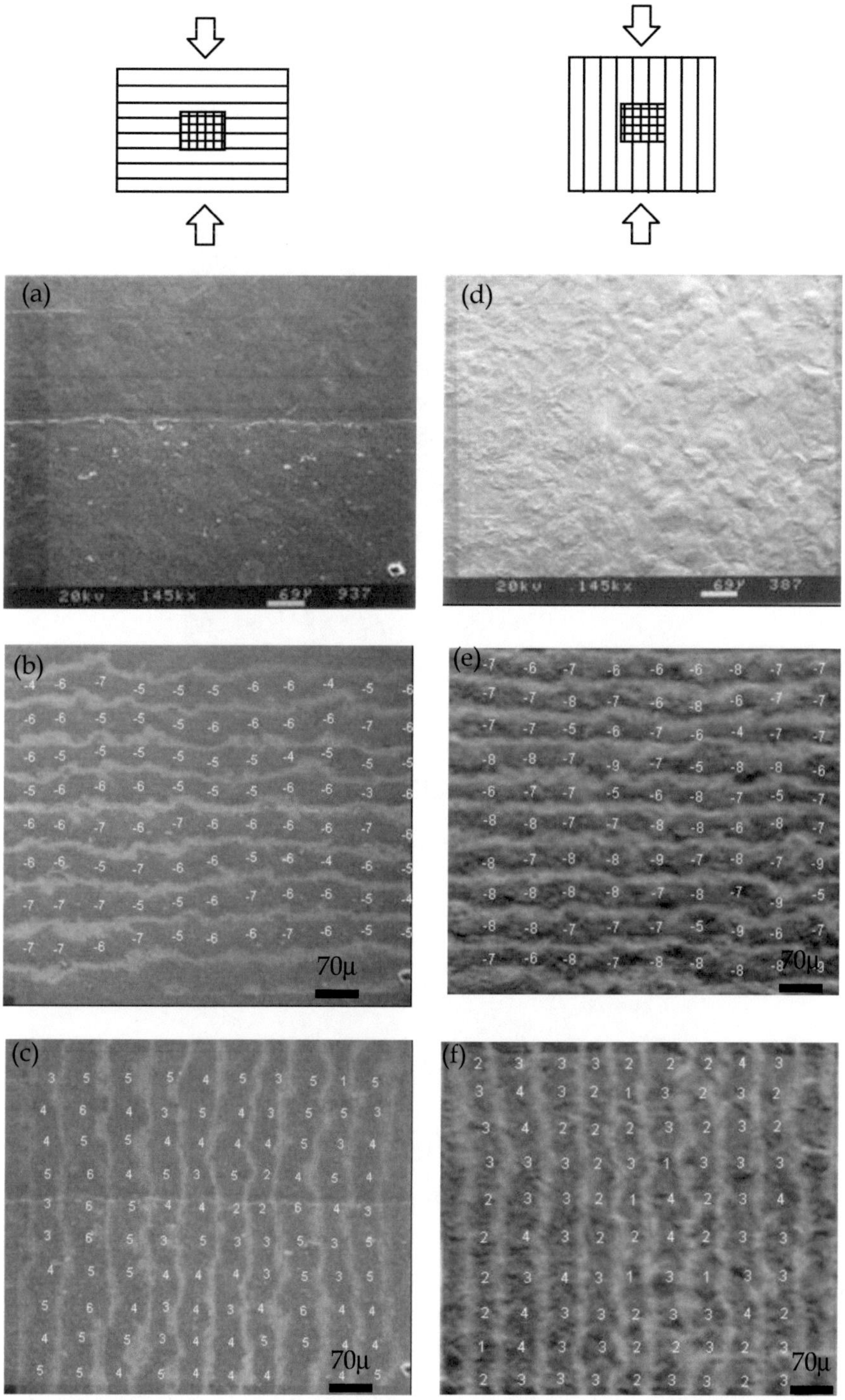

Fig. 8. SEM images; (a) and (b), electron moiré fringe in the direction perpendicular to the loading axis; (c) and (d), and electron moiré fringe in the loading direction; (e) and (f)after about 4.5% compressive deformation. (Numbers are distributed strain (%) at same point)

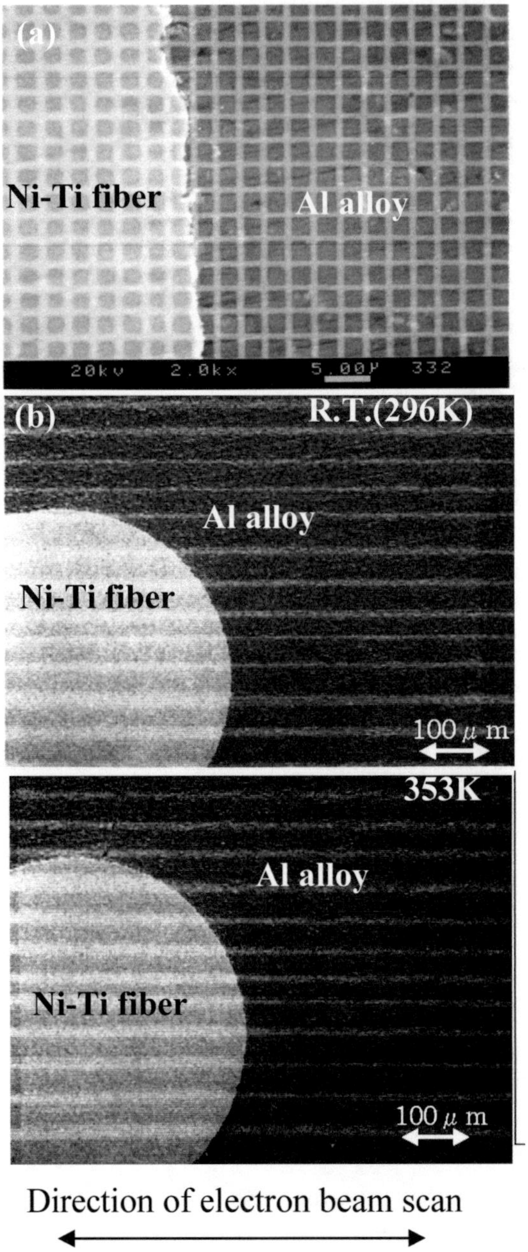

Fig. 9. Higher magnification SEM image of model grid; (a), electron moiré fringes around the Ti-Ni fibers in Al alloy at room temperature; (b) and electron moiré fringes around the Ti-Ni fibers in Al alloy at room temperature at 353K; (c).

The strains in Figure 8 (b) and (c), ε_x, and ε_y, changed from +1% to +6% and -3% to -7%, respectively. The strains in Figure 8 (e) and (f), ε_x, and ε_y, changed from +1% to +4% and -5% to -9%, Compare with the strains ε_x, and ε_y, in the SUS304 area and WT-780C area absolute value of ε_x, and ε_y, is almost same. However, the strain was distributed largely.

3.4 Deformation by thermal expansion

Al alloy with Ni-Ti fiber which had a model grid was set on the heating stage in an SEM chamber and the electron moiré fringes before and after heating were observed. Figure 9 (a), (b) and (c) show an SEM image of the morel grid with 2.9 μm spacing, electron moiré fringes before (room temperature (296K)) and after heating (353K), respectively. Electron beam exposure of 2.7 μm spacing Strain on to the model grid (2.9 μm spacing) on the specimen's surface created the electron moiré fringe.

Strain caused by thermal expansion was calculated by using equation (1). The strain of the Ni-Ti fiber part and Al alloy part caused by the heating was 0.23% and 0.15%, respectively. From these values, coefficient of thermal expansion was also calculated. The coefficient of thermal expansion of the Ni-Ti fiber and Al alloy in this material was 4.0×10^{-5}/K and 2.6×10^{-5}/K, respectively.

4. Conclusion

The strain distribution of the fiber and the matrix in the fiber reinforced plastic, around the interface of the laminated steel, the thermal strain in or around the metallic fiber in Al alloy were observed by electron moiré method. After initiation of cracksi FRP non-uniform deformation was observed. Non-uniform deformation near the boundary of laminated steel of brittle martensitic steel (WT-780) and ductile austenite stainless steel (SUS304) during compressive test was observed. Also, thermal expansion could be observed and coefficient of thermal expansion was calculated.

5. Acknowledgment

This work was partly supported by the LISM (Layer-Integrated Steel and Metal) Project by the Ministry of Education, Culture, Sports, Science and Technology (MEXT), Japan and the Failsafe hybrid composite project of the National Institute for Materials Science, Japan.

6. References

Weller, R and Shepard, B.M., (1948) Displacement measurement by mechanical interferometry, *Proc. Soc. for Exp. Stress Anal.*, 6 35–38

Morse, S.A., Durelli, J., and Sciammarella, C.A., (1960),Geometry of moiré fringes in strain analysis, *J. Eng. Mech. Div.*, 86 105-126.

Sciammarela, C.A., Durelli, A.J, (1961) Moiré fringes as a means of analyzing strains, *J. Eng. Mech. Div.*, 87, 55-74.

Durelli, A.J., Parks, V.J., Moiré, (1970) Analysis of Strain, Prentice Hall..

Theocaris, S., (1969) Moiré Fringe in Strain Analysis, Pergamon..

Post, D. , Han B., Ifju, P., (1994) High Sensitivity Moiré, Springer.

Chiang, F.P., (1982), Moiré method of strain analysis Manual on Experimental Stress Analysis 5th edn, Doyle and Phillips, Society for Experimental Mechanics, 107–35

Post, D., (1988) Sharping and multiplication of moiré fringe, *Exp. Mech.* 28 329

Kishimoto, S., Egashira, M., Shinya, N.and Carolan, R.A., (1991) Local Micro-deformation Analysis by Means of Micro-grid and Electron beam Moiré Fringe Method, *Proc. 6th Int. Conf. on Mech. Behavior of Materials*, Pargamon Press, pp.661-666,

Kishimoto, S., Egashira, M. and Shinya, N., (1993) Micro-creep deformation measurement by a moiré method using electron beam lithography and electron beam scan, *Opt. Eng.*, 32 522-526 .

Read D.T and Dally, J.W., (1994), Electron Beam Moiré Study of Fracture of a Glass Fiber Reinforced Plastic Composite, *J APPL MECH-T ASME*, 61, 402-409.

Dally, J.W., Read, D.T., (1993), Electron-Beam Moiré, *Exp. Mech.* 33, 270-277.

Read D.T and Dally, J.W., Szanto, M., (1993) Scanning moiré at high magnification using optical methods, *Exp. Mech.*, 33 110

Xie, H., Wang, Q., Kishimoto, S., Dai, F., (2007) Characterization of planar periodic structure using inverse laser scanning confocal microscopy moiré method and its application in the structure of butterfly wing, *J. Applied Physics*, 101, 103511.

Part 2

New Materials with Unique Properties

12

Joining of C_f/C and C_f/SiC Composites to Metals

K. Mergia
National Centre for Scientific Research "Demokritos",
Institute of Nuclear Technology and Radiation Protection
Greece

1. Introduction

Composite is a combination of a reinforcing material within a matrix material that act together to provide enhanced properties unobtainable in either single material. Often the matrix provides protection from damage and gives mechanical stability to the reinforcement under loading; at the same time the reinforcement can enhance the mechanical properties of the matrix [Messler, 2004].

Ceramic Matrix Composites (CMCs) combine reinforcing ceramic phases within a ceramic matrix to overcome the intrinsic brittleness and lack of reliability of engineering ceramics (monolithic ceramics) while retaining their most favourable properties. The ceramic matrix composites (CMCs) considered in this study consist of carbon ceramic fibers (arranged in multidirectional preforms) embedded in a C- or SiC-matrix. These CMC materials have the potential for being used at high temperatures, as thermo-structural materials, in different fields including heat exchangers used in severe environments such as rocket and jet engines [Schmidt et al., 2005], gas turbines for power plants, heat shields for space vehicles, inner walls of the plasma chamber of nuclear fusion reaction [Nozawa et al., 2009; Schlosser et al., 2005], aircraft brakes, heat treatment furnaces, etc. The main advantages of CMCs lie in the fact that they present superior mechanical properties and resistance against high temperatures and at the same time they are lightweight and cost-effective [Lamouroux et al., 1999; Naslain, 2004].

The key property of the CMCs is achieved through a proper design of the fiber/matrix interface arresting and deflecting cracks formed under load in the brittle matrix and preventing the early failure of the fibrous reinforcement [Evans & Marshall, 1989]. Crack deflection is controlled via the deposition of a thin layer of a compliant material with a low shear strength, on the fiber surface, referred to as the interphase and acting as a mechanical fuse (to protect the fiber) [Naslain, 2004].

Carbon fiber reinforced carbon composites (C_f/C) have excellent high temperature properties; they maintain high strength and high toughness even at temperatures above 2273 K [Goto et al., 2003; Schmidt et al., 2005]. The strength to weight, in particular at high temperatures, compared with metal materials is just one special feature that make them attractive as a construction materials. On the basis of these attributes C_f/C composites are used in the nose cone and leading edges of the space shuttle, solid propellant rocket nozzles and exit cones, heat shield, and other components for ballistic missiles, and aircraft braking systems. C_f/C composites will be part of the divertor, a complex and critical component of International Thermonuclear Experimental Reactor (ITER) [Tivey et al., 2005].

Carbon fiber reinforced SiC ceramic matrix composites (C_f/SiC) are lightweight, hard, and wear resistant and stable in oxidizing environment up to a high temperature. Owing to the embedded carbon fibers, they have an excellent combination of mechanical properties. Therefore, C_f/SiC composites are promising new structural materials for a variety of high-temperature burner environments, including hypersonic aircraft thermal structures, advanced rocket propulsion thrust chambers, cooled panels for nozzle ramps, turbo pump blisks/shaft attachments, and brake disks [Ishikawa et al., 1998].

Most such applications in aerospace and nuclear industry require joining CMCs to metals, ceramics or composites. Fabrication of complex large-scale structural components requires robust integration technologies capable of assembling smaller, geometrically simple CMC parts. Thus, advances in joining science and technology are important in order the benefits of these advanced materials to be realized. Generally the requirements for CMC-metal joints are high mechanical strength at high temperatures and in aggressive environments, high oxidation resistance, high stability under service and low production cost.

In general, there are four basic joining methods, mechanical, adhesive, diffusion bonding and brazing (or combination of them) which rely on mechanical, chemical and physical forces. The mechanical joining, although simple in its realization, can present a plethora of weak points which can make it unreliable. The adhesive method is mainly used for joining CMC to CMC or CMC to polymer. The two latter techniques have been developed for application in extreme environmental conditions.

For a successful fabrication of CMC-metal joining that performs well at elevated temperatures and under the application of loads there is a series of aspects which have to be addressed and successfully coped with. Both materials to be bonded are inhomogeneous and have very different coefficients of thermal expansion (CTEs) which lead to high residual stresses. Moreover, the difference in Young's modulus may introduce stress concentrations at the interface between the dissimilar materials. To cope with the mismatch of the usually widely differing CTEs, a number of approaches could be used [Razzell, 2004]:

(i) Functionally graded joints in which the strain differences are spread gradually through intermediate layers by altering ratios of constituents.

(ii) Strain isolation systems in which a flexible (e.g., metallic) interlayer accommodates the strain mismatch by elastic deformation.

(iii) Variation of geometry (angled or domed shapes) to minimize stress concentrations.

The inhomogeneous surface structure can be addressed by the application of a coater which bonds well with all the constituents or by an appropriate surface modification. In some cases the rough surface may be advantageous if it is made to act as an interlocker.

A strong bonding can not rely on the weak Van der Waals forces but to chemical reactions which involve the main constituents of the work pieces. This can be achieved by either direct bonding or using interlayers. Whatever the joining process is, chemical equilibrium at the joint interface is needed to form a stable CMC/metal bond for operation at high temperature [Nicholas & Peteves, 1994]. Interfacial reactions can affect every stage of the formation of a joint from the onset of bonding through the development of equilibrated microstructure and to the optimization of the mechanical properties [Peteves et al., 1998; Ashworth et al., 2000].

Another important aspect is the diffusion of species from the metal to the CMC and vice-versa. The diffusion has implications in both the fabrication of joints, if high temperatures are employed and also in the service conditions. The diffusion paths can be very complex as many species are involved but also porosity and microcracking would play an important factor. These effects are expected to be detrimental to both the joining and the joined materials and they can be addressed by choosing a joining technique which can be applied at lower temperatures and creates homogeneous, dense and free of porosity layers. The diffusion of species, in high temperature service conditions, can be handled by the application of a diffusion barrier. A diffusion barrier would also be needed when a combination of techniques are employed e.g. sputtering and diffusion bonding. For a successful metal-ceramic joining the ability to grow both metallic and ceramic layers is of most importance.

In the current review paper first are discussed the main aspects of mechanical joining, adhesive and diffusion bonding and these are presented in sections 2, 3 and 4, respectively. Since brazing is the most common method for ceramic-metal joining and there is an increasing interest for its application in CMC/metal joining, special emphasis is given and its scientific principles as well as the most interesting applications of CMC/metal brazed joints are presented in section 5. A short description of the most significant aspects concerning the wettability and the different brazing alloys used for C_f/C and C_f/SiC joining to metals are discussed in paragraph 5.1. The various brazing parameters, such as temperature, holding time, surface preparation and mostly important the difference in the thermal expansion coefficient (CTE) in the metal/filler/ceramic structure, as well as ways to cope with the different limitations are examined in paragraph 5.2. The chemical interaction of the brazing alloys with the CMC substrate and the formation of different compounds are discussed in section 5.3. In section 5.4 examples of successful joining of C_f/C and C_f/SiC composites to metals for mainly aerospace and nuclear applications are given. Finally, in section 6 characterization techniques assessing the structure and quality of the joint are briefly described.

2. Mechanical joining

Mechanical joining is the oldest and the simplest way to join ceramic matrix composites (CMCs) to metals using either fastening or mechanical attachment. In this type of joining only forces of mechanical origin enable the joining. The materials that comprise the joined pieces remain distinct and separate at the atomic or molecular level.

The principle of fastening in mechanical joining is based on drilling holes in the two materials to be joined and then placing a mechanical fastener through the holes and fixing the fastener in place. This type of joining, although simple in its realization, can present a plethora of weak points such as a) deformation of the CMC materials due to the compressive forces at the joints, b) delamination due to drilling, c) failure of the CMC material due to localized forces (stress concentration) and d) fatigue after long term use of the mechanical joint.

In mechanical attachment, no supplemental devices (or fasteners) are needed. Joining is accomplished by interference or interlocking between designed-in or fabricated-in geometric features integral to the parts being joined. Examples include dove-tails, tongues-and-grooves, T-slots, wedges, and integral snap-fit features molded into plastic parts, as well as elastic thermal shrink-fits, elastic or plastic mechanical interference fits, and plastic staking, crimping, folding and hemming all also found in plastic (especially thermoplastic) parts [Messler, 2004]. Snap-fit integral features are beginning to appear in metal extrusions and castings (to allow the snap assembly of the space-frame in aluminum-intensive automobiles) and in composites (to allow the snap assembly of power transmission towers (Goldsworthy et al., 1994)).

In general, mechanical joining may present a number of disadvantages which can make it unreliable. Issues such as stress concentration, positions of bores, performance under high temperature of joining parts (for instance, bolts) and their match properties with base materials etc, renders the application of mechanical CMC/metal joining, limited [Zhang et al. 2006]. Another issue is the possible intrusion of a fluid which may be prevented by using appropriate seals or gaskets. In addition, it is difficult to bore through the ceramic matrix composites, and the machining process often produces defects, degrading the mechanical behaviour of the composites.

In aerospace applications, and specifically for the re-entry demonstrators currently being developed in Europe thermal protection systems (TPS) based on C_f/SiC mechanically fastening is used for their joining to the metallic panels [Pichon et al., 2009]. TPS of CMC material may be connected to the space vehicle by CMC fasteners [Böhrk & Beyermann, 2010]. However, examination after the re-entry tests showed that these CMC fasteners may present loss of torque which can be mission critical [Reimer, 2006].

One method of making a joint with a metallic bolt and a CMC is by arranging the thermal expansion to cause sliding between the two components, rather than loosening the joint [Razzell, 2004]. A 45° conical head metallic fastener can be used in a matching countersunk hole in the CMC. The arrangement can be adapted for linear joints as well, with metallic inserts used to provide the angled surface. The above joint is limited to moderate temperatures by the capabilities of the bolt, a typical upper temperature being ~800 °C for typical superalloys. Fretting between CMCs and between CMCs and metals is a component specific problem.

3. Adhesive bonding

Adhesive bonding is a process of joining materials with the aid of a substance able to hold those materials together by surface attachment. The forces that enable the surface attachment are a combination of substantial secondary (and, occasionally, primary) chemical bonding, often the result of some chemical reaction, and, at least for rough-surfaced or porous adherends, some mechanical locking at the microscopic level between the adhesive and the adherends [Messler, 2004].

The most common structural adhesives are: epoxies, polyurethanes, acrylics, cyanoacrylates, anaerobics, silicones, and phenolics. The adhesives to be successful have to ensure good wetting and chemical compatibility with the surfaces to be bonded. Their advantages rely on the weight saving, the more homogeneous distribution of forces compared to mechanical joining, ability to join complex shapes and capability for sealing, insulating, damping, and resisting fatigue. However, it is problematic that the bonded joint cannot be disassembled without damage. Their use and performance can be limited or degraded by a number of environmental factors, such as temperature, humidity, UV-light exposure and attack by solvents and fungus. The most serious problem for adhesive bonding, however, is that design engineers are uncertain regarding long-term structural integrity and, further, bonded joints tend to fail instantaneously, not progressively [Kweon et al., 2006]. Even if a bonded joint meets strength requirements in laboratory testing, it is doubtful whether the bonding will work with the designed target performance throughout the whole life span of the joint.

Also hybrid joints have been proposed which combine mechanical fastening and adhesive bonding [Kweon et al., 2006]. It is pointed out that hybrid joining is effective when the mechanical fastening is stronger than the bonding. On the contrary, when the strength of the bolted joint is lower than the strength of the bonded joint, the bolt joining contributes little to the hybrid joint strength.

Currently used thermal protection systems (TPS) in aerospace applications utilize epoxy for the composite matrix. Epoxy matrices are known to be unstable and consequently can decompose during re-entry. Another problem with currently used TPS is the adhesives that are used to bond the composite to the metal panels of spacecraft. The adhesives which have good behaviour at high temperature are brittle, and the adhesives with better mechanical behaviour are not adequate for high temperatures and can degrade and lead to devastating consequences.

Different adhesives have been used to join C/C-SiC composites to Ti-6Al-4V alloys [Srivastana, 2003]. The author analysed the effect of glueline length, glueline thickness, etching time, temperature, exposure time in distilled water and strain rate on adhesive joints. The bond strength was found to decrease about 40-50% with increase of temperature and exposure time simultaneously. Also, they lose their strength in hot water and even their strain rates decrease.

4. Diffusion bonding

Diffusion bonding is a solid state welding process in which two materials can be bonded together through the interdiffusion of atoms across the interface. The two materials are pressed together at an elevated temperature usually between 50 and 70% of their melting point. In most cases the surfaces of the materials to be joined have to be of very low roughness. The pressure is used to relieve the void that may occur due to the different surface topographies. The diffusion of atoms at the interface is carried out by several mechanisms, such as the replacement of near atoms, movement of vacancies etc. Diffusion bonding can be achieved with or without the use of an interlayer. The interlayer can reduce cracking, relax the thermal residual stress and improve the joining strength [Zhang et al., 2006].

In literature there is limited information about the use of diffusion bonding in joining CMCs to metals, whereas there is able information in the case of joining monolithic ceramics, especially SiC, to metals [Li J. & Xiao, 2004; Matsuo et al., 2011; Zhong et al., 2011; Liu H.J. et al., 2000a]. W was diffusion bonded to SiC at 1500°C for 0.5 h under high purity Ar flow atmosphere [Matsuo et al., 2011]. The reaction phase was amorphous with a gap from 500 to 600 nm in which precipitates having a size of several tens of nanometer were formed. The shear strength of the W-SiC joint strongly depends on its amorphous layer existing between SiC and W and the shear strength of the each interface.

Reaction-bonded Si-SiC (RBSC) was diffusion bonded to Inconel 600 superalloy at temperatures from 900 to 1080°C [Li J. & Xiao, 2004]. It was shown that the interfacial reaction products at 900 and 950°C were various silicides with some voids formed in the RBSC. As the bonding temperature increased to 1000 °C, the superalloy/RBSC reactions become more intensive, although some pores in the RBSC were filled by the reaction products. With the bonding temperature increasing to 1080 °C, a thin layer of $CrSi_2$ was formed at superalloy/SiC interface without formation of any pores in the RBSC and of excessive brittle silicides. The shear strength of this joint was measured as 126 MPa.

SiC was diffusion bonded to F82H steel using a double W/Cu and a multiple W/Ni/Cu/Ni interlayer to reduce the residual stress [Zhong et al., 2011]. Diffusion bonding was used to join SiC to a Ti-43Al-1.7Cr-1.7Nb (at %) cast alloy at 1473–1573 K for 15–240 min under a pressure of 35 MPa [Liu H.J. et al., 2000b]. Two kinds of reaction products or new phases have formed during the diffusion bonding, a face-centered cubic TiC and an hexagonal $Ti_5Si_3C_x$.
Successful bonding of three-dimensional C_f/SiC to a niobium alloy has been reported using a Ti–Cu bi-foil interlayer by a two-stage joining process: at 800°C for 30 min under 6 MPa (solid state diffusion bonding) and at 1020°C for 8–120 min under 0.01–0.05 MPa (transient liquid phase diffusion bonding) [Xiong J.H. et al., 2006]. These results showed that the residual Cu layer at the joining interface relaxed the thermal stress of the joint effectively, and the Ti–Cu eutectic liquid, formed by the contact melting of Ti–Cu, not only infiltrated into C_f/SiC, but also reacted with the SiC coating of C_f/SiC. Liquid-phase diffusion bonding was also used for the successful joining of two-dimensional C_f/C to Nb alloy using again a Ti/Cu interlayer [Li J. et al., 2008]. The shear strength of the joint reached a value of 28.6 Mpa.

5. Brazing

Brazing, principally used for metal-metal joining, is now increasingly used for the fabrication of metal-ceramic and ceramic-ceramic couples. It employs a brazing alloy with a low melting point which as it melts wets the two solid surfaces and during the solidification the bond is established between the two parts.
There is a broad knowledge base on brazing monolithic ceramics to metals but the information in the literature concerning the brazing of CMCs to metals is rather limited. The scientific principles involved in the brazing of monolithic ceramics to metals have been discussed by Akselsen where the mechanical properties of directly brazed ceramic-metal joints are discussed as a function of key parameters such as temperature and time [Akselsen, 1992].
For a ceramic-metal brazed joint to be successful, a crucial point is wettability of the ceramic by the liquid metals since in the majority of ceramic–metal systems direct wetting is not observed [Naidich et al., 2008]. The limitation of poor wettability can be overcome by the incorporation of an "active metal" into the braze or to precoat the ceramic with a wetting promoter or the metallization of the ceramic prior to the brazing process [Dixon, 1995; Moutis et al., 2010]. The issue of wettability and the different brazing alloys used for C_f/C and C_f/SiC joining to metals are discussed in paragraph 5.1. Brazing parameters as temperature, holding time, surface preparation and mostly important the CTE difference between metal/filler/ceramic and ways to cope with are examined in paragraph 5.2. The introduction of an active metal has as a result the strong chemical interaction of the filler alloy with the ceramic substrate and the formation of different compounds and locally different distribution of the alloy elements and these effects are discussed in section 5.3. Finally in section 5.4 examples of successful joining of C_f/C and C_f/SiC composites to metals for mainly aerospace and nuclear applications are given.

5.1 Brazing alloys

During brazing the liquid alloy must come into complete contact with the solid surfaces and must be able to enter all cavities and porosities or in other words it should "wet" the solids to be joined. The contact angle and work of adhesion are the primary parameters which define the interaction of a brazing alloy with the surface of the materials to be joined as well as the quality of the joint. Wettability is affected by many factors, such as atmosphere, temperature, thermodynamic stability of liquid and solid, alloying constituents, and the surface condition of the solid (surface roughness, surface stress and crystallographic orientation). Thus, in order to obtain high quality joints controlling of the wettability is necessary. The typical method to promote the wettability in metal/ceramics is to introduce in the liquid metal a so-called adhesion-active element -titanium, hafnium and others- which are characterized by high chemical affinity to the atoms of the solid phase. The liquiditus temperature of the braze is important as, for work pieces to be used at lower temperatures, a high processing temperature might result to the ceramic deterioration. On the other hand, for joints to be used at high temperature applications, a high solidification temperature is required. The CTE and yield strength of the filler metal are of importance for the fabrication of a sound joint and some of these aspects are discussed in section 5.2. Brazing alloys in the form of paste, foil or powders can be used. Complex joint configurations are often better accommodated by use of braze foils than pastes [Morscher et al., 2006]. Powders and pastes containing organic binders often leave contaminants (binder residue) at the interface and impair the joint strength.

Brazing C_f/C and C_f/SiC composites to metals, as generally ceramics to metals, has to tackle the problem of poor wettability of metals on ceramics [Akselsen, 1992]. In order to improve the wettability interfacial chemical reactions between the ceramic and braze are put into play by the introduction of an active metal to the braze. Thus, brazing fillers for metal/ceramic joining are obtained by modifying a conventional filler by the addition of an active element promoting the wetting of the ceramic surface [Nicholas, 1998]. It has also been found that the reactive wetting correlates with the final interfacial chemistry of the system, not with the intensity of the interfacial reactions [Eustathopoulos, 1998]. Unfortunately at present, the experimental estimation of reliable equilibrium contact at high temperatures remains a major challenge and a serious obstacle to the development of scientific approaches to wetting phenomena [Sobczak et al., 2005]. The concentration of the active metal is crucial for a successful brazing. High content of the active metal ensures better wetting but might lead to brittle intermetallic phases or could harden the filler alloy which is detrimental to residual stress relief. The active metal interacts with C or SiC to form different compounds. Active metals in the form of foils have been used to join ceramics to ceramics as Ti or Zr foils for joining SiC [Morozumi et al., 1985].

The brazing fillers mainly consist of two groups of alloys, i.e. Cu-X or Ag-Cu-X brazes where X is the active element typically a group IVB element (Ti, Zr, Hf). In many cases, active brazes also contain additions of Ni, Be, Cr, V, In and Co. The addition of tin to copper does not enhance its ability to wet ceramics, but the addition of titanium does, and the simultaneous addition of tin and titanium is particularly beneficial. The commercially developed Cu-10.3at%Sn–12.8at%Ti brazing alloy wets very well carbon [Standing and Nicholas, 1978]. The binary Cu-Ti alloys for Ti content above 10at% solidify over a narrow temperature range which improves the flowability and reduces the risk of liquation and brittleness during solidification. The initial contact angle of the alloy of 145° on carbon is reduced almost instantaneously to 50°, that of pure Cu. Before the final reduction to 0° there is a plateau lasting of about 3 mins. This plateau indicates the presence of an incubation period needed for the initiation of the chemical reactions between the absorbed Ti and carbon [Li J.G., 1992]. The same effect is achieved by the addition of Ti to the eutectic composition of Ag-Cu (78at% Ag -22at% Cu).

For brazing using the Ti-Ag-Cu system the isothermal section of the ternary phase diagram is used [Petzow & Effenberg, 1988]. The miscibility gap divides the liquid braze into solutions of low and high Ti content [Paulasto et al., 1995] and has a significant effect on the mechanisms of active brazing. The very large values of the surface tension of Ag and Cu of the braze at their melting points show that these metals in pure state do not wet the carbon based materials. The high chemical affinity of Ti towards C, which promotes wetting and bonding, results in the segregation of Ti at the composite/braze interface. Also Ti additions to Sn, Ag, Cu, and CuSn alloys sharply decrease the contact angle on carbon. The addition of Ti and/or Si improves the wetting through the formation of carbides TiC and SiC at the interface [Grigorenko et al, 1998].

The very large values of the surface tension of Ag, Ni and Cu of the braze at their melting points show that these metals in pure state will probably not wet the C_f/C. However, the wettability of TiCuNi brazing alloy (70Ti-15Cu-15Ni, wt%) on C_f/C is very high. A joint, of about 30–35 μm thickness formed at the interface between the C_f/C and the brazing alloy, is continuous, free of micro-defects, independently from the carbon fibre orientation (parallel or perpendicular to the surface) [Salvo et al, 2010]. A thin titanium carbide layer of about 2 μm close to C_f/C substrate causes beneficial near-interfacial changes which promote wetting and bonding [Singh et al., 2007].

The operation temperature of C_f/C composites to metal joints fabricated using the fillers discussed above is limited to around 500 or 600 °C. Therefore in order to fully exploit the high temperature potentialities of C_f/C composites high temperature brazes are required. PdNi alloys offer such a possibility and PdNi alloys and with different Cr-active content have been examined and the alloys PdNi-(12—25)Cr and Ni-33Cr-24Pd-4Si (wt%) showed excellent wettability [Chen et al., 2010].

In terms of SiC brazing, research progress has so far been rather slow. According to the reactivity of SiC with metals two groups can be defined: a) metals (Ni, Co, Fe, etc. . . .) which react only with silicon and b) metals (Zr, Ti, Hf, Mn, etc. . . .) able to react with both silicon and carbon. It is known that a banded structure would form at the interface between SiC and elements Ni, Co or Fe, consisting of alternating bands of silicides and of randomly distributed graphite precipitates in the silicide matrix which has as a consequence the weakening of the joints [Chou, 1993; Mehan et al., 1976]. Therefore, taking also into account the presence of carbon fibres in a C_f/SiC composite the second class of metals are the most appropriate for C_f/SiC to metal joining. For example brazing alloys of 22Ti–78Si (wt.%) and 21Ti–74.2Si–4.8Cu (wt.%) exhibit both good wettability and cohesion with SiC after cooling [Li J. et al., 2009]. Cu-Cr and Cu-Ni-Cr alloys wet well SiC and the addition of Si suppresses the formation of a duplex layer of Cr_3C_2 [Xiao & Derby, 1998]. For high temperature applications of SiC joints the addition of V in Co and Pd–Ni based alloys eliminated the periodic band structure of the interfacial reaction layer controlling thus effectively the interface reaction [Xiong H.P. et al., 2007].

5.2 Brazing processes

Once the brazing alloy has been chosen the brazing temperature is more or less defined by the liquidus of the braze. A little higher temperature might improve braze-ceramic substrate interactions as well as the penetration of the braze into the ceramic. The holding time at the brazing temperature controls the reaction layer formation. Both the brazing temperature and holding time should be optimized so as to achieve a controlled thickness of the reaction layer without the development of brittle reaction phases. Heating and cooling rates are important for the thermal stresses minimization but also on the formation of secondary phases. The conventional brazing process is characterized by slow heating and cooling cycles while the whole component is placed in a uniform heat zone. Another important parameter is the composite ceramic surface preparation. Ceramic composite surfaces are generally rough but additional surface roughness and sub-surface are damage may be introduced while vacuum heating and opening-up of partially closed pores during heating under vacuum has been observed [Sobczak & Asthana, 2001]. Decreasing the brazing temperature and time are always recommended with the advantages of decreased interfacial reactions, decreased erosion of substrates and minimum loss of base-metal properties. The application of brazing alloys which contain reactive metals requires that joining is performed in a vacuum with a very low oxygen potential, or in a dry inert-gas atmosphere with a low dew point to prevent the active elements from reacting with the atmosphere. If these requirements cannot be met, the ceramic surfaces should be metallized prior to brazing.

The effect of brazing temperature and time are very important on the quality and shear strength of the brazed joint. This is very nicely demonstrated by the brazing of C_f/C to TC4 alloy [Qin & Feng, 2009]. The maximum shear strength of the joint has a temperature window of about 20 °C while the holding time window is of about 5 min at the best brazing temperature. The temperature controls which phases will occur between the active metal(s) of the filler and the composite whereas temperature and time the thickness of the reaction layer. Formation of brittle phases is also a consideration for the brazing temperature adjustment. Thus, both brazing temperature and time have to be optimized for improved joint strength [Paiva et al., 2000; Tillmann et al., 1996].

Residual stresses developed during the cooling process will be sufficiently high to influence the strength and properties of the brazed joint during subsequent mechanical or thermal loadings, and in some cases will lead to catastrophic failure during the fabrication process. Careful design with, for example, the employment of finite element analysis (FEA) can minimize the residual stresses developed during the brazing. Increasing the braze thickness generally leads to higher peak residual stresses, but plastic behavior within the joint reduces this effect. There is the possibility of introducing an efficient stress-relaxing procedure, in which, during cooling, the temperature is held constant at an intermediate level at which stresses are high but not catastrophic, to allow for stress relaxation through creep [Levy, 1991].

The critical defect size, d_c, at the interface or within the ceramic can be calculated from the equation

$$d_c = \frac{4K^2}{\pi^3 \sigma^2} \tag{1}$$

where K is the fracture toughness of ceramics and σ is the residual stress (no external load applied) which depends on the CTE difference given by [Ning et al., 1989]

$$\sigma = \frac{E_1 E_2}{E_1 + E_2}(a_1 - a_2)(T_B - T_0) \tag{2}$$

where E is Young's modulus, a the CTE, T_B the brazing temperature and T_0 the temperature of the surroundings. For small CTE mismatch the critical crack length is of the order of tenths of mm whereas for large CTE mismatch could be in the order of μm. Defects of this size may be present in ceramics in the form of pores. This strongly suggests that ceramic surface processing prior to brazing is important for adequate joint mechanical properties. Young's modulus and strength [Ryshkewitch, 1953] and hence strain at fracture, are found to be exponentially dependent on the volume fraction of pores [Moulson, 1979]. Therefore, it is not surprising that failure may occur in metal/ceramic joints even without application of an external load. Thus, evaluation of the joints also requires testing for porosity and internal cracks.

Composite surface polishing prior to brazing might promote chemical reactions which the interfacial roughness of the unpolished substrate could reduce by hampering braze spreading and physical contact. Surface roughness could result in an increase of the contact angle in non-wettable systems whereas in wettable systems in a decrease. Such a surface treatment has been proven beneficial in the case of monolithic ceramics, however, its beneficial effects in the case of composites has not yet been established. A disadvantage in the case of composites is that polishing very likely might introduce surface and sub-surface damage such as grain and whisker or fiber pull-out as well as porosity.

One of the main reasons for the metal-ceramic joint to fail either during its fabrication or during service is the different CTE between metal and ceramic (see equation (2) above). The CTEs of C_f/C and C_f/SiC composites are normally much smaller than those of most alloys (e.g. steels or superalloys) with which need to be joined together (CTE mismatch). When constrained to cool together either through the process of joining or work thermal cycle, residual stresses resulting from the differences in thermal expansion can compromise the integrity of the joint. This intrinsic failure mechanism of the joint can be alleviated by the introduction of an interlayer between the metal and the composite. In addition, complex interlayers with incremented thermal expansion coefficients may be used which, through a gradual transition of CTE, relieve the thermal stresses.

Ductile metallic interlayers as Cu (yield strength of about 50 MPa) provide significant reduction of residual stresses, but their applicability in real systems is limited due to their low resistance to corrosion and oxidation at elevated temperatures. Mo thermal expansion coefficient is very low (about 4.8×10^{-6} K^{-1}), very close to that of the ceramic and can be used as CTE matching interlayer. Using a double foil of Mo adjacent to the metal and Cu adjacent to the C_f/C thermal stresses can be minimized and cracks during brazing or cooling from a working temperature to be avoided. When the thermal stress of the joint is over the yield strength of Cu, the Cu foil yields releasing part of the stress. Moreover, part of stress diverts to the Mo foil resulting to a decrease of stress at the composite and the filler metal interface.

Stress release can also be accomplished by the incorporation of few interlayers (compensated interlayer) as in the case of a triple layer Kovar/tungsten/nickel used for brazing SiC to a nickel-based superalloy [Xiong H.P. et al., 2007] or by a layer with linearly or non-linearly varying CTE (functionally graded interlayer). Another approach is that of a flexible interlayer which uses a thin, dimpled or corrugated sheet of metal brazed to the metal and ceramic parts. In this context it has been proposed the use of metal foam for joining ceramics to metals. This has been applied in brazing aluminium nitride to Inconel 600 and it has been demonstrated that the joints are tolerant to severe thermal cycling tests [Shirzadi et al., 2008]. Relief of the residual stresses could also be accomplished by the combination of reinforced and un-reinforced filler alloy layers and for example

Incusil®ABA™ ($Ag_{59}Cu_{27.25}In_{12.5}Ti_{1.25}$) brazing filler was reinforced with SiC particles [Galli et al., 2006].

Properly designed multiple interlayers can reduce the strain energy in the ceramic-metal joining more effectively than a single interlayer [Park J.W. et al., 2002] and the most desirable gradation of interlayer properties is not a simple linear decrease from one base material to the other. The use of rigid layers with increasing CTE away from the ceramic interface and insertion of ductile layers between each rigid layer can reduce the strain energy most effectively Also a very effective way to decrease the strain energy at the ceramic/metal interface and to provide a better solution for larger area joints is through the control of porosity at the interlayer [Park J.-W. et al., 2004]. This has been confirmed by shear tests which proved that the joint strength is increased up to a certain level of porosity due to relaxation of residual stresses. Generally multiple interlayers play an important role in reducing the strain energy in the ceramic and in increasing the strength of the joint and their designing should be based on material properties and numerical or analytical calculations.

5.3 Surface Interactions

The brazing alloy at the brazing temperature is in liquid form thus the liquid-metal/ceramic interfaces play a central role in the brazing of metals to ceramics. As mentioned above, the decisive parameters determining the chemical and physical interactions that take place at the interface is the wettability of the solid substrate by the liquid alloy, characterised by the contact angle, and the interface adhesion strength. The width, w, of the reaction zones can be estimated by the following relationship:

$$w = c_0 t^n \exp(-Q/RT) \qquad (3)$$

where $n \approx 0.5$, t the holding time, T is the brazing temperature, R the gas constant and Q the activation energy for diffusion. The activation energy for diffusion depends on the diffusing species and the chemical and structural form of the substrate (e.g. porosity, fibre versus matrix). If the constituents of a filler alloy have different activation energy for diffusion (which usually will be the case) the ceramic composite will assist in their separation and different activation zones will be observed. Also faster or slower diffusion can occur dependent on the path of the melt which could be through or around a fiber or through an open porosity channel. In addition, some of the elements of the filler metal might react with the ceramic and as the content of the filler alloy changes the activation energy will also change. Thus, the filler ceramic surface interactions can be very complicated and the phases formed will depend not only on the phase diagram but very strongly on the changing concentration of the penetrating filler versus depth. The surface structure, both chemical and morphological, is what will determine the joint quality and strength and has thus to be controlled by the filler and process choice. In fact, interfacial reactions can affect every stage of the formation of a joint from the onset of bonding through the development of equilibrated microstructure to the optimization of the mechanical properties. An important point is that all investigations related to interface compound formation are carried out at room temperature and it could be contemplated whether the compounds are formed when the metal filler is in the liquid state or are formed as a result of crystallization when the more refractory component of the alloy solidifies. Notwithstanding that it is thought that most of the compounds are formed at the brazing temperature the importance of the cooling from the brazing temperature to room temperature should not be ignored. Further work in this area is needed as sometimes holding the cooling at some specific temperature improves the joints and this could be connected with the formation of beneficial phases [Naidich, 2005]. Also it is not quite clear the mechanism of the spreading of a liquid on a ceramic at high temperature which is usually accompanied by chemical reactions and diffusion [Saiz & Tomsia, 2005]. Theoretical calculations on the energetics of liquid metal/ceramic interface based on statistical mechanics [Novakovic et al., 2003] will be of great help for understanding brazing interface phenomena and guiding developments in controlling the interface interactions and thus the fabrication of sound joints for demanding applications.

5.3.1 Carbon and filler metal interactions

Carbon strongly interacts with the active metals (Cr, Ti) in the filler to form carbides. Different stochiometric or sub-stochiometric carbides can be formed of which their chemical formula and mixture will depend on the Gibbs free energy. Bulk thermodynamics of different alloys and carbon could be a guide as to what compound might be formed. However, there are severe limitations of bulk thermodynamics as the interaction occurs in surfaces and the surface energy has to be taken into account. In addition, the filler alloy concentration changes in time and in depth and during cooling, phases formed at higher temperatures might be transformed. Further, it should be remembered that we are far away from thermodynamic equilibrium and the products of the interactions are mostly diffusion controlled.

An approach to study surface interactions is to deposit an active element on the surface of carbon and by X-ray diffraction measurements to determine the phases formed at different temperatures or by imitation of the brazing process thermal cycle [Moutis et al., 2009]. Cr deposited on carbon was completely transformed into a mixture of the Cr_7C_3 and Cr_3C_2 phases after the thermal cycle of brazing. Generally, when Cr is the active element in a brazing filler it reacts strongly with C to form different carbides $Cr_{23}C_6$, Cr_4C, Cr_7C_3 and Cr_3C_2 and spreading kinetics are limited by the Cr diffusion [Voitovitch et al., 1999]. The Cr carbide formation promotes the reaction of brazing fillers with C_f/C composite, resulting in the improvement of wettability on carbon matrix. Cu poorly wets carbon but the liquid Cu-Cr alloys (Cr 0.5 and 2 at.%) show improved wettability on vitreous carbon substrates which is due to the formation of a 1-5 μm thick layer of the compound Cr_7C_3 [Voitovitch et al., 1999]. The high temperature filler Ni-33Cr-24Pd-4Si forms a 15-20 μm in width homogeneous diffusion layer at the C_f/C composite interface. Close to the C_f/C surface Cr reacts with carbon to form a Cr-C reaction layer. Subsequently, Pd and Si participate in the reactions and form Pd_2Si and Pd_3Si phases, and in this reaction zone, the residual brazing alloy became Ni-rich and Pd-depleted [Chen et al., 2010].

Wetting by the Ti-Ag-Cu braze is due to the formation of Ti carbide. Only one stable intermediate phase is formed in the Ti-C system: TiC, with a face centered NaCl type structure. The TiC has a wide composition range which extends from 29 at% C to 50 at% C with the lattice parameter increasing from 4.303 to 4.330 Å with increasing carbon content. As the Ti is depleted from the alloy the wetting behaviour will change as Ag and Cu do not wet carbon or TiC [Frage et al., 2002]. Whenever carbon fibers are used for CTE matching or for increasing joint strength, Ti in the braze alloy reacts with the fibers bonding them to the structure [Zhu & Chung, 1994]. Usually different interaction zones are observed in the C_f/C substrate when Ti-Ag-Cu braze is used. The closest to the substrate zone is divided into a

TiC layer and a layer which mainly consist of Ti and Cu [Qin & Feng, 2007]. The second layer consists of Cu-rich and Ag-rich phases. In the next zone a mixture of Ag-rich phase and a phase mainly consisting of Ti and Cu are observed. The formation of the TiC phase near the C_f/C interface is also observed in the brazing of the C_f/C composites with a Ni superalloy [Moutis et al., 2010] and the second layer consists of separated Ag and Cu rich phases. Ti carbide formation is not only observed on the composite surface but also around the carbon fibers. The Ti-Ag-Cu melt and composite interactions are mainly determined by the TiC formation at the composite/filler interface. As the Ti interacts with the carbon its concentration in the melt adjacent to the composite surface decreases. Ti from the bulk melt diffuses to the depleted zone and interacts with the underlying carbon increasing the carbide layer thickness. As the melt is depleted from Ti and probably during cooling separation of Ag and Cu occurs. Also TiCu phases have been observed and probably their formation would be possible in cases in which the formation of TiC is not the prevailing compound. In addition, the mutual exclusion between the Ag and Ti elements have a significant influence on the joint microstructure [Wang et al., 2010].

The nickel in TiCuNi also has a higher affinity for carbon than Cu, Au and Ag, and has been found to segregate at the carbon/ metal interface. There are no stable Ni carbides however there is some evidence on the formation of metastable ones. However, as the principal constituent of TiCuNi is Ti (~70%), which strongly reacts with the carbon, formation of nickel carbides is less probable.

A foil filler metal Ti–37.5Zr–15Cu–10Ni (wt%) forms a 2 μm reaction layer with a C_f/C composite which consists probably of Zr and Ti carbides [Qin & Feng, 2009]. When Cu is used as a stressed reliever three layers between C_f/C composite and Cu are formed. A $Cu_{51}Zr_{14}$ compound is formed at the Cu-filler interface (10 μm). At the composite site (Ti,Zr)C reaction layer is formed which changes from discontinuous at lower temperatures (850 °C) to continuous at higher brazing temperatures (920 °C) and its thickness increases. In the third layer (50 μm) many phases ($Ti_2(Cu,Ni)$ + Ti(Cu,Ni) + TiCu + Cu_2TiZr) coexist. As the temperature increases the plethora of phases is reduced and only Cu and $Ti(Cu,Ni)_2$ remain. The phases present and the joint microstructure determine the shear strength of the joint which increases from 7 MPa for the brazing fabricated at the lowest temperature to 21 MPa for that fabricated at the highest temperature.

5.3.2 SiC and filler metal interactions

The brazing of C_f/SiC to metals is mainly based on processes and methodologies developed for brazing monolithic SiC to metals. Therefore, in order to put the discussion of the brazing of C_f/SiC to metals in the right perspective and assist in the expected future intense activity in this area the interaction of SiC to filler metals is outlined in this section. Designing and controlling the chemical reactions between SiC and metals is an important issue in the fabrication of the C_f/SiC to metal joints. Formation of various reaction products at the SiC/metal interface during the joining process may lead to completely different mechanical properties of the joints. Thus, the understanding of reactions between SiC and different metals is substantial for the realization of a successful joining between C_f/SiC and metallic alloys. In addition, the control of interface reactions between the SiC matrix and the brazing alloys is a significant matter and should be considered in the design of new brazing alloys for joining SiC based materials.

The reaction kinetics of SiC/metal can be indentifed into two modes i) formation of silicides and free carbon (type I) and ii) formation of carbides and silicides (type II) [Park J.S. et al.,

1999]. Ni, Co, Fe, etc. metals react with SiC to form brittle silicides which are accompanied by carbon precipitation in the form of graphitic layers which weaken the joint of the ceramic to a metal. Zr, Ti, Hf, Mn, etc., that react with both silicon and carbon, can lead to a high reactivity with SiC that must be controlled in order to select the nature of the reaction products and the thickness of the reaction zone. Bhanumurthy & Schmidt-Fetzer studied the interface reactions of SiC with Ni, Cr, Pd and Zr [Bhanumurthy & Schmidt-Fetzer, 2001]. It was observed that none of these metals are in thermodynamic equilibrium with SiC. The interface reactions lead to the formation of complex structures in the reaction zone. The interface reactions in SiC/Ni, SiC/Pd form periodic bands and SiC/Cr, SiC/Zr form layered structures. C and metal atoms are the most dominant diffusing species in SiC/Cr, SiC/Zr and C is almost immobile in SiC/Ni and SiC/Pd. Therefore development of brazing alloys will rely on employing combinations of these metals in order to control the intensity of the interactions and chemical composition of the products. Other elements (Si, Ag, Al, etc.) may be added in order to control the strength of the interactions and the final products. Nickel's strong interaction with SiC can be controlled by the addition of, for example, Ag or Si. An Ag coating, about 2 μm thick, significantly reduces the reaction of SiC with Ni and this approach has been used in joining SiC to Ni super-alloys for high temperature applications [Hattali et al., 2009]. A Ni–56Si filler alloy was used to join SiC to Kovar alloy (Fe–32at%Ni–15at%Co) [Liu G.W. et al., 2010].

McDermid & Drew used solution thermodynamic theory to compute an optimum composition of Ni–Cr–Si alloy for brazing SiC ceramics [McDermid & Drew, 1991] and they conducted brazing experiments in order to assess the effect of changing the Si content away from the optimum composition on the joint microstructures. They pointed out that the alloys containing less than 36 at% Si lead to the formation of a porous reaction zone at the brazing alloy/SiC interface due to the excessively vigorous joining reaction between brazing alloy and ceramic. The best microstructures were attained for 40 at% Si alloy joints, closely agreeing with the thermodynamic model, whereas higher Si content alloys exhibited localized debonding of the brazing alloy from the SiC.

The AgCuTi filler poorly wets SiC [Südmeyer et al., 2010]. Titanium and silicon carbide in the temperature range between 1250 and 1500 °C form the ternary phase Ti_3SiC_2 [Gottselig et al., 1990] which could improve the wettability if the chemical potential is appropriate for its formation. Good wettability with a contact angle less than 30° on SiC is observed when using as a braze filler SnAgTi pellets with a Sn content above 30 wt%. Standing and Nicholas pointed out that the solubility of Ti in ternary systems was reduced by the addition of an element with a low surface energy like Sn [Standing and Nicholas, 1978]. In that manner the activity of Ti, which is necessary for the diffusion process at the ceramic–braze interface, can be increased so that the ceramic–braze interface is strengthened. Moreover the low solidus temperature (200 °C) and the small Young's modulus of Sn leads to a reduction of residual stresses, which has a positive effect on the compound strength. Ti reacts with the carbon in the C_f/SiC composite substrate to form stochiometric carbide (TiC) as well as sub-stochiometric carbides (e.g., $TiC_{0.95}$, $TiC_{0.91}$, $TiC_{0.80}$, $TiC_{0.70}$, $TiC_{0.60}$ and $TiC_{0.48}$). In addition, formation of silicide phases at the interface from the reaction of Ti and Si (from the SiC) is a distinct possibility.

5.4 Fabrication of brazed joints of C_f/C and C_f/SiC to metals

5.4.1 C_f/C-metals brazed joints

Semi-3D C_f/C has been brazed to TC4 (Ti-6Al–4V (wt. %)) aerospace alloy [Donachie , 1982; Roger et al., 1993] using a Ag-26.7Cu-4.6Ti (wt.%) 50 μm thick foil [Qin & Feng,

2007] or Ti–37.5Zr–15Cu–10Ni (wt. %) 50 μm thick foil [Qin & Feng, 2009]. Notwithstanding the thickness of AgCuTi foils is 50 μm, joining zones of about 60–80 μm are formed indicating inter-diffusion of the filler metal and base materials. Within the three zones (metal/braze, intermediate, braze/composite) different layers are formed. In the braze/composite zone TiC+C and TiCu layers are formed. In the intermediate zone there is the formation of Ag (s.s) phase and the TiCu + Ti_3Cu_4 phases. Whereas in the metal/braze zone the layers of phases Ti_3Cu_4, TiCu, Ti_2Cu and Ti_2Cu + Ti (s.s), are observed. Brazing at 900 °C for 5 min using Ti–Zr–Cu–Ni foil results in the formation of a reaction layer of 2 μm thickness which is probably (Ti,Zr) carbides. The formation of (Ti,Zr)C between C_f/C composite and the filler metal speeds the wettability of the liquid brazing alloy. A good joining is observed but there is also a crack arising from the thermal stresses. Thus, a pure Mo foil (0.1 mm thick) next to the TC4 and the pure Cu foil (0.3 mm thick) next to the C_f/C were used as composite interlayer in order to release stresses (Cu) and provide a matching (Mo) CTE to the composite. Samples brazed at 900 °C for 5 min have the maximum shear strength of 21 MPa, a four fold increase in comparison to samples (5 MPa) without the Mo/Cu interlayer. This strength corresponds to that of the Cu/filler/C_f-C system which is the weakest and fracture occurs in the composite when carbon fibers are parallel to the joining surface and the fracture surface lies in the composite/filler metal interface when the fibers are vertical. In order to reduce the CTE mismatch between the C_f/C composite and the TC4 metal, SiC particles were introduced in the Ag-26.7Cu-4.6Ti (wt.%) filler alloy. SiC particles as reinforced phase had an average particle diameter of 4.6 μm. The best shear strength was attained for volume fraction of SiC particles of about 15%. SiC particles have reacted with the brazing alloy, forming a new phase, probably a Ti–Si–C compound [Qin & Yu, 2010].

Ti-metal/C_f-C composite joints were formed by reactive brazing with three commercial brazes, namely, Cu-ABA, TiCuNi, and TiCuSil [Singh et al., 2005]. Ti-rich phases such as TiC_{1-x}, which bond well to both the carbon and the braze, were formed at the C_f-C/TiCuNi and C_f-C/TiCuSil interface. At the braze/Ti interface some dissolution of the metal in the molten braze has occurred. The highest joint strength was obtained by the Cu-ABA braze material. Joining of Ti tubes to C_f/C composite plates provide a test of the joined structure in tension. The Cu-ABA braze composition had the highest load-carrying ability in comparison to TiCuNi and TiCuSil. Fracture always occurred within the surface ply of the C_f/C composite which indicates a stronger bond between the braze material and the Ti-tube or C_f/C composite than the strength within the outer ply itself. The interlaminar tensile strength of these C_f/C composites was an order of magnitude higher than the bond strength measured based on the failure load and bonded area of the surface ply. Probably the brazing process introduces defects in the outer ply reducing its interlaminar tensile strength or the loading condition induces stress-concentrations and other fracture modes. The fiber orientation is also a factor that dictates the load-bearing capability of the joint [Morscher et al., 2006]. Titanium aluminide is regarded as an excellent material for high-temperature applications [Li Y.L. et al., 2006] and its joining to C_f/C could further elevate the operating temperature of thermal structures. The interfacial structure of the C_f-C/AgCuTi/TiAl braze joint was greatly influenced by the mutual exclusion between the Ag and Ti elements with typical interface structure TiAl/Ti_3Al + AlCuTi/$AlCu_2Ti$/Ag(s.s)/TiC/C_f-C [Wang et al., 2010].

Copper-clad-molybdenum and copper-clad-invar because of their thermal conductivity and low thermal expansion are used for thermal management applications. A joined Cu-clad-Mo/C_f-C composite system can provide excellent heat dissipation capability at reduced weight. The C_f/C composites made from T-300 C fibers and resin-derived carbon matrix have been joined to Cu-clad-Mo using four brazes (Cu–ABA, Cusin-1 ABA, Ticuni, and Ticusil) [Singh et al., 2007]. Cu–ABA (92.8Cu–3Si–2Al–2.25Ti) microstructure consists of needle-like precipitates that are a Cu(Ti,Si) phase, and a Cu-rich homogeneous matrix. Mo is dissolved in the molten braze and diffuses within 20 μm in the composite. Redistribution of alloying elements because of dissolution and interdiffusion was also observed with the other brazes and the surface active element Ti preferentially segregated at the C_f-C/braze interface in all joints. Also 3D C_f/C composites having CVI carbon matrices were joined to Cu-clad-Mo for heat rejection applications using two Ti-containing Ag–Cu active braze alloys (Ticusil and Cusil-ABA) [Singh & Asthana, 2008]. Both Cusil-ABA and Ticusil have infiltrated the inter-fiber regions in the 3-D C_f/C composite of several hundred micrometers. The TiC reaction layer that forms is known to be discontinuous and this permits extensive infiltration of porous carbon by the melt even in a short time interval.

C_f/C composites (CARBOTEX) have been joined to a Nimonic alloy using Ticusil (Ti-Cu-Ag) filler metal [Moutis et al., 2010]. In order to accommodate the different linear coefficients of thermal expansion between ceramic composite and metal as well as to provide compatibility between the surfaces to be joined, the C_f/C surface was metallized through the deposition of a chromium layer. Heat treatment at 700 °C for 1 h results in the transformation of part of the deposited Cr to chromium carbide Cr_7C_3. During brazing with Ticusil filler it is found that the Ti penetrates into C_f/C in depth of about 100 μm and coats the carbon fibers. This deep Ti penetration in the carbon substrate has also been observed in C_f/C composites brazed to Cu clad Mo [Singh & Asthana, 2008], and in graphite brazed to Nimonic alloy [Moutis et al., 2009]. Layered structures of different compounds are observed and these arise from the Ti carbon interaction to form TiC and the separation of Ag and Cu in the remaining filler melt. The surface metallization with chromium improves the reactivity of the elements of the filler metal and the carbon.

The use of metallic glass brazes to join ceramic composites, notwithstanding their high strength and excellent corrosion resistance, has been limited. C_f/C made from T-300 C fibers and resin-derived carbon matrix CVI SiC matrix reinforced with T-300 carbon fibers was brazed to Ti using two Ni-base metallic glass braze foils (MBF-20 Ni–6.48Cr–3.13Fe–4.38Si–3.13B–0.06C–0.07Co–0.01Al and MBF-30 Ni–4.61Si–2.8B–0.02Fe–0.02Co–0.01(Al, P, Ti, Zr)) [Singh et al., 2008]. The very large Ti concentrations in the vicinity of the C_f/C composite surface in both MBF brazes suggest that Ti from the substrate had actually dissolved in molten braze during joining, and segregated at the C_f/C surface. The joints show large internal stresses and inter-laminar shear failure within the composite matrix was also noted in some joints.

C_f/C composites as low neutron activation materials are considered as excellent materials for nuclear fusion reactors because they also have very low atomic number and very high melting and sublimation temperatures. For fusion applications the joining material and process must satisfy certain criteria as not using high pressures, the braze should contain low activation materials and present thermomechanical stability at least 600 °C [Salvo et al., 1997].

The general configuration of the ITER high heat flux components consists of armour material, which directly faces the thermonuclear plasma, and the heat sink, which transfers the heat loads from the armour to the water coolant [Merola et al., 2002; Liu J.Y. et al., 1994]. C_f/C is the reference design solution for the armour material as it can withstand high heat loads, has high thermal shock and thermal fatigue resistance as well as high thermal conductivity. For this application C_f/C material has to be joined to Cu interlayer. The main

problems in the development of the C_f-C/Cu joints are the large thermal expansion mismatch and the fact that Cu does not wet carbon. Plansee AG developed a methodology which consists in the casting of pure copper on the CFC laser machined surface which previously has been activated by titanium and Ansaldo Ricerche developed a brazing alloy with good wetting characteristics on CFC surface [Merola et al., 2003].
Also direct joining of the CFC to Cu was performed by modifying the surface of the composite by a solid-state reaction at high temperature of a transition metal of VI B group (chromium and molybdenum) deposited on the CFC surface by a simple slurry technique (metal powder suspension in ethanol) [Appendino et al., 2004]. The direct joining of copper (in form of slurry or foil) to CFC was performed at 1100 °C for 20 min. The direct joining could be performed as the wetting angle of copper on the modified CFC was lower than 20° at 1100 °C [Appendino et al., 2003].

5.4.2 C_f/SiC -metals brazed joints

Brazing is the main method for joining SiC ceramic to metal [Liu H.J. et al., 2000a]. Carbon fiber-reinforced SiC matrix (C_f/SiC) composites present higher fracture toughness than SiC monolithic ceramics and also combine the merits of the SiC ceramics such as low density, excellent oxidation resistance and high-temperature strength. Therefore, they are excellent potential candidates for application in the aerospace industry [Jian et al., 2005]. Reliable joining of the composite to a metal, especially to the Ti-alloys widely used in aerospace field, is essential for full exploitation of the composite properties and saving weight of the overall structure. As these composites are reaching maturity a recent effort has been devoted in the development of brazing them to metals [Lin G. et al., 2007; Liu Y.Z. et al., 2011; Singh et al., 2008; Singh et al., 2010, Xiong J.H. et al. 2010a; Xiong J.H. et al., 2010b; Xiong J.H. et al., 2006].
For aerospace applications joining of C_f/SiC composite to aerospace alloys, as Ti–6Al–4V, is required. For the brazing of the Ti rich alloys is apparent that Ti as an active metal in the braze would be the best choice. Also Ti reacts with both the matrix (Si, C) and the fibers (C) and therefore stronger joints are expected. Cu-Ti or Ag-Cu-Ti alloys have been the base of brazing C_f/SiC to Ti alloys and in some cases carbon or carbon fibers have been introduced in the braze. Lin G. et al. used carbon fiber-reinforced brazing material (67.6Ag-26.4Cu-6Ti, wt%) to braze C_f/SiC to Ti alloy [Lin G. et al., 2007]. The volume fraction of the carbon fibers in the braze defines the strength of the joint, controls the reaction between the Ti element and the brazed composite and it is associated with the brazing parameters. The reactive products include TiC_x thin layers covering the fibers and TiC_x small particles distributed near these fibers in the brazing layer. With the increase of the brazing temperature and dwell time, more Ti dissolves into the brazing layer from the Ti-alloy and the extent of interfacial reaction of Ti with the composite increases. However, excessively high brazing parameters and dwell times incur the formation of pores in the brazing layers.
Using brazing alloys in a powder form, being cheaper in principle, could also permit the mixing of different elements for which brazing foils or pastes are difficult to be fabricated. As an interlayer a mixed powder of Ag-Cu-Ti-W was used for joining C_f/SiC to a Ti alloy [Lin G.B. et al., 2006]. The results showed that W grains mainly distributed in Ag phase in the brazing layer provide the effects of reinforcement and lowering residual thermal stress at the joint. The room temperature and 500 °C shear strengths of the joints performed at 500°C for 30 min with Ag–Cu–Ti–50W (vol.%) are remarkably higher than the optimal strengths of the joints brazed with Ag–Cu–Ti. Successful joining of C_f/SiC composite to TC4

alloys (Ti–6Al–4V (wt%)) was realised using 94(72Ag–28Cu)–6Ti (wt.%) alloy powder with particle size of 320 mesh [Xiong J.H. et al., 2010b]. The joint interfaces were microstructurally sound, well bonded, and without cracks and voids. Ti_3SiC_2, TiC and Ti_5Si_3 phases were formed in the reaction layers between the composite and the interlayer whereas TC4 dissolved in the braze and Cu diffused into the TC4. The best strength of the joints, 102 and 51 MPa at room temperature and 500 °C, respectively, was obtained for brazing at 900 °C and with dwell time of 5 min.

Also mixed powders of Cu, Ti and graphite were used for brazing C_f/SiC composite to a Ti alloys [Ban et al., 2009; Xiong J.H. et al., 2010a]. Formation of TiC around surplus graphite and TiC particles in the bonding layer reduced the thermal stress significantly. The reaction rate was controlled by the diffusion rate of C from graphite particles to the liquid bonding layer. The shear strength was remarkably higher than the optimal shear strengths of the joints brazed with pure Cu-Ti.

A Ti-Ni-Nb (39.4-39.4-21.2 at%) brazing alloy [Liu Y.Z. et al., 2011] or a Ti–Cu bi-foil interlayer [Xiong J.H. et al., 2006] were used to join C_f/SiC composites and Nb alloy (rocket propulsion). The brazing using Ti-Ni-Nb was performed at 1220 °C for 20 min and the ductile filler metal released the thermal stress in the joint. Both Ti and Nb elements in the filler reacted with C_f/SiC during the brazing process, and a well bonded C_f/SiC-Nb joint was obtained with shear strength of 149 MPa. The Ti–Cu bi-foil interlayer was used together with a two-stage joining process: at 800 °C for 30 min under 6 MPa and at 1020 °C for 8–120 min under 0.01–0.05 MPa. The results showed that the residual Cu layer at the joining interface relaxed the thermal stress of the joint effectively, and the Ti–Cu eutectic liquid, formed by the contact melting of Ti–Cu, not only infiltrated into C_f/SiC, but also reacted with the SiC coating of C_f/SiC. These characteristics were beneficial to the joint, of which the shear strength was as high as 34.1 MPa.

C_f/SiC composites were vacuum brazed to Ti and a Ni-base superalloy using Ni-base metallic glass braze foils (MBF-20 and MBF-30) [Singh et al., 2008]. For the Ti joint, the results showed that the braze/composite interfacial contact, in both braze foils, is intimate. However, significant cracking through the braze region was observed in the case of MBF-20 braze foil, whereas in the case of MBF-30 there was no evidence of interfacial microvoids and cracks in the joint region. The cracking in MBF-20 was attributed to the higher boron content of MBF-20 (3.13%) than MBF-30 (2.8%), and the presence of ~6.48%Cr in MBF-20 as opposed to MBF-30 that does not contain any Cr. For the case of the Ni-base superalloy the joints for both braze foils are sound, but in the case of MBF-20 shrinkage cavities had been formed. It is pointed out that compositional changes due to substrate dissolution led to secondary-phase precipitation which aided interfacial bonding although inter-laminar shear failure occurred within some composites. Residual thermal stresses in the joint led to hardness gradients; however, stress accommodation by the braze prevented interfacial cracking.

C_f/SiC composites reinforced with T-300 carbon fibers in a CVI SiC matrix were joined to Cu-clad Mo using two Ag-Cu active braze alloys, Cusil-ABA (1.75% Ti) and Ti-Cu-Ag (4.5% Ti) [Singh & Asthana, 2010]. The brazed joints revealed good interfacial bonding, preferential precipitation of Ti at the composite/braze interface. The Knoop microhardness distribution across the joints revealed hardness gradients at the interface, and a higher hardness in Ti-Cu-Ag than in Cusil-ABA. The effect of composite surface preparation revealed that joints made using ground samples did not crack whereas un-ground samples cracked due conceivably to amplification of residual stress at surface imperfections.

Theoretical predictions of the effective thermal resistance suggest that composite-to-Cu-clad-Mo joints may be promising for lightweight thermal management applications.
Also interlayers have been used in order to achieve bonding and relieve strain mismatch between the composite and steel or Ti or Ni alloys. Brazed joints using an interlayer which accommodates strain mismatch have been successfully produced between a SiC_f/coridierite CMC and Ti-6Al-4V alloy, and between SiC_f/coridierite and Fe-18Cr-8Ni stainless steel (Dixon, 1995). Ductile interlayers of Ni, Cu, W, and SiC/Ti-6Al-4V metal matrix composite were used between the metals and the CMC. The CMC was coated with 1μm Ti in order to induce wetting and bonding and a Ag-28Cu eutectic braze or aluminium braze was used. Joints brazed normal to the CMC layup plane were generally much stronger than those brazed parallel to the layup plane. Li S. et al. used Cu/W/Cu/W/Cu multiple interlayers to bond C_f/SiC to Ni-based superalloy [Li S. et al., 2003]. The bending strength of the joints has been improved by using the active metallic filler contacted with C_f/SiC. The strength of the joints was remarkably affected by the welding temperature. The maximum value 102.1 MPa has been obtained at 970°C (dwell time: 10 min, welding pressure: 34.3 MPa).

6. Characterization techniques

The characterization techniques can be categorized into two groups, those assessing the joint structure, thermal and mechanical properties and are in general destructive in nature and those assessing the quality of a specific joint and for a specific application which have to be non destructive (NDT). The first type can be performed at the different stages of joint fabrication or after a service cycle, real or laboratory. NDT on composite/metal joints has an inherent difficulty as the response of these complex structures to the excitations used are difficult to be appraised with regard to the component performance in service conditions. However, NDT techniques calibrated against metallographic examinations and mechanical tests can form a basis for reliable NDT test in manufacturing components for critical applications as in the aerospace and nuclear industry [Escourbiac et al., 2007; Ezato et al., 2002; Mergia et al., 2009; Merola et al., 2002].
The meatallographic type examinations of the joints as well as mechanical tests are well known and will not be referred here [Kim & Lee, 1998; Park J.-W et al., 2002; Jadoon et al., 2004; Takahashi et al., 2003; Galli et al., 2009; Lee et al., 1995; Serizawa et al., 2006; Serizawa et al., 2007; Serizawa et al., 2008;]. However, some advanced techniques not widely used and not well known need to be mentioned. Laboratory X-ray diffraction is well suited in identifying different phases in the joint. Taking into account that there are cases that we need locally to identify phases or that the amount of a phase present is below the laboratory X-ray detection limit, synchrotron X-ray techniques would be of an advantage. X-ray diffraction measurements near the ceramic interface can be used to determine the residual stress distribution [Eigenmann et al., 1987] as indentation fracture [Larsson 2011] or combination of them with finite-element analysis [Lee et al., 1997]. Residual stresses measured by X-ray diffraction in conjunction with finite element models predict very well the joint strength assessed by four-point bend tests [Gali et al., 2009]. Also neutron diffraction could be of great advantage for identifying phases and internal stresses in the joint. Neutron diffraction can be used as a NDT technique even for large components. Neutron diffraction has been applied in determining the stresses of C_f-C/Mo structures to be used for the ITER divertor [Ceretti et al., 1994]. Also strain field both in CFC graphite and in the brazing metal of Cu/C_f-C joints produced for fusion applications has been

determined by neutron diffraction in temperatures up to 600 °C, temperature at which these joints are going to be employed [Ceretti et al., 1998].
An interesting method for detecting weak or defective bonding areas within the brazed adjoining contact surfaces of composite ceramic-metal plates is the measurement by dynamic holographic interferometry of the response of transient flexural waves transmitted through the joint structure [Conrad & Sayir, 2001]. The quality of joints between CFC composites and Cu or CuCrZr to be used as high heat flux components of fusion machines has been tested by lock-in thermography, ultrasonic inspection, microtomography and micro-radiography [Casalegno et al., 2008]. The thermal diffusivity at the material interface of a C_f/C composite joined to Cu has been determined by the laser flash method. Subsequent calculation of the thermal conductivity and thermal contact resistance can be used as a qualitative assessment of the ceramic/metal joint integrity [Casalegno et al., 2010]. In conclusion, these NDT test need further evaluation and development in order to routinely screen the quality of joints.
In order to understand the filler/ceramic interactions and its wetting process is important its activity to be observed "in situ" as the filler metal interacts with the ceramic at the brazing temperature. A technique sensitive to the different interface formation with resolution of few Å is neutron reflectivity. In addition this technique can be used at high temperature and due to the high penetration of neutrons into the matter real interacting metal/ceramic sandwiches can be used [Derby et al., 1998; Xiao et al., 1997].

7. Conclusions

Ceramic-matrix composites (CMCs), and especially C_f/SiC and C_f/C composites, are attractive materials mainly for aerospace, thermonuclear fusion and other applications that require heat and wear resistance. Most of such applications require joining CMCs to metals which has been done using a number of approaches: mechanical joining, adhesive and diffusion bonding and active alloy brazing. Brazing is the method that presents the highest perspectives for the use of CMCs in advanced applications and demanding service conditions. The most critical issues related to the brazing process of CMC/metal are the poor wettability of ceramics by metals and the mismatch in the thermal expansion coefficient which induce residual thermal stresses at the joint. Different approaches to address the wettability issues include the development of brazing alloys, surface modification of the CMC by a wetting promoter of by direct metallization of its surface, and the promotion of certain chemical reactions in combination with the restriction of others through diffusion barriers. The relief of thermal strains at the joint is coped with by the use of ductile interlayers, the employment of carbon fibers or SiC particles reinforced brazing material and the use of an efficient stress-relaxing procedure, such as intermediate holding times during cooling for stress relaxation through creep.
However, notwithstanding sound joints between CMCs and metals have been produced, a lot of improvement in brazing is required in order to address the standards imposed by aerospace and nuclear applications. For this a better understanding of the liquid-metal/ceramic wetting and the underlying physics and chemistry has to be achieved. This will help in the development of predictive models of liquid-metal/ceramic substrate interactions and also of new brazing alloys. The brazing thermal cycles need to be optimized by finite element analysis for the induced thermal stresses and by incorporating predictive models determining the compounds to be formed at elevated temperatures through the

liquid-metal/ceramic interactions. Advanced characterization techniques such as synchrotron X-ray and neutron based techniques will be valuable for the joint characterization. In addition, standardization of existing and development of new non-destructive techniques are needed in order to asses the quality of the joints for the intended application.

8. References

Akselsen, O. M. (1992). Advances in brazing of ceramics. *Journal of Materials Science*, Vol. 27, No. 8, (April 1992), pp. 1989-2000

Appendino, P.; Casalegno, V.; Ferraris, M.; Grattarola, M.; Merola, M. & Salvo, M. (2003). Joining of C/C composites to copper. *Fusion Engineering and Design*, Vol. 66–68, (September 2003), pp. 225-229

Appendino, P.; Ferraris, M.; Casalegno, V.; Salvo, M.; Merola, M. & Grattarola, M. (2004). Direct joining of CFC to copper. *Journal of Nuclear Materials*, Vol. 329–333, Part.2, (August 2004), pp. 1563–1566, ISSN 00223115

Ashworth, M.A.; Jacobs, M.H. & Davies, S. (2000). Basic mechanisms and interface reactions in HIP diffusion bonding. *Materials & Design*, Vol.21, No.4, (August 2000), pp.351-358

Ban, Y.H.; Huang, J.H.; Zhang, H.; Zhao, X.K.& Zhang, Z.Y. (2009). Microstructure of Reactive Composite Brazing Joints of C_f/SiC Composite to Ti-6Al-4V Alloy with Cu-Ti-C Filler Material. *Rare Metal Materials and Engineering*, Vol. 38, No. 4, (April 2009), pp. 713-716, ISSN: 1002-185X

Barlak, M.; Piekoszewski, J.; Werner, Z.; Pakiela, Z.; Sartowska, B.; Składnik-Sadowska, E.; Walis, L.; Kierzek, J.; Starosta, W.; Kolitsch, A.; Gröetzchel, R. & Bochenska, K. (2009). The influence of distribution of titanium alloyed into carbon ceramics by the intense plasma pulses on their surface wettability with liquid copper.*Vacuum*, Vol. 83, Suppl. 1, (May 2009), pp.S81-S85

Bhanumurthy, K. & Schmid-Fetzer, R. (2001). Interface reactions between silicon carbide and metals (Ni, Cr, Pd, Zr). *Composites Part A: Applied Science and Manufacturing*, Vol. 32, No.3-4, (March 2001), pp.569-574, ISSN 1359835X

Böhrk, H. & Beyermann, U. (2010). Secure tightening of a CMC fastener for the heat shield of re-entry vehicles. *Composite Structures*, Vol.92, No.1, (January 2010), pp.107-112

Casalegno, V.; Salvo, M.; Ferraris, M.; Smeacetto, F.; Merola, M. & Bettuzzi, M. (2008). Non-destructive characterization of carbon .ber composite/Cu joints for nuclear fusion applications. *Fusion Engineering and Design*, Vol.83, No.5-6, (October 2008), pp. 702–712

Casalegno, V.; Vavassori, P.; Valle, M.; Ferraris, M.; Salvo, M. & Pintsuk, G. (2010). Measurement of thermal properties of a ceramic/metal joint by laser flash method. *Journal of Nuclear Materials*, Vol. 407, No.2, (December 2010), pp. 83–87

Ceretti, M.; Coppola, R.; Di Pietro, E.; Lodini, A.; Perrin, M.; Piant, A. & Rustichelli, F. (1994). Neutron diffraction study of internal stresses in brazed CFC/Mo divertor structures for NET/ITER. *Journal of Nuclear Materials,* Vol.212-215, Part 2, (September 1994), pp. 1617-1620

Ceretti, M.; Coppola, R.; Di Pietro, E. & Nardi, C. (1998). High-temperature residual strain measurements, using neutron diffraction, in brazed Cu/CFC graphite divertor structures. *Journal of Nuclear Materials*, Vol.258-263, Part 1, (October 1998), pp. 1005-1009

Conrad, M. & Sayir, M. (2001). Composite ceramic-metal plates tested with flexural waves and holography. *Experimental Mechanics* Vol.41, No.4, December 2001, pp. 412-420

Chen, B.; Xiong, H.-ping; Mao, W. & Cheng, Y.-yong. (2010). Wettability and interfacial reactions of PdNi-based brazing fillers on C-C composite. *Transactions of Nonferrous Metals Society of China*, Vol. 20, No.2, pp. 223-226

Chou, T.C. (1993). Interfacial Debonding by Solid-State Reactions of SiC with Ni and Co. *Scripta Metallurgica et Materialia*, Vol. 29, No.2, (July 1993), pp. 255-260

Derby, B.; Edwards, R.; Webster, J. & Xiao, P. (1998). Characterisation of liquid metal/solid ceramic interfaces by neutron reflection. *Acta Materialia*, Vol.46, No.7, (April 1998), pp. 2387–2392

Dixon, D. G. (1995). Ceramic matrix composite-metal brazed joints. *Journal of Materials Science*, Vol. 30, No.6, (January 1995), pp. 1539-1544

Donachie, M.J. (1982). *Titanium and Titanium Alloys Source Book*, American Society of Metals, Metals Park, OH, pp. 265–269

Eigenmann, B.; Scholtes, B. & Macherauch, E. (1987). Residual stresses in ceramic and ceramic-metal composites by X-ray diffraction methods. *Materials Science and Engineering A*, Vol.118, (October 1989), pp. 1-17

Escourbiac, F.; Constans, S.; Courtois, X. & Durocher, A. (2007). Application of lock-in thermography non destructive technique to CFC armoured plasma facing components. *Journal of Nuclear Materials*, Vol.367-370, Part 2, (August 2007), pp. 1492–1496, ISSN 00223115

Eustathopoulos, N. (1998). Dynamics of wetting in reactive metal/ceramic systems. *Acta Materialia*, Vol.46, No.7, (April 1998), pp. 2319–27

Evans A.G. & Marshall D. (1989). The mechanical behavior of ceramic matrix composites. *Acta Metallurgica*, Vol. 37, No.10, (October 1989), pp. 2567–83

Ezato, K.; Dairaku, M.; Taniguchi, M.; Sato, K. & Akiba, M. (2002). Non-destructive testing of CFC monoblock divertor mock-ups. *Journal of Nuclear Materials*, Vol. 307–311, Part 1, (December 2002), pp. 144–148, ISSN 00223115

Frage, N.; Froumin, N. & Dariel, M.P. (2002). Wetting of TiC by non-reactive liquid metals. *Acta Materialia*, Vol.50, No.2, (January 2002), pp. 237–245

Galli, M.; Botsis, J. & Janczak-Rusch, J. (2006). Relief of the Residual Stresses in Ceramic-Metal Joints by a Layered Braze Structure. *Advanced Engineering Materials*, Vol. 8, No.3, (March 2006) pp. 197-201

Galli, M.; Botsis, J.; Janczak-Rusch, J.; Maier, G. & Welzel, U. (2009). Characterization of the residual stresses and strength of ceramic-metal braze joints. *Journal of Engineering Materials and Technology, Transactions of the ASME*, Vol.131, No.2, (April 2009), pp. 0210041-0210048

Goldsworthy, W.B.; Johnson, D.W. & Korzeniowski, G. (1994). Bifurcated column joint system for electrical transmission tower, US Patent 5, 319, 901.

Goto, K.; Hatta, H.; Oe, M. & Koizumi, T. (2003). Tensile Strength and Deformation of a 2D Carbon-Carbon Composite at Elevated Temperatures. *Journal of the American Ceramic Society*, Vol.86, No.12, (December 2003), pp. 2129-2135

Gottselig, B.; Gyarmati, E.; Naoumidis A., & Nickel, H. (1990). Joining of ceramics demonstrated by the example of SiC/Ti. *Journal of the European Ceramic Society*, Vol. 6, No.3, pp.153-160

Grigorenko, N.; Poluyanskaya, V.; Eustathopoulos, N. & Naidich, Y., (1998). Interfacial Science of Ceramics Joining, In: Bellosi A, Kosmac, Tomsia AP, (Ed.), pp. 69–78, Kluwer Academic Publishers, Boston

Hattali, M.; Valette, S.; Ropital, F.; Stremsdoerfer, G.; Mesrati, N. & Treheux, D. (2009). Study of SiC–nickel alloy bonding for high temperature applications. *Journal of the European Ceramic Society*, vol. 29, No.4, pp.813-819

Jadoon, K.; Ralpha, B. & Hornsby, P. R. (2004). Metal to ceramic joining via a metallic interlayer bonding technique. *Journal of Materials Processing Technology*, Vol.152, No.3, 30 (October 2004), pp. 257-265

Jian K.; Chen Z.H.; Ma Q.S. & Zheng W.W. (2005). Effects of pyrolysis processes on the microstructures and mechanical properties of Cf/SiC composites using polycarbosilane. *Materials Science Engineering* A, Vol. 390, No.1-2, (January 2005), pp. 154–158

Ishikawa, T.; Kajii, S.; Matsanaga, K.; Hogani, T.; Kohtoku, Y. & Nagasawa, T. (1998). A Tough, Thermally Conductive Silicon Carbide Composite with High Strength up to 1600°C in Air. *Science*, Vol. 282, (November 1998), pp. 1295–1297, ISSN 0036-8075

Kim, J. H. & Lee, S. B. (1998). Stress intensity factors and crack initiation directions for ceramic/metal joint, *Theoretical and Applied Fracture Mechanics*, Vol.30, No.1, (September 1998), Pages 27-38

Kweon, J.; Jung, J.; Kim, T.; Choi, J. & Kim, D. (2006). Failure of carbon composite-to-aluminum joints with combined mechanical fastening and adhesive bonding. *Composite Structures*, Vol. 75, No.1-4, (September 2006), pp. 192-198

Lamouroux, F. ; Bertrand, S. ; Pailler, R. ; Naslain, R. & Cataldi, M. (1999). Oxidation-resistant carbon-fiber-reinforced ceramic-matrix composites. *Composites Science and Technology*, Vol. 59, No.2, (May 1999), pp. 1073-1085

Larsson, P.-L. (2011). On the mechanical behavior at sharp indentation of materials with compressive residual stresses. *Materials & Design*, Vol.32, No.3, (March 2011), pp. 1427-143

Lee, S.-B.; Kobayashi, H. & Huh, J. -W. (1995). Fatigue strength and fracture mechanism of ceramic-metal joints under cyclic bending. *International Journal of Fatigue*, Vol.17, No.6, pp. 427-435

Lee, S.-B. & Kim, J.-H (1997). Finite-element analysis and X-ray measurement of the residual stresses of ceramic/metal joints. *Journal of Materials Processing Technology*, Vol. 67, pp. 167-172

Levy, A. (1991). Thermal Residual Stresses in Ceramic-to-Metal Brazed Joints. *Journal of the American Ceramic Society*, Vol. 74, No.9, (September 1991) pp. 2141-2147

Li, J. G. (1992). Kinetics of wetting and spreading of Cu-Ti alloys on alumina and glassy carbon substrates. *Journal of Materials Science Letters*, Vol. 11, No.23, (January 1992), pp. 1551-1554

Li, J.; Liu, L.; Wu, Y.; Li, Z.; Zhang, W. & Hu, W. (2009). Microstructure of high temperature Ti-based brazing alloys and wettability on SiC ceramic. *Materials and Design*, Vol.30, No. 2, (February 2009), pp. 275–279

Li, J. & Xiao, P. (2004). Fabrication and characterisation of silicon carbide/superalloy interfaces. *Journal of the European Ceramic Society*, Vol. 24, No.7, (June 2009), pp. 2149–2156

Li, J.; Xiong, J. & Zhang, F. (2008). Transient liquid-phase diffusion bonding of two-dimensional carbon–carbon composites to niobium alloy. *Materials Science and Engineering A*, Vol.483-484, (June 2008), pp. 698-700

Li, S.; Zhang, J.; Liang., X.; Duan, H. & Zhang, Y. (2003). Joining of Carbon Fibre Reinforced SiC (Cf/SiC) to Ni-Based Superalloy with Multiple Interlayers. *International Journal of Modern Physics B*, Vol. 17, Nos. 8-9, (April 2003), pp. 1777-1781

Li, Y.L.; He, P. & Feng, J.C. (2006). Interface structure and mechanical properties of the TiAl/42CrMo steel joint vacuum brazed with Ag–Cu/Ti/Ag–Cu filler metal. *Scripta Materialia*, Vol. 55, No.2, (July 2006), pp. 171–174

Lin, G.; Huang, J. & Zhang, H. (2007). Joints of carbon fiber-reinforced SiC composites to Ti-alloy brazed by Ag-Cu-Ti short carbon fibers. *Journal of Materials Processing Technology*, Vol. 189, No. 1-3, (July 2007), pp. 256-261

Lin, G. B.; Huang, J. H.; Zhang, H.; Liu, H. Y. (2006). Microstructure and mechanical performance of brazed joints of Cf/SiC composite and Ti alloy using Ag–Cu–Ti–W. *Science and Technology of Welding & Joining*, Vol. 11, No. 4, pp. 379-383

Liu H.J.; Feng J.C. & Qian Y.Y. (2000). Microstructure and strength of the SiC/TiAl joint brazed with Ag–Cu–Ti filler metal. *Journal of Materials Science Letters*, Vol. 19, No.14, (July 2000), pp. 1241–1242

Liu, H.J.; Feng, J.C. & Qian, Y.Y. (2000). Interface structure and formation mechanism of diffusion-bonded joints of SiC ceramic to TiAl-based alloy. *Scripta Materialia*, Vol. 43, No.1, (June 2000), pp. 49-53

Liu, G. W.; Valenza, F.; Muolo, M. L. & Passerone, A. (2010). SiC/SiC and SiC/Kovar joining by Ni–Si and Mo interlayers. *Journal of Materials Science*, Vol. 45, No.16, (August 2010), pp. 4299-4307

Liu, J.Y.; Chen, S. & Chin, B.A. (1994). Brazing of vanadium and carbon-carbon composites to stainless steel for fusion reactor applications. *Journal of Nuclear Materials*, Vol. 212-215, Part.2, (September 1994), pp. 1590-1593, ISSN 00223115

Liu, Y.Z.; Zhang, L.X.; Liu, C.B.; Yang, Z.W.; Li, H.W. & Feng, J.C. (2011). Brazing C/SiC composites and Nb with TiNiNb active filler metal, In: *Science and Technology of Welding & Joining*, vol.16, No.2, (February 2011), pp. 193-198

Evans A.G. & Marshall D. (1989). The mechanical behavior of ceramic matrix composites. *Acta Metallurgica*, Vol. 37, No.10, (October 1989), pp. 2567–83

Matsuo, G.; Shibayama, T.; Kishimoto, H.; Hamada, K. & Watanabe, S. (2011). Micro-chemical analysis of diffusion bonded W-SiC joint. *Journal of Nuclear Materials*, (in press), ISSN 00223115

Mcdermid, J. R. & Drew, R. A. L. (1991). Thermodynamic Brazing Alloy Design for Joining Silicon Carbide. Journal of the American Ceramic Society, Vol. 74, No. 8, (August 1991), pp. 1855-1860.

Mehan, R.L. & Mckee, D.W. (1976). Interaction of metals and alloys with silicon-based ceramics. *Journal of Materials Science*, Vol. 11, No.6, (June 1976), pp. 1009-1018

Mergia, K.; Grattarola, M.; Messoloras, S. & Gualco, C. (2009). Residual Stress Measurements In Mo / CuCrZr Tiles Using Neutron Diffraction. *Advanced Materials Research*, Vol.59, pp. 299-303, ISSN: 10226680

Merola, M.; Akiba, M.; Barabash, V. & Mazul, I. (2002). Overview on fabrication and joining of plasma facing and high heat flux materials for ITER. *Journal of Nuclear Materials*, Vol. 307–311, Part 2, (December 2002), pp. 1524–1532, ISSN 00223115

Merola, M.; Dänner, W.; Palmer, J.; Vielder, G.; Wu, C.H. & EU-ITER Participating Team (2003). European contribution to the development of the ITER divertor. *Fusion Engineering and Design*, Vol. 66-68, (September 2003), pp. 211-217

Merola, M.; Chappuis, P.; Escourbiac, F.; Grattarola, M.; Jeskanen, H.; Kauppinen, P.; Plöchl, L.; Schedler, J. Schlosser, B.; Smid, I.; Tähtinen, S.; Vesprini, R.; Visca, E. & Zabernig, A. (2002). Non-destructive testing of divertor components. *Fusion Engineering and Design*, Vol.61–62, (November 2002), pp. 141–146

Messler, R. (2000). Trends in key joining technologies for the twenty-first century, In: *Assembly Automation* Vol. 20 . No.2, pp. 118-128, MCB University Press, ISSN 0144-5154

Messler, R. (2004). Joining Composite Materials and Structures: Some Thought-provoking Possibilities. *Journal of Thermoplastic Composite Materials*, Vol. 17, No. 1, (January 2004), pp. 51-75

Morozumi, S.; Endo, M. & Kikuchi M. (1985). Bonding mechanism between silicon carbide and thin foils of reactive metals, In: *Journal of Materials Science,* Vol.20, (November 1985), pp. 3976–3982

Morscher, G.N., Singh, M., Shpargel, T.P. & Asthana, R. (2006). A simple test to determine the effectiveness of different braze compositions for joining Ti tubes to C/C composite plates. *Materials Science and Engineering A*, Vol. 418, No.1-2, (February 2006), pp. 19–24

Moulson, A.J. (1979). Review: Reaction-bonded silicon nitride: its formation and properties. *Journal of Materials Science*, Vol. 14, No.5, (May 1979), pp. 1017-1051

Moutis, N. V.; Jimenez, C. ; Speliotis, T. ; Azpiroz, X. & Mergia, K. (2009). Graphite-Nimonic alloy brazing. *Advanced Materials Research*, Vol.59, pp. 209-213

Moutis, N. V.; Jimenez, C.; Azpiroz, X.; Speliotis, T.; Wilhelmi, C.; Messoloras, S. & Mergia, K. (2010). Brazing of carbon–carbon composites to Nimonic alloys. *Journal of Materials Science*, Vol. 45, No.1, (January 2010), pp. 74-81

Naidich, Y. (2005). About liquid metal/ceramic interface interaction mechanism and mode of a new intermediate compound formation. *Current Opinion in Solid State and Materials Science*, Vol.9, No.4-5, (August-October 2005), pp. 161–166

Naidich, Y.; Zhuravlev, V.; Gab, I.; Kostyuk, B.; Krasovskyy, V.; Adamovskyy, A.A. & Taranets N.Y. (2008). Liquid metal wettability and advanced ceramic brazing. *Journal of the European Ceramic Society*, Vol. 28, No. 4, pp. 717-728

Naka, M. & Maeda, M. (1991). Application of ultrasound on joining of ceramics to metals. *Engineering Fracture Mechanics*, Vol. 40, No.4-5, pp. 951-956

Naslain, R. (2004). Design, preparation and properties of non-oxide CMCs for application in engines and nuclear reactors: an overview. *Composites Science and Technology*, Vol. 64, No.2, (February 2004), pp. 155-170

Nicholas, M. G. (1998). Joining Process: Introduction to Brazing and Diffusion Bonding, M.G. Nicholas, (Edi), Kluwer Academic Publishers, Dordrecht, The Netherlands

Nicholas, M.G. & Peteves, S.D. (1994). Reactive joining; chemical effects on the formation and properties of brazed and diffusion bonded interfaces. *Scripta Metallurgica et Materialia*, Vol. 31, No.8, (October 1994), pp. 1091-1096

Ning, X. S.; Okamoto, T.; Miyamoto, Y. & Koreeda, A. (1989). Effect of oxide additive in silicon nitride on interfacial structure and strength of silicon nitride joints brazed with aluminium. *Journal of Materials Science*, Vol.24, No.8, (August 1989), pp. 2865-2870

Novakovic, R.; Ricci, E.; Muolo, M.L.; Giuranno, D. & Passerone, A. (2003). On the application of modelling to study the surface and interfacial phenomena in liquid alloy–ceramic substrate systems. *Intermetallics*, Vol. 11, No.11-12, pp. 1301–1311

Nozawa, T.; Hinoki, T.; Hasegawa, A.; Kohyama, A.; Katoh, Y.; Snead, L. L., Henager, Jr. C.H. & Hegeman, J.B.J. (2009). Recent advances and issues in development of silicon carbide composites for fusion applications. *Journal of Nuclear Materials*, Vol. 386-388, (April 2009), pp. 622-627, ISSN 00223115

Paiva, O.C. & Barbosa, M.A. (2000). Brazing parameters determine the degradation and mechanical behaviour of alumina/titanium brazed joints. *Journal of Materials Science*, Vol. 35, No.5, (March 2000), pp. 1165–1175

Park, J. S., Landry, K., & Perepezko, J. H. (1999). Kinetic control of silicon carbide / metal reactions. *Materials Science and Engineering A*, Vol. 259, No. 2, (January 1999), pp. 279 - 286

Park, J.-W. & Eagar, T.W. (2004). Strain energy release in ceramic-to-metal joints with patterned interlayers. *Scripta Materialia*, Vol. 50, No. 4, (February 2004), pp. 555–559

Park, J.-W.; Mendez, P. F. & Eagar, T. W., (2002). Strain energy distribution in ceramic-to-metal joints, In: *Acta Materialia*, Vol.50, No.5, (March 2002), pp. 883-899

Paulasto, M.; Loo, F.J.J. van & Kivilahti, J.K. (1995). Thermodynamic and experimental study of Ti-Ag-Cu alloys. *Journal of Alloys and Compounds*, vol. 220, No.1-2 , (April 1995), pp.136-141

Peteves, S.D.; Paulasto. M.; Ceccone, G. & Stamos, V. (1998). The reactive route to ceramic joining: Fabrication, interfacial chemistry and joint properties. *Acta Materialia*, Vol. 46, No.7, (April 1998), pp. 2407-2414

Petzow, G. & Effenberg, G. (Eds.) (1988). Ternary Alloys, Vol. 2, p. 55, VCH, Verlagsgesellschaft, Weinheim, Germany

Pichon, T.; Barreteaua, R.; Soyris, P.; Foucault, A.; Parenteau, J.M.; Prel, Y. & Guedron, S. (2009). CMC thermal protection system for future reusable launch vehicles: Generic shingle technological maturation and tests. *Acta Astronautica*, Vol. 65, No.1-2, (July-August 2009), pp.165–176, ISSN: 00945765

Qin, Y. & Feng, J. (2007). Microstructure and mechanical properties of C/C composite/TC4 joint using AgCuTi filler metal. *Materials Science and Engineering A*, Vol. 454–455, (April 2007), pp. 322–327

Qin, Y. & Feng J. (2009). Active brazing carbon/carbon composite to TC4 with Cu and Mo composite interlayers. *Materials Science and Engineering A*, Vol. 525, No.1-2, (November 2009), pp.181–185

Qin Y. & Yu Z. (2010). Joining of C/C composite to TC4 using SiC particle-reinforced brazing alloy. *Materials Characterization*, Vol. 61, No.6, (June 2010), pp. 635-639

Reimer, T. (2006). The KERAMIK thermal protection system experiment on the FOTON-M2 Mission, *Proceedings of the 5th European workshop on thermal protection systems and hot structures* (SP-631), Noordwijk, The Netherlands, May 17-19, 2006

Razzell, A.G. (2004). Joining and Machining of Ceramic Matrix Composites Comprehensive Composite Materials, In: *Carbon/Carbon, Cement, and Ceramic Matrix Composites*, Vol.4, pp. 689-697, (November 2004), ISBN 0080437222

Roger, R.; Collings, E.W. & Welsch, G. (1993). Materials Properties Handbook: Titanium Alloys, ASM International, Materials Park

Ryshkewitch, E. (1953). Compression Strength of Porous Sintered Alumina and Zirconia. *Journal of the American Ceramic Society*, (February 1953), Vol. 36, No.2, pp. 65-68

Saiz, E. & Tomsia, A.P. (2005). Kinetics of high-temperature spreading. *Current Opinion in Solid State and Materials Science*, Vol.9, No.4-5, (August-October 2005), pp. 167–173

Salvo, M.; Casalegno, V.; Vitupier, Y.; Cornillon, L.; Pambaguian L. & Ferraris, M. (2010). Study of joining of carbon/carbon composites for ultra stable structures. *Journal of the European Ceramic Society*, Vol. 30, No.7, (May 2010), pp. 1751–1759

Salvo, M.; Lemoine, P.; Ferraris, M. & Montorsi, M. (1997). Joining of Carbon-Carbon Composites for Thermonuclear Fusion Applications. *Journal of the American Ceramic Society*, Vol. 80, No. 1, pp. 206-212

Serizawa, H.; Fujita, D.; Lewinsohn, C.A.; Singh, M. & Murakawa, H. (2006). Finite element analysis of mechanical test methods for evaluating shear strength of ceramic composite joints using interface element. *Ceramic Engineering and Science Proceedings*, Vol.27, No.2, pp. 115-124

Serizawa, H.; Katayama, K.; Lewinsohn, C.A.; Singh, M. & Murakawa, H. (2008). Effect of residual stress on fracture behavior in mechanical test for evaluating shear strength of ceramic composite join. *Ceramic Engineering and Science Proceedings,* Vol.28, No.2, 2008, pp. 503-511

Serizawa, H.; Lewinsohn, C.A.; Singh, M. & Murakawa, H. (2007). Numerical analysis of test methods for evaluating shear strength of ceramic composite joints using interface elemen. *Materials Science Forum*, Vol.539-543, Part 3, pp. 2143-2148

Shirzadi, A.A.; Zhu, Y. & Bhadeshia, H.K.D.H. (2008). Joining ceramics to metals using metallic foam. *Materials Science and Engineering A*, Vol. 496, No.1-2, (November 2008), pp. 501–506

Schlosser, J.; Escourbiac, F.; Merola, M. ; Fouquet, S. ; Bayetti, P. ; Cordier, J.J. ; Grosman, A.; Missirlian, M.; Tivey, R. & Rödig, M. (2005). Technologies for ITER divertor vertical target plasma facing components. *Nuclear Fusion*, Vol. 45, No.6, (June 2005), pp. 512–518

Schmidt, S.; Beyer, S. & Immich, H. (2005). Ceramic Matrix Composites: A Challenge in Space-Propulsion Technology Applications. *International Journal of Applied Ceramic Technology*, Vol.2, No.2, (March 2005), pp. 85–96

Schmidt, S.; Beyer, S.; Immich, H.; Knabe, H.; Meistring, R. & Gessler, A. (2005). Ceramic Matrix Composites: A Challenge in Space-Propulsion Technology Applications. *International Journal of Applied Ceramic Technology*, Vol. 2, No.2, (March 2005), pp.85-96

Singh, M. & Asthana, R. (2010). Joining and Integration of Advanced Carbon-Carbon and Carbon-Silicon Carbide Composites to Metallic Systems, In: *Ceramic Materials and Components for Energy and Environmental Applications,* D. Jiang, Y. Zeng, M. Singh and J. Heinrich, (Eds.), pp. 493-504, ISBN 9780470408421, John Wiley & Sons, Inc., Hoboken, NJ, USA

Singh, M. & Asthana, R. (2008). Characterization of brazed joints of C/C composite to Cu-clad-Molybdenum. *Composites Science and Technology,* Vol. 68, No.14, (November 2008), pp. 3010–3019

Singh, M., Asthana, R. & Shpargel, T.P. (2007). Brazing of C/C composites to Cu-clad Mo for thermal management applications. *Materials Science and Engineering A,* Vol. 452–453, (April 2007), pp. 699–704

Singh, M.; Asthana, R. & Shpargel, T.P. (2008). Brazing of ceramic-matrix composites to Ti and Hastealloy using Ni-base metallic glass interlayers. *Materials Science and Engineering A*, Vol.498, Nos.1-2, (December 2008), pp.19-30

Singh, M.; Shpargel, T.P.; Morscher, G.N. & Asthana, R. (2005). Active metal brazing and characterization of brazed joints in titanium to carbon–carbon composites. *Materials Science and Engineering A*, Vol.412, Nos.1-2, (December 2005), pp. 123–8

Sobczak, N. & Asthana, R. (2001). The role of wetting phenomenon in the structure and bonding of metal-ceramic interfaces: some recent developments. *Recent Research Developments in Materials Science,* Vol. 2, pp. 135–160, Research Signpost, Trivandrum, India

Sobczak, N.; Singh, M. & Asthana, R. (2005). High-temperature wettability measurements in metal/ceramic systems - Some methodological issues. *Current Opinion in Solid State and Materials Science,* Vol. 9, No. 4-5, (August-October 2005), pp. 241–253

Srivastava, V. (2003). Characterization of adhesive bonded lap joints of C/C–SiC composite and Ti–6Al–4V alloy under varying conditions. *International Journal of Adhesion and Adhesives,* Vol. 23, No. 1, pp.59-67, ISSN: 01437496

Standing, R. & Nicholas, M.G. (1978). The wetting of alumina and vitreous carbon by copper–tin–titanium alloys. *Journal of Materials Science,* Vol.13, No.7, (July 1978), pp. 1509–1514.

Südmeyer, I.; Hettesheimer, T. & Rohde, M. (2010). On the shear strength of laser brazed SiC–steel joints: Effects of braze metal fillers and surface patterning. *Ceramics International,* Vol. 36, No.3, (April 2010), pp. 1083-1090

Takahashi, M.; Okabe, N.; Zhu, X. & Kagawa, K., (2003). Strength estimation of ceramic-metal joints with various interlayer thicknes. *Fatigue and Fracture of Engineering Materials and Structures,* Vol.26, No. 5, (May 2003), pp. 391-398

Tillmann, W.; Lugscheider, E.; Xu, R. & Indacohea, J.E. (1996). Kinetic and microstructural aspects of the reaction layer at ceramic/metal braze joints. *Journal of Materials Science,* Vol. 31, No.2, (January 1996), pp. 445-452

Tivey, R.; D'Agata, E.; Chuyanov, V. & Heidl, H. (2005). Overview of the engineering design of the ITER divertor improvements towards manufacture. Fusion Engineering and Design, Vol.75–79, (November 2005), pp. 447–450

Trehan, V.; Trickey, S.; Indacochea, J.E.; Lugscheider, E.; Buschke, I. & Tillmann, W. (1999). Brazing of silicon nitride with reactive filler metals. *Science and Engineering of Composite Materials*, Vol. 8, No.2, 107-112, ISSN: 0334-181X

Voitovitch, R.; Mortensen, A.; Hodaj, F. & Eustathopoulos, N. (1999). Diffusion limited reactive wetting: study of spreading kinetics of Cu–Cr alloys on carbon substrates. *Acta Materialia*, Vol.47, No.4, (March 1999), pp. 1117–1128

Wang, H.; Cao, J. & Feng, J. (2010). Brazing mechanism and infiltration strengthening of C/C composites to TiAl alloys joint. *Scripta Materialia,* Vol. 63, No.8, (October 2010), pp. 859–862

Xiao, P. & Derby, B. (1998). Wetting of silicon carbide by chromium containing alloys. *Acta Materialia*, Vol.46, No.10, (June 1998), pp. 3491–9

Xiao, P.; Derby, B.; Webster, J. & Penfold, J. (1997). The characterization of metal/ceramic interfaces using specular neutron reflection. *Acta Materialia*, Vol.45, No.1, (January 1997), pp.273–279

Xiong, J.H.; Li, J.; Zhang, F. & Huang, W. (2006). Joining of 3D C/SiC composites to niobium alloy. *Scripta Materialia*, Vol. 55, No.2, (July 2006), pp. 151-154

Xiong, J. H. ; Huang, J. H. ; Wang, Z. P. ; Ban, Y. H. ; Zhang, H. ; Zhao, X. K. (2010). Brazing of carbon fibre reinforced SiC composite and Ti alloy using Cu—Ti—C filler materials. *Materials Science and Technology*, vol. 26, No.3, (March 2010), pp. 356-360, ISSN 0267-0836

Xiong, J. H.; Huang, J. H.; Zhang, H. & Zhao, X. K. (2010). Brazing of carbon fiber reinforced SiC composite and TC4 using Ag–Cu–Ti active brazing alloy. *Materials Science and Engineering: A*, Vol. 527, Nos. 4-5, (February 2010), pp. 1096-1101

Zhang, Y.; Feng, D; He, Z.Y. & Chen, X.C. (2006). Progress in Joining Ceramics to Metals. *International Journal of Iron and Steel Research*, Vol.13, No.2, (March 2006), pp. 1-5

Zhong, Z.; Hinoki, T. & Kohyama, A. (2011). Microstructure and mechanical strength of diffusion bonded joints between silicon carbide and F82H steel. *Journal of Nuclear Materials*, Vol. 10-14 (in press), ISSN 00223115

Zhu, M. & Chung, D. D. L. (1994). Active Brazing Alloy Containing Carbon Fibers for Metal-Ceramic Joining. *Journal of the American Ceramic Society*, Vol.77, No.10, (October 1994), pp. 2712-20

13

Optical and Structural Studies of Binary Compounds by Explosive Laser Irradiation and Heat Treatment

S. Kar
Department of Physics, R. K. Mission Sikshanamandira,
Belur Math, Howrah,
India

1. Introduction

The chalcogens are the name for the periodic table group 16 (old-style: VIB or VIA) in the periodic table. It is sometimes known as the oxygen family. It consists of the elements oxygen (O), sulfur (S), selenium (Se), tellurium (Te), the radioactive polonium (Po), and the synthetic ununhexium (Uuh). Their compounds, particularly the sulfides, selenides and tellurides are collectively known as chalcogenides. Binary compounds of the chalcogens are called chalcogenides (rather than chalcides, which breaks the pattern of halogen / halide and pnictogen / pnictide).

Chalcogenide compounds have the structure mostly composed of segments, chains and molecules. The primary covalent type structure prevailing within the chains of segments in particular a weak one. The materials of low covalent bond strength shows optical transparency in the mid to far IR. While the chalcogenide glasses (Boer & Ovshinsky, 1970; Collins, 1970; Zope M & Zope J, 1984; Afifi et al., 1988; Fadel & El-Shair , 1992 & Fadel et al. ,1992) are regarded as weakly bonded materials in contact to oxide glasses containing both types of heteropolar and homopolar covalent bond viz. Ge-Se, Ge-S etc. and Se-Se, S-S etc. Substantial number of the works on chalcogenide glasses were pointed towards device making games in memory thin film electronic switches (von Allmen, 1987 & Fadel, 1993). Although the applications are manifold yet the necessary structures are still to be understood in the sense of transition between amorphous to crystalline states (Brice,1986). A tendency towards attaining glassy states for ternary chalcogenide glasses increases in the following order

$$S > Se > Te \quad \text{Gr. (VI)},$$

$$As > P > Sb \quad \text{Gr. (V)},$$

$$Si > Ge > Sn \quad \text{Gr. (IV)}$$

This principle adopted is an indicator that the thin films of Ge-As-S system are comparatively easier than Ge-P-S system. The softening temperature decreases with the increase of atomic number and with the addition of any other element to the host of Ge- Se and Ge-Te (Katsuyama & Matsumura, 1992) may increase the temperature of softening thereby shifting of the crystalline temperature on to much higher order.

Chalcogenides have wide range of applications in materials science domain. Several structural or photochemical changes have been observed in amorphous elements and compounds viz. through the route of laser treatment for photo- crystallization (Katsuyama & Matsumura, 1992; Mott & Davis, 1979; Andonov,1982) and heat treatment. Selenium is one such characteristic element, which undergoes unstable morphological transitions from amorphous to crystalline phases. Generally Se thin films grown by evaporation technique, crystallize to pattern resembles like a spherulitic one in hexagonal form compounds of helical chains (Kawarada & Nishina, 1975). During transitions, the different phases of the a-Se films reveal that the structure, growth processes and other properties depend on the method of preparation, deposition conditions (vapor temperature, rate and angle of deposition), physical condition of the substrates and vacuum environment (Kim & Turnbull, 1973 & Gross et al., 1977).

The present work deals with the synthesis of thin films of binary chalcogenide compounds of Ge, Se & Sn elements under explosive laser shots and heat treatments. A number of experimental results were used to explain a new phase generation of the composite materials of chalcogenide. The present attempt is mainly confined with the formation of chalcogenide films and their characterization through Optical Microscopy, SEM, XRD, UV-VIS-NIR etc. In general the chalcogens are viewed as two fold coordinated atoms and can easily be obtained in the form of thin films. Melting, quenching and vapour deposition process are the main tools to fabricate the chalcogenide glasses and alloys. The analysis of the micrographs accumulated by means of SEM and optical Microscopy from the experiment, which are carefully monitored and analyzed.

2. Experimental details

2.1 Importance of vacuum in film preparation

For the preparation of thin film of a material, it is required (a) suitable source to heat the material and convert it from solid to vapour phase (b) suitable chamber at a reduced pressure (c) a supporting material on which the vapour will condense to form a film. If the pressure inside the chamber is not in the order of 10^{-5} to 10^{-6} torr, the deposition is not good. As a result the film often peels out after deposition.

Substrate is the material on which the film is to be deposited. Different materials have been used as a substrates. The common substrate materials are used like glass, quartz, plastic etc. To prepare a good type of cleanliness of substrate is more essential. Temperature of the substrate during deposition is also another factor.

2.2 Cleaning of substrates

To prepare a high quality thin film made of CIS cleaning procedure of substrates glass is more essential (Chopra,1969). The step by step cleaning procedure is given below.

i. clean with dilute HCl (1-5%) and rinse the water,
ii. rub the substrate with brush and a mild detergent,
iii. wash in double distilled deionnized water several times,
iv. place in ultrasonic cleaner with the glass slide immersed in water,
v. boiled them in an electric heater,
vi. degrease in vapour of isopropyl alcohol in a degreasing chamber fitted to a condenser system.

2.3 Film preparation

Thin films of different specimens have been prepared by adopting thermal evaporation technique through a vacuum coating unit (modified Edward 12EA - 784) as shown in **(Fig.1.)** under a stable vacuum of the order of $5x10^{-6}$ torr. on a good quality
E-glass substrates. The vacuum system consists of a rotary pump followed by silicone oil (704 grade oil) diffusion pump. The vacuum measurement is conducted through a Pirani gauge followed by a Penning gauge. The circulation of the chilled water is maintained all throughout the course of the experiment. There is a provision in the evaporator that takes care of the substrates thermal state. The distance between the source and the substrate is optimized to the order of 10 cm in order to prepare good quality uniform thin films.

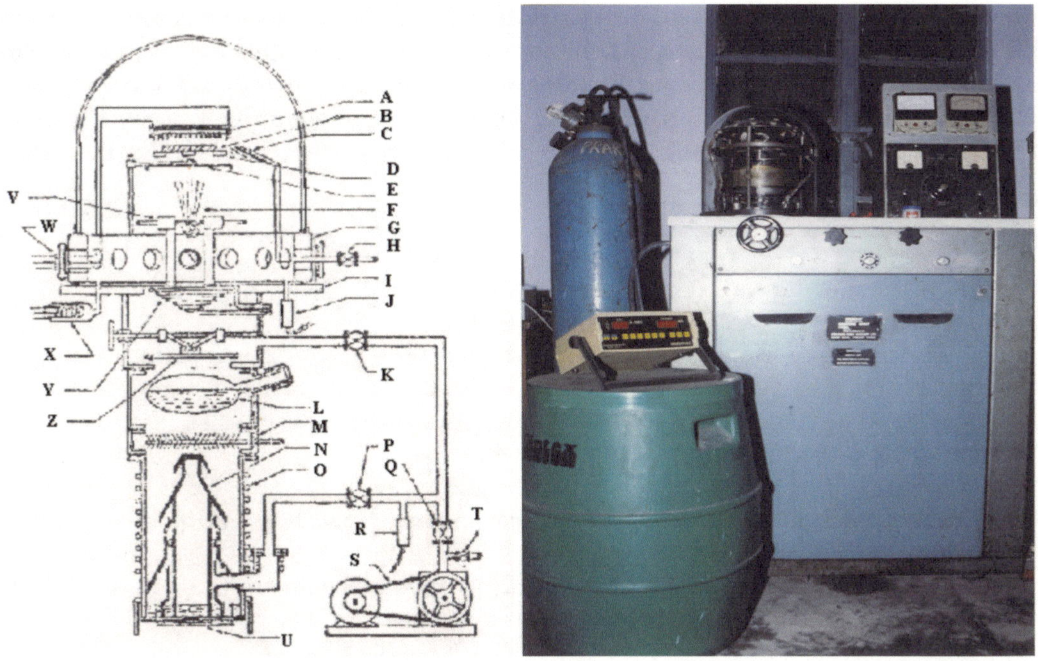

Fig. 1. A typical oil-diffusion pump evaporation station used in the laboratory. The notation stand for A, quartz iodine lamp heater; B, substrate; C, quartz crystal rate controller and deposition monitor; D, substrate mask; E, shutter (mechanical); F, vapour from evaporation source; G, adapter collar between the bell jar and the pump base plate flange; H, air inlet valve; I, base plate flange; J, Pirani or thermocouple gauge; K, roughing valve; L, liquid air trap; M, cooled chevron baffles; N, diffusion pump; O, cooling coils; P and Q , backing valves; R, Pirani gauge; S, fore-pump with air inlet valve T; U, diffusion pump heater; V, filament holders; W, multiple feedthrough; X, ionization gauge; Y, Miessner trap; Z, baffle valve.

The substrates were initially washed in warm soap solution followed by chemical cleaning with anhydrous grade tri-chloro-ethyline, acetone and alcohol followed by CIS ultrasound cleaning. These initial treatments render the occluded materials to be wiped away before the specimen is put in the evaporation chamber for deposition. Different specimens from the chalcogenide groups of various thickness have been deposited on the substrates for both individual and stacked bi-layer samples. The adhesion of the film and its quality is highly depending on the cleanliness, temperature of the substrate and order of the vacuum during deposition. If the substrate is not properly cleaned or the residual gas inside the chamber is not being sufficiently low during deposition, it has been found that the film often peels out after deposition. After cleaning, the substrate has been baked inside the vacuum chamber to remove the absorbed water vapor and further cleaned with ion bombardment in presence of argon atmosphere. Argon ion cleaning was carried out before the evaporation process started and also the initial sputtering of the source materials are ignored with the help of a movable umbrella placed over the tungsten boat.

Materials used for thermal evaporation of purity grade 5N (99.999 %) from Johnson & Matthey Co. Ltd. The shutter so used is important by which the unwanted atomic depositions were filtered and the heating has been raised with immediate removal of the shutter. The thickness of the film, the rate of deposition and the pre-factors associated with the elements used were monitored with the help of Edward FTM5 quartz crystal monitor unit. Deposition rate of the material on the substrate was in the order of 5 to 8 nm sec^{-1}. After the completion of the deposition, the films were exposed to high purity dry argon atmosphere for quite sometime and then taken out from the chamber, for the photo-inducement and heat treatment.

- Coating Unit : Edward 12EA-784
- Thickness monitor Unit : Edward FTM5 Quartz Crystal
- Materials : Se, Ge, Sn
- Purity grade : 5N (99.999 %)(Johnson and Mathay Co. Ltd.)
- Substrates : Glass slides
- Substrates are cleaned with tri – chloro - ethyline, acetone and alcohol.
- Order of Vacuum maintained : ~10^{-3} Pa.
- Substrate to film material gap : ~ 10 cm

  Shutter arrangement is being interposed between source and the substrate to avoid unwanted material deposition.
- Ar ion cleaning was initially carried on achieving the requisite vacuum prior to deposition, in order to improve the ultimate vacuum.
- Deposition rate : 5 ~ 8 nm sec^{-1}

Table 1. Summary of deposition Conditions

Films of amorphous Ge, Sn & Se on E-glass substrates have been prepared singularly by vacuum evaporation technique under a vacuum of the order of $5x10^{-6}$ torr. Out of such individual films the two as deposited front surfaces (one combination of Ge & Se and other Sn & Se) are sandwiched by applying small mechanical pressure in order to prevent air pockets at the film interface. The composite films are then placed on a micro-positioner facing Se-substrate interface towards the laser beam. The sandwiched films are then irradiated under pulsed shot through Nd : YAG laser beam for fundamental and 3rd harmonics to form compounds of Ge-Se and Sn-Se.

Several specimens of different films (Se, Ge, Sn, Ge-Se, Sn-Se etc.) are grouped in various ways following their depositing thickness. The deposition rate and thickness were recorded by Edward FTM5 thickness monitoring system housed in the vacuum chamber. All the films were heat-treated and laser treated according to their needs in the experimental process.

2.4 Laser treatment

For laser treatment individual films were sandwiched by applying mechanical pressure to make the interface free from air pockets and then irradiated by pulsed laser output from a third harmonics generator (KDP) of 355 nm UV wavelength pumped by Nd:YAG laser of Spectra Physics Model from one side. Depending upon the requirement one to hundred shots of 20-40 mJ cm^{-2} were applied on the composite films.

The full power of the laser pulse is around 50W and the reason for using this amount of power during the irradiation process is to make approximately equivalent heat treatments, otherwise the quality of equivalence under the two different processes may be confused.

In laser-assisted reactions generally, the problems encountered during laser shots on multilayer thin films are the following:

a. Precise knowledge of absorbance of the optical energy by the films remains unknown. This is because the film absorbance and refractivity are time-dependent phenomenon during the phase transition period.
b. Conduction of heat depends on time during reaction.
c. Sticking co-efficient (Bhadra et al., 1994), roughly the diffusion of reactant in the substrate is unknown. The reaction kinetics due to pulsed laser irradiation generally synthesize thin chalcogenide films (Antioniadis & Joliet, 1984). In this way laser pulse shots may have the advantage of making tailored one-dimensional multilayer systems.

2.5 Heat treatment

Individuals and composite bi-layer films were heated under constant temperature for particular time with the help of a temperature controlled furnace and they were quenched to room temperature. For heat treatment, bi-layer of Ge-Se and Sn-Se films and individuals of Ge, Sn, Se films were heated under constant temperature of 100 ^{0}C, 150 ^{0}C etc. for 30 minutes, 45 minutes with the help of a temperature controlled furnace and they were returned to room temperature by decreasing the temperature in steps of about 1/ 2 ^{0}C per minute.

2.6 Measurements

Microstructures of treated and untreated films have been studied through polarizing optical microscope [OLYMPUS - KH (MLX - TR) model]. Some representative SEM micrographs of treated samples have been carried out with the help of Jeol SEM, Model JSM 5200. X-ray diffraction pattern of different specimens were carried out with model BRUKER-aXS-D8 Advance of Cu K-α irradiation of monochromatic wavelength 1.5406 A^0. UV-VIS-NIR transmission spectra were yielded by a spectrophotometer (Shimadzu UV - 3101PC).

3. Results and discussions

The present attempt is mainly confined with the formation of chalcogenide films and their characterization through Optical Microscopy, SEM, XRD UV-VIS-NIR etc. In general the chalcogens are viewed as two fold coordinated atoms and can easily be obtained in the form of thin films. Melting, quenching and vapour deposition process are the main tools to fabricate the chalcogenide glasses and alloys. The analysis of the micrographs accumulated by means of SEM and optical Microscopy from the experiment, which are carefully monitored and analyzed. It was known in the 80's and authenticated by Mott & Davis, 1979 that under rapid cooling of chalcogenides on melting by Laser bombardment can yield material, which can be achieved in amorphous form.

3.1 Morphological study

Microstructures **(Fig. 2 & 3)** of treated and untreated films have been studied through polarizing optical microscope [OLYMPUS - KH (MLX - TR) model].

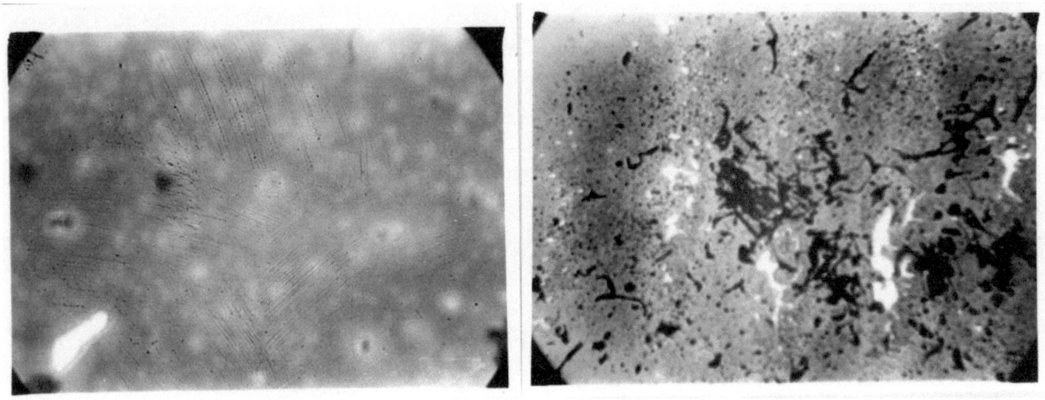

(a) (b)

Fig. 2. Optical Micrographs (a) GeSe Film (stacked), (b) GeSe Film (sandwiched).

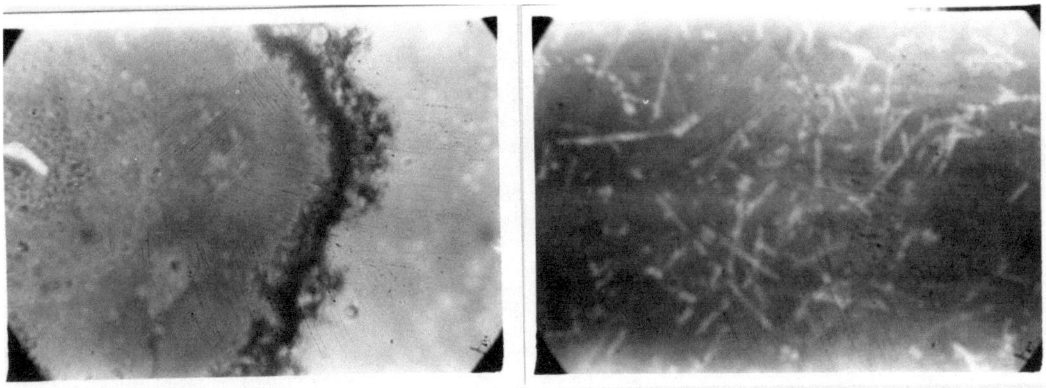

(a) (b)

Fig. 3. Optical Micrographs (a) SnSe Film (stacked) and (b) SnSe Film (sandwiched)

The dendritic feature observed in **Figs. 4 - 8** arises from the differences in the solute content at a faster rate than the movement of the solidification isotherms (Bhadra et al., 1994: Griffiths et al., 1969 & Bhadra et al., 1998). The flow initially dispersed in many paths since the flowing warm material starts to dissolve the already solidified materials. The face growth rate lags and the corner growth increases to maintain the growth rate more or less steady. In order to attain the forced growth rate, the corner penetrates further while the faces grow slowly in depleted part of the material in the film (Brice, 1986; Antioniadis & Joliet, 1984; Griffiths et al., 1969; Bhadra et al., 1998 & Pamplin, 1980) when Se film heat-treated at 150 ^{0}C and returned to room temperature, the warm Se cools on a highly polished substrate, the stem of the dendrite and the branches grow from its sides which are directed along the fast growing crystallographic axes. The symmetry of each flake may be due to hexagonal symmetry of Se and the substrate topology. Morphological instability occurs when the film subsequently heat-treated at higher temperature.

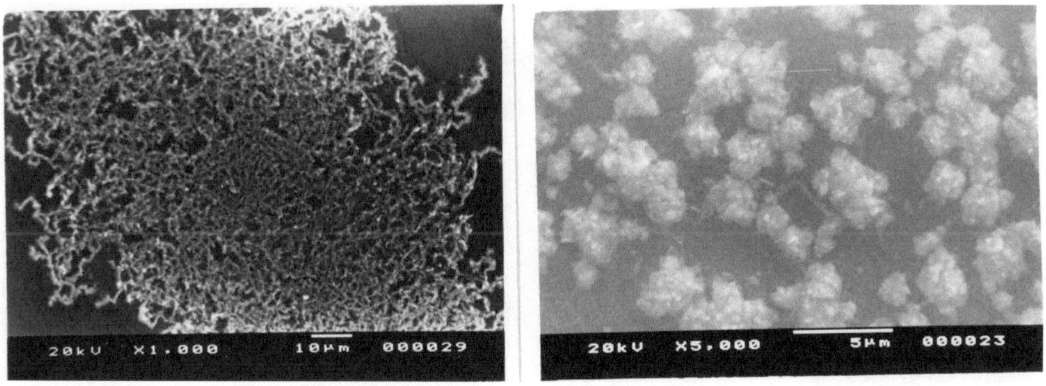

(a) (b)

Fig. 4. Scanning Electron Micrographs (SEM) (a) Se Film (Heat treated) and(b) Se Film (as deposited).

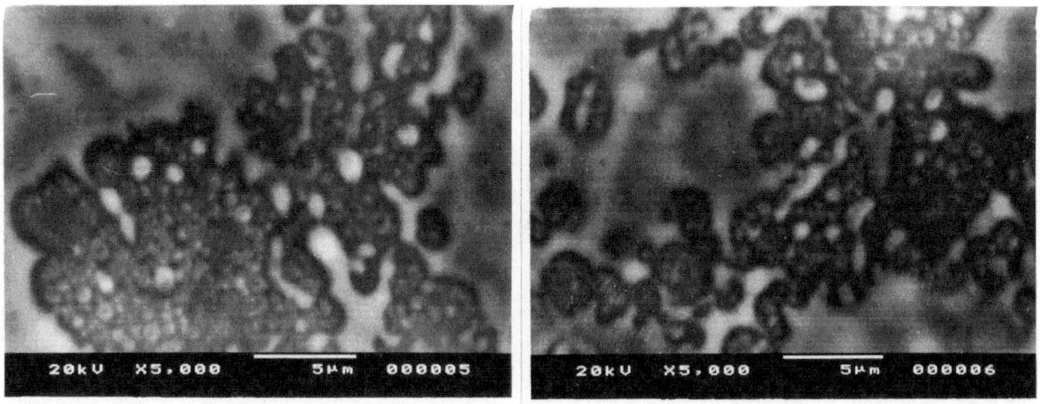

(a) (b)

Fig. 5. Scanning Electron Micrographs (SEM) (a) & (b) GeSe Film (Heat treated)

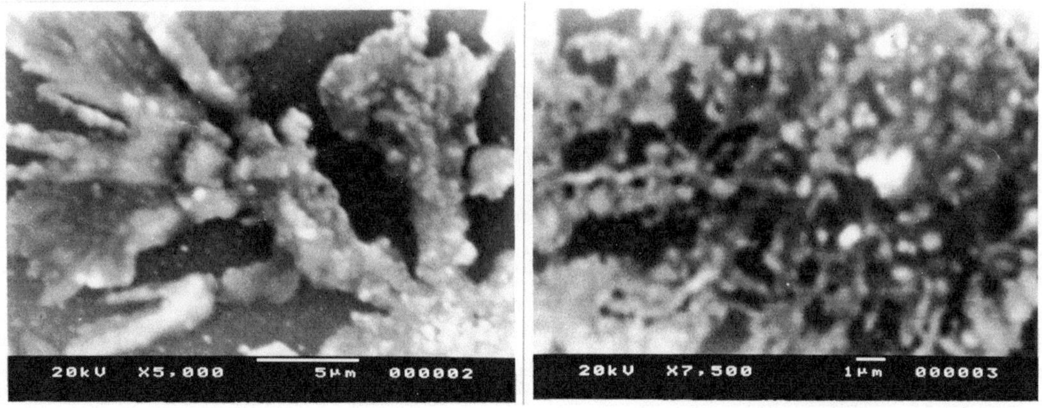

(a) (b)

Fig. 6. Scanning Electron Micrographs (SEM) (a) & (b) SnSe Film (Heat treated).

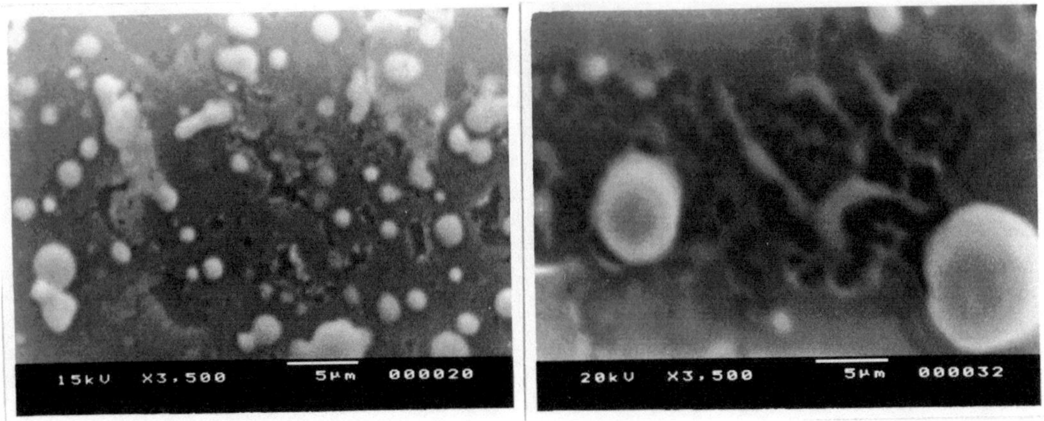

(a) (b)

Fig. 7. Scanning Electron Micrographs (SEM) (a) & (b) GeSe Film (Laser treated)

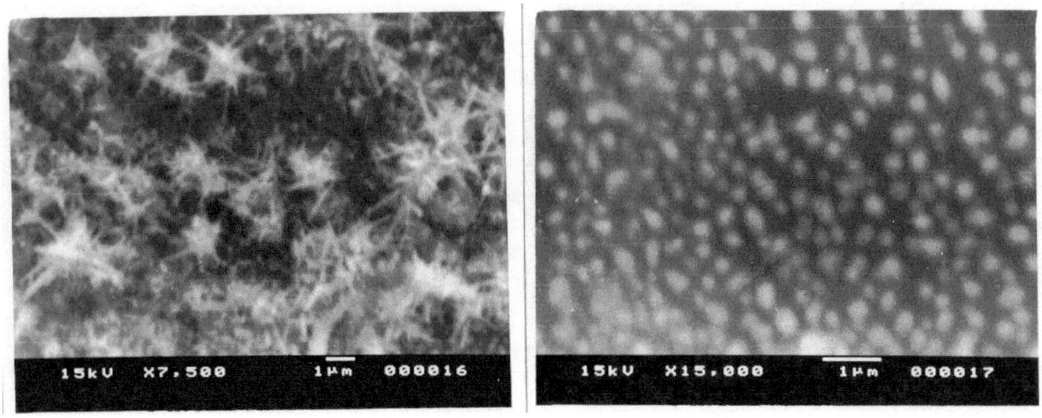

(a) (b)

Fig. 8. Scanning Electron Micrographs (SEM) (a) & (b) SnSe Film (Laser treated).

It has been observed that if the materials of Ge be deposited over Se film or Sn be deposited over Se the photo micrographs show practically a little change in structure. In the event of singular deposition of Sn & Se and sandwiched and then laser shots be made through Se in front some changes are found even the shots are not optically compatible that is to say fundamental beam. The difference in the explosive system is clearly observed through SEM studies and it is concluded that third harmonics generated a picture, which is optically compatible and the rest is equivalent to the simple heating process. The work on such explosive laser irradiation is found to be very important towards the appropriate formation of compounds.

3.2 Crystallographic study

The structural changes on various types of vapour grown amorphous films during both laser and heat treatments have been analyzed by XRD. Two significant peaks have been noticed at 2θ values 23.43^0 and 41.37^0 for both the films under treatments as mentioned above in the XRD

(Fig. 9) peaks corresponds to (100) and (110) diffraction planes and are resembling to that of hexagonal patterns of Se. In case of heat treated binary films the diffraction peaks corresponds to Ge-Se (201), Sn-Se {(042) & (212)} were clearly observed. Experimental results from XRD pattern were given in **table - 2.** The corresponding XRD data i.e. the inter-planar distance *d* and *(hkl)* plane are verified to the standard JCPDS data [JCPDS data. ID nos. 42-1425 (Se), 18-0549 (Ge), 19-1365 (Sn), 15-0404 (GeSe) and 35-1042 (SnSe)].

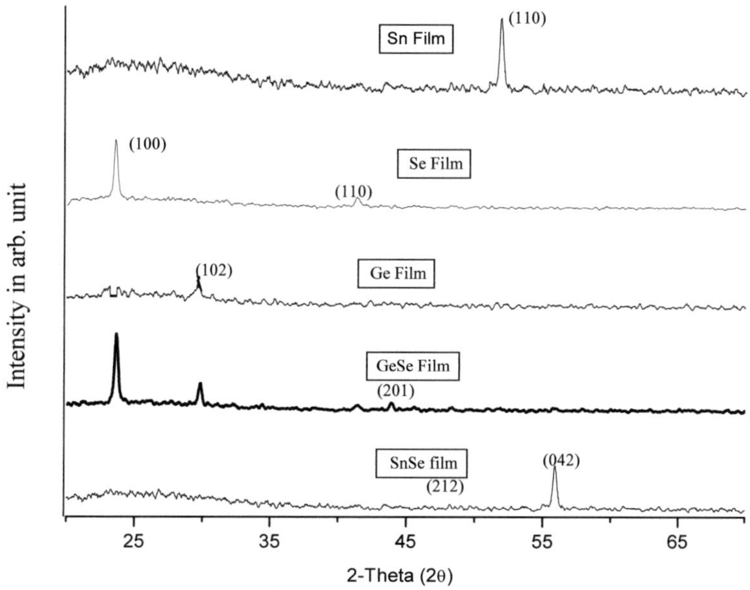

Fig. 9. XRD pattern of different films

Specimen	d-values from JCPDS (A^0)	2θ in degree	d-values observed (A^0)	(hkl)
Se	3.77 2.18	23.43 41.37	3.79 2.17	(100) (110)
Ge	3.01	29.84	3.01	(102)
Sn	2.61	51.96	2.62	(110)
GeSe	2.14	43.96	2.13	(201)
SnSe	1.85 1.73	49.17 55.61	1.81 1.70	(212) (042)

Table 2. Experimental results from XRD pattern

The changes in the thermally treated specimens are much faster than those observed in case of laser treatment (Kar et al.,2007). This is due to the fact that the optical matching of these materials are tantamount with the third harmonics of the Nd : YAG pulsed laser in that zone when the energy is mostly utilized under lasing action. The behaviors of the laser treated with third harmonics irradiation in Ge-Se are the example of yielding faster results. Thus, it is imperative to activate the specimens as per the optical fundamental peaking or else the process will not generate the expected results.

3.3 Optical study

The optical transmission spectra before and after laser irradiation of Ge-Se films and Sn-Se films are shown Fig. 10-12. The transmission plot has been grouped into transparent, weak and medium absorption zones to find the optical constants. Using Manificier et al.,1976 and Swanepoel, 1983, the transmission T is a complex function

$$T = f(\lambda, s, n, d, \alpha) \tag{1}$$

Where s is the refractive index (RI) of the substrate, n is the RI of the film, d is the film thickness and α the absorption coefficient.
The refractive index value n is given by the expression

$$n = [N + (N - s)^{0.5}]^{0.5} \tag{2}$$

where

$$N = 2s\frac{(T_M - T_m)}{T_m T_M} + \frac{(s^2 + 1)}{2} \tag{3}$$

and $T_m(\lambda)$ and $T_M(\lambda)$ are the envelopes of the minima and maxima in the transmission spectra (Fig.10).
By using Equation 2 the refractive index as well as the thickness were calculated and verified with the experimental data.

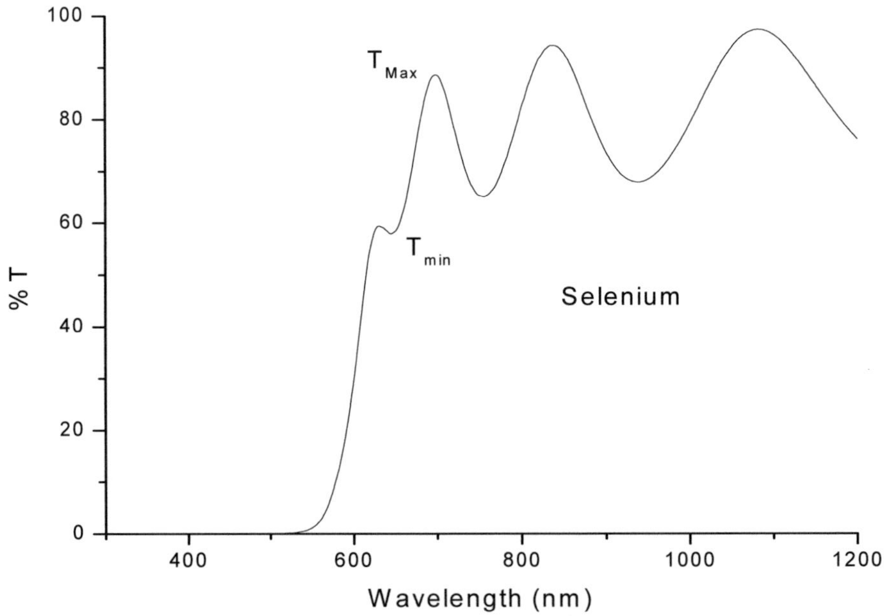

Fig. 10. UV-VIS-NIR transmittance (% T) versus wavelength plot of typical Se film

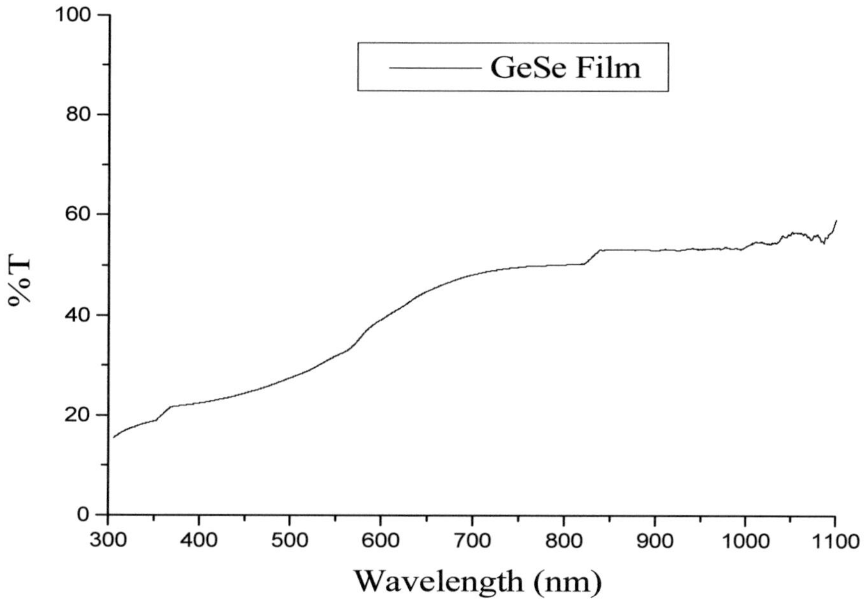

Fig. 11. UV-VIS-NIR transmittance (% T) versus wavelength plot of typical GeSe film

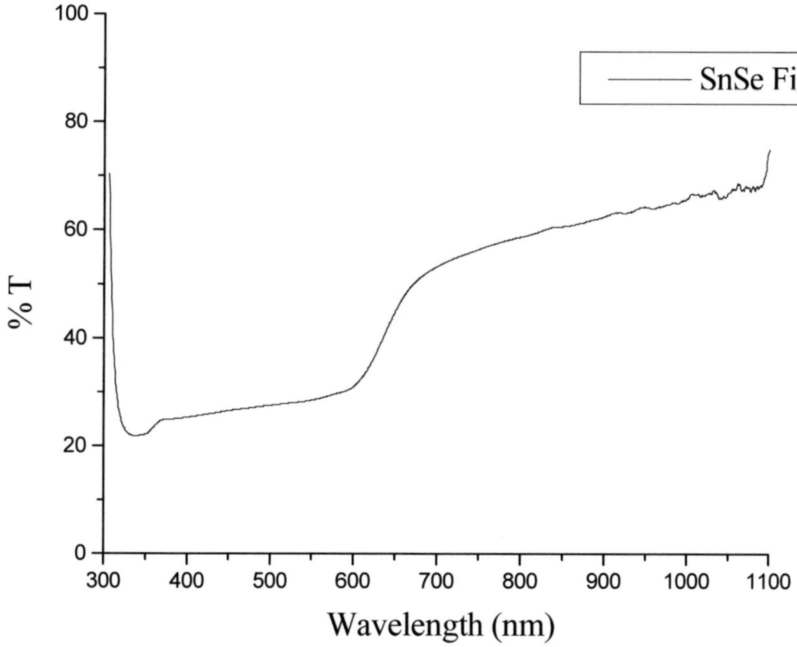

Fig. 12. UV-VIS-NIR transmittance (% T) versus wavelength plot of typical SnSe film

The optical absorption coefficient (α) of the films was calculated using the formula

$$\alpha = d^{-1} Ln\left(\frac{I'_0}{I}\right) \tag{4}$$

Where d and I are the sample thickness and transmitted intensity, respectively, and I'_0 is the effective incident photon intensity (Bhadra et al., 1999).

The UV–VIS–NIR transmission spectra were carried out through a spectrophotometer (Shimadzu UV-3101PC) for the films deposited on glass substrates. The optical band gap was calculated from the Tauc, 1974 relation;

$$(\alpha h\nu)^{1/2} = A\,(h\nu - Eg) \tag{5}$$

where A is a constant and hn is the photon energy, and α the absorption coefficient. The graph showing $(\alpha h\nu)^{1/2}$ versus $h\nu$ is shown in **Fig. 13**. From the intercept on the $h\nu$ axis $(\alpha h\nu)^{1/2} = 0$ gives the band gap energy Eg which comes out for Ge–Se to be **1.45 eV** and for Sn–Se **1.62 eV**. The variation of absorption coefficient (α) versus $h\nu$ plot is shown in **Fig. 14**. The optical absorption edge (Tauc, 1974 & Connell, 1979) is usually described as having three different regions. At high energies, and, for $\alpha > 10^4$ cm^{-1} Eq. (4) is valid for evaluating the optical band gap. In Eq. (4), A contains an average matrix element, constant with energy, and joint density of states for the conduction and valence bands. The equation is formulated on the basis of a square root dependence of the density of states on energy in both bands. The selection rule for the crystal momentum is broken by disorder and so the value of A is much greater than in the crystal. The energy gap in an amorphous material is empirically defined as E_0 of the above-mentioned state.

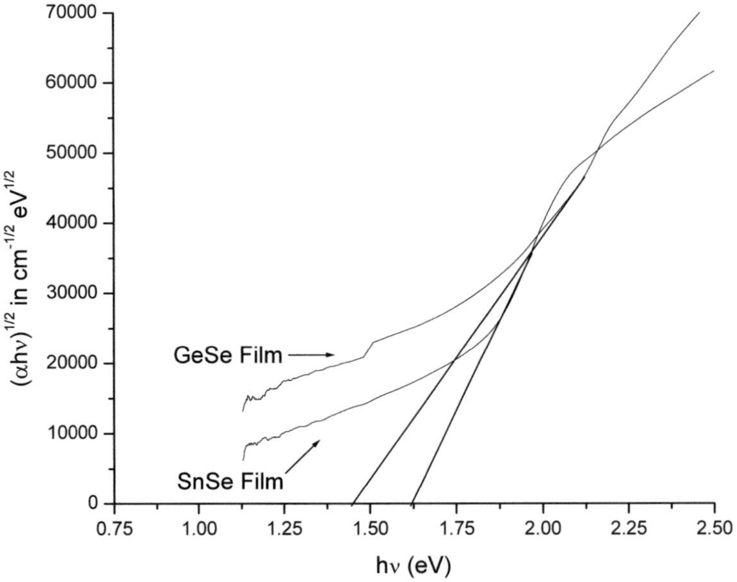

Fig. 13. Variation of $(\alpha h\nu)^{1/2}$ versus $h\nu$ plot

Experimental data on heat-treated films were recorded in the higher absorption region, and then the best fit least squares method were made to fit in the Tauc plot. The transition from the amorphous to crystalline state was realized on heat treatment. Thus the extrapolation is made on a very small region of the absorption energy without incorporating any possible error.

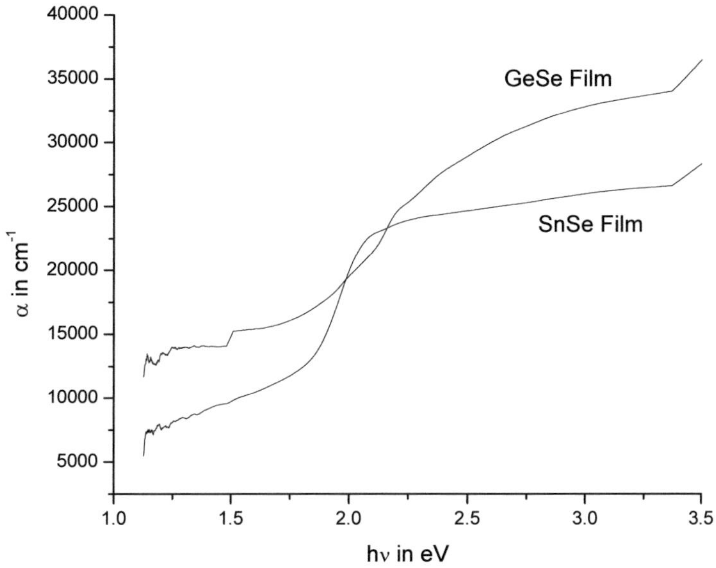

Fig. 14. Variation of absorption coefficient (α) versus $h\nu$ plot

The changes in the thermally treated specimens are much faster than those observed in case of laser treatment. This is due to the fact that the optical matching of these materials is related to the third harmonic of the Nd:YAG pulsed laser in that zone where the energy is mostly utilized under lasing action. The behavior of the laser treated with third-harmonic irradiation in Ge-Se is an example of yielding more rapid transitions. Thus, it is imperative to activate the specimens as per the optical fundamental peaking or the process will not generate the expected results.

4. Conclusions

The process of preparation of thin films of mono-system and also binary chalcogenide alloys and there various structural and morphologies have thoroughly been studied and lots of interesting information was gathered to explain a new phase generation of the composite materials. The typical molecular nature of chalcogenide alloys give rise to a subtle changes in optical properties by photo induction. As the semiconductor Se is photo - sensitive with a lower melting temperature than Ge and Sn; so it is possible to diffuse Se into these elements by an adiabatic impulsive energy. The short range ordering structure of chalcogens when mixed up with four fold-coordinated atoms like Ge, Sn etc. produce alloy or compound with certain specific signature. The optical band gaps calculated from the Tauc plot were found to be 1.45 eV for Ge–Se and 1.62 eV for Sn–Se and they behave like semiconductors. The laser impulse on those materials proves an advantage over the other adopted methods because of the fact that the instant quenching under laser irradiation pave the path for the materials to be alloyed properly. The disadvantage in laser induction is that the large-scale compound preparation may be a different proposition.

5. Acknowledgements

The author wishes to thank Dr. A. K. Maiti of Bidhannagar Govt. College, Prof. K Goswami of Jadavpur University and S. K. Bhadra of CGCRI for helping in characterization.

6. References

Afifi M. A., Labib H. H. A., El-Ela A. H. Abou, and Sharaf K. A. , (1988). Appl. Phys., A46, p-113.
Andonov P. (1982). J. Non-cryst. Solids; 47, p-297.
Antioniadis C., Joliet M. C. (1984). Thin Solid Films;115, p-75.
Bhadra S. K., Bandyopadhyay T., Maiti A. K. and Goswami K., (1999). Surface Review and Letters 6, 2, pp- 219-223.
Bhadra S. K., Maiti A. K. , Goswami K. (1998). Indian J. Phys. 72 A (3),pp-171–87.
Bhadra S. K., Maiti A. K. and Goswami K. (1999). J. Mater. Sci. Lett.; 18, pp-1543-1545.
Bhadra S. K., Maiti A. K., Bhar R., Talapatra D., Goswami K. (1994). J. Mater. Sci. Lett. ; 13, pp-525–527.
Boer K. W. and Ovshinsky S. R., (1970). J. Appl. Phys., 41, p- 2675.
Brice J. C. (1986). Crystal growth processes. New York: Wiley.
Chopra K. L.(1969). Thin Film Phenomena, McGraw-Hill Book Company (New York).
Collins F. M., (1970). J. Non-crystal. Solids, 2 , p-496.
Connell G. A. N. (1979). In: Brodsky M. H., editor. Topics in applied physics: amorphous semiconductors, vol. 36. New York: Springer; p- 73.
Fadel M. (1993). Vacuum ; 44, p-855.
Fadel M. and El-Shair H. T., (1992).Vacuum, 43, p-253.
Fadel M., Negem A., Metwally H. and Afifi M. A., (1992). J. Appl. Phys., A54, p- 288.
Griffiths C. H. , Fitton B. (1969). In: Cooper W. C., editor. Physics of Se and Te. Oxford: Pergamon; p-163.
Gross G, Stephens R. B., Turnbull D. (1977). J Appl Phys; 48, p-1139.
JCPDS data. ID nos. 42-1425 (Se), 18-0549 (Ge), 19-1365 (Sn),15-0404 (GeSe) and 35-1042 (SnSe).
Kar S., Maiti A. K. and Goswami K., (2007). Vacuum 82, pp- 45-49.
Katsuyama T, Matsumura H. (1992). J. Non-cryst. Solids; 139,p-177.
Kawarada M., Nishina Y.; (1975). Jpn. J. Appl. Phys.;14, p-1519.
Kim K. S, Turnbull D. (1973). J. Appl. Phys.;44, p-5237.
Manifacier J. C., Gaslot J. and Fillard J. P., (1976). J. Phys. E : Sci. Instrum. 9, p-1002.
Mott N. F., Davis E A. (1979). Electronic processes in non-crystalline materials. Oxford: Clarendon; p. 518.
Pamplin B. R.(1980). Crystal growth. 2nd ed. Oxford: Pergamon Press.
Swanepoel R., (1983). Ibid.; 16,p-1214.
Tauc J.(1974). Amorphous and liquid semiconductors. New York: Plenum Press; p- 159.
von Allmen M. (1987). Laser-beam interactions with materials. Germany; Springer.
Zope M. J. and Zope J. K. ,(1984). J. Mater. Sci. Lett., 3, p-850.

Part 3

Applications of New Materials

14

Development Liquid Rocket Engine of Small Thrust With Combustion Chamber from Carbon - Ceramic Composite Material

Alexander A. Kozlov[1], Aleksey G. Vorobiev[2],
Igor N. Borovik[3], Ivan S. Kazennov[4], Anton V. Lahin[5],
Eugenie A. Bogachev[6] and Anatoly N.Timofeev[7]
[1,2,3,4]*MAI*
[5,6,7]*«Kompozit» corporation*
Russia

1. Introduction

Advances in the development of high-temperature composite materials and coatings based on glass and ceramics were the basis for their application in aerospace and other branches of engineering. Carbon-ceramic composite materials (CCCM) have unique heat-shielding, erosion-resistant performance and durability at a low density.

At present, composite material used as protective coatings of elements of aircraft gas turbine engines, turbopump assembly of liquid rocket engines, hypersonic technology products, tile protection of spacecraft, inserts the nozzle into a block of solid fuel engines and other technical fields, where the most acute problem of protection structure at high temperature in an oxidizing environment.

In the MAI at the department #202 investigations in the development of small rocket engine thrusts. Ability to use CCCM as a material of the combustion chamber is considered as one of the most promising ways to improve small engines thrusts.

2. Overview of problems and problem statement

Intention to apply the composite material (CM) to create a combustion chamber LRE of small thrust (LREST) associated with essential growth engine specific impulse, which characterizes its effectiveness. The first attempts to introduce carbon-carbon composites as a material for the combustion chamber LREST undertaken in the early-middle of the last decade [6]. However, the resulting designs are generally not meet one or more of the requirements for chamber LREST. Addressing the application of CM based on the following tasks:

- availability of technology forming a thin-walled shell profile, characteristic of combustion chamber of LREST;
- protect material from a high temperature in an oxidizing environment possible conditions;
- develop design a secure connection composite combustion chamber with metal mixing head;

- ensure zero gas-tightness of the wall;
- the possibility of mechanical machining of parts from CM;
- ensuring the strength of the material with a sharp change in pressure and the presence of thermal stresses, typical for pulsed operation of LREST.

As a result of the development and improvement of production technology CM related to the choice of optimal process parameters, the technical level of equipment and tooling availability of reliable methods for nondestructive testing of composite structures and semi-finished products for their production, managed to develop a scientific basis and the basis for developing an extensive list of composite materials and technologies of their production [1]. Currently, there are all prerequisites for a successful practical application CCCM as a material for LREST.

Long time the main materials used for combustion chamber of LREST in Russia and other country were niobium alloys with protective silicide coating. They are able to withstand temperatures up to 1200°C, although the temperature of the combustion products can reach 3,500°C. To reduce the temperature of the wall of the combustion chamber the mixing fuel and oxidizer is organized with sub-optimal ratio of components. This reduces fuel efficiency, which in general is reflected in the perfection of the aircraft. Russian serial thruster (KBHM, Engineering Research Institute) and currently as the main material used alloys based on niobium. To date, the value of specific impulse for Russian LREST on the components of the nitrogen tetroxide (NTO) + unsymmetrical dimethylhydrazine (UDMH) / monometilgidrazin (MMG) is less than 310 (1, 2).

Foreign LREST as the material of the combustion chamber and the nozzle using alloys based on niobium (engine TR-308, TR-312-100MN, Northrop Grumman) (LEROS 1R, LEROS 1C American Pacific Corporation, AMPAC, USA), platinum (S400 - 12, S400 - 15, EADS Astrium, Europe), iridium (engines R-4D, R-4D-15 (HiPAT) (Fig. 3), Aerojet, USA) with protective coatings. To reduce temperature effects on the wall using the hanging film. Operating temperature chamber walls with precious metals of platinum may reach 2200 ° C. Specific impulse of modern foreign engine components NTO + UDMH / MMG) reaches 327 sec.

With the advent of composite materials that are not inferior in its characteristics, but at a price much cheaper than the above alloys, foreign manufacturers have switched to the development of the combustion chambers LREST with the CM. The use of non-metallic composite is promising, because, being comparable in price to conventional niobium alloys, it has a lower density, which is important in terms of reducing the mass of the engine, a substantially lower cost compared to the platinum group metals.

In Russia, the issues involved in the development of KM "Kompozit" [9], VIAM [7, 8], Company "Iskra" and several other organizations. In Russia the use of KM in rocket engines is reduced to the use of CCC for the attachment of radiative cooling of the engine 11D58M, but understanding the prospects for the application of CM in the elements of missile technology exists [4].

Abroad, there are many organizations dealing with composite materials (ULTRAMet, SNECMA, DuPont). In some countries there is a separate program for the development of aerospace industry with abroad application of advanced CM. Some foreign developers LREST already implementing the combustion chamber of CM in their propulsion systems.

One example of successful introduction of advanced composite technology is the development corporation EADS apogee thruster, called the European Apogee Motor. Engine European Apogee Motor, thrust 500N, in which the combustion chamber and nozzle are made at the same time (4, 5), is lightweight and high specific impulse, which amounts to more than 325 seconds [10]. European Apogee Motor will be the main engine for the platform AlphaBus.

Along with the progressive combustion chamber from CM, which can resist high temperatures, vibration and shock loads, to achieve this level of specific impulse possible to optimize the mixing head.

Various methods of nondestructive testing have been investigated and applied, including ultrasound, thermography and tomography. European Apogee Motor can be used for various tasks in the commercial and military satellites, Transfer Vehicle, vehicles reusable. Low weight and high specific engine performance save fuel, which positively affects the value of the payload in comparison with other engines. Reasonable price of production and procurement composite material allows the engine to compete successfully in the market.

Features LREST domestic and foreign production with the designation of the material used are given in Table 1.

Fig. 1. Engine KBChM DST-100A

Fig. 2. Engine Engineering ResearchInstitute 11Д428A

Fig. 3. Engine HiPAT with c iridium- rhenium combustion chamber.

Fig. 4. Combustion chamber and nozzle of European Apogee Motor.

Fig. 5. Engine European Apogee Motor

	11Д428A NIIMash (Russia)	11D458M NIIMash (Russia)	DST-200A KMChM (Russia)	S400 - 15 EADS	R-4D-15 (HiPAT) Aerojet	500 H European Apogee Motor
Components:	UDMH NTO	UDMH NTO	UDMH NTO	MMG NTO, MON-1, MON-3	NTO MMG	MMH, NTO, MON-1, MON-3
Vacuum thrust (N):	130,5	392.4	200	425	445	500 +/- 20N
Vacuum specific impulse (sec):	308	302	307	321	325	>/= 325 sec
O/F:	1,85±0,05	1,85±0,05	1.85	1.65	1	1.65
Pressure chamber (bar):			7	10	9.4	10
Valve current (V)	27	27	34			50
Length (mm)	274	461		669		803
Weigh (kg):	1.5	3	1.7	4.3	5.2	< 5
Expansion ratio	150		100		300	
Combustion chamber materisl	Niobium alloy	Niobium alloy	Niobium alloy	Platinum alloy	Iridium alloy with rhenium coating	Composite material

Table 1. LREST performance.

Thus, in terms of technology development of production structures made of composite materials, an effort the developers of space vehicles and platforms to increase the mass of the payload, the task of creating the combustion chamber of LREST from carbon-ceramic material is urgent.

3. Development of the COP from KM to LREST MAI-202

202 Department of the Moscow Aviation Institute has long been working in the design and creation of experimental LREST [3, 11, 12]. On this subject holds a number of contracts, some contracts are in work. The construction mixer head engines LREST MAI-202 is the use of separate plates welded together the components and the presence of low-pressure film cooling layer with the ability to regulate its relative expense.

The main LREST for which is developed by the combustion chamber from CM are the engines: MAI-202-200 thrust of 200 N on the components of the NTO + UDMH (fuel film cooling), MAI-202-500-HTHP-Kerosene thrust 500 N on the components of high tests hydrogen peroxide (96%) + kerosene (oxidizer film cooling), MAI -202-200-Oxygen-Kerosene thrust of 200 N on the components of gaseous oxygen and kerosene (oxidizer film cooling). The expansion ratio for all engines 70, the pressure in the combustion chamber 9-12 atm.

To reduce the cost of production engines combustion chamber were made of heat resistant alloy EP-202 and HN60VT with protective antioxidant coating based on chromium oxide. Maximum temperature of the wall of the combustion chamber in the tests did not exceed 1200 K.
As a result of cooperation with NPO "Kompozit", on the basis of the developers at that time, technology for the production of similar products [9, 5] have developed a program to develop the experimental combustion chambers of ceramic based composite material for the above engines.
Table 2 shows the characteristics of the developed material CCCM C-SiC for the combustion chamber of LRE compared with traditional materials - niobium alloys 5VMTS and similar materials C-SiC, used in non-Russian LREST.
NPO "Kompozit" has several recipes and opportunities for applying the method of oxide coatings on the surface CCCM nanotechnology to improve the radiating capacity of the outer wall of the chamber and the reflectivity of the inner wall. These activities are aimed at raising the temperature boundary layer of combustion products without increasing the wall temperature of the combustion chamber.

Parameters	Performance		
	LREST of dep. 202 MAI	Russian engines	Foreign engines
	Combustion chamber from CM C-SiC	Serial combustion chamber from alloy 5VMC with disilicide molybdenum coating	Novoltex C-SiC-, (SNECMA, France)
Working temperature, ^oC	1600	1450	1600
Density, gr/cm^3	1,75	~ 8,7	2,1
Combustion chamber weigh decreasing , %	300-400	-	300

Table 2. Comparison of characteristics of the developed material for the COP LREST with the characteristics of traditional material and foreign analogues

The technology of obtaining ceramic combustion chambers clean, does not require large expenditures for tooling and expensive manufacturing equipment, unlike their foreign counterparts. Method of forming a matrix does not introduce damage in the reinforcing components. Reagent has not previously been used for construction and materials and protective coatings.
The existing technology of the matrix provides the opportunity to connect with the metal wingtips - flanges due to the formation of metal-composite junction, which gives a strongtight ceramic-composite mount the camera to the metal parts of the engine - mixing head and nozzle attachment.
In the design (Fig. 6) new combustion chambers, the following conditions:

- preservation of the internal geometric profile of the combustion chamber and the nozzle;

- the availability of ready-mixing heads of the engines;
- conservation sectional design of the engine in the main parts (heads, the combustion chamber, the nozzle attachment) for testing individual units with the possibility of not folding design;
- the ability to install fittings for measuring pressure in the combustion chamber when developing a workflow.

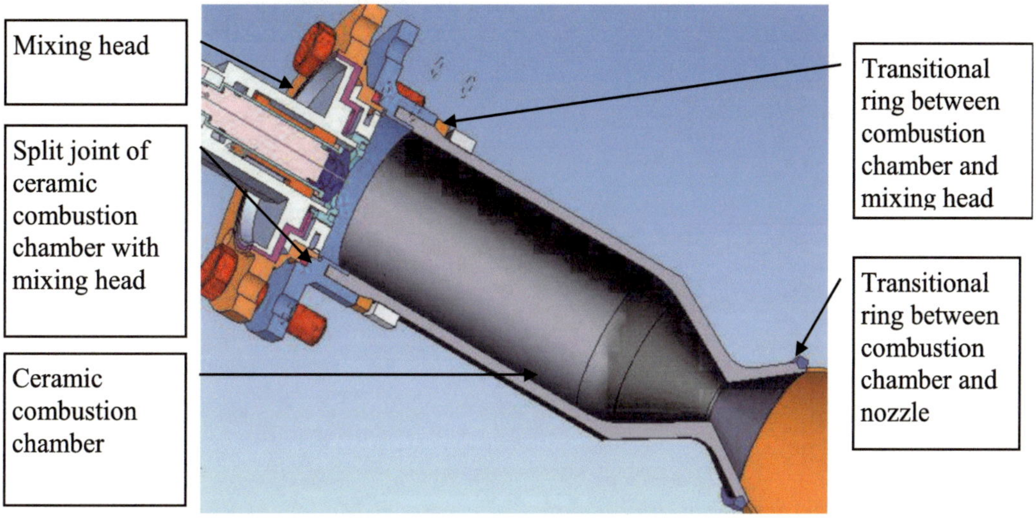

Fig. 6. Model LREST MAI-202-200-Oxigen-Kerosene with the composite combustion chamber

The samples of the combustion chambers (Fig. 7) underwent the following operations:
- formation of carbon fiber frame blanks;
- pre-machining;
- carbonation and high-temperature treatment;
- formation of silicon carbide matrix by chemical vapor saturation;
- forming a composite coating for gas isolation.

As a result, studies were identified and addressed a number of problems:
- improving technology, laying patterns to create a complex profile of the combustion chamber with small diametral dimensions of the throat;
- designed connectors connect the camera to the mixing head and nozzle of the heat-resistant steel.

Currently, new engines with the designation of MAI-202C, where the letter C means the application ceramic composition combustion chamber (Fig. 8), in preparation for firing tests.

In addition to weight reduction designs and improve specific impulse by raising the temperature of combustion products, the use of composite materials with an antioxidant coating will in future go to oxidative veil of low-flow that will positively affect the efficiency of the engine.

Fig. 7. Work material of ceramic combustion chamber.

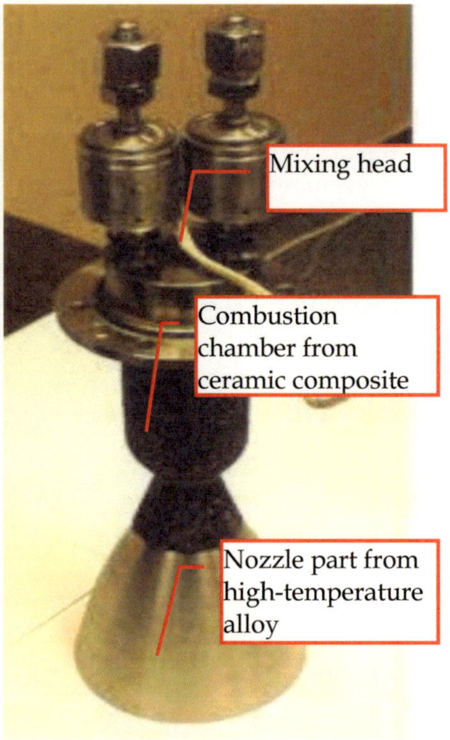

Fig. 8. Model of LREST MAI-202-200-OK.

4. Analysis of the efficacy of CM for the combustion chambers LREST

By raising the temperature of combustion products and the operating temperature of the walls of the combustion chamber due to changes in mixing and reduce the consumption component in the veil is possible to obtain high specific impulse in the application of CM in the construction of the combustion chamber.

On the basis of firing experiments on the engine MAI-202-200 (NTO UDMH) [11] analysis was conducted to increase the specific impulse in the case of the combustion chamber of CM [2]. The calculations for the experimental-theoretical model of thermalstate LREST it was shown that the use of new material for the engine MAI-202-200, operating at 1800 K allows to achieve a specific impulse of 325 sec. and for the engine MAI-202-500-HTHPK specific impulse will be 326 seconds. that at the world's leading manufacturers LREST (Fig. 9, Fig. 10).

The calculation results show that increasing the specific impulse apogee LREST for 5 seconds increases the weight of the payload of 7 kg for the model geostationary satellite weighing 4800 kg, equivalent can be replaced by extending the life of the device [13]. A more detailed analysis of the gain in weight of the payload from an increase in specific impulse LREST requires binding to a specific machine.

For engines MAI-202-200, MAI-202-500K-HTHPK preparing fire tests to assess their energy efficiency with ceramic composition of the combustion chamber. Also planned for the study of composite combustion chamber engines MAI-202 with pulse mode operation to prove the efficiency of the material under cyclic thermal and mechanical stresses.

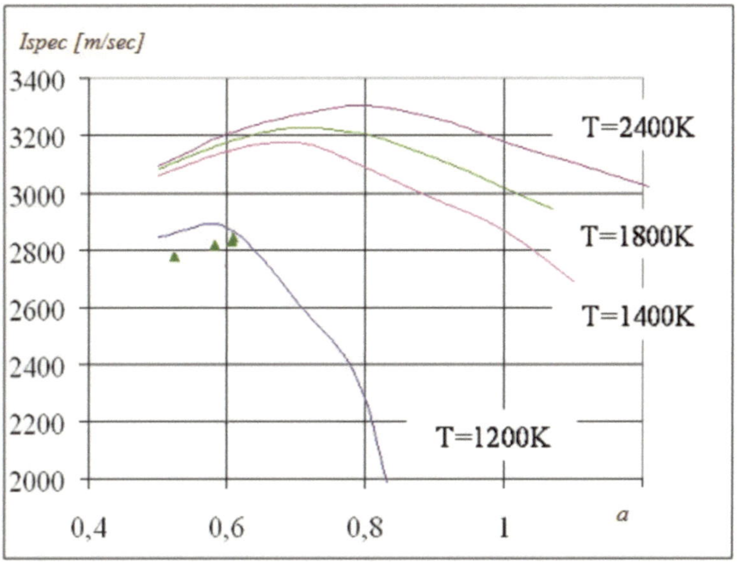

Fig. 9. Calculated dependence of the pulse motor MAI-202-200 depending on the coefficient of oxidizer for different maximum wall temperature of the combustion chamber (for the single-injector head). Green triangles - experimental data/

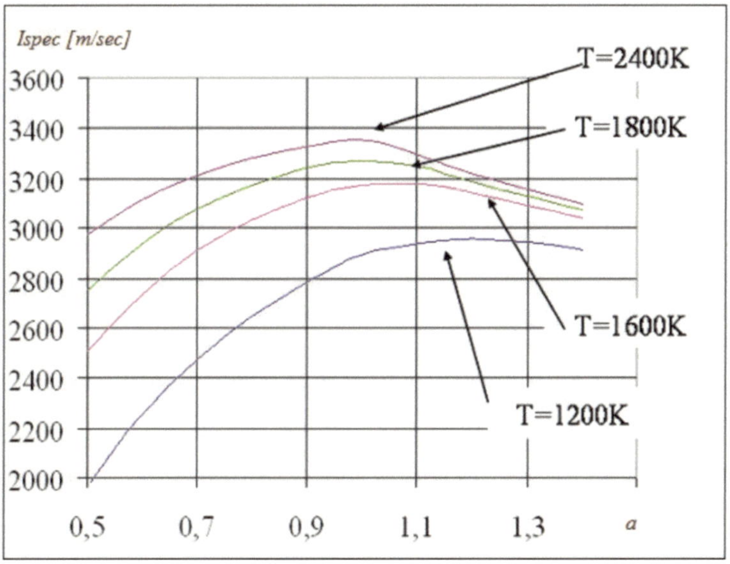

Fig. 10. Calculated dependence of the pulse motor MAI-202-500-VPVK depending on the coefficient of excess oxidizer for different maximum wall temperature of the combustion chamber.

5. Conclusions

At the Department of MAI 202 together with "Kompozite" is under active development of small rocket engines with thrust combustion chambers of the carbon-ceramic composite

materials. The analysis shows that the use of CM achieves a specific impulse in excess of Russian flight models, and corresponding foreign equivalents developed.
Detailed information can be found at www.mai202.ru.

6. References

[1] Bulanov I.M., Vorobey V.V. Tekhnologiya raketnykh i aerokosmicheskikh konstruktsiy iz kompozitsionnykh materialov: Ucheb. dlya vuzov. M.: Izd-vo MGTU im. N.E.Baumana, 1998, 516 s.

[2] Vorob'ev A.G. Matematicheskaya model' teplovogo sostoyaniya ZhRDMT. Vestnik MAI. T14, №4. Moskva. 2007. - S. 42-49.

[3] Kozlov A.A., Abashev V.M. Raschet i proektirovanie zhidkostnogo raketnogo dvigatelya maloy tyagi. Moskva, MAI, 2006.

[4] Koshlakov V.V., Mironov V.V. Perspektivy primeneniya kompozitsionnykh materialov v raketnykh dvigatelyakh. Raketno-kosmicheskie dvigatel'nye ustanovki: sbornik materialov Vserossiyskoy nauchno-tekhnicheskoy konferentsii. M.: Izd-vo MGTU imeni N.E. Baumana , 2008. - 10-11 s.

[5] Lakhin A.V. Protsessy polucheniya kompozitsionnykh materialov i pokrytiy na osnove karbida kremniya khimicheskim gazofaznym osazhdeniem iz metilsilana pri otnositel'no nizkikh temperaturakh i davleniyakh: Dis. kand. nauk. - Moskva, 2006. - 140 s.

[6] Pavlov S.V., Grachev V.D., Tokarev A.S. Rezul'taty razrabotki i issledovaniy rabotosposobnosti kamer sgoraniya ZhRDMT iz UUKM // Raketno-kosmicheskaya tekhnika, vyp. 3 (136). NII teplovykh protsessov, 1992 g. 30-33 c.

[7] Solntsev S.S., Isaeva N.V. Keramicheskiy kompozitsionnyy material teplonagruzhennykh uzlov i detaley. Pervyy mezhvedomstvennyy nauchno-tekhnicheskiy seminar po problemam nizkoemissionnykh kamer sgoraniya gazoturbinnykh ustanovok. Opyt razrabotki, problemy sozdaniya i perspektivy razvitiya nizkoemissionnykh kamer sgoraniya GTU. 14-16 dekabrya. Moskva, TsIAM

[8] Solntsev S.S. Vysokotemperaturnye keramicheskie kompozitsionnye materialy i antiokislitel'nye resursnye pokrytiya. // 75 let. Aviatsionnye materialy. Izbrannye trudy «VIAM» 1932-2007. Pod red. Kablova E.N. - M.: «VIAM», 2007. - 438 s.

[9] [9] Timofeev A.N., Bogachev E.A., Gabov A.V., Abyzov A.M., Smirnov E.P., Persin M.I. Sposob polucheniya kompozitsionnogo materiala. - Patent RF №2130509 ot 20.05.1999, prioritet ot 26.01.1998.

[10] Astrium.EADS Web Page: http://cs.astrium.eads.net/sp/ /SpacecraftPropulsion /BipropellantThrusters.html

[11] Kozlov A.A., Abashev V.M., Denisov K.P. ets. Experimental finishing of bipropellant apogee engine with thrust 200 N. 51st International Astronautical Congress. Rio de Janeiro, Brazil. October 2-6, 2000.

[12] Kozlov A.A., Abashev V.M., Hinckel J.N. Organization of the working process in the small thrust engine LRESTH МАИ-200. 52nd International Astronautical Congress. Toulouse, France. October 1-5, 2001.

[13] Liou, Larry, C. Advanced Chemical Propulsion for Science Missions. NASA/TM-2008-215069.

15

New Routes to Recycle Scrap Tyres

Xavier Colom, Xavier Cañavate,
Pilar Casas and Fernando Carrillo
Universitat Politècnica de Catalunya
Spain

1. Introduction

Most means of transport need tyres as an essential component to work. Tyre composition is complex and is designed according to high technical standard requirements of adhesion, flexibility, forces and pressures, which are necessary for their functioning. However, tyres suffer from wear and have a limited lifetime due to its use. After their replacement, the unusable tyres are known as End of Life (EOL) tyres.

The crosslinked chemical structure of the rubber, the high amount of stabilizers and other additives present in tyres formulations make them a non-biodegradable, non-environmentally friendly material. The growing environmental awareness linked to the development of new european and national regulations have instigated the research for recovering EOL tyres for other applications.

A common industrial procedure prior to any form of recovery consists in grinding the tyre, in order to obtain a powder called ground tyre rubber (GTR). This powder has been proposed as a suitable reinforcement for composite materials as a way to reduce the amount of EOL tyres in added value applications. However, one of the main issues of working with GTR is its low compatibility with most of polymeric matrices used, mainly due to highly crosslinked rubber structure. Another drawback is related with the high particle size of GTR obtained by standard industrial grinding process (between 400 and 600 µm approximately) that results in brittle composite materials. Different methods have been already tested to compatibilize the polymer blends and to reduce GTR's particle size, turning out to be too expensive procedures to provide an economically competitive material.

In this chapter, several ways to blend GTR with commodity polymers were discussed. The proposed alternatives are based on the preparation of new composite materials using GTR as a reinforcement, thereby providing another way of adding value and reducing the stock of used tyres.

For the development of the proposed GTR based composites, it will be necessary to improve adhesion between matrix and rubber reinforcement. In this regard, the following methods have been carried out, which was discussed in this chapter: 1) acid pre-treatment of GTR materials, 2) use of wetting additives and waxes into the mixture and 3) use of a ternary blend to prepare a thermoplastic elastomer.

The effect of each proposed method on the mechanical and morphological properties of the GTR based composites was studied. Also, microstructural and chemical characterization of the composites was provided. Moreover, the materials have been optimized in order to

obtain composites reinforced with the highest GTR amount by keeping the most appropriated mechanical properties, an adequate processability and minimum cost.
In particular, the aforementioned methods proposed for the development of the GTR based composites will cover the following aspects:

1. A treatment with H_2SO_4 which improves the rubber's ability to interact with high density polyethylene matrix (HDPE). This treatment provides a greater stiffness to the GTR reinforcement, which is a consequence of rubber rigidification after the acidic treatment due to the extraction of additives and degradation;
2. Treatments with wetting additives and waxes which produce an increase in the mechanical properties of the polymer matrix (i.e. Young's modulus and tensile strength).
3. The use of ethylene propylene diene monomer rubber (EPDM) and peroxides as a third component for the production of thermoplastic elastomers materials based on reused tyres. A material containing up to 30% by weight (w/w)of GTR can be obtained by this method, which shows a combination of strength, toughness and elongation appropriate for industrial use.

2. Approach

Composites made out of ground tyre rubber and other polymers would have good adhesion if the ground tyre rubber would have pores, holes, crevices or other irregularities, unfortunately they do not. Ground rubber obtained by the cryogenic and room temperature methods are quite different.
Figure 1 shows that the major difference between particles of rubber generated by room temperature and cryogenic processing systems is their shape. Particles derived from the cryogenic process have a smooth surface, akin to shattered glass while the particles derived from the room temperature process have a rough surface, giving it greater surface area relative to the cryogenically produced particle [Adhikari 2000]. In later sections, it will be seen how the multilobed morphology of ambient GTR is not enough for an acceptable mechanical adhesion.

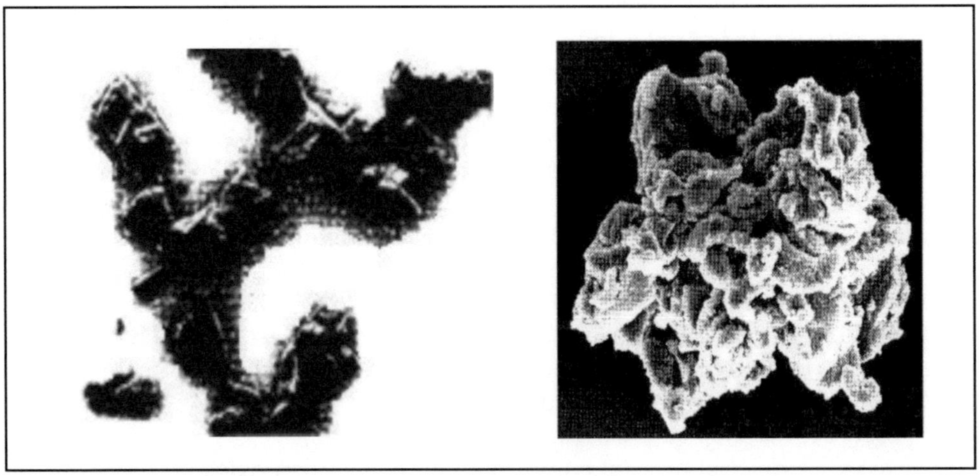

Fig. 1. Cryogenic ground rubber [Burford 1982] (x100 – left) and room temperature GTR [Erickson Materials] (x325 right).

2.1 Previous studies of composites based on GTR and thermoplastic matrices

Composites based on GTR and thermoplastics have not been deeply studied. Even so, some interesting and successful studies can be found in the literature.

Shojaei et. al. [Shojaei 2007] studied the behaviour of HDPE after degradation by re-extrusion. In order to improve the HDPE modulus and tensile strength, Polypropylene (PP) filled with 30 wt % of glass fibber was added. In addition, they studied the effect of GTR particles in the final product where they observed a lower stiffness due to its role (in that specific composite) as soft filler and an undesirable slightly increase in elongation. Unfortunately, ground tyre rubber did not produce any improvement on impact properties probably because of the low interfacial adhesion beetween the matrix and GTR particles. They finally concluded that the addition of ground tyre rubber particles into the final composition should be kept at low amounts (less than 10%wt).

Oliphant and Baker [Oliphant 1993] precoated the cryogenic GTR (CGTR) with ethylene acid acrylic copolymer and mixed it afterwards with LLDPE and HDPE matrixes. They found that the deleterious effects of the mixture could be overcome while still retaining composite processability. For example, a blend with 40 wt % of ethylene acid acrylic (EAA) coated CGTR particles with LLDPE had an impact and tensile strengths of 90% compared with those of pure LLDPE. However, very poor mechanical properties of the CGTR/HDPE composites were found with HDPE composites. This was believed to be because particles were too large to induce a brittle to ductile transition. The failure occurred largely through crack propagation and the large CGTR particles (even with moderate adhesion), acted as serious flaws, providing an easy path for the crack to follow. They concluded that the addition of CGTR to a semi-brittle matrix such as HDPE, requires much higher levels of adhesion (to retard the crack growth at the particle/matrix interface) or much lower particle sizes.

Another way to modify adhesion between these two incompatible phases is via radiation. Sonnier et. al. [Sonnier 2006] studied the influence of γ radiation on GTR/rHDPE (recycled) composites. Gamma irradiation allows achieving in situ compatibilisation, leading to an improvement of the mechanical properties. With irradiation doses of 25-50kGy the elongation at break and Charpy impact strength significantly increased. Only Young's modulus slightly decreased due to the fact that radiation induced also crosslinking of the rHDPE matrix.

The grafting of compatibilizers on GTR surface is another method to improve compatibility. Fuhrmann and Karger-Kocsis [Fuhrmann 1999] functionalized GTR with methacrylic acid and glycidyl methacrylate by photoiniciated polymerization. The process consists in UV radiation of GTR in presence of air and afterwards a process of grafting compatibilization is carried out. The presence of reactive carboxy and epoxy groups was demonstrated. The idea behind these studies is the creation of polar functional groups on the GTR surface that should result in increased reactivity and improved interfacial adhesion between the GTR and matrix polymers. Kim et. al. [Kim 2000] studied the mixture of HDPE filled with surface modified GTR particles round tyre rubber particles surface modified with acrylamide (AAm) using UV radiation. Ground tyre rubber particles and HDPE were extruded using a single-screw extruder and maleic anhydride-grafted polypropylene was added as a compatiblizer to improve adhesion between phases. They demonstrate an improvement in tensile stress, strain and impact strength.

2.2 Materials and general process for samples preparation

The basic materials used in this study were HDPE and GTR. Properties of those materials are described below. Other specific materials used in the different purposed methods will be described in the corresponding section.

The high density polyethylene (HDPE) used as matrix was supplied by REPSOL-YPF (ALCUDIA® 4810-B), with a density of 960Kg/m^3. The HDPE was characterized by melt flow index ($MFI_{190/2.16}$) of 1.0g/10min, Young's modulus of 927.90MPa, tensile strength of 17.17MPa, elongation at break of 390.80% and toughness of 38.4J.
Two different suppliers provided the GTR particles: Gestión Medioambiental de Neumáticos S.L. (GMN) in Maials (Lleida, Spain) and Alfredo Mesalles (Barcelona, Spain), both with a average particle size between 0.4 and 0.6mm.
The mixing process for the composites was carried out in a two roll mill heated at 150-155ºC. The HDPE matrix is melted for a minute and then, the GTR particles are added and mixed for another 5 minutes.
Composite sheets (150x150x2mm^3) were prepared by hot press moulding at 100kN and 170ºC for 10 minutes. After that step, the sample sheets are cooled for 5 minutes under pressure using - water. After that, the materials were mechanically shaped as test specimens according to ASTM-D-412-98 specifications.

2.3 Oxidant treatments on GTR surface in order to improve composites compatibility

Sulphuric and nitric acids have been used previously to modify GTR surface from tyre wastes resulting in a great porosity development [Manchón 2004][Sonnier2006]. According to these previous results the chemical attack produced by the acid on surfaces creates an appropriate morphology to improve interlocking between matrix and the GTR particles. On the other hand, it is worth to mention that acid treatments seem a financially worthwhile way to achieve a suitable material, due to the fact that their application does not require any specific equipment or complex technical processes.
These results lead to consider the possibility of using the main features of the pre-treated oxidised rubber to improve the mechanical adhesion of GTR/HDPE composites. The first method to improve adhesion between HDPE and ground tyre rubber consists on using three chemical acids, such as H_2SO_4, HNO_3 and a sulphuric-nitric solution (50/50% v/v), as GTR surface modifiers. The effect of the chemical and physical modifications on the GTR particles surface and the effect of this modification on the composites performance were monitored by determining its mechanical and structural properties, by the use of FTIR-ATR spectroscopy, and SEM.
From the mechanical properties [Fig.2] it was concluded that all of them are reduced significantly. Only with lower particle size (<200μm, after sieving) an increase of the mechanical properties was observed due to the etching caused by the acid pre-treatment (Figure 3). The effect on Young's modulus is clearly visible in [Fig.3]. The etching of the acid acts promoting mechanical adhesion, but also rigidizes the rubber by extracting the substances that may act as a plasticizer. The acid treated GTR particles have a rigid nature as it has been also observed in the studies by other authors [Cepeda 2000a] [Cepeda 2000b]. Then, GTR obtained with previous acid etching are usually stiffer than not treated ones. The sulpho-nitric mixture seems to provide the best results, but in general all acids have an improving effect. On the other hand, acidic pre-treatments were found to increase the tensile strength too. The effect of the acidic pre-treatment is related to the etching of the GTR surface, the elimination of moieties and additives and the achievement of a microporous surface more suitable for mechanical adhesion. The acid pre-treated samples showed similar behaviour, obtaining for the 50% sulphuric-nitric solution better results than the nitric or sulphuric acid pre-treatment, for the lowest particle size and highest GTR content. Low values for elongation at break and toughness were found in this case. The decrease of elongation at break is related to the imperfect interfacial adhesion between components. The

incidence of the poor adhesion between phases on this property is especially important. On the other hand, the compatibilizing effect achieved by the acid pre-treatments is counteracted by the increase of the stiffness of the rubber. The extraction of additives, oligomers or plasticizers of the GTR particles by the acid produces a rigid material. The reduction of the ability of deformation of the rubber influences the decrease of elongation and subsequently the decrement of the toughness.

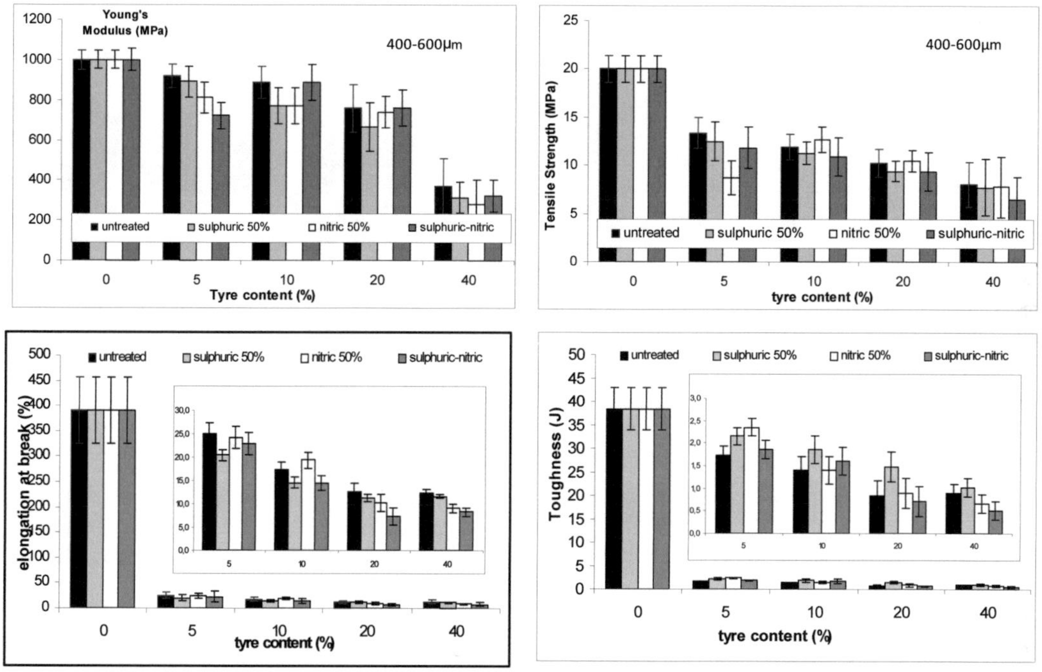

Fig. 2. Young's Modulus, tensile strenght, elongation at break and toughness of composites based on treated GTR with particle size between 400-600μm and HDPE.

Chemical changes produced by acid treatments on GTR particles were studied by means of FTIR-ATR spectroscopy. Figure 4 shows a FTIR-ATR spectral area of 400-1800cm^{-1} for the samples treated with H_2SO_4 and HNO_3 and compares these values to those for untreated GTR particles. To obtain this spectrum, it was chosen the band at 1026cm^{-1} assigned to carbon black [Delor 1998] [Cañavate 2000] as a reference. The spectral analysis of untreated GTR shows a weak band at 1739cm^{-1} that is associated to the thermal oxidation that occurs as a result of the exposure of the surface to oxygen (specially during the grinding handling), and which induces the formation of an oxidation skin that includes carbonyl groups. The strong band at 1640cm^{-1} is associated to the C=C of polyisoprene, the weak band at 1540cm^{-1} to zinc sterarate (an anti-adherent compound), bands at 1430cm^{-1} with the scissoring vibrations of =CH_2 (on butadiene), the band at 875cm^{-1} with the trans isopropenyl unit (-C(CH_3)=CH-) and the band at 470cm^{-1} with S-S.

The study shows that treating GTR particles with H_2SO_4 acid produces several chemical and degradative modifications on the tyre surface, mainly the formation of sulphonic acid, a decrease in double bonds (1640cm^{-1}) due to the degradation process of polybutadiene and other unsaturated components of the tyre, and a decrease in content in minor components.

This phenomena may be related to the disappearance of the weak bands at 1739 and 1540cm^{-1}, the decreasing of bands at 1430 and 875cm^{-1} and the appearance of new bands at 1402, 1121 and 618cm^{-1}, which is indicative of the O=S=O stretching of the absorptions of sulphonic acid. These results are in agreement with previous studies [Cepeda 2000a] [Cepeda 2000b].

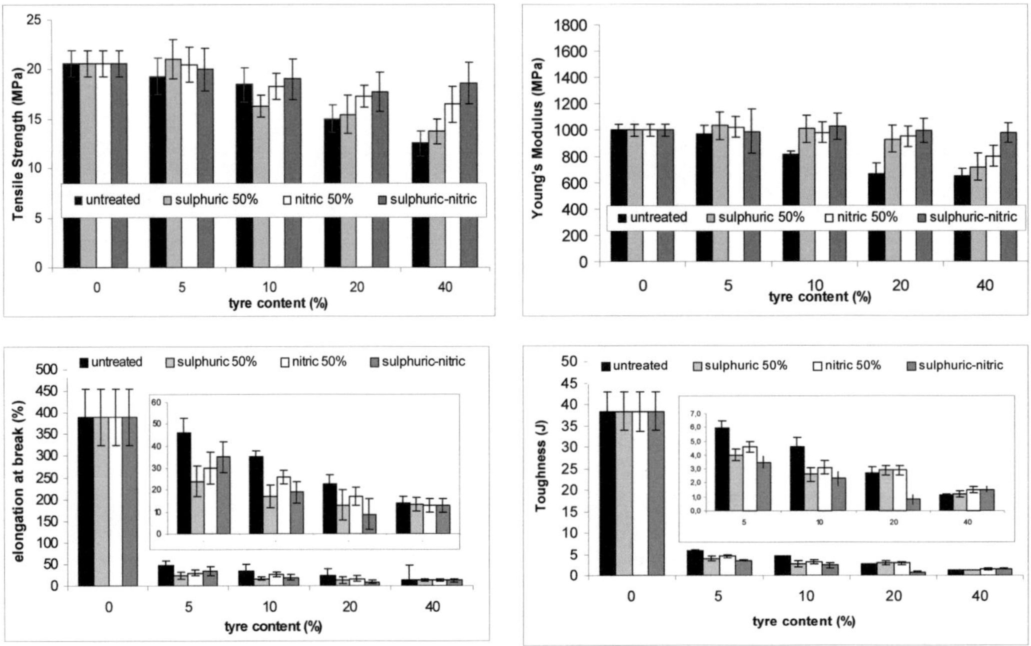

Fig. 3. Young's Modulus, tensile strenght, elongation at break and toughness of composites based on treated GTR with particle lower than 200μm and HDPE.

The treatment of GTR particles with nitric acid shows similar results to the sulphuric acid treatment except for the sulphur-oxygen derivatives. As shown in Figure 4, spectra of samples exposed to nitric acid show a decrease of the same bands as those exposed to sulphuric acid (1739, 1640, 1540, 1430 and 875cm^{-1}), plus an increase of the band at 1382cm^{-1} assigned to N-N=O. The bands assigned to O=S=O do not appear in this case. The observed chemical modifications agree with those presented in precious studies [Figovslq 1996] [Dierkes 2003]. Sulphuric acid acts as a strong dehydrating substance that can take up hydrogen and oxygen from organic matter and cause carbonisation. HNO_3 is also a powerful oxidizing agent when used in concentrated solutions. They modify the surface of the material introducing sulphur and nitrogen surface groups as - SO_3 and - NO_2. There is also an increase of groups as O=S=O and C-SO_2-OR, a decrease of C-H and an increase of C=C.

By the study of the spectrophotometry FTIR-ATR spectral bands, the main results obtained were following: i) every treatment studied produces a specific chemical modification on the ground tyre rubber particles, thus inducing the formation of specific groups; ii) several degradative effects appear in a similar way and do not depend on the acid used in the treatment (mainly the decrease in double bounds and the elimination of minor components and moieties).

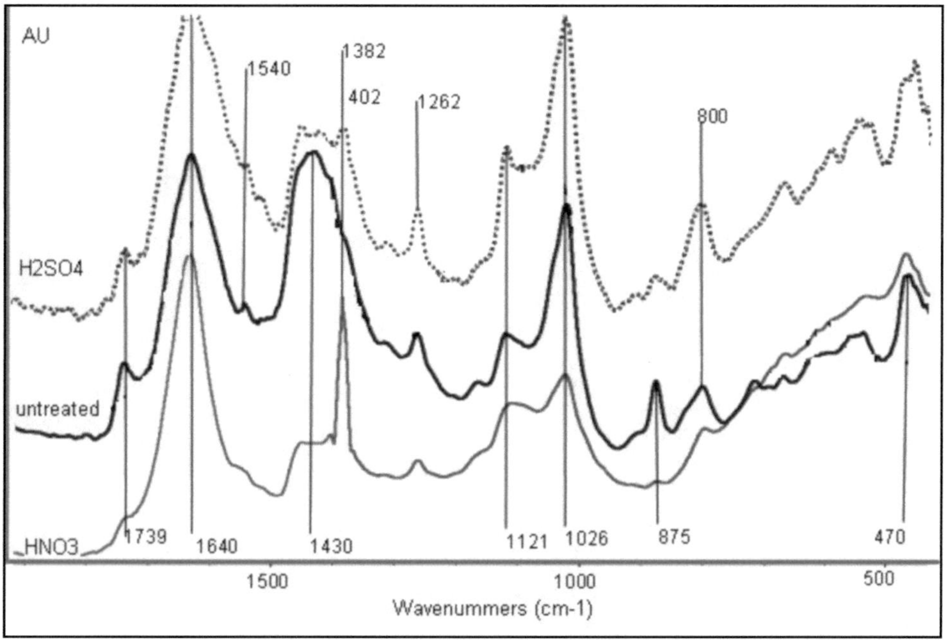

Fig. 4. Spectra of GTR particles: 1) treated with HNO_3, 2) untreated and 3) treated with H_2SO_4.

Some SEM microphotography pictures depicting the fracture surface of the GTR particles composites are shown in Figure 5. The images show different levels of magnification, which allows the comparison of the different particle sizes. The picture "a" shows a blend sample containing sulphuric treated GTR particles sizes between 400-600µm. In the centre appears a big particle [see arrow 1], showing some cracks and pores big enough to be observed at this level of magnification. The particle is unliked to the matrix, as it can be observed by the deep voids around its contour [see arrow 2]. The GTR seems to be resting on the HDPE, without being properly attached to it. On the other hand, the matrix has been strained and deformed plastically [see arrow 3]. Microphotography "b" from a sample including particles with sizes between 200-400µm pre-treated with HNO_3 shows similar features. The interaction between both components of the blends is not good and there are many cavities [see arrow 4] around the ground tyre rubber particle. Image "c" shows a different situation, since magnification is 10 times higher, particle is much smaller. The particle seems much more integrated in the matrix, and there is an area on the right with a clean cut that indicates that the particle itself has been broken instead of detached [see arrow 5], which proves the good performance of the interfacial contact, when the particles has been treated with H_2SO_4. The contour of the particle does not show voids around, instead some fragments of HDPE sprout from the ground tyre rubber showing points of good attachment between both components [see arrow6]. The high magnification allows the appreciation of the roughness achieved by the pre-treatment. Picture "d" shows particles (size between 200 and 400 µm) treated with H_2SO_4-HNO_3 (50%). Several medium size particles appear showing different levels of attachment.

The results of SEM and FTIR-ATR which are summarized above suggest that the surface modification of GTR particles influences their mechanical properties, and in some cases may also improve the interaction and compatibility between the HDPE matrix and the GTR.

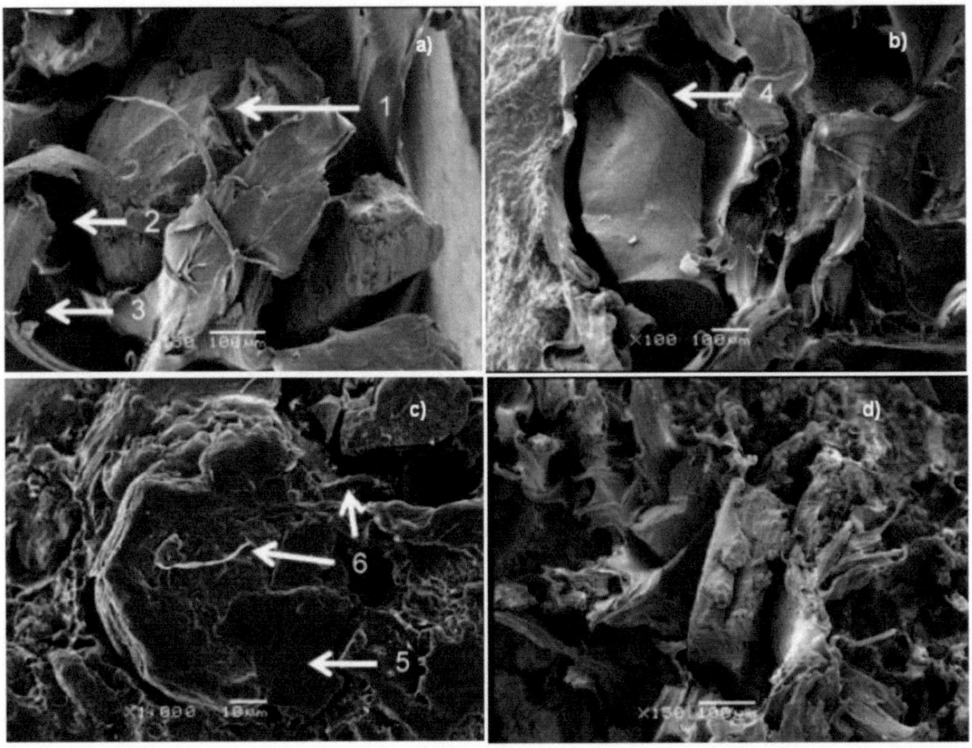

Fig. 5. SEM microphotographies of different fracture surfaces of GTR/HDPE components. a) particle size between 400-600μm treated with H_2SO_4, b) particle size between 200-400μm treated with HNO_3, c) particle size lower than 200μm treated with H_2SO_4, d) particle size between 200 and 400μm treated with H_2SO_4-HNO_3 (50%).

2.4 Use of waxes and wetting agents

K. Oliphant and W. E. Baker [Oliphant 1993] studied the influence of a coating layer process upon cryogenically GTR particles (CGTR). As a coating, they used Dow Primacor 3460, an ethylene acrylic acid (EAA) copolymer. They pre-treated the CGTR particles with the EAA copolymer. Afterwards the "coated" particles were added onto the melt LLDPE and HDPE matrixes. They observed that the deleterious effects of the CGTR particles is more pronounced in composites with HDPE than LLDPE. For example, the impact failure for pure LLDPE was seen to be a ductile yielding process in which the dart drawn the material out as it passes through. In contrast, the failure of the pure HDPE, although it involves some plastic deformation, is observed to occur through catastrophic propagation of a crack through the impact zone. This difference in impact failure was believed to be responsible for the poorer properties of the CGTR/HDPE composites. For LLDPE, where failure is ductile, large particles with moderate adhesion were easily tolerated but in CGTR/HDPE composites, the failure remained semi-brittle because particles were too large to induce brittle-to-ductile transition. Failure then occurs largely through crack propagation, and the large particles, even with moderate adhesion, act as serious flaws providing an easy path for the crack to follow. They concluded that the addition of CGTR to a semi-brittle matrix requires much higher level of adhesion to retard crack growth at the particle/matrix interface, or much lower particle sizes to lower the brittle-ductile transition temperature. In their study they concluded that composites of 40 to 50% of precoated CGTR with LLDPE have impact properties approaching those of the pure LLDPE and retain adequate processability. This property improvement was believed to be due to an interaction between the carboxylic acid groups on the EAA copolymer with functional groups on the CGTR particles surface, which result in an increase on adhesion and greater ductibility. However, poor mechanical properties were obtained with CGTR/HDPE composites due to the semi-brittle nature of HDPE.

The mechanical properties of different composites of untreated GTR/HDPE are shown in Table 1. The increase of the GTR amount produces a decrease of the mechanical properties of the final composite. As commented before, this is due to a very weak adhesion between the two phases. The big particle size of the GTR and its crosslinked structure, which avoid any compatibility with the thermoplastic matrix, are the main reasons of this behaviour. The brittle response of the material is related to poorly adhering large particles present in the matrix [Bartczak 1999]. According to previous studies [Colom 2006], the composite with 20% of untreated GTR is a good compromise between mechanical properties and GTR content. Samples with higher percent of GTR are more difficult to process and show an excessive decrease in Young's modulus and tensile strength. The composition with 20% of GTR was used to study the effect of additives.

% GTR	Young's modulus (MPa)	Std. Dev.	Tensile strength (MPa)	Std. Dev.	Elongation at break (%)	Std. Dev.	Toughness (J)	Std. Dev.
0	927,90	27,95	20,67	2,22	390,08	22,20	38,4	3,00
10	889,00	24,94	16,44	1,53	17,31	1,31	1,57	0,23
20	759,65	33,11	14,34	2,13	12,55	1,46	1,13	0,21
40	370,25	12,84	8,74	1,72	12,00	1,73	0,88	0,15

Table 1. Mechanical properties of untreated GTR/HDPE composites.

The HDPE/Additives blends were prepared in order to analyze the influence of the additives in the mechanical properties of the matrix. Additive dosages are the same as those used in the mixture of HDPE with GTR particles. Table 2 shows the mechanical properties of HDPE with the different additives. In general, it is observed that additives decrease elongation at break and toughness but increase Young's modulus of HDPE. Tensile strength only increases with ester wax additive (Ceridust 5551) while it is lower with all the other additives. Hato and Luyt [Hato 2007] observed similar behaviour in terms of mechanical properties when they studied a blend of paraffin wax and HDPE. They explained the results in terms of morphology, suggesting a possible cocrystallization of the mixture HDPE/wax. According to their interpretation, wax chains are short enough (9nm compared to an approximate polyethylene lamellar thickness of 10nm) to be incorporated as straight chains into the HDPE lamellae. This incorporation occurs only at low dosages of wax. The cocrystallization would be responsible for the increase in Young's Modulus and the reduction in elongation at break. Blends of polyethylene (LDPE or LLDPE) with paraffin and oxidized paraffin waxes show similar properties [Krupa 2002].

%Aditive	Young's modulus (MPa)	Std. Dev.	Tensile strength (MPa)	Std. Dev.	Elongation at break (%)	Std. Dev.	Toughness* (J)	Std. Dev.
0	927,90	27,95	20,67	2,22	390,80	22,01	38,41	3,00
2.5% Dis-108	1.075,43	32,17	20,3	1,62	290,91	27,21	33,41	3,52
5% Byk-9077	1.087,04	24,13	19,53	2,61	317,42	36,01	32,62	4,44
2% Byk-Syn 2100	1.111,27	33,85	17,48	1,82	342,21	25,22	35,61	3,12
3% Byk-P 105	1.049,24	15,25	16,46	2,37	357,32	25,01	35,51	2,71
5% LPE520	1.088,26	17,48	16,94	3,14	337,66	15,90	34,47	1,61
5% LCPP1502	1.051,86	20,78	16,55	4,14	362,01	30,90	36,11	5,12
2.5%Ceridust 5551	1.248,99	36,01	22,19	2,15	266,21	16,61	33,00	3,12

Table 2. Formulations showing the influence of the additives on the mechanical properties of the matrix.

Table 3 shows the mechanical properties of the HDPE/GTR composites with the different additives. Disperbyk-108 and Byk-9077 are both designed to interact with acid groups and, as can be seen, do not improve any of the properties of the blend. It could be assumed then that GTR particles have a basic or rather a non polar surface. In order to improve interaction between a non polar surface and basic additives (such as Disperbyk-108 or Byk-9077) a synergic additive with acids groups can be used. As mentioned in section 4.3 Byk-Synergist 2100 has acid groups able to interact with the basic additives and with the non polar surfaces of the GTR particles, acting as a kind of "bridge". It can be observed that this additive causes a significant increase in elongation at break (13.2%). Byk-P 105, designed to interact with basic groups, is the wetting additive that most improves the mechanical properties of the composite. This result is related to carbon black characteristics. Carbon black is present on tyre formulations in form of furnace black pigments in amounts between 20 to 40%. As a result of the manufacturing process, furnace black pigments always contain small amounts of basic surface oxides (pyrone-like structures) [Buxbaum 1993]. The presence of these basic groups on the GTR particles surface could provide interaction with Byk-P 105. The wetting additive is anchored onto the ground tyre rubber particles surface and its non polar part is extended in the HDPE matrix. This wetting effect decreases the interfacial energy between the two phases and leads to a better adhesion. These results, obtained with an acidic wetting additive (Byk-P 105) are in concordance with the improvement in compatibility achieved by the ethylene-acrylic acid (EAA) in GTR/LLDPE composites attributed to the interaction between the acid groups on the EAA copolymer and CGTR particles surface [Oliphant 1993].

Licowax PE520 and Licocene PP1502 waxes also offer good mechanical properties, whereas the ester wax Ceridust 5551 reduces elongation at break and toughness. As discussed previously, there is an increase of Young's modulus and tensile strength when the ester wax is mixed with the matrix (Table 3). When GTR particles are included in the composite (Table 4), the same effect can be noticed. Despite this improvement, elongation at break and toughness drastically decrease. Therefore, the effect of Ceridust 5551 seems mainly related to the matrix, being unable to create the thin coating onto the GTR particles to improve adhesion. The effect of Licowax PE520 and Licocene PP1502 is also related to the zinc derivatives present in tyre formulations. Pastor et al. [Pastor 1994], Romero et. al [Romero 2001]. Monteiro et al. [Monteiro 2002] and Colom et al. [Colom 2009] observed a diffusion and migration of different tyre additives, like zinc derivatives from the bulk of different rubbers and reused tyres to their surfaces. These additives lead to a poor adhesion between the GTR particles and the HDPE matrix due to its low free energy (unpolar properties). As mentioned before, waxes cover the particles reducing slightly the viscosity in the layers adjacent to the matrix. During calendaring at 150-155ºC, zinc derivatives can be solubilised into the matrix due to the lower viscosity obtained by the waxes. This effect achieves a "cleaner" GTR particles surface that is easier to wet.

%Additive	Young's modulus (MPa)	Std. Dev.	Tensile strength (MPa)	Std. Dev.	Elongation at break (%)	Std. Dev.	Toughness* (J)	Std. Dev.
0	759,65	33,11	14,34	2,13	12,55	1,46	1,13	0,21
2.5% Dis-108	760,49	16,44	13,97	4,38	12,71	1,27	1,02	0,13
2.5% Dis-108 + 2% Byk-Syn 2100	737,05	18,69	13,98	2,88	14,19	2,33	1,24	0,11
5% Byk-9077	724,03	21,06	12,89	4,20	11,81	3,24	0,89	0,25
5% Byk-9077 + 2% Byk-Syn 2100	712,05	21,02	13,22	2,52	14,20	1,96	1,12	0,29
3% Byk-P 105	796,47	26,22	15,08	3,48	13,19	2,01	1,25	0,26
5% LPE520	797,10	31,44	15,55	1,84	13,69	2,33	1,32	0,19
5% LCPP1502	753,07	29,44	14,59	4,27	13,51	1,72	1,24	0,19
2.5% Ceridust 5551	846,16	27,23	14,66	2,80	11,04	1,56	0,94	0,14

Table 3. Influence of the additives on the mechanical properties of composites of 20% of GTR and 80% of HDPE.

Different percentages of GTR particles were tested with the wetting additive and waxes that gave best results (Byk-P 105, Licowax PE520 and Licocene PP1502). Table 4 shows the mechanical properties of the different composites.

Fracture surfaces of the composites containing the additives (Byk-P 105, Licowax PE520 and Licocene PP1502) that gave better mechanical properties were examined by SEM. Figure 6 contains SEM images of the composite with 20% of GTR particles and selected additives. Figure (a) shows the composite with 20% of GTR particles and 80% of HDPE without any additives, where GTR particles do not display any signs of adhesion to HDPE. Figures (b), (c), and (f) corresponding to the additives combination of 2.5% Disperbyk-108 + 2% Byk-Synergist 2100, 5% Byk-9077 + 2% Byk-Synergist 2100 and 5% Licocene PP1502 respectively, show also poor interaction. Figure (d) corresponds to the composite including 3% of Byk-P 105: the particle is surrounded by the HDPE matrix but the picture does not show a perfect interaction. On the other hand, photograph (e) with 5% of Licowax PE520 shows good adhesion between the two phases. The existence of some filaments protruding from the GTR particles indicates a strong interaction.

Four samples of each specimen were examined under the microscope for dispersion quality study. The mechanical energy provided by the two roll mill during the milling stage breaks the GTR agglomerates and creates smaller particles with larger interfaces in contact with the thermoplastic matrix. Wetting additives and waxes improve dispersion stability of single GTR particles; consequently, better mechanical properties can be achieved.

% of GTR	% of additive	Young's modulus (MPa)	Std. Dev.	Tensile strength (MPa)	Std. Dev.	Elongation at break (%)	Std. Dev.	Toughness* (J)	Std. Dev.
10%	No additive	889,00	24,94	16,44	1,53	17,31	1,31	1,57	0,23
10%	3% Byk-P 105	914,94	36,54	16,86	2,58	17,61	2,71	1,65	0,26
10%	5% LPE520	933,27	17,48	16,91	2,80	14,86	2,59	1,59	0,16
10%	5% LCPP1502	906,25	30,78	16,44	3,91	17,62	3,41	1,60	0,32
20%	No additive	759,65	33,11	14,34	2,1	12,55	1,4	1,13	0,21
20%	3% Byk-P 105	796,47	26,22	15,08	3,48	13,19	2,01	1,25	0,26
20%	5% LPE520	797,10	31,44	15,55	1,84	13,69	2,33	1,32	0,19
20%	5% LCPP1502	753,07	29,44	14,59	4,27	13,51	1,72	1,24	0,19
40%	No additive	370,25	12,84	8,74	1,72	12,00	1,73	0,88	0,15
40%	3% Byk-P 105	427,10	15,92	8,46	2,70	13,92	1,25	0,90	0,05
40%	5% LPE520	421,39	22,41	9,3	1,82	14,97	2,04	0,92	0,14
40%	5% LCPP1502	405,25	17,64	8,29	2,95	11,81	1,61	0,54	0,16

Table 4. Influence of the additives on the mechanical properties of composites of 20% of GTR and 80% of HDPE.

Figure 7 shows four pictures. (a) and (b) correspond to the composites without additives and with 2.5% Disperbyk-108 + 2% Byk-Synergist 2100 respectively. Both pictures show the existence of agglomerates, which can be seen as GTRr particles in contact along their edges with its interstitial spaces filled with the HDPE matrix. On the other hand, pictures (c) and (d) correspond to 3% of Byk-P 105 and 5% of Licocene PE520 respectively; both show single GTR particles surrounded by the HDPE matrix. This lack of agglomerates means a better dispersion and stabilization of the particles, and consequently an improvement on the mechanical properties of the final blend.

From the study of the mechanical properties, the following conclusions can be drawn:

i. the increase of the GTR particles amount in neat composites produces a decrease of the mechanical properties in the final composite, especially in elongation at break and toughness. The mixture with 20% of GTR particles gives a balanced compromise between the amount of GTR particles and good mechanical properties;
ii. the wetting additive Byk-P 105 and the wax Licowax PE520 give the best performance in the blends with 10, 20 and 40% of ground tyre rubber particles;
iii. all additives produce an increase in the Young's modulus and tensile strength of the HDPE matrix. This increase is attributed to cocrystalization of the waxes with HDPE.

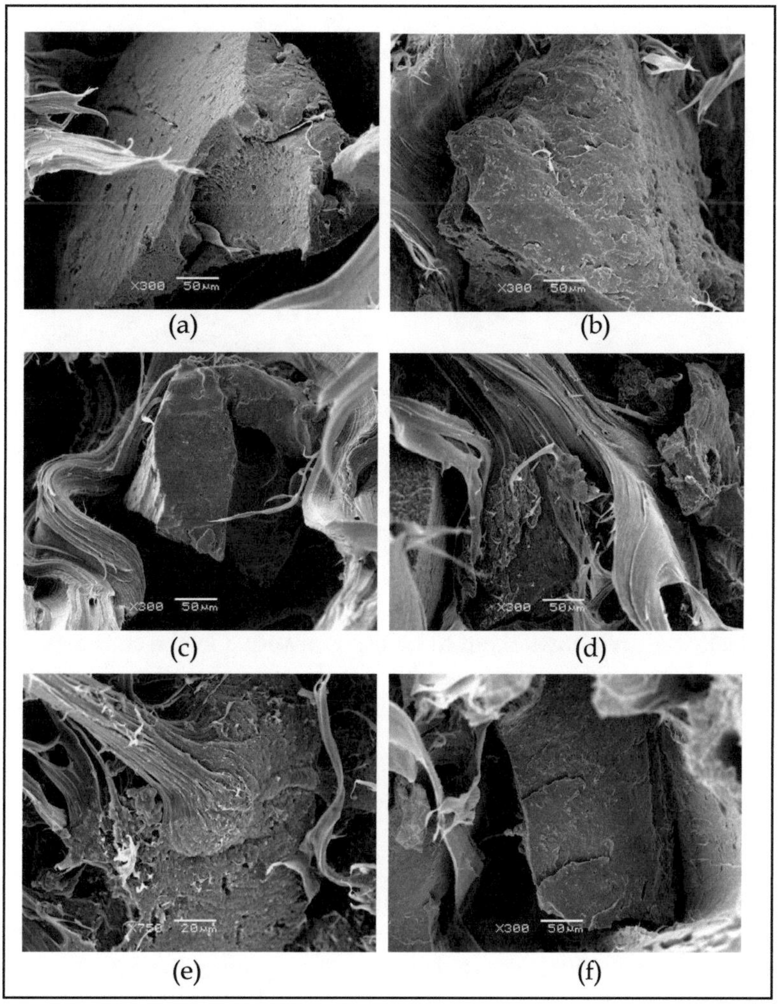

Fig. 6. SEM microphotographs of the composites 20% of GTR and 80% of HDPE with different additives. a) neat composite, b) 2.5% Dis-108 + 2% Byk-Syn 2100, c) 5% Byk-9077 + 2% Byk-Syn 2100, d) 3% Byk-P 105, e) 5% LPE520 and f) 5% LCPP1502.

Licowax PE520 acts adequately promoting adhesion. Fracture surface reveals fragments of HDPE attached to the particles. On the other hand the Byk-P 105 image shows particles better wrapped in thermoplastic than with any other wetting additive. Optical microscope's photographs show that samples including Byk-P 105 and Licowax PE520 contain fewer agglomerates. The homogeneous dispersion of the GTR particles in the matrix produces

2.5 Formulations of thermoplastics vulcanizates

TPVs based on EPDM and PP or HDPE are the most representative example of this kind of materials. They are used for automotive applications, like bumpers or hoses, and in other uses such as covering wires, pipes, boots, handle tools, etc. [Van Duin 2006].

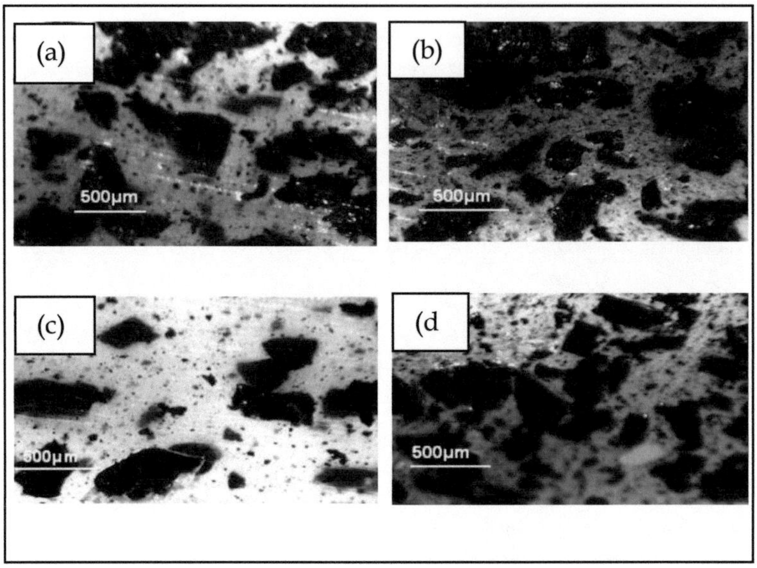

Fig. 7. Microscope pictures of the composites with 20% of GTR and the following additives: a) neat mixture, b) 2.5% Dis-108 + 2% Byk-Syn 2100, c) 3% Byk-P 105 and d) 5% LPE520.

Their thermoplastic nature allows them to be reprocessed. On the contrary, rubber articles, like tyres, are not recyclable or re-processible due to their vulcanized structure. However, the increasing price of EPDM over the last decade has led to its substitution by metallocene-based ethylene-octene and ethylene-butene elastomers [Sherman 2008]. Therefore, the attempt to substitute as much EPDM as possible by GTR in a TPV formulation could be considered a good option for cost reduction of the final formula and as an upcycling application for GTR.

Over recent years, some attempts to use GTR in TPV's formulations have been made. [Kumar 2002]studied the feasibility of producing TPE with Low Density Polyethylene (LDPE), fresh rubber (SBR, NR or EPDM) and mechanical degraded GTR (with and without processing oil). [Punnarak 2006] reported the study of composites based on reclaimed tyre rubber (RTR) and HDPE, dynamically vulcanized with sulphur, maleic anhydride (MA) and dicumyl peroxide (DCP), [Naskar 2001] presented a TPE made of cryogenic GTR, EPDM and ethylene-co-acrylic acid, vulcanized with DCP. Other studies have been published with GTR pre-treated with bitumen, which according to the authors, contributes to the GTR devulcanization, also acting as a plasticizer and compatibiliser in the composites. From these published works, it can be concluded that the addition of fresh EPDM with good matrix compatibility (similar surface energies) produces an effect of encapsulation of the GTR particles, creating a co-continuous phase and improving the adhesion between phases. There are several factors that influence the final properties, like plastic/rubber composition, mixing conditions, and type of curing agent used to crosslink the elastomeric phase. This section reports the study of a TPV based on EPDM, standard injection HDPE and GTR, dynamically crosslinked with peroxides. Peroxides are fundamental in the final TPV properties. A new peroxide, 3,3,5,7,7-pentamethyl-1,2,4-trioxepane, developed by Akzo Nobel Polymer Chemicals (Trigonox 311) was used in the present study. Unique about Trigonox 311 is its decomposition temperature, which is significantly higher compared to any other commercially available crosslinking peroxides, allowing processing temperatures of 180-190°C with no scorch problems.

The most important factors that affect thermoplastic vulcanizates morphology and mechanical properties are: a) type of curing agents used to crosslink the elastomeric phase, b) plastic/rubber composition (including EPDM/GTR proportion on the rubber) and c) degree of crosslinking. Four different experimental sets were designed in order to perform an accurate study of the thermoplastic vulcanizates mechanical properties. The first set was carried out on composites 40/30/30 (HDPE/EPDM/GTR) with peroxide contents from 0.1 to 3% using Trigonox 311 (Table 6). The influence of the Trigonox 311 content on thermoplastic vulcanizates mechanical properties was studied. In order to assess the previous results, a second set of samples containing only DCP was tested. Different temperatures of curing, related to peroxides activity, were examined too (Table 7). Moreover, the influence of the plastic/rubber composition of the composites was tested at constant EPDM/GTR ratio and peroxide content (Table 7). Finally, GTR was substituted at different concentrations by EPDM in a fixed plastic/rubber and peroxide content composition (Table 8). These last sets of samples were tested in order to determine the amount of EPDM that can be replaced by GTR particles maintaining a good balance of mechanical properties of the thermoplastic vulcanizates.

Table 5 shows the influence of the peroxide content on the mechanical properties of composites where the plastic/rubber and EPDM/GTR ratios were kept constant at 40/60 and 50/50 respectively. A reference sample without EPDM and without peroxide (only HDPE/GTR) is also included showing a brittle performance (composite nº1, Table 5). When EPDM is incorporated to the composite, elongation at break and toughness increase immediately, however, Young's modulus and tensile strength decrease due to the incorporation of a rubber-like phase. As peroxide content increases, all properties show higher values due to crosslinking, but at higher contents of Trigonox 311 (2 and 3%) elongation at break start decreasing. This could be related to the reaction of the peroxides with the thermoplastic matrix, because they are not selective for unsaturated elastomers. Similar results were reported in previous studies [Kumar 2002]. In polymer composites, an increase in elongation at break is directly related to a better compatibility between phases. Therefore it can be concluded that composite 6, containing 1% of Trigonox 311, shows maximum elastomeric thermoplastic properties. In other words, from this experiment, 1% of Trigonox produces the best results to crosslink EPDM and seems to be effective creating a network to encapsulate ground tyre rubber particles.

Processing conditions are important issues regarding final properties of the cured composite. DCP, as standard peroxide for TPV's studies, was also tested in combination with Trigonox 311, in order to study synergetic effects and compare processing and physical properties of the composites. Although 1% of Trigonox showed maximum performance in the first study, the content was reduced to 0.5% when adding DCP as additional peroxide, in order to avoid excess of peroxide. That would be also beneficial for cost savings. Table 6 shows the mechanical properties of the DCP alone and in combination with Trigonox 311, at different processing temperatures. Composite 9, with 0.1% of DCP, was consolidated at 170°C and did not show difficulties in handling. However, its properties were not as good as those of composite 6. Composites with higher DCP contents lead to premature crosslink in the two roll mixer (working temperature 150-155°C) and those pressed at higher temperatures became very sticky and difficult to remove from the press being impossible to measure. The synergy effect of both peroxides at 210, 190 and 180ºC was studied in composites 10, 11 and 12. Consolidation temperature seems to have an important role in the final properties. Different conditions for composites containing 0.5% of Trigonox and 0.1% of DCP were tested. Composite 10 was difficult to take off the press too but composites 11 and 12 showed no handling problems. Composite 11 shows better elongation at break and toughness than 6. Therefore a combination based on 0.5% of Trigonox and 0.1% of DCP respectively seems to give the maximum elastomeric thermoplastic properties. Nicolini et al. [Nicolini 2006] also found that the mixture of two curing agents, DCP/bismaleimide, in TPVs based on PP/EPDM 35/65 w/w gave the best mechanical properties and much better ones than DCP alone.

The plastic/rubber composition has also an important influence in the final material properties. Table 7 shows the mechanical properties of composites where EPDM/GTR ratio was kept constant at 1:1, and peroxide combination of 0.5% Trigonox 311 and 0.1%DCP. As HDPE content increases from 10 to 40%, Young's modulus, tensile strength and toughness gradually increase too. Elongation at break decreases due to the incorporation of a plastic phase. With even higher contents of HDPE (more than 50%), Young's modulus and tensile strength increase while elongation at break and toughness decrease. The higher the HDPE content the less toughening effect of the elastomeric phase. Therefore the mixture with 40% of HDPE and 60% of rubber phase shows the highest desired properties. Similar results were found by A. K. Naskar et al. [Naskar 2001]. The last set of samples study mechanical properties when fresh rubber is replaced by GTR particles. Up to this point, the best properties have been obtained with the 40/60 composition of plastic/rubber phase. Therefore, the HDPE phase and peroxide content are maintained while EPDM content is varied from 60% to 0%. Table 8. shows the results. When GTR particles content increased the loss in mechanical properties due to the incompatibility between GTR particles and HDPE is clear. From 60% to 30% of EPDM, the composite becomes gradually less toughened but when 40% of GTR particles is included, EPDM seems to loose its encapsulating and compatible effect leading to an important decrease of elongation at break and toughness. It's interesting to notice that the composite with 60% of GTR particles is even more brittle than the one containing 10% of EPDM. It seems that peroxides only vulcanizes EPDM and do not react with the possible existing active sites of GTR.

After this study an overall vision of composites behaviour is acquired. In order to assess the substitution of EPDM for GTR, two composites are selected for comparison. Composite 23 does not contain GTR (40%HDPE + 60%EPDM + 0.5%Trig + 0.1%DCP) and composite 26 has 30% of EPDM substituted by GTR (40%HDPE + 30%EPDM + 30%GTR + 0.5%Trig + 0.1%DCP). Results show (*Table 8.*) that even with the considerable amount of 30% in GTR particles, the 297% value in elongation at break fulfils the major criteria for a thermoplastic elastomer: which this value must be higher than 100%. However, if the GTR particles amount is increased to 40% the elongation at break drastically drops down to 121% which is almost in the limit. Therefore, 30% of GTR in the composite is the optimum amount for the desired balance: maximum GTR particles quantity with acceptable mechanical properties.

Figure 8 shows an increasing hardness tendency when peroxide dosage (Trigonox) is increased in the blends where the plastic/rubber and EPDM/GTR ratios were kept constant at 40/60 and 50/50 respectively. Figure 9 shows the effect on hardness, when EPDM is substituted by GTR (Table 8). At high GTR dosages, from 40 to 60%, the hardness value is maintained constant at values of 89-90, in those cases, EPDM (less than 20%) is not able to form the co-continuous phase to encapsulate the dispersed GTR particles and give the rubber-like property on the final blend. This is in concordance with the results obtained in the mechanical properties, where composites with less than 30% of EPDM suffer an important decrease in elongation at break and toughness.

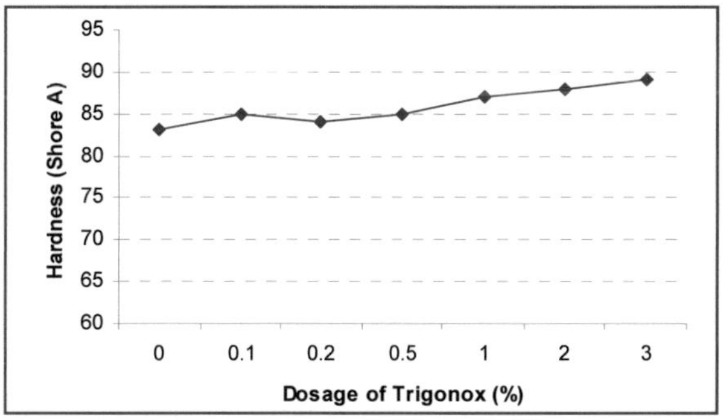

Fig. 8. Hardness as function of Trigonox dosage

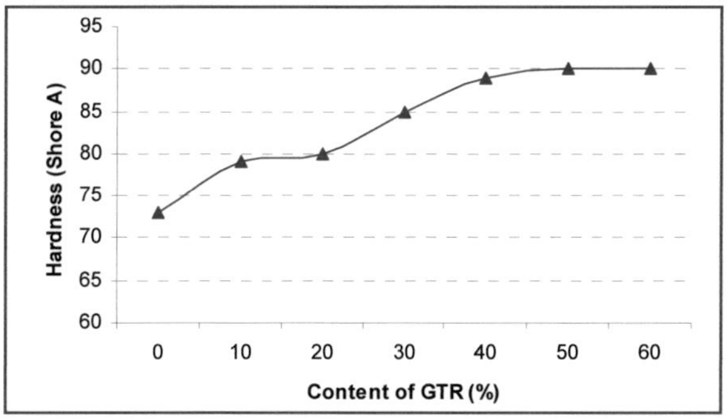

Fig. 9. Hardness as function of GTR dosage.

ATR-FTIR spectroscopy was used to study the chemical changes involved after dynamic vulcanization. The main peroxides reaction consists on the hydrogen atoms abstraction from the polymer chain and formation of the corresponding macroradicals. The reactivity of the generated free radicals depends on the hydrogen bond dissociation energy [Naskar 2004]. Trigonox 311 major decomposition products are methane, acetone, isopropyl acetate, 3-hydroxy-1,3-dimethylbuthyl acetate and 3-oxy-1-methylbuthyl acetate, while, major DCP decomposition products are methane, acetophenone, 2-phenylpropanol-2, α-methylstyrene and water. Peroxide cross-linking of unsaturated rubbers and polymers is achieved via free-radical mechanism that involves three steps: the first is the generation of radicals by thermal decomposition of the peroxide, on second place the radical attack on the polymer chain via hydrogen abstraction to generate polymer radicals, and third and last, the combination of two polymer radicals to form carbon-carbon crosslinks . In order to investigate the chemical changes on composites with Trigonox 311, the sample's spectra with constant 40/30/30 ratio of HDPE/EPDM/GTR and an increasing peroxide dose from 0 to 3% were examined.

Composite	% of HDPE	% of EPDM	% of GTR	% of Trigonox 311	Young's Modulus (MPa)	Std. dev %	Tensile Strenght (MPa)	Std. dev %	Elongation* at break (%)	Std. dev %	Toughness* (J)	Std. dev %
1	40	0	60	0	440.45	14.80	9.56	0.23	12.88	1.92	1.03	0.12
2	40	30	30	0	98.99	6.00	5.12	0.25	153.03	21.63	8.93	2.13
3	40	30	30	0.1	116.91	10.80	5.1	0.23	145.98	34.81	8.69	2.84
4	40	30	30	0.2	126.71	9.70	5.32	0.42	170.14	40.03	8.90	2.01
5	40	30	30	0.5	120.99	4.60	5.6	0.22	190.19	32.72	9.35	2.32
6	40	30	30	1	126.71	4.45	5.85	0.12	268.36	29.15	11.34	1.31
7	40	30	30	2	128.26	7.61	6.22	0.15	260.93	29.21	11.08	1.33
8	40	30	30	3	130.83	6.43	6.55	0.95	228.53	48.43	10.73	5.32

Table 5. Mechanical properties of the composites containing HDPE/EPDM/GTR ratio of 40/30/30 and different contents of Trigonox 311. The table includes the composite of 40% HDPE with 60% GTR.

Composite	% of HDPE	% of EPDM	% of GTR	% of Trigonox 311	% of DCP	Temperature (°C)	Young's Modulus (MPa)	Std. dev %	Tensile Strenght (MPa)	Std. dev %	Elongation at break (%)	Std. dev %	Toughness* (J)	Std. dev %
9	40	30	30	0	0.1	170	116.26	6.12	6.98	0.23	221.94	21.52	13.91	1.22
10	40	30	30	0.5	0.1	210	94.98	10.01	4.99	0.81	156.24	24.74	7.03	3.93
11	40	30	30	0.5	0.1	190	97.51	5.32	7.06	0.67	297.33	30.62	17.17	3.52
12	40	30	30	0.5	0.1	180	90.34	8.49	6.74	0.42	281.41	28.91	15.86	4.01

Table 6. Mechanical properties of the composites containing HDPE/EPDM/GTR ratio of 40/30/30 with 0.1% of DCP and mixtures of 0.1% DCP + 0.5% Trigonox 311, at different temperatures.

Composite	% of HDPE	% of EPDM	% of GTR	% of Trigonox 311	% of DCP	Young's Modulus (MPa)	Std. dev %	Tensile Strenght (MPa)	Std. dev %	Elongation at break (%)	Std. dev %	Toughness* (J)	Std. dev %
13	10	45	45	0.75	0.15	9.78	0.21	2.45	0.21	360.00	35.02	8.33	1.9
14	20	40	40	0.67	0.13	23.75	1.43	4.00	0.42	337.00	44.81	12.72	2.1
15	30	35	35	0.58	0.11	55.49	4.52	6.04	0.51	293.42	48.23	15.31	2.3
16	40	30	30	0.5	0.1	97.51	5.32	7.06	0.67	297.33	30.62	17.17	2.5
17	50	25	25	0.42	0.08	209.00	12.23	7.10	0.31	135.00	12.31	9.04	0.7
18	60	20	20	0.33	0.07	263.15	18.81	8.25	0.22	107.70	24.21	7.91	4.9
19	70	15	15	0.25	0.05	385.52	16.62	10.44	0.41	154.63	45.01	5.68	4.9
20	80	10	10	0.17	0.03	637.47	14.61	14.59	0.51	42.26	10.83	3.51	1.3
21	90	5	5	0.08	0.01	783.07	60.61	16.70	0.44	30.80	7.82	2.70	1.4
22	100	0	0	0	0	927.90	27.91	17.17	0.52	390.80	22,01	38,41	3.0

Table 7. Mechanical properties of formulations showing variation of the rubber/plastic composites at fixed peroxide dose and constant EPDM/GTR ratio of 1:1.

Composite	% of HDPE	% of EPDM	% of GTR	% of Trigonox 311	% of DCP	Young's Modulus (MPa)	Std. dev %	Tensile Strenght (MPa)	Std. dev %	Elongation at break (%)	Std. dev %	Toughness* (J)	Std. dev %
23	40	60	0	0.5	0.1	41	2.32	14.22	0.53	751.21	18.84	54.61	3.46
24	40	50	10	0.5	0.1	65.59	6.95	10.92	0.53	592.65	19.14	39.42	2.36
25	40	40	20	0.5	0.1	89.19	11.84	7.26	1.14	417.68	32.68	22.89	3.98
26	40	30	30	0.5	0.1	97.51	5.32	7.06	0.67	297.33	30.62	17.17	2.52
27	40	20	40	0.5	0.1	132.93	10.19	6.14	0.16	121.38	22.27	6.3	1.61
28	40	10	50	0.5	0.1	161.37	14.36	6.23	0.42	33.54	11.75	1.72	0.72
29	40	0	60	0.5	0.1	172.42	10.32	6.33	0.45	25.78	4.46	1.24	0.32

Table 8. Mechanical properties of the 40:60 plastic:rubber composites at different rubber (EPDM:GTR) ratios.

In Figure 10 can be seen, the presence of different C=O stretching bands (1744, 1592, 1586 cm^{-1}) associated to carbonylated products such as ketones, aldehydes or other products that can be generated by the Trigonox 311 decomposition. The C=O peak intensity is observed to be higher for the composites cured with higher peroxide amount, indicating an oxidative degradation of the matrix polymer due to the higher peroxide concentration.

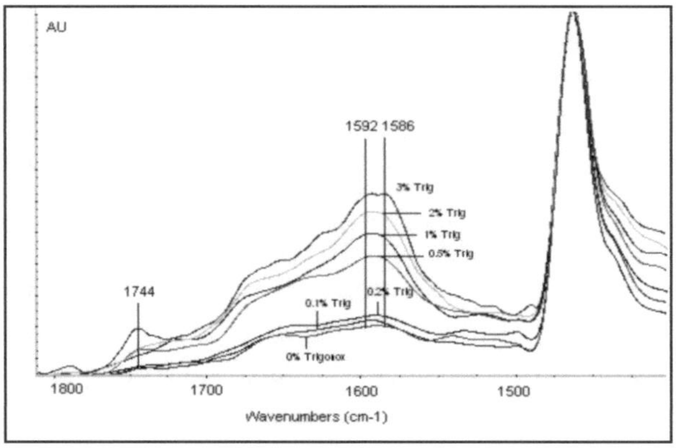

Fig. 10. ATR-FTIR spectres of samples with HDPE/EPDM/GTR ratio of 40/30/30 and an increasing peroxide dose (from 0 to 3%) for the area between 1800 and 1400 cm^{-1}.

Figure 11 shows the spectra between 1400 and 1000cm^{-1}. Bands at 1374, 1258, 1090 and 1025cm^{-1} are assigned to the asymmetric stretching vibrations of C-O-C bonds of esters and ethers linkages during the curing reaction between peroxide free radicals and the polymeric macroradicals. The C-O-C band absorbance increases up to 1% of Trigonox but it decreases at higher peroxide dose (2 and 3%) due to the preferred oxidative degradation reaction of the peroxide with the matrix. The results are in concordance with the mechanical properties, where the composite with 1% of Trigonox showed the more interesting mechanical behaviour.

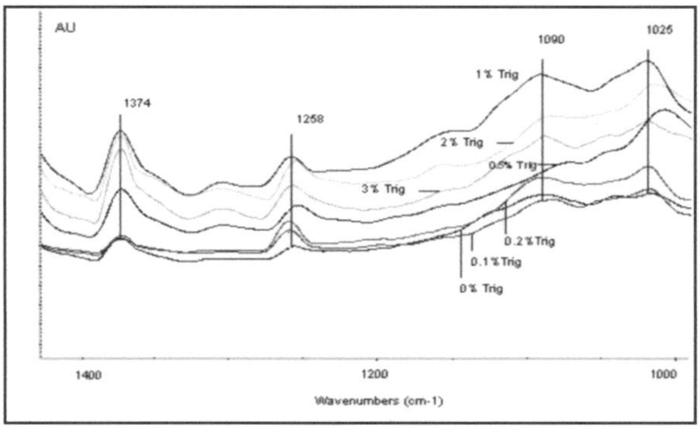

Fig. 11. ATR-FTIR spectres of samples with HDPE/EPDM/GTR ratio of 40/30/30 and an increasing peroxide dose (from 0 to 3%) for the area between 1400 and 1000 cm^{-1}.

Figure 12 shows SEM micrograph of fracture surface of the composite without EPDM and peroxides, only HDPE and GTR. These two materials are incompatible due to their different chemical composition as can be seen in picture 12.a where GTR particles appear almost isolated without being wet by the HDPE matrix. Without any kind of compatibility between the two materials, very little adhesion is possible, leading to very poor mechanical properties as already described.

The influence of EPDM addition into the HDPE-GTR composite is shown in picture 12.b. The addition of an elastomeric component which, at the same time contains high amount of

polyethylene should help to improve compatibility between HDPE and GTR particles. Particles are not as isolated as in the composite without EPDM. However, can be observed as some GTR particles parts are still not surrounded neither by HDPE nor EPDM. In other words, EPDM still is not able to encapsulate all GTR particles.
As expected, the encapsulation effect is seen in Figures 12c and 12d with peroxides addition. In both pictures can be observed how GTR particles are encapsulated by EPDM. In the case of Figure 12d corresponding to the composite with 0.5%Trigonox + 0.1%DCP, the encapsulating effect is much more visible. The fact that GTR particles are completely encapsulated means that EPDM has formed a three-dimensional network within them, and at the same time, EPDM creates a co-continuous phase with HDPE. This effect leads to a much better adhesion and better mechanical properties as already observed.

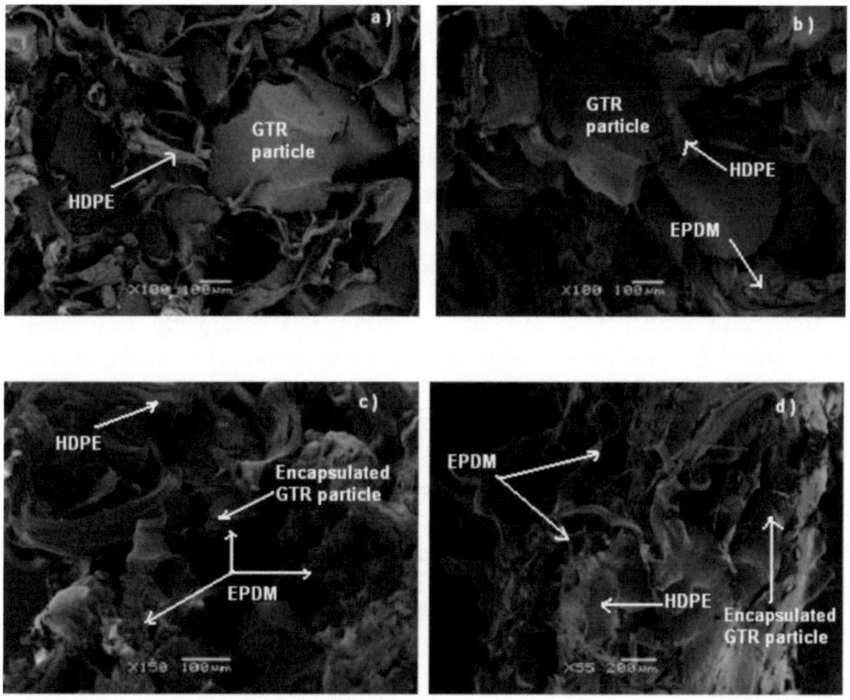

Fig. 12. Shows the SEM micrograph of the following composites: (a) 40%HDPE 60%GTR, (b) 40%HDPE 30%EPDM 30%GTR, (c) 40%HDPE 30%EPDM 30%GTR 3%Trigonox and (d) 40%HDPE 30%EPDM 30%GTR 0.5%Trig 0.1%DCP.

The following conclusions can be drawn from this section: i) The composition consisting on 40% HDPE 30% EPDM 30% GTR 0.5% Trigonox 311 and 0.1% DCP gives the best balance regarding GTR amount (30%) and mechanical properties. It is proved that process conditions, such as press temperature, have an important role in the final mechanical properties. The mixture of two different peroxides types give the best synergist effect. High peroxide amount has a negative impact on crosslinking due to a possible thermooxidative degradation of the thermoplastic matrix as observed by FTIR-ATR analysis. ii) The encapsulation phenomena of EPDM over GTR particles can be observed in SEM pictures. An optimization of composite components for maximum adhesion between GTR particles and HDPE matrix is studied during the mechanical properties test. Good adhesion is only possible if GTR particles are encapsulated by fresh rubber, i.e. EPDM. Good adhesion between phases is directly translated as good mechanical properties in the final composite. GTR can be up-cycled if is used as a rubber component in TPV production. In the present study, 30% of fresh rubber can be substituted by this recycled material, still fulfilling the major criteria for this kind of materials.

3. Conclusion

After the complete study of the three compatibilization methods and the comparison of the mechanical properties, estimation costs, maximum amount of GTR particles and easiness of production of the final compound, the following final conclusions can be taken:

- A detailed study of the composites obtained by each compatibilization method was done. In the three methods there has been a mechanical properties study, a chemical and morphological study.
- The effectiveness of the compatibilization method can be check with mechanical properties of the blends. Mechanical properties are method dependant.
    - Acid treatment does not improve any of the properties with the chosen particle size. The acid causes etching on the surface of GTR particles improving, in this way, mechanical adhesion. A brittle performance of the compound can be observed at lower particle size (< 200μm).
    - Addition of waxes and wetting additives has a slightly improvement of all mechanical properties. The best results were obtained with the wetting additive Byk-P 105 and the wax Licowax PE 520. Byk-P 105 interacts with the basic groups present in the GTR particles surface. Licowax PE 520 acts covering the GTR particles and reducing slightly the viscosity in the layers adjacent to the matrix.
    - Peroxides are use for dynamic vulcanization and the blend is transformed into a thermoplastic vulcanizate. A mixture of peroxides is used. Trigonox 311, which is designed to work at high temperatures and DCP, which is standard peroxide used for TPV's, were used synergetic effects and compare processing and physical properties of the composites. The composition consisting on 40% HDPE 30% EPDM 30% GTR 0.5% Trigonox 311 and 0.1% DCP gives the best TPV mechanical properties.
- The maximum amount of GTR particles in the final blend was desired in order to help the global reduction of EOL tyres. The blend consisting on 40% HDPE 30% EPDM 30% GTR 0.5% Trigonox 311 and 0.1% DCP gives the best balance regarding GTR amount (30%) and mechanical properties without deleterious defects.
- Due to cost savings, it was a premise to have a compatibilization method that does not need many complications to be applied (e.g. especial machines, etc…). Besides the acid treatment, which need an step more in order to apply the pre-treatment onto the GTR particles, the rest of the methods do not need any special handling or special equipment as the raw materials used can be add directly to the mixture.

4. Acknowledgment

Financial support from the Spanish Ministry of Science and Technology (MAT 2007-64569 Project) is gratefully acknowledged.

5. References

B. Adhikari, D. De and D. Maiti. Reclamation and recycling of waste rubber. Progress in Polymer Science 2000; 25, 909-948.

R. P. Burford and M. Pittolo. Rubber Chemistry and Technology 1982; 55, 1233

A. Shojaei, H. Yousefian, S. Saharkhiz. Performance characterization of composites materials based on recycled high density polyethylene and ground tyre rubber reinforced with short glass fiber for structural applications. Journal of Applied Polymer Science 2007; 104, 1- 8.

K. Oliphant, W. E. Baker. The use of cryogenically ground rubber tires as a filler in polyolefin blends. Polymer Engineering and Science 1993; 33, 166.

R. Sonnier, E. Leroy, L. Crerc and J. M. López Cuesta. Compatibilization of polyethylene/ground tyre rubber blends by γ irradiation. Polymer degradation and stability 2006; 91, 2375-2379.

I. Fuhrmann, J. Karger-Kocsis. Promising approach to functionalisation of ground tyre rubber - photochemically induced grafting. Plastics, rubber and composites 1999; 28.

J. I. Kim, S. H. Ryu, Y. W. Chang. Mechanical and dynamic mechanical properties of waste rubber powder/HDPE composite. Journal of Applied Polymer Science 200; 77, 2595-2602

S. H. Lee, M. Basasubramanian, J. K. Kim. Dynamic reaction incide co-rotating twin screw extruder. II. Waste ground rubber tire powder/polypropylene blends. Journal of Applied Polymer Science 2007; 106, 3209-3219.

S. Coiai, E. Passaglia, F. Ciardelli, D. Tirelli, F. Peruzzotti, E. Resmini. Modification of crosslinked rubber particles by free radical polymerization. Macromol. Symp. 2006; 234, 193-202.

E. Manchón, E. Macías, A. Nadal, C. Fernández and V. Gómez. Preparation of mesoporous and macroporous materials from rubber of tyres wastes. Microporous and Mesoporous Materials 2004; 67, 35-41.

C. M. Cepeda, M. M. Pastor, T. P. Ferrrándiz and J.M. Martín. Surface characterization of vulcanized rubber treated with sulfuric acid and its adhesion to polyurethane adhesive. The Journal of Adhesion 2000; 73, 135-160.

C. M. Cepeda, M. M. Pastor, T. P. Ferrrándiz and J.M. Martín. Polymer surface modification: relevance to adhesion. Ed.; V S P Internacional Science Publishers: Zeist, 2000; 2, 305-334.

J. Cañavate, P. Pagès, J. Saurina, X. Colom and F. Carrasco. Determination of small interactions in polymer composites by means of FTIR and DSC. Polymer Bulletin 2000; 33, 294-300.

O. Figovslq, D. Beilin, N. Blanck, J. Potapo and V. Chernyshe. Development of polymer concrete with polybutadiene matrix. Cement and Concrete Composites 1996; 18, 437-444.

W. K. Dierkes. Rubber recycling. In: Pandalai SG, editor. Recent research developments in macromolecules, vol. 7, Research Signpost, Trivandrum 2003; 265-292.

M. Hato, A. Luyt. Thermal fractionation and properties of different polyethylene/wax blends. Journal of Applied Polymer Science 2007; 104: 2225-2236.

X. Colom, F. Carrillo, F.J. Cañavate; "Composites Reinforced with Reused Tyres: Surface Oxidant Treatment to Improve the Interfacial Compatibility". European Polymer Journal 42, 2369-2378, 2006.

I. Krupa A. Luyt. PE/wax blends: interesting observations. Macromolecular Symposium 2002; 178: 109-116.

G. Buxbaum. Industrial Inorganic Pigments. VCH Verlagsgesellschaft GmbH, Weinheim. 141-176.

M. Romero, M. Pastor and J. Martín. Adhesión improvement of SBR rubber by treatment with trichloroisocyanuric acid solutions in different esters. International Journal of Adhesion and Adhesives 2001; 21: 325-337.

P. Monteiro, N. Segre and G. Espósito. Surface characterization of recycled tire rubber to be used in cement paste matrix. Journal of Colloid and Interface Science 2002; 248: 521-523.

X. Colom, , F.J. Cañavate, F. Carrillo; J.J. Suñol " Effect of the particle size and acid pre-treatments on compatibility and properties of recycled HDPE plastic bottles filled with ground tyre powder". Journal of Applied Polymer Science 2009; 112: 1882-1890.

16

A Review of Thermoplastic Composites for Bipolar Plate Materials in PEM Fuel Cells

Rungsima Yeetsorn[1], Michael W. Fowler[2] and Costas Tzoganakis[2]
King Mongkut's University of Technology North Bangkok
University of Waterloo
[1]Thailand
[2]Canada

1. Introduction

Polymer composite materials have particular properties that meet special requirements. A conductive polymer composite is positioned to play an increasingly significant role in industry and academia, specifically in the area of electrical conductivity. Even general knowledge about electrically conductive composites has been available for many years, less attention has been given in the literature to the use of conductive composites for alternative energy production.

Why is the use of composite materials for energy production interesting? With a continued growth in the worldwide demand for energy, there is increasing interest in alternative technologies of energy generation such as fuel cells, for various stationary and mobile applications. In this chapter, the authors are mainly interested in a fuel cell as an energy generator, since a fuel cell is expected to play a major role in the economy of this century and for the foreseeable future. A number of factors provide the incentive for fuel cells to play a role in future energy supplies and for transportations, including climate change, oil dependency and energy security, urban air quality, and growth in distributed power generation [1].

A polymer electrolyte membrane fuel cell (PEMFC) is a good contender for portable and automotive propulsion applications because it provides high power density, solid state construction, high chemical-to-electrical energy conversion efficiency, near zero environmental emissions, low temperature operation (60 - 120 °C), and fast and easy start-up [2,3, and 4]. The U.S. Department of Energy (DOE) has also identified the polymer electrolyte membrane fuel cells as the main candidate to replace the internal combustion engine in transportation applications [2]; however, barriers to commercialization remain. Fundamental technical challenges facing the commercialization of PEM fuel cells are manufacturing and material costs; material durability and reliability; and hydrogen storage and distribution issues [4, 5, and 6]. One of the major factors limiting fuel cell commercialization is the development of bipolar plates, which are one of PEMFC's key components. Bipolar plate characteristic requirements are a challenge for any class of materials, and none fits the profile characteristics exactly. Therefore, research on materials, designs and fabrications of bipolar plates for PEMFC applications is a vital issue for scientists and engineers wanting to achieve the appropriate PEMFC for global commercialization. Several types of materials are currently used in bipolar plates, including non-porous graphite plates, metallic plates with or without coating and a number of composite plates. Thermoplastic composite bipolar plates are an attractive option for PEMFC use. They do not only offer advantages of low cost, lower weight and greater ease of manufacturing than traditional graphite, but their properties can also be tailored through changes of reinforcements and the resin systems. The weakest point of thermoplastic composite bipolar plates is their low electrical conductivity compared to conventional graphite or metallic bipolar plates. To increase the electrical conductivity of the plates, electrically conductive polymers or fillers have been used as bipolar plate materials.

This chapter, based on, reviews and extends existing polymer composite material (focusing on thermoplastic composites) and manufacturing literature within the context of a bipolar plate material for a PEMFC application. Throughout this chapter the authors also critique and broaden composite types and alternative manufactures based on functions of bipolar plates and the applications of process selection techniques with respect to a marketable thermoplastic bipolar plate production. Beside an electrically conductive application, the chapter documents interesting information for the design of conducting composite formulations in other applications such as thermal conductivity and mechanical applications. It also permits evaluation and comparison of processing characteristics of different composites, fundamental phenomena for the composite fabrication.

2. Background of a PEMFC and bipolar plates

The successful conversion of chemical energy into electrical energy in a primitive fuel cell was first demonstrated over 160 years ago by lawyer and scientist Sir William Grove in 1843[2,4]. These early devices, however, had very low current density. General Electric (GE), a more efficient design in the late 1950s for NASA's Gemini and Apollo space missions, and in addition fuel cell system provided electricity and drinking water for the crew. In developing its fuel cell technology, NASA funded more than 200 research contracts that finally brought the technology to a level that was viable for commercial applications [2, 4]. The types of fuel cells under active development include hydrogen fuelled ones such as alkaline fuel cell (AFC), polymer-electrolyte-membrane fuel cell (PEMFC) and phosphoric-acid fuel cell (PAFC)[7]. The first development of PEMFC was by GE for the Gemini space project, but after that, the PEMFC development became dormant. The improvement of PEMFC programs was reactivated in the 1980s, by Ballard Power Systems, subsequently recognized as the world leader in fuel cell technology [2]. The company and spin-off company Automotive Fuel Cell Corporation (AFCC) has dominated the developing automotive market and has started productions of stationary and portable power applications. For example, the largest fleet of hydrogen fuel cell buses in service was in Whistler, BC, Canada for the 2010 Olympic and Paralympic Winter Games. The twenty transit buses, powered by Ballard's FCvelocity-HD6 power module, provide a 62% reduction in greenhouse gas emissions compared to diesel buses [8]. There are now many companies, for instance General Motors, Toyota, and Honda, involved in the growth of the PEMFC technology, especially in the automotive fuel cell market. Siemens and some Japanese companies have particularly focused on portable and residential fuel cell systems, where potentially high-volume markets are expected. Companies such as Plug Power and Hydrogenics have made significant advances in the lift truck and back-up power market with their low pressure PEMFC technology.

2.1 The operation of a PEMFC [2, 4, and 9]

An expanded view of PEMFC components is shown in Figure 1 below. Key components include the membrane-electrode assembly (MEA), bipolar plates (flow fields or separators), and seals. The traditional PEMFC has a polymer electrolyte membrane placed between two gas diffusion electrodes, an anode and a cathode respectively, each usually containing a metal catalyst, such as Pt, supported by an electrically conductive material. The gas diffusion electrodes are exposed to the respective reactant gases: the reduction gas (hydrogen) and the oxidant gas (oxygen/air). An electrochemical reaction occurs at each of the two junctions (three phase boundaries) where one of the electrodes, the electrolyte polymer membrane and the reactant gas interface.

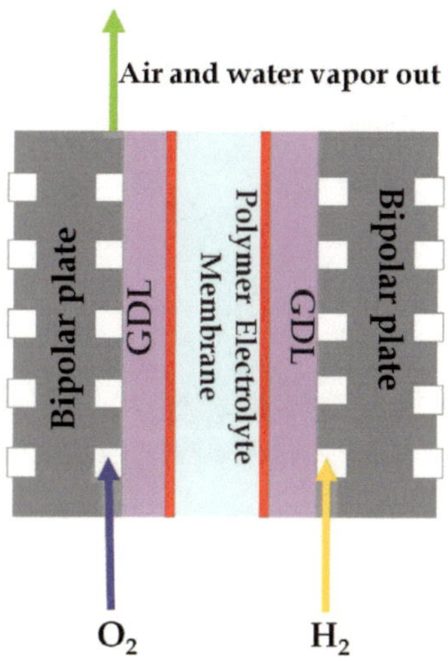

Fig. 1. Exploded view of PEMFC components

The fuel cell is an electrochemical energy device that converts chemical energy, from typically hydrogen, directly into electrical energy. The electrochemical reactions in fuel cells happen simultaneously on both sides of a membrane: the anode and the cathode. The basic PEMFC reactions are shown as follows.

At the anode,

$$H_2 \longrightarrow 2H^+ + 2e^- \tag{1}$$

At the cathode,

$$\tfrac{1}{2} O_2 + 2H^+ + 2e^- \longrightarrow H_2O + 286\ \text{kJ/mol} \tag{2}$$

Overall,

$$H_2 + \tfrac{1}{2} O_2 \longrightarrow H_2O + 286\ \text{kJ/mol} \tag{3}$$

Figure 2 shows a schematic of a PEMFC operation to gain a fundamental understanding of the polymer electrolyte membrane fuel cell technology. During PEMFC operations, hydrogen permeates through the anode and interacts with the noble metal catalyst, producing electrons and protons (1). The electrons are conducted via an electrically conductive material through an external circuit to the cathode, while the protons are simultaneously transferred via an ionic route through a polymer electrolyte membrane to the cathode. This polymer membrane also serves as a gas barrier so that the reactant species cannot freely combine. At the cathode, oxygen permeates to the catalyst sites where it reacts with the protons and electrons when properly hydrated, producing the reaction (2). Consequently, the products of the PEMFC reactions are water, electricity and heat.

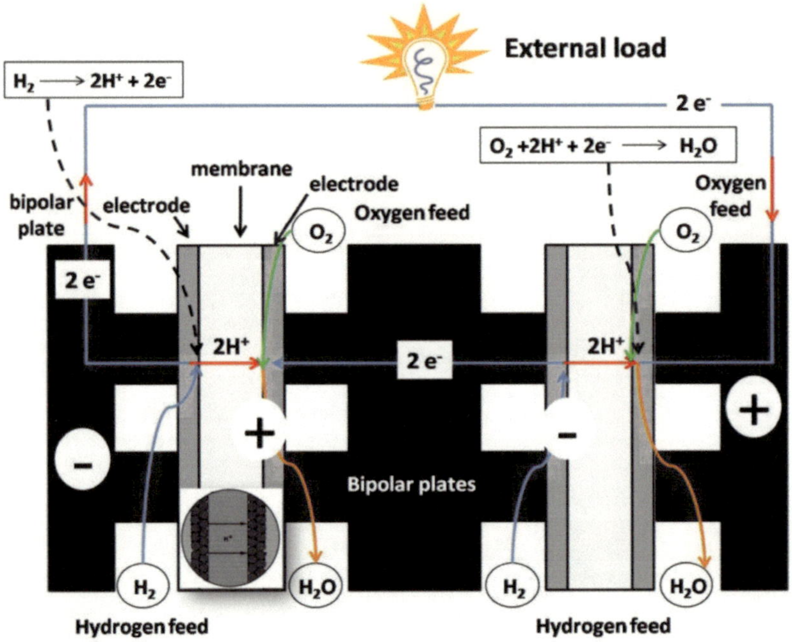

Fig. 2. Schematic of PEMFC operation [2]

2.2 Bipolar plates [2, 4, and 9]

The bipolar plate performs a number of functions within the PEMFC as described below.

1. Conducting electrons to complete the circuit, including by collecting and transporting electrons from the anode and cathode, as well as, connecting individual fuel cells in series to form a fuel cell stack of the required voltage (i.e., fuel cells are typically arranged in a bipolar configuration);
2. Providing a flow path for gas transport to distribute the gases over the entire electrode area uniformly;
3. Separating oxidant and fuel gases and feeding H_2 to the anode and O_2 to the cathode, while removing product water and un-reacted gases;
4. Providing mechanical strength and rigidity to support the thin membrane and electrodes and clamping forces for the stack assembly;
5. Providing thermal conduction to help regulate fuel cell temperature and removing heat from the electrode to the cooling channels.

The materials of the bipolar plate must have particular properties because of its multiple responsibilities and the challenging environment in which the fuel cell operates. Material's properties must be considered for achievable design for a fuel cell application, specifically, electrical and thermal conductivity, gas permeability, mechanical strength, corrosion resistance and low weight. An ideal material should combine the following characteristics that are defined by Department of Energy (DOE) as shown in table 1.

Property	Value
Electrical conductivity	> 100 Scm^{-1}
Contact resistance	0.1 - 0.2 $ohmcm^2$
Thermal conductivity	> 10 $W(mK)^{-1}$
Weight	< 0.4 $kgkW^{-1}$
Flexural strength	$\geq$ 25 MPa
Flexibility	3-5% deflection at mid-spa
Compression strength	$\geq$ 50 MPa
Tensile strength	$\geq$41 MPa
Gas permeability	< 2×10^{-6} $cm^3\ s^{-1}\ cm^{-2}$
Cost	$25/kW or < $10/plate.
Corrosion resistance	< 1 $\mu A\ cm^{-2}$

Table 1. US DOE technical targets for composite bipolar plates [10, 11]

Currently, efforts to improve the PEMFC cost and reliability for the industry, including the automotive industry, are comprised of reducing the cost and weight of the fuel cell stack, the goal being a 50 kW system of $35/kW and <133 kg in mass [12].The bipolar plates in the stack require significant improvement, since bipolar plates account for approximately 55% of the PEMFC weight, and 37% of the stack manufacturing and materials cost [13, 14] as shown in Figure 3. Accordingly, the development of bipolar plates may present opportunities for cost and weight reductions in PEMFCs. Moreover, bipolar plate characteristic requirements are a challenge for any class of materials, and none fits the profile characteristics exactly. Therefore, research on materials, designs and fabrications of bipolar plates for PEMFC applications is a vital issue for scientists and engineers wanting to achieve the appropriate PEMFC for global commercialization.

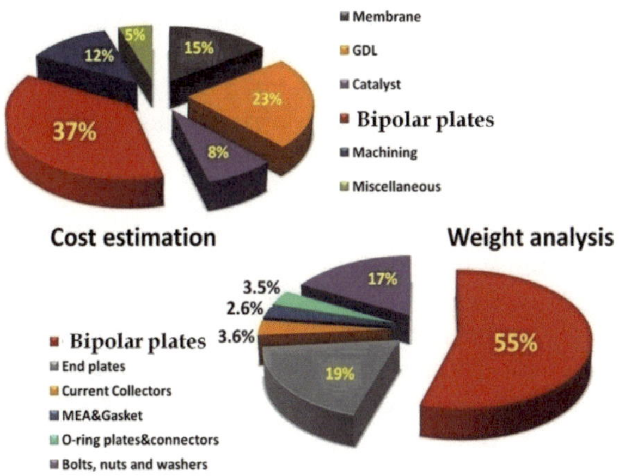

Fig. 3. Relative cost and weight components from a PEMFC using graphite bipolar plate [11]

3. Bipolar plate materials

Several types of materials are currently used in bipolar plates, including non-porous graphite plates, metallic plates with or without coating and a number of composite plates.

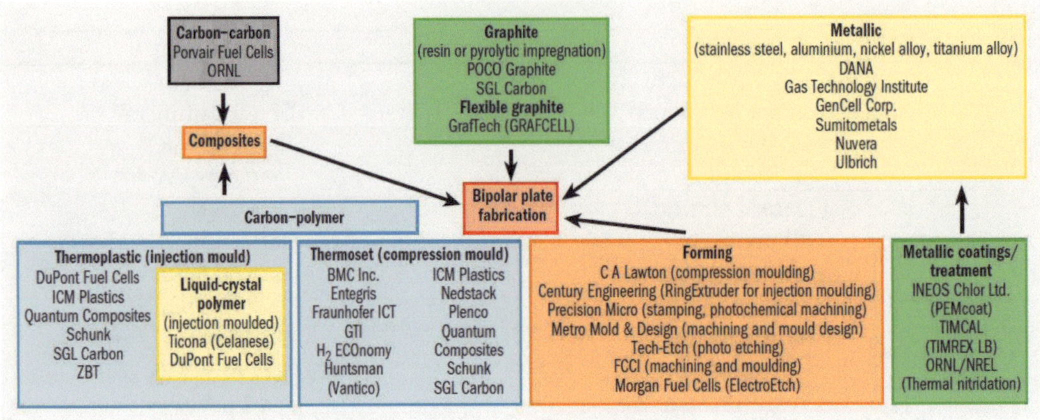

Fig. 4. Bipolar plate materials and names of manufacturers [3]

3.1 Metallic bipolar plates [15, 16, and 17]

Metals, as sheets, are potential candidates for bipolar plate material since they have good mechanical stability, electrical and thermal conductivity and gas impermeability. Probably the most important benefit is that the resultant stack can be smaller and lighter than graphite bipolar plates. Two advantages to metallic plates that they can be stamped to accommodate flow channels and that the resultant plate can be varied thick, for example 100 µm. However, the main disadvantage of metal plates is their susceptibility to corrosion and dissolution in the fuel cell operating environment of 80ºC and a pH of 2-3. This corrosion is harmful to fuel cell performance for the following reasons [18]. First, surface oxide creation significantly enlarges the contact resistance between the plate and the GDL. Second, the corrosion process changes the morphology of the surface, potentially reducing the contact area with the GDLs. Lastly, when the metal plate is dissolved, and the dissolved metal ions diffuse into the PEM membrane and become trapped in the ion exchange sites [19].This trapping results in ionic conductivity diminution, leading to increased membrane degradation. To solve these issues, researchers have considered of non-coated metal alloys, precious non-coated metals, and coated metals with a protective layer.

3.2 Graphite bipolar plates [15, 16]

Bipolar plates in the PEMFC have traditionally been made from graphite, since graphite has excellent chemical stability to survive the fuel cell environment. Other advantages of graphite are its excellent resistance to corrosion, low bulk resistivity, low specific density, and low electrical contact resistance with electrode backing materials. This low contact resistance results in high electrochemical power output. The disadvantages of graphite plates are its high costs, the difficulty of machining it, its porosity, and its low mechanical strength (brittleness). Bipolar plates have traditionally been created from graphitic carbon impregnated with a resin or subject to pyrolytic impregnation. A thermal treatment is used in the process to seal the pores. This seal renders the bipolar plates impermeable to fuel and oxygen gases. Due to the brittle nature of graphite, graphite plates used in fuel cell stacks must typically be several millimetres thick, which add to the volume and weight of the stack.

In order to solve this problem, flexible graphite was considered the material of choice for bipolar plates in PEMFC. Flexible graphite is made from a polymer/graphite composite, in which the polymer acts as a binder. The graphite principally used for the composite is expanded graphite (EG), produced from graphite flakes intercalated with highly concentrated acid. The flakes can be expanded up to a few hundred times their initial volume [20] produced from graphite flakes intercalated with highly concentrated acid. The expansion leads to a separation of the graphite sheets into nano-platelets with a very high aspect ratio. This layered structure gives higher electrical and thermal conductivity. The expanded form is then compressed to the desired density and pressed to form the bipolar plate. In comparison to conventional graphite bipolar plates, the bipolar plates produced from EG are thinner.

3.3 Polymer composite bipolar plates [10, 17]

Thermoplastic or thermosetting composites are beneficial over metallic and traditional graphite materials with regard to corrosion resistance, flexible and low weight. In addition, polymer composites may be produced in economical processes, such as compression, transfer or injection moulding processes, depending on the number of units to be manufactured. The main drawback to polymer composite is the lack of electrical conductivity, which is critical independently of final applications. To enhance the electrical conductivity of the bipolar plates, electrically conductive polymeric materials have been used as bipolar plate materials. Electrically conductive polymeric materials are organic based materials that permit electron transfer.

According to the electrically conductive structure, conductive polymeric materials can be divided into two categories: intrinsically conducting polymers (ICPs) and conductive polymeric composites (CPCs). ICPs are organic polymer semiconductors. Electrical conductivity is realized by the presence of chain unsaturation and electron delocalization effects. Much research effort and interest has therefore been devoted towards the development of polymers with intrinsic electrical conduction characteristics brought about by the presence of the conjugated group and by doping techniques. ICPs can be used for few applications due to their poor productivity and the high manufacturing costs, although they own terrific conductive performance. In terms of CPCs, composite materials for bipolar plates can be categorized as metal or carbon-based. The combination of conventional polymers (ABS, PC, PP, and etc.) with conductive loads of fillers (e.g. carbon black or carbon fibers, metallic or metallic fibers, metallic powders) allows the creation of new polymeric composite materials with unique electrical properties. CPCs are advantageous over ICPs with regards to because of the large-scale variation in the conductivity, the favorable processability and low costs. However, CPCs would not be unsuccessful to improve conductive and mechanical performance simultaneously for the reason that high filler concentration improves on conductive performance but it deteriorates on mechanical performance.

In spite of the well created commercial products available to the fabrication of a polymer composite bipolar plate (Figure 4), there is a continuous attempt for the development of better composites with maximized electrical conductivity.

4. Electrically conductive thermoplastic composites

Typically, polymer composites are created by incorporating a conductive material into a polymer matrix. The preference for the polymer binder is governed by the chemical compatibility with the fuel-cell environment, mechanical and thermal stability,

processability when loaded with conductive filler, and cost. Two different main types of resins have been used to fabricate composite plates: thermoplastic and thermosetting. Among the thermosetting resins, such as phenolics, epoxies, polyester, and vinyl ester, etc., the epoxy resin is a popular choice for a polymer composite bipolar plate production [6]. The thermosetting resins have low viscosity, and thereby contain a higher proportion of conductive fillers. During the moulding process, the thermosetting resin allows for moulding of intricate details. Moreover, the resins can be highly cross-linked through a proper curing process, and the cross-linked structure gives good chemical resistance [3].
Thermoplastic resins, such as polypropylene (PP), polyethylene (PE), poly (vinylidene fluoride) (PVDF), liquid crystalline polymer (LCP), poly (phenylene sulfide) (PPS), and fluoropolymer[6]are used less in bipolar plate fabrication than thermosetting resins for various reasons. These reasons are: thermoplastics are generally less chemically stable as thermosets and a fuel cell must be operated at lower temperatures to avoid plates melting. On the other hand, they can be injection-moulded and are therefore more beneficial in automated manufacturing. This process is suitable for mass production for future markets. However, high filler loadings limit possibilities for the injection moulding process because of the higher viscosity of the composite material. As a consequence, thermoplastic composites generally have lower electrical conductivity than other technologies. The pathway to modify such advanced materials is to incorporate conductive fillers in the thermoplastic matrix with a proper conductive network structure. This review mainly discusses about the development of thermoplastic/conductive carbon filler composites which is CPCs. A comparison between a thermoplastic/filler composite and a thermoset/filler composite for the bipolar plate material is shown in Table 2.

	Thermoset/ filler composite	*Thermoplastic/filler composite*
Advantage	- Higher temperature operation than thermoplastic - Fast cycle-time - Flow-field introduced during moulding - Low contact resistance	- Injection moulding lends itself to manufacturing automation - Fast cycle-time - Flow-field introduced during moulding - Low contact resistance
Disadvantage	- Relatively low electrical conductivity	- Low electrical conductivity when using standard thermoplastics - Limited to low-temperature operation - Injection moulding difficult at high filler loading - Generally less chemically stable than thermoset resin-
Processing option	- Compression moulding - Post-moulding CNC milling of blank	- Injection moulding - Compression moulding - Post-moulding CNC milling of blank

Table 2. A comparison between a thermoplastic/filler composite and a thermoset/filler composite for the bipolar plate material [3]

An important theory for understanding conductivity within thermoplastic composite materials, especially where the thermoplastic matrix and the fillers have very different characteristics, is the concept of percolation. Percolation processes were developed by Flory (1941) and Stockmayer (1943) to describe how small branching molecules react and form very large macromolecules [22]. In a view of electrical conduction in a polymer matrix, electrons are free to flow through conductive filler particles. If these filler particles contact one another, a continuous path is formed through the polymer matrix, which is an insulating material, for electrons to travel through. This path is called a ***conductive network***, and the material with the conductive network turns into a conducting material, as illustrated in Figure 5.

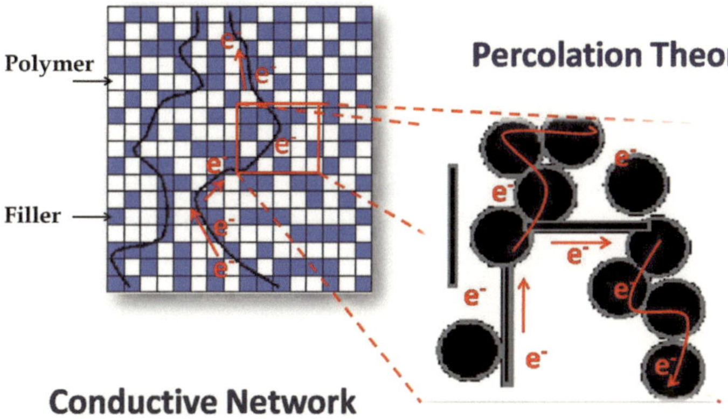

Fig. 5. Schematics of percolation pathway[11]

For the increase in conductive filler loading, three main regions define the relationship to the conductivity of conductive-filled-polymer composites as shown in Figure 6. At low filler loadings (region A), the electrical conductivity value equals zero, since no path exists for electron transport. The conductivity of the composite is still very close to that of the pure polymer matrix. At a certain critical loading, known as the ***percolation threshold***, enough filler has been introduced so that it begins to form a continuous conductive network through the composite. Following the percolation threshold is a region that produces a significant increase in conductivity with very little increase in filler amount, as displayed by region B. After this region of drastic increase, the conductivity slows its increase, and approaches that of the filler material as increase happens because the conductive network through the sample is complete. This is depicted in area C of Figure 6. Eventually, P_{max} is reached, at which point the addition of more filler does not increase the ease of electron movement. The addition of more filler will not enhance the conductivity to any significant degree.

In this chapter, different aspects of percolation phenomena of several polymer systems are reviewed. The information in this chapter is closely related to the electrical performance of CPCs and it may be useful for the improvement of thermoplastic composite bipolar plates.

To investigate the possibility of using conductive composite blends as bipolar plates in PEMFC, electrical conductivity and fuel cell performance must be characterized. Prior to discussing the development of electrically conductive thermoplastic composites, it is essential to understand the concepts of electrical conductivity and fuel cell performance measurements.

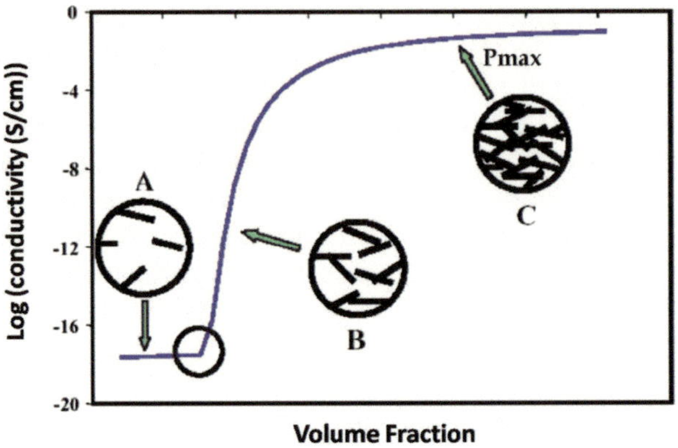

Fig. 6. Percolation S-Curve[23]

4.1 Electrical conductivity measurements

According to PEMFC operation, electrons must transfer through bipolar plates to complete a circuit (Figure 7), so one critical fuel cell performance factor is the electrical conductivity of the bipolar plate. The better the electrical transport of a bipolar plate, the fewer plates are required to produce a given power output.

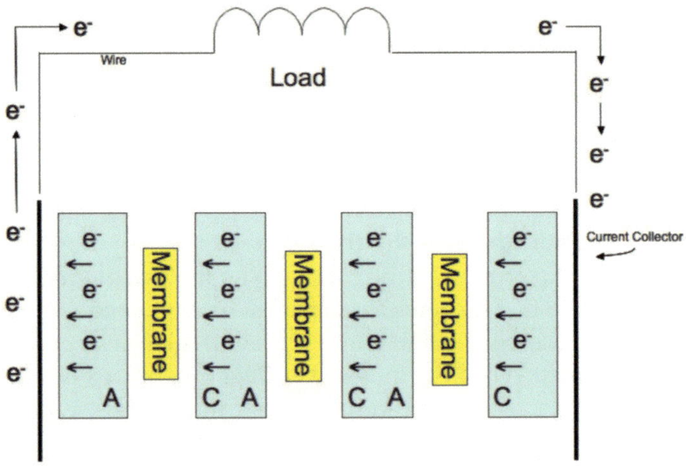

Fig. 7. Schematic of electron transport in the cell [11]

Two main methods, through-plane and in-plane conductivity measurements, are employed to measure the electrical conductivity of composite plates. The through plane test method [24], developed by the US Fuel Cell Council, can measure bulk resistance and contact resistance. Both result in a voltage drop across bipolar plates. Bulk resistance is that of the bipolar plate material, and contact resistance is that of the bipolar plate and gas diffusion layer interface. To measure the through-plane conductivity, a composite specimen is placed between two conducting plates (gold-nickel-copper plates). A current source is applied through those conducting plates and the voltage drop across two conducting plates is measured with a multimeter. In a typical fuel cell, the stack of bipolar plates is held under pressure to seal the interfaces. Stress affects the electrical characteristics of the bipolar plate material; therefore, to measure the conductivity of a bipolar plate, a sample must be placed under pressure (Figure 8).

The conductivity can be calculated using Ohm's law (Equation 4)

$$Electricalconductivity = \frac{I}{V} \times \frac{(samplethickness(cm))}{(width(cm) \times length(cm))} = Scm^{(-1)} \tag{4}$$

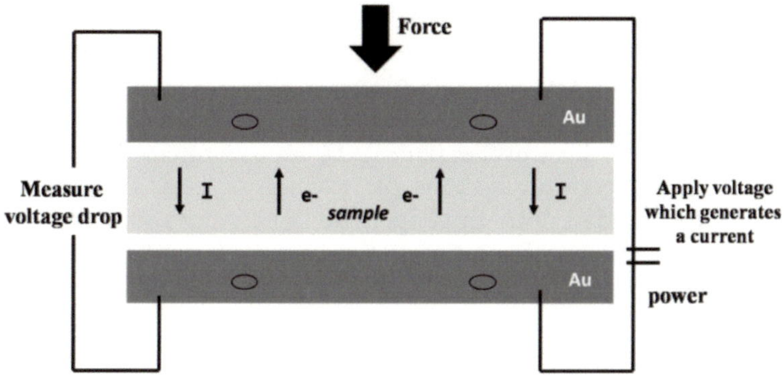

Fig. 8. Basic setup for through-plane conductivity test method[11]

Typically, a four-point probe is used to measure the in-plane electrical conductivity [25] of bipolar plates. This tool supplies excellent measurement results for near-surface and surface-related electrical properties of bipolar plates. However, it does not provide the same insight into electrical characteristics deep inside the bipolar plate (Figure 9). The four probes are arranged in a linear fashion, where the two outer probes are connected to a voltage supply, and the inner probes to a volt meter. As current flows between the outer probes, the voltage drop across the inner probes is measured. The relationship of the current and voltage values is dependent on the resistivity of the material under test, and the geometrical characteristics of the probes.

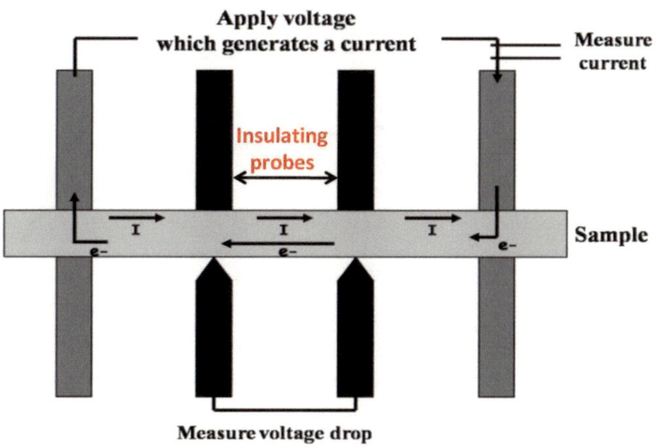

Fig. 9. Four point probe tester[11]

Using a rectangular sample, Mironov [25] calculated the resistivity relating to two geometric correction factors: a rectangular specimen of thickness (F1) and a rectangular specimen of length (F2).

$$\rho = \frac{\pi T}{\ln 2}\left(\frac{V}{I}\right) F_1 F_2 \tag{5}$$

4.2 Fuel cell performance measurement [2, 5]

The quality and functionality of fuel cells and their components can be evaluated in terms of conductivity, mechanical strength, permeability, reliability, durability, and power output. A polarization curve is the important indicator of overall fuel cell performance (Figure 10). The plotted curve relates the cell voltage to a changing current density. The maximum cell voltage, which occurs when the current density is zero, is referred to as the open circuit voltage (OCV). In electrochemical systems the voltage decreases as the current density increases and can be divided into three general areas: activation polarization, ohmic polarization, and concentration polarization. As shown in Figure 10, at the OCV, no power is produced. The power then increases with increasing current density up to a maximum, the position of which depends on the design and quality of the fuel cell components employed. Beyond the maximum, the drop in cell voltage is stronger than the increase in current density. The overall fuel cell performance is given in Equation (6), which signifies the cell voltage as the ideal Nernst voltage (E^o) minus different sources of voltage loss (η).

$$V_{Cell} = E^{\circ} - \eta_{activation} - \eta_{ohmic} - \eta_{concentration} \tag{6}$$

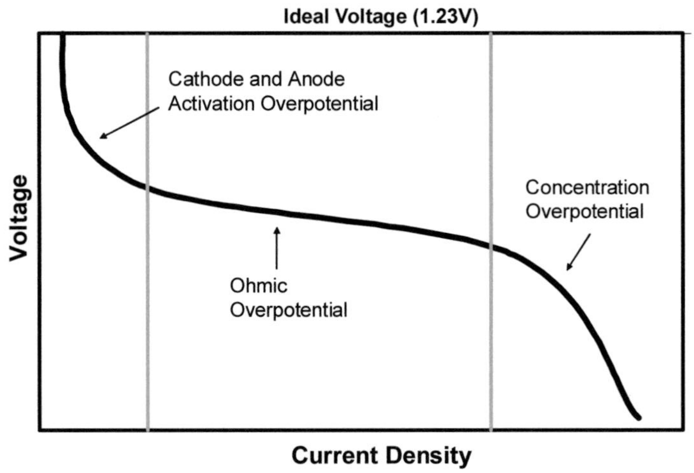

Fig. 10. Regions of a polarization curve

According to percolation phenomena, the intrinsically insulating polymer resin must be filled with conductive particles to meet the conductivity requirements for bipolar plate use. For example, Cokturk et al. [26] prepared composite materials by incorporating various nickel fillers of different shapes into polyethylene. The nickel-fillers used in this project were nickel powders, nickel filamentary powders, nickel flakes, and nickel-coated graphite fibres. It was found that the volume resistivity values of the composite materials, at maximum filler volume fraction (67 vol %), decreased more than 17 orders of magnitude with respect to neat polyethylene. A composite bipolar plate made from polyvinylidene fluoride (PVDF) as the binder and titanium silicon carbide (Ti_3SiC_2) as the conductive filler was fabricated by a compression moulding technique. The metallic filler was able to improve electrical conductivity to 28.83 Scm^{-1}, but it did not meet all the criteria [27]. Instead, poor dispersion and wetting of nickel coated-graphite in the polypropylene matrix were observed [28]. The poor dispersion and incompatibility of materials lowered the electrical conductivity when compared to the blend with synthetic graphite. This result implied that the procedure for the polymer-metal filler system needs improving to produce better dispersion composites. Moreover, the polymer-metal filler composites have the potential for ion dissolution into the fuel cell membrane, degrading performance.

To solve this problem, researchers mixed a polymer with corrosion-resistant conductive particles that do not leach and thus met the conductivity target. In summary, conductive carbonaceous fillers have proven to be interesting fillers for polymer composite bipolar plates.

4.3 Electrically conductive carbonaceous fillers

Graphite has been incorporated into polymers for bipolar plate purposes since it is able to enhance electrical properties and may offer attractive electrochemical, physical, mechanical, and economic possibilities. To reach target electrical conductivity, a high load of graphite (60-80 wt%) was must be formulated for the bipolar plate material. For example, the graphite composite bipolar plates in the research of Heo and his collaborators were composed of expanded graphite, graphite flake and a novolak-type phenol resin, with a 75:25 wt% ratio of graphite to phenol [29]. The production involved pre-curing and compression moulding processes. The results indicated an electrical conductivity of 250 Scm^{-1}, which met the requirements for PEMFC bipolar plates. According to the requirement of high graphite loading, thermosets are a popular polymer matrix used in graphite/ polymer bipolar plates, since they can accommodate a higher proportion of conductive graphite [30, 31].

Although thermoset/graphite composites are more preferable for using as a bipolar plate material than thermoplastic/graphite composites, scientists and engineers have been attempting to employ thermoplastics for the graphite composite bipolar plates. The main advantage of using thermoplastics is that they can be injection-moulded and are more favourable to automated manufacturing. High temperature thermoplastic graphite composites were therefore prepared using polyphenylene sulfide (PPS) and polyether sulfone (PES) containing natural graphite powder [32]. A low resistance, in the order of 0.1 ohm, was obtained for a graphite concentration of 50 wt%. PPS/G composite with 20% PPS resin content could be produced under a moulding temperature of 380 ^{o}C [33]. The bending strengths of the composites are 52.4MPa and 55.7MPa, when the electrical conductivities of the composites are 118.9 S cm^{-1} and 105 S cm^{-1}, respectively. The values can meet the requirements of DOE standard at the same time (Figure 11). The plate had an electrical conductivity of 118.9 Scm^{-1}. Xiao et al. [20] first prepared a poly (arylenedisulfide)/graphite nanosheet composite via a direct ring opening polymerization, and then graphite nanosheets were prepared by ultrasonic bathing. The plate produced was a good candidate for the bipolar plates of a PEMFC.

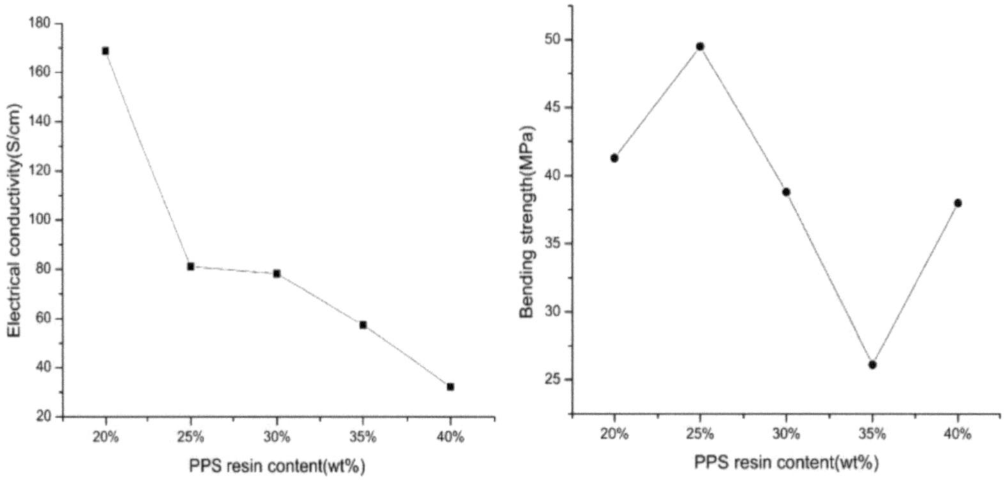

Fig. 11. Effect of PPS content on electrical conductivity and bending strength of the PPS/graphite composites[33]

Carbon black (CB)-filled thermoplastic composites are widely used as antistatic and electrostatic dissipative and semi-conductive materials. To increase the electrical conductivity, polymers with high carbon powder loading have been explored. Carbon black compounds are typically loaded with 30% to 40% carbon black because CB-filled polymers have relatively low-bulk electrical conductivities, most of which were below 1 Scm^{-1} [34]. Del Rio et al. used polyvinylidene (PVDF) and carbon black to achieve a conductivity of 2.36 Scm^{-1}, corresponding to PVDF with 40wt%CB at 30°C, for example[35]. When the CB content changed from 30 to 40 wt%, a sudden rise in conductivity was observed. The conductivity increased because the percolation threshold appeared just at this range. The best results from a performance test of a conventional fuel cell using these composite bipolar plates were obtained with the highest CB concentration (100 $mWcm^{-2}$ at 200 $mAcm^{-2}$). Although carbon black can increase conductivity, the addition of carbon black diminishes processability of the material. An additional problem associated with high CB concentrations in polymers is the substantial reduction in the strength and ductility of the polymer composites [36]. To address such problems, numerous studies were directed at reducing the carbon black concentration and the percolation threshold. Gubbels and his co-workers [37] proposed a concept of a triple-continuous structure to provide carbon-filled polymer blends with high electrical conductivity and tensile strength simultaneously. They found that the conductive carbon is preferentially located in one phase and forms a triple-continuous structure which has the advantage of achieving conductive composites at lower carbon concentrations. It is because only the percolation threshold in one phase needs to be exceeded, rather than the entire polymer blend (Figure 12).

Even though fibres are frequently used to improve the strength and rigidity of polymers, several studies have considered the reinforcement of thermoplastic polymers using carbon fibres to achieve improve the thermal and electrical properties. Fibres with finite aspect ratios can enhance the conductivity of a composite by aligning the long and more conductive axis of fibre particles in the current-flow or "through-thickness" plate direction. The fibre particles make smaller and fewer insulating gaps in the composite, resulting in a material resistivity reduction and, in turn, an increase in conductivity [38]. To consider the influence of fibre loading on electrical and mechanical properties, some studies have been performed on fibre orientation. Clingerman and his team [39] performed compounding runs followed by injection moulding of test specimens of carbon fibres filled nylon 6, 6 and polycarbonate based polymer. They found that the difference in electrical conductivity was likely due to the higher length, aspect ratio, and orientation of the carbon fibres in the composites. Carbon nanotubes are one type of carbon fibres that have been used to enhance an electrical property of bipolar plates. Liao, S.H. and his co-workers [40] fabricated thin nanocomposite bipolar plates (the thickness <1.2 mm) including multiwalled carbon nanotubes, graphite powder and polypropylene (PP). They used three types of PP with different crystallinities including high crystallinity PP, medium crystallinity PP, and low crystallinity PP to investigate the influence of crystallinity on the dispersion of MWCNTs in PP matrix. The results showed that the dispersion of MWCNTs in low crystallinity PP is favoured. Since highly crystalline polymers have strong van der Waals forces between intermolecular chains, MWCNTs tend to aggregate within the matrices.

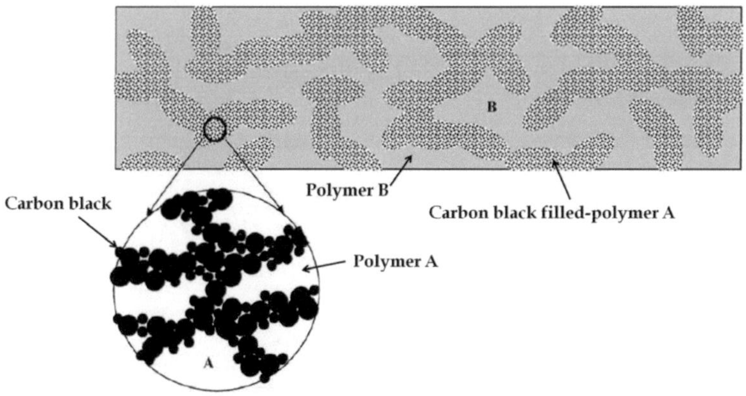

Fig. 12. Schematic of the microstructure of the proposed carbon black-filled polymer blend composite[35]

The addition of various carbon fillers with different physical and electrical properties, including natural flaky graphite, synthetic graphite, carbon black and carbon fibres, were found to be essential for the development of bipolar plates with the desired properties [31]. The single filler gives typical blend characteristics as follows. Carbon black has a high surface area and aggregation behaviour, which are needed to achieve the desired conductivities and are also responsible for relatively low production cost. It is difficult, however, to prepare well dispersed polymer/carbon black composites containing more than 35 wt% carbon black. As a result, graphite is generally added to the composite to combine the electrical properties of graphite with ease of processing. High-aspect-ratio carbon fibre has also been added to polymers to increase both electrical conductivity and mechanical reinforcement. Therefore, ***the synergistic effect***, which results from combining various types of conductive fillers, on electrical conductivity is a topic of growing interest to researchers. In previous works, the use of two filler types in polyethylene composites combined the conducting features of carbon black and carbon fibre [41].While fibres provide charge transport over large distances, carbon black particles improve inter-fibre contacts. The results showed that for composites in which the segregated carbon black-polyethylene component was laid above the percolation threshold, the electrical inter-fibre contacts were activated through carbon black particle bridges, leading to a rise in conductivity. The addition of sufficient amounts of carbon black must be controlled to obtain good processability. Therefore, graphite was used instead of carbon black to overcome this disadvantage. Radhakrishnan[32] stated that the addition of carbon black component led to further improvement in electrical conductivity over and above that achieved at high graphite contents (Figure 13).

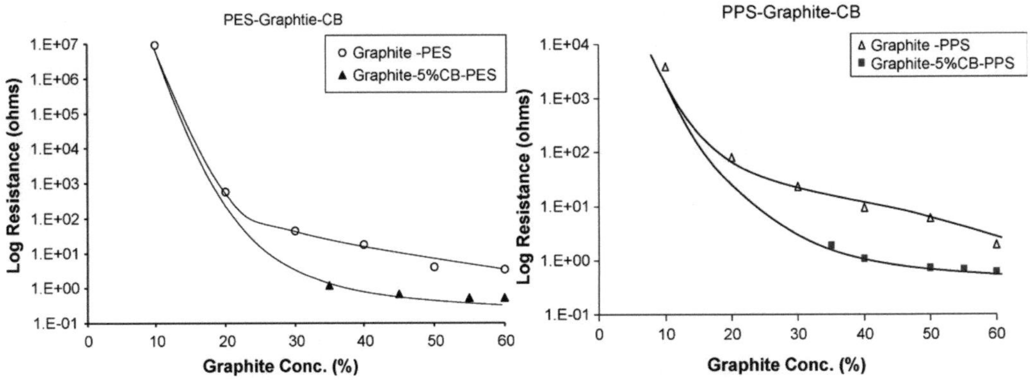

Fig. 13. Compositional dependence of electrical resistance of graphite composites [32].

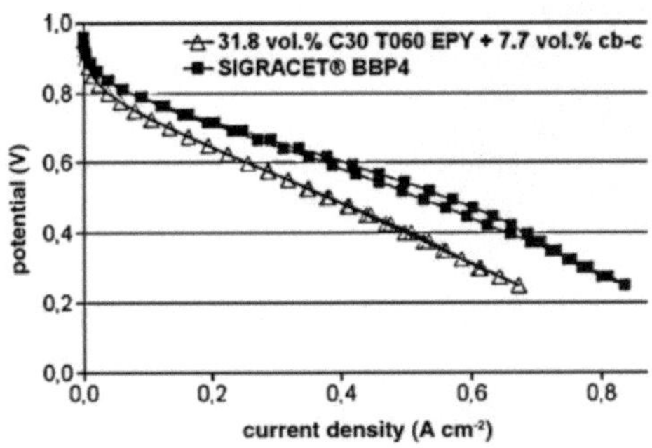

Fig. 14. Current–voltage characteristic curves of a bipolar plate of composite bipolar plate.

Wolf and Willert-Porada[42] blended Ticona (Liquid crystal polymer (LPC)) with 31.8 vol% of carbon fibre and 7.7 vol% of carbon black. The current-voltage characteristic curve (Figure 14) of the composite was compared with a commercial composite material (SIGRACET® BBP4). The maximum current density of the assembly containing the composite bipolar plates was 20% below the maximum current density of the assembly containing the commercial product, which consists of more than twice the amount of carbon. Ternary composites were also used as electrically conductive materials to prepare bipolar plates. Huneault formulated a new material using polypropylene (PP) and polyphenylene sulfide (PPS) as matrices and carbon black, graphite, and carbon fibres as conductive additives [43]. The University of Waterloo has investigated the synergistic effect of two and three filler systems on bipolar plate properties [11, 23, and 28]. Conductive networks in composites can be associated by the combination of carbon fillers as shown in Figure 15.
These formulations have properties suitable for bipolar plate manufacturing. The University of Duisburg-Essen together with the Zentrum für BrennstoffzellenTechnik GmbH (ZBT)

produced carbon fillers–thermoplastic composite bipolar plates using injection moulding process. The bipolar plates had density of 1.6 gcm^{-3}, specific bulk conductivities 150 Scm^{-1} and material prices between 2 and 10 Euro kg^{-1}. Even though, the investigations showed that the electrical conductivity and ability of injection moulding process can be improved significantly, three carbon fillers always resulted in an increased viscosity [44].

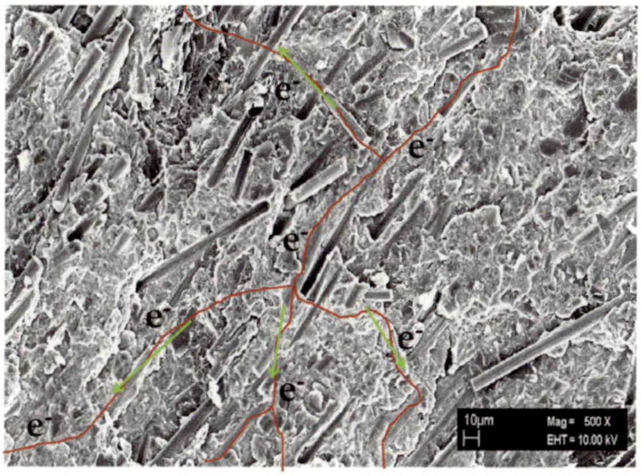

Fig. 15. SEM micrograph of composite with carbon fibre, synthetic graphite, and carbon black using magnification of 500X[11]

In composite bipolar plates, the filler should extend continuously along the cross sectional area and through the thickness of the plate. In other words, the composite needs appropriate conductive networks to have enough electron transport paths. However, several obstacles exist, such as adhesion failure, agglomeration, and poor distribution of fillers in the matrix. Furthermore, filler distribution may increase contact resistance if the conductive filler does not fully disperse to reach the surface. The introducing of coupling agents is one technique for formulating well dispersed polymer/filler composites. A coupling agent can act as a molecular bridge at the interface between the two substrates; hence, it improves the interfacial adhesion between fillers and the matrix [45]. It also plays the role of a dispersion agent; thus, it increases the probability of conductive network formation throughout the matrix. Improvement in the electrical conductive network will diminish the percolation threshold. For example, the electrical percolation of polyethylene/carbon black composite decreases when CB is treated by a titanate coupling agent [46]. Polymer composites prepared as a mixture of a conjugated conducting phase and classical nonconducting phase can exhibit good electrical conductivity with relatively low conducting phase content. They normally show a significant jump in the electrical conductivity at a certain critical concentration of the conducting polymer. The possibility of using conductive polymers to produce bipolar plates is an interesting notion for fuel cell development. For example, an exploration was made of the electrical properties of polypropylene/conductive filler (i.e., graphite, carbon black, and polyaniline) composites as potential replacements for the traditional graphite bipolar plates [47]. Two methods were mainly adopted in this work for the preparation of composites: melt compounding and solution blending. The results indicated that the solution blending led to further improvement in electrical conductivity compared with the melt compounding. Yeetsorn and collaborators [48] found that using polypyrrole (PPy) as an additive was also successful in further improving electrical properties of the composites through better dispersion, as well as contributing to the conductive network. The results indicated that PPy improved PEMFC performance by improving the electrical conducting network within the composites (Figure 16). PPy acts as a conducting additive dispersing on the interface between fillers and the polymer matrix, and it can also coat some parts of filler surfaces. These functions of PPy improve the number of conductive paths [11].

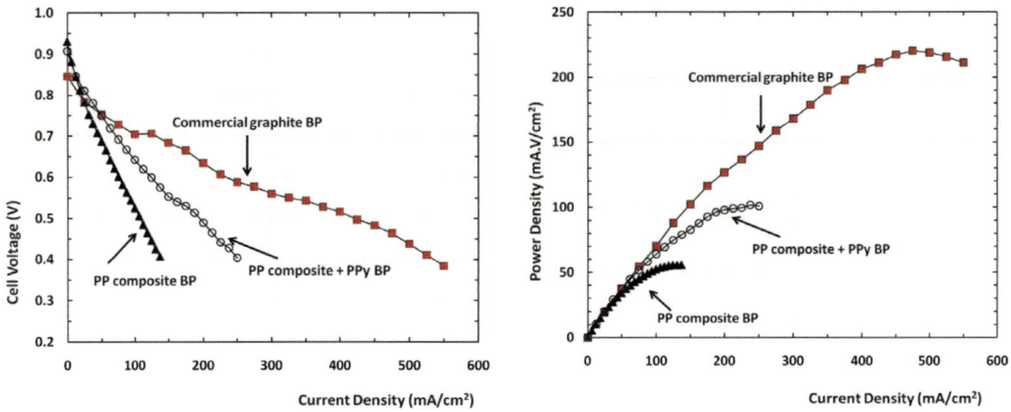

Fig. 16. Initial performance of the single cells assembled with different bipolar plates[11]

5. Thermoplastic composite bipolar plate manufacture [15, 49, and 50]

From a processing point of view, the polymer composites offer ease of manufacturing. For instance, flow fields can be moulded directly into these composites, thereby eliminating the costly and difficult machining steps required for graphite or metal hardware. The composites are used to form bipolar plates with flow channels by conventional compression or injection moulding or compression-injection moulding.

Compression moulding and injection moulding are two main processes for current manufacturing of bipolar plates. ***The compression moulding*** is favoured for both thermoplastic and thermosetting matrix composites. However, if compression moulding is used, the thermoset has to be cured, and the thermoplastic material has to be cooled. Both of these processes greatly increase production time. If compression moulding is used, cost-effective mass production would be more readily achievable with thermosets rather than thermoplastics because of the shorter cycle times for thermosets. With a suitable combination and operating condition, a compression moulded thermoset composite can be cured comfortably in less than 10 minutes, resulting in cycle times less than those required for thermoplastics. ***Injection moulding*** machines make manufacturing automation and mass-production easier, as well as ensuring short processing times, of as little as 30 seconds per plate, and with high dimensional tolerance. However, in order to accomplish practical electrical conductivities, the thermoplastic has to be highly loaded with conductive fillers, which causes the melt to flow poorly. Higher viscosity of materials has become the main concern for composite bipolar plate manufacturing.

Currently, the injection-moulding industry is well ascertained and responding positively to the bipolar-plate challenge. SGL Technologies GmbH has been developing different moulding technologies for graphite composite bipolar plates since 1997 [51, 52]. Injection moulding of polypropylene and phenolic-bonded graphite compounds has been identified as one of the most promising and cost-effective production processes. A range of low-cost carbon–polymer compounds with specific bulk conductivities between 5 and 150 Scm^{-1} was developed in 2005, and a production line for bipolar plates by injection moulding has been set up.

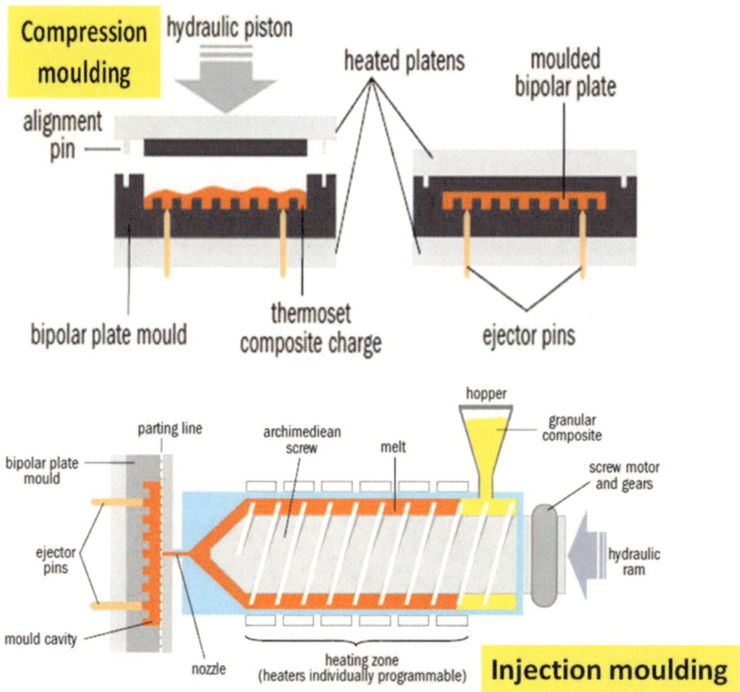

Fig. 17. Compression and injection moulding processes for bipolar plates [3]

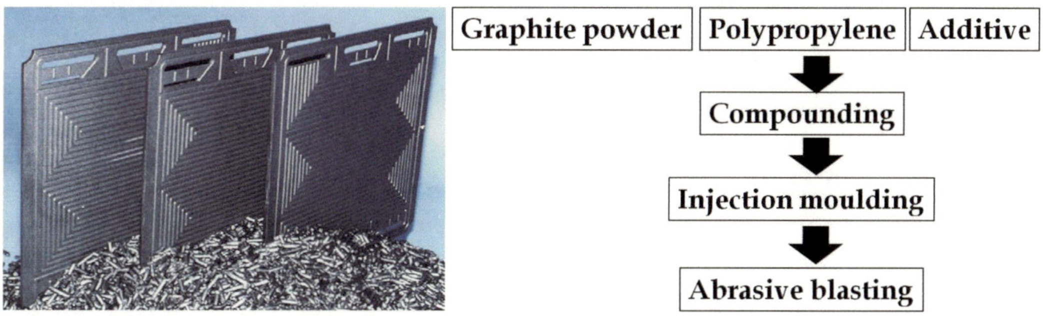

Fig. 18. ®Sigracet PPG86 bipolar plates and their production [52]

Quantum Composites (Ohio, US) and Ferromatik Milacron (Germany) were the first to demonstrate injection moulding for a thermoset-based carbon-polymer composite based on vinyl ester. Century Engineering (Michigan, US) has developed RingExtrude (12-screw) which is a technology that modernizes the method by which bipolar plates are produced [3]. An Injection moulding is a low cost high-productivity manufacturing method; however, other processes, such as extrusion with appropriated die, rolling and thermoforming may be alternative progressions to producing composite bipolar plates.

Although the injection moulding process favours the massive production rates that are essential to reduce production costs, the electrical conductivity of injected composites is negatively influenced because of the high shearing typical of the process [6].The high shearing promotes the rupture of conductive network structure in the composites. In recent years, some new techniques for prototyping bipolar plates for PEMFC have been reported to avoid the handicaps of injection moulding but faster than conventional compression moulding. Huang and his collaborators used the wet-lay method to make bipolar plates. This method produced wet-lay sheets or mats consisting of graphite particles, glass or carbon fibres, and a base made of poly(phenylene sulfide) [53, 54]. The process was comprised of a slurry-making process and then compression moulding to form conductive composite plaques. The use of the wet-lay composite in the formation of bipolar plates has multiple advantages. First, it increases in-plane conductivities of the composite plate significantly, 200 - 300 S cm^{-1}, while maintaining through-plane conductivity. Second, it dramatically increases the mechanical properties of the composites. Third, one can use different components, including polymers, graphite particles, and reinforcement, for the core and outer layers of the plate, respectively, to optimize the properties and/or reduce the cost of the plate. Kuo and Chen created bipolar plates using a composite material composed of Nylon-6 and S316L stainless steel alloy fibres by the injection moulding process [55].

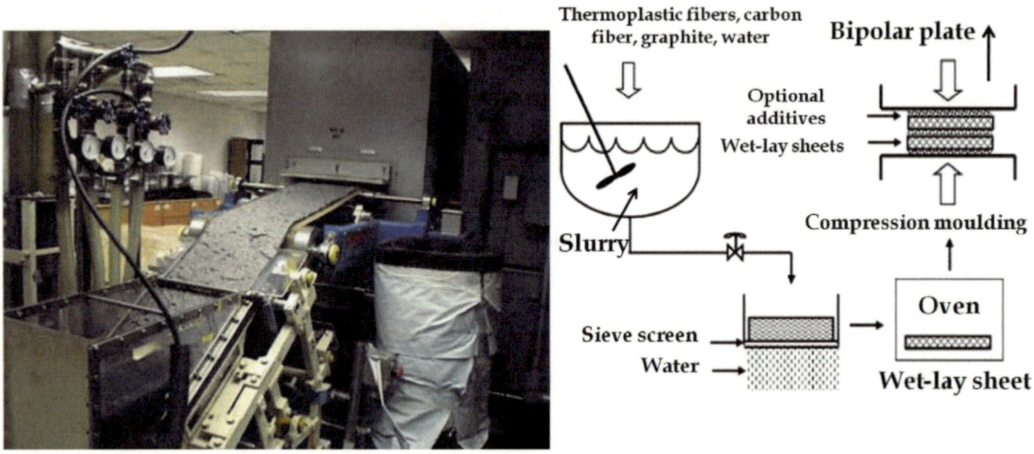

Fig. 19. Manufacture of wet-lay composite sheets and bipolar plates [54]

Although this technique provided a possible alternative to graphite for bipolar plates, the performance of current PEMFC failed to match that of conventional PEMFC with graphite bipolar plates, and metal fibres are likely to impact the durability of moulds and moulding equipment. Inserting metal sheets into bipolar plates is another idea aimed at increasing the electrical conductivity of plates. Hsiao and his team prepared metal mesh hybrid polymer composite bipolar plates by inserting a copper or aluminium mesh in polymer composites [56]. Poly(oxyalkylene)-diamines composites consist of 70 wt% graphite powder and 0-2 wt% modified multi-walled carbon nanotubes. Results indicated that the in-plane electrical conductivity of m-MWCNTs/polymer composite bipolar plates increased from 156 Scm^{-1} (with 0 wt% MWCNT) to 643 Scm^{-1} (with 1 wt% MWCNT). However, the aluminium mesh hybrid polymer composite bipolar plate exhibits an obvious decrease in through-plane conductivity (22.9 Scm^{-1}) due to the passive film layer on the aluminium surface. Prototype plates were successfully fabricated using a two-shot injection over-moulding procedure over aluminium plates [36].

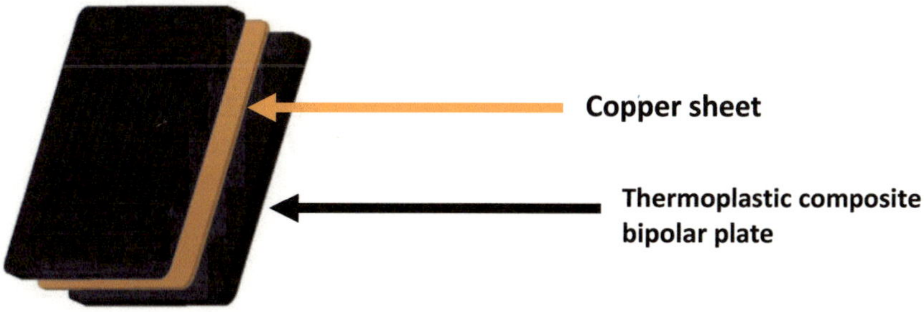

Fig. 20. Copper insert thermoplastic composite bipolar plate

The over-moulding approach is promising, but in order to reduce the contact resistance, improved polymer-metal adhesion and better matching of the aluminium and polymer blend thermal expansion coefficient is required. Direct injection without the metallic core leads to better resistivity results. In the conventional graphite plates, the ribs and channels in the flow fields are formed while removing part of the material to create channels. Thermoplastic composite bipolar plates were fabricated using a two-component moulding process developed by NedStack[57]. The NedStack bipolar plate comprises of two parts which were a highly conductive compound used for the gas-flow-channel area and non-conductive injection moulding grade polymer used for the border area. The two-component moulding process had moulding cycle less than 10 seconds which is 100 times faster than other plate moulding processes. Therefore this short cycle process is considered a break trough in productivity and cost reduction.

6. Effect of a process condition on electrically conductive network

The electrical conductivity or other properties of insulating polymer matrices can be increased with relatively high filler compositions of a conductive fill. However, the non-uniformity of material properties is a major problem in a highly filled polymer system due to poor dispersion of fillers in the polymer matrix. Besides filler concentration, the electrical conductivity also depends upon filler dispersion, interaction between composite constituents, and process conditions. Manufacturing processes and processing conditions of composites influence the filler orientation, filler dispersion, and inter-particle distance within the polymer matrix. Furthermore, the processes would change the degree of crystallinity of the matrix and the aspect ratio of conductive fillers [58]. Different systems give different tendencies of changes in electrical conductivity because in reality, the fibre orientation and filler distribution could be due to the complex interactions among many factors during injection moulding [59. The filler orientation and dispersion typically affect conductive network association in the blends. The injection parameters, such as injection speeds and melt temperature, strongly affect filler orientation. Chandra and collaborators [60] observed the effects of melt temperature and screw speed on electrical conductivity of polycarbonate/carbon nanotube composites. They stated that in order to achieve uniformly distributed high electrical conductivity, the composites should be processed at high melt temperatures and low injection speeds to certify appropriate and uniform electrical conductivities.

Additionally, higher electrical conductivities were found at locations farthest from the gate (Figure 21). Yang and Shi [61] investigated the effect of moulding time on the in-plane electrical conductivity of compression moulded PPS-mesocarbon microbeads composite bipolar plates. The in-plane conductivity increased with moulding time from 10 min to 30min. The optimum moulding time and moulding temperature may form a three-dimensional network that favours the transport of electrons between the conductive fillers.

To accomplish a suitable conductive network structure the processing conditions should be carefully characterized through specific experimental research, regarding both polymer and filler types. Using the simulation of flow injection to describe the conductive network pattern may be useful, as the combination of fillers in a hybrid filled system and various process parameters used in the injection moulding process significantly affect the formation of conductive networks [11, 62]. Yeetsorn and team found the electrical conductivity values of injected bipolar plate cut from different areas of an injected plaque were different, since the electrical conductivity depends largely on the carbon fibre alignment, filler dispersion, and filler distribution.

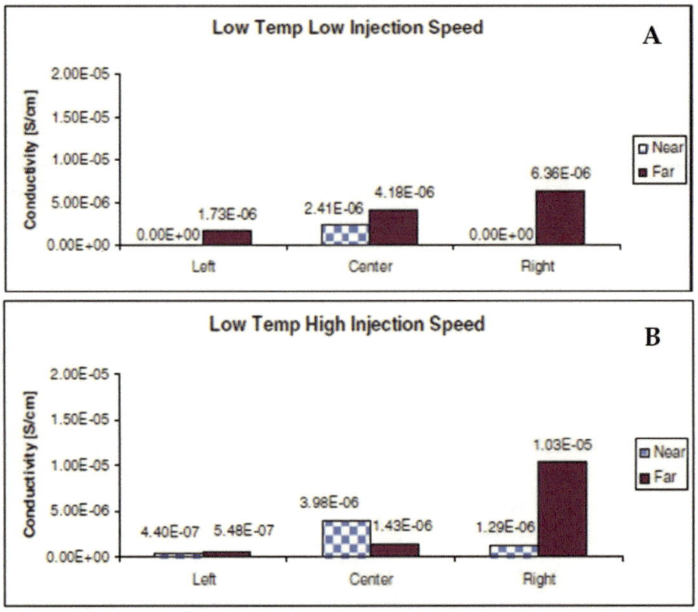

Fig. 21. Electrical conductivities of PC/CNT nanocomposites injection moulded at low melt temperatures (A) low injection speed, (B) high injection speed[61]

Regarding to electrical conductivity values, in-plane and through-plane electrical conductivity of the composite containing only carbon fibres as filler had a positive correlation with the simulation output (Figure 22); however, the final distribution and dispersion of the filler were caused by complex interactions among many factors during injection moulding. Therefore, the simulation output could not explain the orientation of carbon fibres in the composites containing hybrid fillers.

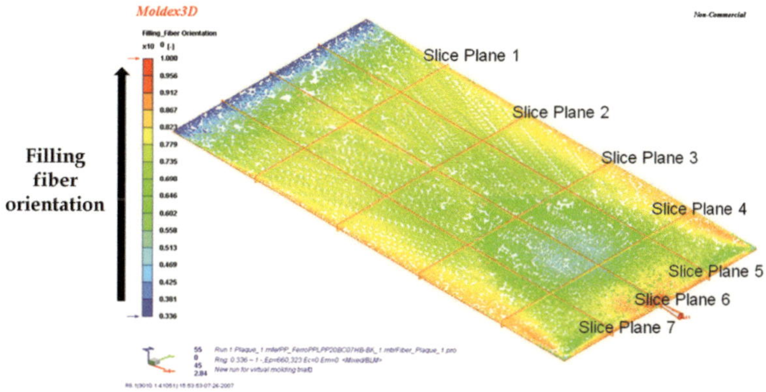

Fig. 22. Simulation output fibre orientation-skin[11]

7. Conclusion

Polymer based formulations using thermoset matrix with carbon conductive carbon filler were successful in a PEMFC market, because their electrically conductive performance achieves the DOE target. However thermoset is not suitable for the injection moulding industry which is well established and responding positively to the bipolar plate challenge. Injection moulding has been typically used for thermoplastic products, therefore; thermoplastic composites are promising for future opportunities composite materials within the PEMFC commercialization. An electrical conductivity performance of thermoplastic composite bipolar plates can be satisfactorily modified through the acceptable materials and manufacturing processes. The fundamental concept of conductive composite improvement and the information of material properties are the basic tool to start a successful bipolar plate development for commercialization.

Although, the thermoplastic composites with conductive carbon fillers and conducting additives were successful in further improving electrical properties of the composites, it is important to understand how different carbon fillers behave in thermoplastic matrices due to the percolation theory. The knowledge of the filler and polymer characteristics will be a guideline for the selection of inherent high electrically conductive fillers and less resistive polymeric matrices. Thermoplastic composites with conductive carbon fillers and conducting additives were successful in further improving electrical properties of the composites. These successes are promising for future opportunities composite materials within PEMFCs, even though the developed thermoplastic composite are currently inferior to the performance of commercial bipolar plates. Therefore, there are many factors of thermoplastic composite bipolar plate development that must be concerned and improved as discussed in previous sections to achieve the required properties of bipolar plates. The scientific literatures in this chapter contribute to the development of materials processes and main characterizations of bipolar plates, especially electrical conductivity and fuel cell performance. Ultimately, all information from this article provides a data base for conductive composites and bipolar plate development in academic and commercial fields.

Future directions in the development of thermoplastic composite bipolar plates may be guided by the following recommendations:

- Investigation of multi filler electrical and rheological percolation thresholds of the composites in multi filler system
- The incorporation of small amount conductive polymer into thermoplastic composites
- Using the simulation of bipolar plate production processes to investigate the effects of operating conditions on conductive network association
- Design an appropriate mould (for injection or compression moulding process) must be provided for conductive network association.

8. References

[1] Weaver, G. (2002). *World Fuel Cells - An Industry Profile with Market Prospects to 2010*, 1st ed., Elsevier, New York, pp. 234.

[2] Barbir, F. (2005). *PEM Fuel Cell*. 2005, Elsevier Inc, UK:

[3] Brett, D. and Brandon, N., Bipolar Plates: The Lungs of The PEM Fuel Cell. *The Fuel Cell Review*, Vol. 2, No. 1, 2005, pp. 15-23.

[4] Vishnyakov, V. M. Proton exchange membrane fuel cells. *Vacuum*, Vol. 80, (2006), pp. 1053-1065.

[5] Larmininie, J., and Dicks, *A. (2003). Fuel cell systems explained*. Wiley and Sons Ltd., West Sussex, England, John.

[6] Hermanna, A., Chaudhuria, T., and Spagnol, P., Bipolar Plates for PEM Fuel Cells: A review. *International Journal of Hydrogen Energy*, 2005, Vol. 30, pp. 1297 – 1302.

[7] Steele, B. C. H., and Heinzel, A. Materials for fuel-cell technologies. *Nature*, 2001, Vol. 414, pp. 345-352.

[8] Kusie, J., Knight, J., and Morton, J. (2010). News Release. *Ministry of Transportation and Infrastructure*, BC, Canada.

[9] O'Hayre, R., Cha, S.W.,Colella, W.,Prinz, F.B. (2009). *Fuel Cell Fundamentals*, ed. 2nd., Jonhn Wiley and Sons Inc., New York.

[10] Antunes, R.A., Oliveira, M.C.L., Ett, G., Ett, V. Carbon materials in composite bipolar plates for polymer electrolyte membrane fuel cells: A review of the main challenges to improve electrical performance. *Journal of Power Sources*, 2011, Vol. 196, No.6, pp. 2945-2961.

[11] Yeetsorn, R. (2010). *Development of Electrically Conductive Thermoplastic Composites for Bipolar Plate Application in Polymer Electrolyte Membrane Fuel Cell*, Department of Chemical Engineering, University of Waterloo: Waterloo, Canada.

[12] Besmann, T.M., Klett, J. W., Henry, J. J., Lara, C.E., Carbon/Carbon Composite Bipolar Plate for PEM Fuel Cells. *Journal of The Electrochemical Society*, 2000, Vol 147, No.11, pp. 4083-4086.

[13] Jayakumar, K., Pandiyan, S., Rajalakshmi,N., Dhathathreyan, K.S., Cost-benefit Analysis of Commercial Bipolar Plates for PEMFC's. *Journal of Power Sources*, 2006, Vol. 161, pp. 454–459.

[14] On, I.B., R. Kirchain, and Roth, R., Technical Cost Analysis for PEM Fuel Cells. *Journal of Power Sources*, 2002, Vol 109, pp. 71-75.

[15] Mehta, V. and Cooper, J.S., Review and Analysis of PEM Fuel Cell Design and Manufacturing. *Journal of Power Sources*, 2003., Vol. 114, pp. 32-53.

[16] Dihrab, S.S., K. Sopian , Alghoul, M.A., Sulaiman, M.Y., Review of The Membrane and Bipolar Plates Materials for Conventional and Unitized Regenerative Fuel Cells. *Renewable and Sustainable Energy Reviews*, 2009, Vol 13, pp. 1663–1668.

[17] Tawfika, H., Hunga, Y., and Mahajan, D., Metal Bipolar Plates for PEM fuel cell - A review. *Journal of Power Sources*, 2007, Vol 163, pp. 755-767.

[18] Antunes, R.A., Oliveira, M.C.L., Ett,G., and Ett,V., Corrosion of Metal Bipolar Plates for PEM Fuel Cells: A Review. *International Journal of Hydrogen Energy*, 2010, Vol. 35: pp. 3632-3647.

[19] Hung, Y., El-Khatib, K.M., and Tawfik, H., Corrosion-resistant Lightweight Metallic Bipolar Plates for PEM Fuel Fells. *Journal of Applied Electrochemistry*, 2005 vol. 35, pp. 445-447.

[20] Xiao, M., Lub,Y. Wang ,S.J., Zhaoa,Y.F., Menga,Y.Z., Poly(arylene disulfide)/graphite Nanosheets Composites as Bipolar Plates for Polymer Electrolyte Membrane Fuel Cells. *Journal of Power Sources*, 2006, Vol. 160, No. 1, pp. 165-174.

[21] Ku, C.C. and Liepins, R. (1987). Electrical Properties of Polymers: Chemical Principles., Hanser Publishing, New York

[22] Stauffer, D. (1985). *Introduction of Percolation Theory.*, Tylor and Francis, London.

[23] Mali, T.J. (2006). *Thermoplastic Composites for Polymer Electrolyte Membrane Fuel Cell Bipolar Plates*, Department of Chemical Engineering, University of Waterloo: Waterloo, Canada.

[24] Cunningham, N., Lefèvre, M., Lebrun, G. and Dodelet, J., Measuring The Through-plane Electrical Resistivity of Bipolar Plates (Apparatus and Methods). *Journal of Power Sources*, 2005, Vol. 143, pp. 93-102.

[25] Mironov, V.S., Kim J.K., and M. Lefevre, Lebrun,G., Dodelet, J.P., Comparison of Electrical Conductivity Data Obtained by Four-Electrode and Four-Point Probe Methods for Graphite-Based Polymer Composites. *Polymer Testing*, 2007, Vol. 26, pp. 547–555.

[26] Cokturk, H.S., Fiske, T.J., and Kalyon, D.M., Effects of Particle Shape and Size Distributions on the Electrical and Magnetic Properties of Nickel/ Polyethylene Composites. *Journal of Applied Polymer Science*, 1993, Vol. 50, pp. 1891-1901.

[27] Bin, Z. and Bingchu, M., Study on The Electrical and Mechanical Properties of Polyvinylidene Fluroide/titanium Silicon Carbide Composite Bipolar Plates. *Journal of Power Sources*, 2006, Vol. 161, pp. 997–1001.

[28] Wang, Y. (2006). *Conductive Thermoplastic Composite Blends for Flow Field Plates for Use in Polymer Electrolyte Membrane Fuel Cells (PEMFC)*, Department of Chemical Engineering, University of Waterloo: Waterloo, Canada.

[29] Heo, S.I., K.S. Oh, Yun,J.C., Jung, S.H., Yang,Y.C., and Han, K.S., Development of Preform Moulding Technique using Expanded Graphite for Proton Exchange Membrane Fuel Cell Bipolar Plates. *Journal of Power Sources*, 2007, Vol. 171, pp. 396-403.

[30] Kakati, B.K., Deka,D., Differences in physico-mechanical behaviors of resole and novolac type phenolic resin based composite bipolar plate for proton exchange membrane (PEM) fuel cell Electrochim. Acta, 2007, Vol. 52, pp. 7330–7336.

[31] Mathur,R.B., Dhakate,S.R., Gupta,D.K., Dhami,T.L., Aggarwal, R.K., Effect of different carbon fillers on the properties of graphite composite bipolar plate. Journal of Materials Processing Technology, 2008, Vol. 203, pp.184–192.

[32] Radhakrishnan, S., Ramanujam, B.T.S., Adhikari, A., Sivaram,S., High-temperature, Polymer–graphite Hybrid Composites for Bipolar Plates: Effect of Processing Conditions on Electrical Properties. *Journal of Power Sources*, 2006, Vol. 163, pp. 702–707.

[33] Xia, L. and Li, A., Effects of Resin Content and Preparing Conditions on The Properties of Polyphenylene Sulfide Resin/graphite Composite for Bipolar Plate. *Journal of Power Sources*, 2008. Vol.178 pp. 363–367.

[34] Wu, M. and Shaw, L.L., A novel Concept of Carbon-filled Polymer Blends for Applications in PEM Fuel Cell Bipolar Plates. *International Journal of Hydrogen Energy*, 2005, Vol. 30, pp. 373-380.

[35] Rio, C.D., Rio,C. D., Ojeda, M. C., Acosta, J. L., Escudero, M. J., Hontan, E., and Daza, L., New Polymer Bipolar Plates for Polymer Electrolyte Membrane Fuel Cells: Synthesis and Characterization. *Journal of Applied Polymer Science*, 2002, Vol. 83,No.13, pp. 2817-2822.

[36] Mighri, F., Huneault, M.A., and Champagne, M.F., Electrically Conductive Thermoplastic Blends for Injection and Compression Molding of Bipolar Plates in The Fuel Cell Application. *Polymer Engineering and Science*, 2003, Vol. 44, No.9, pp. 1755-1765.

[37] Gubbels, F., Jerome, R., Teyssie, P., Vanlathem, E., Deltour, R., Calderone, A., Parente, V., Bredas, J., Selective localization of carbon black in immiscible polymer blends: a useful tool to design electrical conductive composites. *Journal of Macromolecules*, 1994, Vol. 27, pp.1972–1974.

[38] Blunk, R.H.J., Daniel, J. L., Yoo,Y.E., and Tucker III, C.L., Enhanced Conductivity of Fuel Cell Plates Through Controlled Fiber Orientation. *AIChE Journal*, 2003, Vol. 49, No.1, pp. 18-27.

[39] Clingerman, M.L, Weber, E.H., King, J., and Schulz, K.H., Synergistic Effects of Carbon Fillers in Electrically Conductive Nylon 6,6 and Polycarbonate Based Resins. *Polymer Composites*, 2002, Vol. 23, No. 5 pp.911-924.

[40] Liao,S.H., Yen,C.Y., Weng,C.C., Lin,Y.-F., Ma,C.C.M., Yang,C.H., Tsai, M.-C., Yen,M.Y., Hsiao,M.C., Lee,S.H., Xie,X.F., Hsiao,Y.H., Preparation and properties of carbon nanotube/polypropylene nanocomposite bipolar plates for polymer electrolyte membrane fuel cells.Journal of Power Sources, 2008, Vol. 185, pp. 1225–1232.

[41] Calleja, F.J.B., Bayer, R.K., and Ezquerra, T.A., Electrical Conductivity of Polyethylene-carbon-fibre Composites Mixed with Carbon Black. *Journal of Materials Science*, 1988, Vol. 23, pp. 1411-1415.

[42] Wolf , H., Willert-Porada M., Electrically conductive LCP–carbon composite with lowcarbon content for bipolar plate application in polymer electrolyte membrane fuel cell. *Journal of Power Sources*, 2006, Vol. 153, pp. 41–46.

[43] Huneault, M.A., Mighri, F., and Champagne, M.F., Conductive Polymer Blends for Injection Molded Bipolar Plates. *Proceedings of ANTEC2003*, 2003, pp. 1330-1334.

[44] King, J.A., Tambling, T. M., Keith, J. M., Cole, A.J., and Morrison, F. A., Synergistic Effects of Multiple Carbon Fillers on The Rheology of Liquid Crystal Polymer Based Resins. *Polymer Composites*, 2009, Vol. 30, pp. 111-119.
[45] Brovko, Rosso, O.P., and Friedrich, K., Adhesion Between Differently Treated Fibers and A Hybrid Resin System. *Journal of Materials Science Letters*, 2002, Vol. 21, pp. 305-308.
[46] Yu, G., Zhang, M.Q., and Zeng, H.M., Carbon-black-filled Polyolefine as A Positive Temperature Coefficient Material: Effect of Composition, Processing, and Filler Treatment. Journal of Applied Polymer Science, 1998, Vol. 70, pp. 559-566.
[47] Dweiri, R. and Sahari, J., Investigations on The Electrical Properties of Carbon-based Polypropylene Composites for Bipolar Plates in Polymer Electrolyte Membrane Fuel Cell (PEMFC), *Journal of Power Sources*. 2007, Vol. 171, pp. 424-432.
[48] Yeetsorn, R., Fowler, M.,Tzoganakis, C., Influence of Polypyrrole on Electrical Conductivity of Injection-Thermoplastic-Composite Bipolar, *Proceedings of ANTEC2009*, Chicago, USA, 2009.
[49] Cunningham, N., Lefèvre, M., Lebrun, G. and Dodelet, J., Measuring The Through-plane Electrical Resistivity of Bipolar Plates (Apparatus and Methods). *Journal of Power Sources*, 2005, Vol. 143, pp. 93-102.
[50] Li, X. and Sabir, I., Review of Bipolar Plates in PEM Fuel Cells: Flow-field Designs. *International Journal of Hydrogen Energy*, 2005, Vol.30, pp. 359-371.
[51] Baurens, P. and Crouvezier, J.P.P., Innovative Concepts for Bipolar Plates. *CLEFS CEA*, 2005. winter(50/51), pp. 2004-2005.
[52] Muller, A., Kauranen, P., Ganski, A., and Hell, B., Injection Moulding of Graphite Composite Bipolar Plates. *Journal of Power Sources*, 2006, Vol. 154, pp. 467-471.
[53] Huang, J. and Baird, D.G., Compression Molding of Highly Conductive Fuel Cell Bipolar Plates from A Thermoplastic Composite. *Proceedings of ANTEC2003*, 2003, pp. 2151-2155.
[54] Huang, J., Baird, D.G., and McGrath, J.E., Development of Fuel Cell Bipolar Plates from Graphite Filled wet-lay Thermoplastic Composite Materials. *Journal of Power Sources*, 2005, Vol. 150, pp. 110–119.
[55] Kuo, J.K. and Chen, C.K., A novel Nylon-6–S316L Fiber Compound Material for Injection Molded PEM Fuel Cell Bipolar Plates. *Journal of Power Sources*, 2006, Vol. 162, pp. 207–214.
[56] Hsiao, M.C., Liao, S.H., Yen, M.Y., Ma, C-C.M., Lee,S-J., Chen,Y-H., Hung, C.H., Lin, Y.-F., and Xie,X-F., Electrical and Thermal Conductivities of Novel Metal Mesh Hybrid Polymer Composite Bipolar Plates for Proton Exchange Membrane Fuel Cells, Journal of Power Sources, 2010, Vol. 195, No. 2, pp. 509-515.
[57] Middelman,E., Kout,W., Vogelaar,B., Lenssen,J., Wa, E. Bipolar plates for PEM fuel cells. Journal of Power Sources, 2003.118: p. 44–46.
[58] Lee, M.-S., Chen, L.-J., He,Z.-R., Yang,S.-H., The Development of A Heterogeneous Composite Bipolar Plate of A Proton Exchange Membrane Fuel Cell. *Journal of Fuel Cell Science and Technology*, 2005, Vol.2, pp. 14-19.

[59] Lutz, W., Herrmann,J., Kockelmann, M., Hosseini, H.S., Jäckel, A., Schmauder,S., Predak,S., and Busse, G., Damage Development in Short-fiber Reinforced Injection Molded Composites. *Computational Materials Science*, 2009, Vol. 45, pp. 698–708.

[60] Chandra, A., Kramschuster, A.J., Hu,X., Turng, L.S., Effect of Injection Molding Parameters on The Electrical Conductivity of Polycarbonate/carbon nanotube Nanocomposites. *Proceedings of ANTEC2007*, 2007.

[61] Yang,T., Shi,P. Study on the mesocarbon microbeads/polyphenylene sulfide composite bipolar plates applied for proton exchange membrane fuel cells. *Journal of Power Sources*, 2008, Vol. 175, pp 390-396.

[62] Keith,J. M., King,J. A., Barton R. L. Electrical Conductivity Modeling of Carbon-Filled Liquid-Crystalline Polymer Composites. *Journal of Applied Polymer Science*, 2006, Vol. 102, pp 3293–3300.

17

High Voltage Electric Discharge Consolidation of Tungsten Carbide - Cobalt Powder

Evgeny Grigoryev
Moscow Engineering Physics Institute
Russia

1. Introduction

Today WC–Co composites are extensively used to enhance the wear resistance of various engineering components, e.g. cutting tools and dies. Cemented carbides are used throughout industry for high wear, abrasive applications as a result of their extreme hardness. Apart from their exceptional hardness, WC has other unique properties such as high melting point, high wear resistance, good thermal shock resistance, thermal conductivity and good oxidation resistance (Crowson & Chen, 1991). WC with ductile metals such as cobalt as a binding medium, which is known to be helpful in cementing fine WC particles, is used in bulk sintered forms. Matrices of ductile metals, such as cobalt, greatly improve its toughness, hence elimination the possibility of brittle fracture during operation. WC–Co composites are extensively used to enhance the wear resistance of various engineering components, e.g. cutting tools and dies. In this paper, we report the results of studying the macroscopic phenomena occurring under high voltage electric discharge consolidation of tungsten carbide - cobalt powder (Grigoryev & Rosliakov, 2007).
The methods of consolidation of powder materials, based on various techniques of transmission of electric current pulses through a powder under mechanical pressure, are widely studied in many research laboratories (Grasso1 et al., 2009). The interest in these methods was motivated by their ability to consolidate a large variety of powder materials to high densities within short periods of time, without having to increase grain sizes. They are especially important because make it makes it possible to obtain bulk nanomaterials (Groza & Zavaliangos, 2003; Kodasha et al., 2004). These methods include electric-discharge sintering (Raichenko, 1987), field-assisted sintering technique, plasma assisted sintering, spark plasma sintering (Groza & Zavaliangos, 2003; Munir, 2006; Olevsky, & Froyen, 2009) etc. The large number of these methods is related to the wide range of variation in the electrical parameters of the action on a powder. The increasing importance of these methods as a tool for consolidation of powders is demonstrated by the large number of papers published in the recent years (Figure 1).
The efficiency of electric-pulse methods is determined by the multifactor effect on consolidated materials (Baranov et al., 2001). Specific features of plastic deformation of conducting materials under the electric-pulse effect on the microscopic level were considered in (Bataronov, 1999). To obtain materials with required properties, one has also

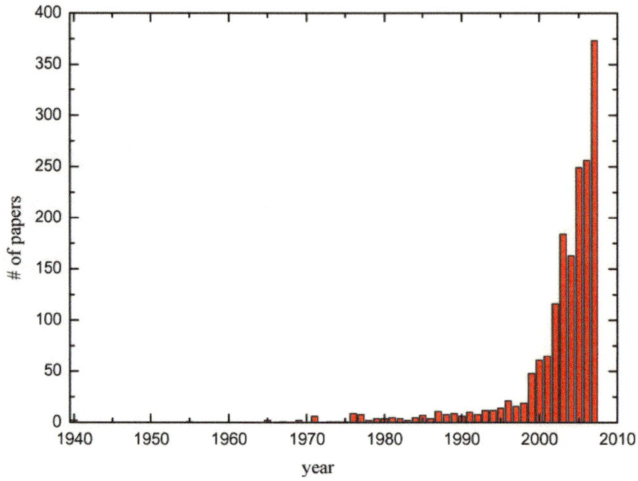

Fig. 1. Number of publications related to Electric Current Activated/assisted Sintering (ECAS) processes (Orrù et al., 2009)

to know the macroscopic processes occurring in the bulk of a consolidated sample. For example, the kinetics of consolidation of powder materials in these methods is significantly different, and their duration changes from several tens of minutes for electric-discharge sintering and spark plasma sintering (Raichenko, 1987; Groza & Zavaliangos, 2003) to several milliseconds for high voltage electric discharge consolidation (Grigoryev & Rosliakov, 2007).

In this paper, we focused on the consolidation of WC–Co powders into a solid bulk without increasing their crystallite sizes by high voltage electric discharge consolidation (HVEDC) (Grigoryev & Rosliakov, 2007; Grigoryev, 2010). The principle of HVEDC is to discharge a high-voltage (up to 30 kV), high-density current (~100 kA/cm^2) pulse (for less than 300 μs) from a capacitor bank through the powders under external pressure, resulting in a temperature rise of more than 2500 K, instantaneously to weld grains of powders together. In this way, a full or near full densification may be achieved with minimal undesirable microstructural changes due to short consolidation time. Furthermore, WC–Co powders could be consolidated into solid bulks by high voltage electric discharge consolidation with densities close to theoretical density. High voltage electric discharge consolidation (HVEDS) (Grigoryev & Rosliakov, 2007) is a developed consolidation method which can produce near-net-shape compacts of high relative density much more rapidly than other conventional process such as pressure-less sintering, hot press and HIP.

Additionally this paper (Grigoryev, 2010) describes the progress that we have made in developing in situ electric discharge consolidation and bonding as an accurate method of joining cemented carbide alloy to steel substrate. Joints between cemented carbide alloy and tool steel are made by the brazing and electron beam welding processes and are applied to various tools. An important need has become evident for joints of high strength, in order to produce tools of lengthened life span and at low cost: thus, the development of innovative joining techniques to replace the brazing and electron beam welding processes is an urgent requirement. Bonding by HVEDS of cemented carbides to steel appears as an attractive complementary technique to conventional joining processes due to its high precision, high process speed, low heat input. This method has demonstrated a potential to provide distinct technological and economic benefits in the consolidation of difficult-to-sinter powders, including short processing times, fewer processing steps, elimination of the need for sintering aids, and near net shape capabilities.

2. Experimental procedure

Dense WC–Co composites were fabricated by an HVEDC method. In this process, the WC–Co mixed powders were poured into an electrically non-conducting ceramic die. The ceramic die was plugged at two ends with molybdenum electrodes-punches and an external pressure up to 300 MPa was applied to the powder on air-operated press. A high voltage capacitor bank was discharged through the powder. The schematic of the high voltage electric discharge consolidation (HVEDC) system is shown in Figure 2.

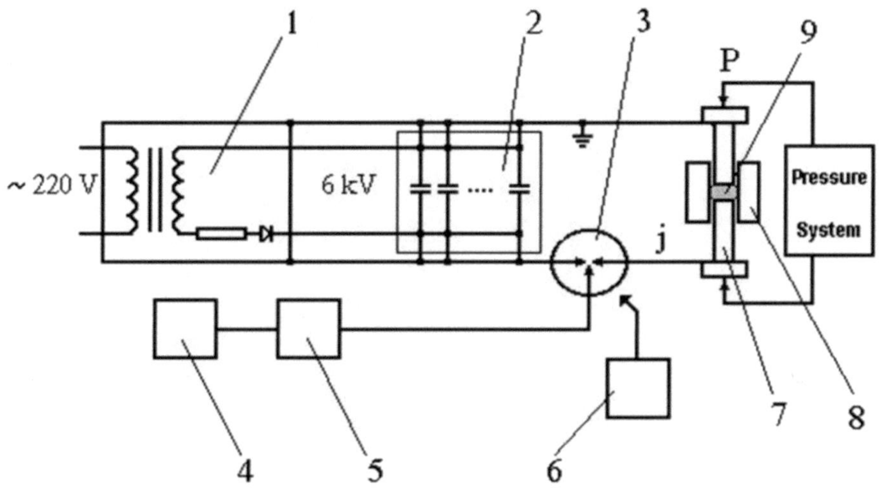

1 - charging unit, 2 - capacitor bank, 3 - trigatron switch, 4 - control system, 5 - electrical discharge ignition system, 6 - pulse electrical discharge registration system, 7 - punch - electrode, 8 - die , 9 - powder.

Fig. 2. Schematic of EDC apparatus

High voltage electric discharge consolidation (HVEDC) apparatus for powder consolidation consists basically of charging unit (1); a bank of capacitors (2) and trigatron switch (3) to connect a powder column (9) suddenly across the charged capacitor bank. The capacitor bank consists of thirty 200 μF capacitors that can store up to 6 kV. HVEDC uses the pulse current generated from the capacitor bank to quickly heat a powder column subjected to constant pressure during the process. Powder column was a circular cross-section rod of diameter ~ 10 mm and length from 10 to 15 mm. In this process the WC–Co powder is poured into an electrically non-conducting ceramic die (8). The ceramic die is plugged at two ends with molybdenum electrodes-punches (7) and an external pressure up to 300 MPa is applied to the powder on air-operated press. Punches - electrodes (7) transfer pressure on the powder from a pneumatic system and simultaneously serve as current leads to apply current pulses from a generator to the pressed powder (9). A high voltage capacitor bank is discharged through the powder.

We used commercial WC–Co powders (grain size WC < 5μm) as starting material for high voltage electric discharge consolidation. Characteristics of the chemical composition of these powders are resulted in Table 1.

Chemical element	WC	Co	free carbon	total oxygen
mass %	~ 80	20	0.101	0.13

Table 1. The chemical composition of WC - Co powders

Table 1 gives the general values before the electric discharge compaction.
The discharge current is measured by a toroidal Rogowsky coil (6) around the powder column. An oscillograph showing a typical output from the Rogowski coil is shown in Figure 3.

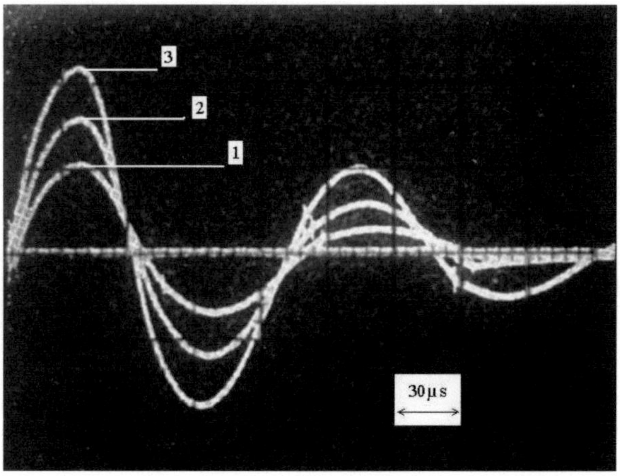

(Peak currents: 1 – 50 kA , 2 – 80 kA , 3 – 110 kA)

Fig. 3. Typical pulse current traces from registration system (Rogowski coil)

The measurements of the current pulse parameters with a Rogowski coil showed that the discharge current pulse time length in all experiments did not exceed 300 μs. This value determines the time of energy injection into the powder. HVEDC applies a high-voltage, high-density current pulse to the powder column under external pressure for a very short period of time. This method uses the passage of the pulse electric current to provide the resistive heating of the powder by the Joule effect. Joule heating occurs at the inter-particle contact to instantaneously weld powder particles, resulting in densification. The achieved WC-Co powder compact density as a result of HVEDC process depends on applied external air-operated pressure, magnitude and waveform of pulse current that depends on RLC – parameters of the electrical discharge circuit. An important feature of high voltage electric discharge consolidation of powders is the high concentration of released energy in the contact zones of powder particles; therefore, the initial state of the surface of powder particles (thickness and structure of oxide films, presence of foreign impurities, etc.), the shape of the powder particles and their sizes, and the external pressure on the powder significantly affect the character of physical processes under HVEDC. The other decisive factors are the injection rate of the electromagnetic field energy into the powder and the character and magnitude of the mechanical pressure applied to the powder during HVEDC.

We determined the density change of consolidated sample versus time by using high-speed movie recording. To study the consolidation kinetics, we used a high-speed "SKS-1M" camera, which makes it possible to record movies with a frequency up to 4×10^3 frames/s. The high-speed consolidation kinetics of the tungsten carbide - cobalt powder was experimentally investigated for different values of the pulse current amplitude and pressure. Consolidation of powder material takes place at constant pressure P, created by the pressure system during all process of the electro discharge compaction. Time dependences of powder density during high voltage electric discharge consolidation process are shown in Figure 4.

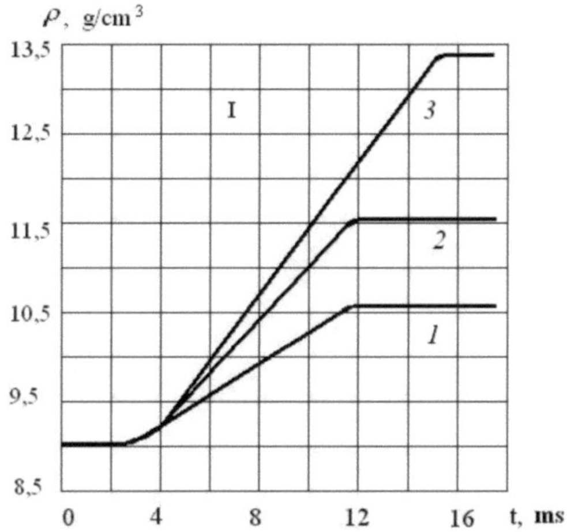

Fig. 4. Variations of powder density as a function of time in HVEDC process at constant external pressure (P = 150 MPa)

Figure 4 shows the dependences of the consolidated WC–Co mean density ρ on the time t, obtained by treating the process cinegrams. In Fig. 4 curves: 1 (75 kA/cm^2), 2 (85 kA/cm^2), 3 (97 kA/cm^2), were determined at constant pressure P = 150 MPa. Duration of process of powder material densification ranged from 6 ms to 16 ms for all our experiments. The results of experiments show that motion of punches in the process of the high voltage electric discharge consolidation takes place with steady speed. The value of speed depends on amplitude of the pulse current and external pressure. The magnitude of external pressure determines initial specific resistance of the powder column and, accordingly, the amount of heat in powder material. With the increase of pressure the specific resistance of powder column goes down sharply, which results in the less heating of powder material. The densification of the consolidated material takes place due to an intensive plastic strain which depends on external pressure P and the yield stress of powder matter $\sigma_T(T)$ (T - temperature). Therefore, speed of plastic flow of the compacted powder material and consequently speed of change of density of the consolidated powder sample is determined by the temperature in HVEDC process. The speed of densification depends on a dimensionless parameter $\beta = \sigma_T(T)/P$. At a constant initial resistance of a WC–Co powder, its temperature increases with an increase in the discharge current pulse amplitude. Hence, the powder consolidation rate increases as well (Fig. 4). At constant amplitude of pulse current a densification speed is determined by temperature dependence of yield strength at different pressures, applied to the powder column. The finishing density (Fig. 4) of a compacted material after the high voltage electric discharge consolidation process is defined by the amount of the external pressure and parameter β. Generally, an increase in the external pressure leads to an increase in the consolidation rate; however, this increase is compensated in the case under consideration by the temperature dependence of the yield strength of the powder material. This circumstance explains the fact that the powder consolidation rate decreases with an increase in the pressure applied to the powder at constant amplitude of the discharge current pulse.

The temperature evolution during high voltage electric discharge consolidation process was measured by means of thermocouple method. We measured the temperature on the powder sample surface, using Chromel–Alumel and tungsten–rhenium thermocouples. A standard temperature versus time curves at electro discharge compaction are shown in Figure 5.

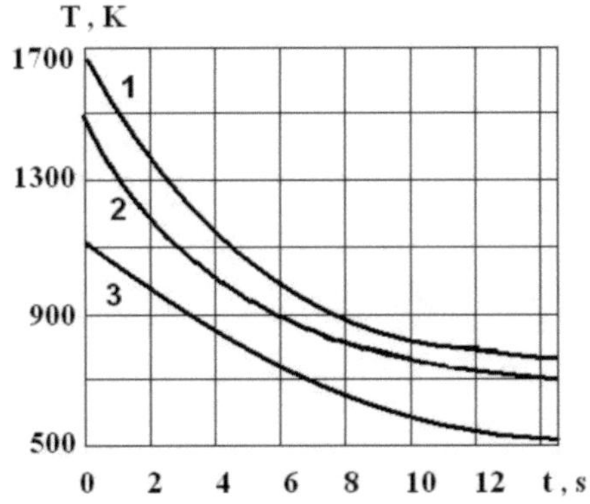

Fig. 5. Standard temperature versus time curves of powder column side surface during HVEDC process (at constant pressure P = 200 MPa)

Temperature variations 1, 2, 3 correspond of powder column side surface during HVEDC process. Temperature dependence 1 was got at amplitude of discharge pulse current equal 95 kA /cm^2, 2 - 90 kA /cm^2, 3 - 85 kA/cm^2. The powder material densification process during of HVEDC takes place at approximately constant temperature. It follows from the measured temperature curves (Figure 5).

Density measurements after HVEDC process were performed using the Archimedes principle in distilled water. X-ray diffraction (XRD) was performed on the as-received powder using a "DRON-3" diffractometer with a Cu target for 2θ from 20° to 120° at a scan speed of 1°/min. XRD was repeated on the consolidated specimens, followed by density measurements.

3. Results and discussion

The high energy density in the particle contact zones causes a change in the aggregate state of the material (from a solid to a liquid and, partially, a dense low-temperature plasma). The physics processes in the contact zones are characterized by spatial inhomogeneity and time dependence. The analysis and finding the main regularities in the behaviour of the material in the contact zones of particles makes it possible to establish the optimal high voltage electric discharge consolidation parameters. Thermal processes under HVEDC of powders were considered previously in (Bazanov et al., 1990). A mathematical model of the physical processes occurring under high voltage electric discharge consolidation in a powder both during compacting as a whole and taking into account the processes in the particle contact zones was proposed in (Grigoryev, 2007). Analysis of the problem parameters made it possible to reveal the hierarchy of the characteristic times of the processes occurring under HVEDC. The approximate general scheme is as follows. A current pulse passing through a powder and punches - electrodes strongly heats only the powder material without significant punches - electrodes heating, because the powder resistivity greatly exceeds that of the electrode material. The intense heating of the powder significantly decreases its resistance to plastic deformation, and, under the action of an external mechanical pressure, it is consolidated with a characteristic rate, dependent on the pneumatic system type. Simultaneously, heat sink from the powder to the punches and matrix occurs due to the thermal conduction. The time of energy injection to the powder is determined by the current pulse width: $\tau_0 < 10^{-3}$ s. The time of formation of a consolidated material from the powder, τ_1, depends on the loading system and lies in the range $2\times10^{-3} < \tau_1 < 2\times10^{-2}$ s. The cooling time of the consolidated material, τ_2, is determined by the thermal conductivity of the materials and the characteristic size of the compacted sample: $\tau_2 \sim 2.5$ s. In this case, the time scales of the processes obey the following relation:

$$\tau_0 < \tau_1 << \tau_2 \tag{1}$$

The system of equations describing the macroscopic processes under high voltage electric discharge consolidation is based on the mass, momentum, and energy conservation laws and the electrodynamics equations for consolidated and powder conductors.

$$\frac{\partial\rho}{\partial t} + div(\rho\vec{v}) = 0 \tag{2}$$

$$\left(\frac{\partial\vec{v}}{\partial t} + (\vec{v},\nabla)\vec{v}\right)_i = \left(\frac{\partial\sigma_{ik}}{\partial x_k}\right) + F_i \tag{3}$$

$$\frac{\partial}{\partial t}\rho\left(\varepsilon + \frac{\vec{v}^2}{2}\right) = -div\left(\rho\vec{v}\left(w + \frac{v^2}{2}\right) - (\vec{v},\hat{\sigma}') - \kappa\nabla T\right) + \frac{\vec{j}^2}{\sigma} \tag{3}$$

$$rot\vec{E} = -\frac{\partial\vec{B}}{\partial t}, \quad rot\vec{H} = \vec{j}, \quad div\vec{B} = 0 \tag{4}$$

$$\vec{F} = \left[\vec{j},\vec{B}\right], \quad \vec{j} = \sigma\left(\vec{E} + \left[\vec{v},\vec{B}\right]\right) \tag{5}$$

where ρ – density, $\vec{v}$ – velocity, $\hat{\sigma}$ – internal stress tensor, ε – internal energy, w – enthalpy, $\hat{\sigma}'$ – viscoplasticity tensor, T – temperature, $\vec{j}$ – electrical current density, $\vec{E}$, $\vec{H}$ – tension of the electrical and magnetic fields, respectively, $\vec{B}$ – magnetic field induction, $\vec{F}$ – Ampere force; k – thermal conductivity; σ – conductivity of the powder material. The system of equations (2) – (6) must be supplemented by the corresponding equations of state for the material. It is assumed that the electrode-punch material obeys Hooke's law. We used viscous-plastic material model analysis (Carroll at al., 1986) for the description of powder compaction process.

$$P = \frac{2}{3}\sigma_T \ln\frac{\alpha}{\alpha-1} - \frac{4}{3}\eta\frac{\dot{\alpha}}{\alpha(\alpha-1)} - \frac{\rho_m a^2}{3(\alpha_0-1)^{2/3}}\frac{d}{d\alpha}\left\{\frac{\dot{\alpha}^2}{2}\left[(\alpha-1)^{-1/3} - \alpha^{-1/3}\right]\right\} \quad (7)$$

where: P - compaction pressure, σ_T - yield stress of powder material (Co), η - viscosity of a powder material (Co), a - the initial size of pores, ρ_0, ρ_m - initial and theoretical density of powder material correspondently (Co), $\alpha = \rho_m / \rho$ ($\dot{\alpha} = d\alpha/dt$, $\alpha_0 = \rho_m / \rho_0$) (Carroll at al., 1986).

The calculation of the macroscopic temperature distribution in the consolidated WC-Co sample is presented in Figure 6.

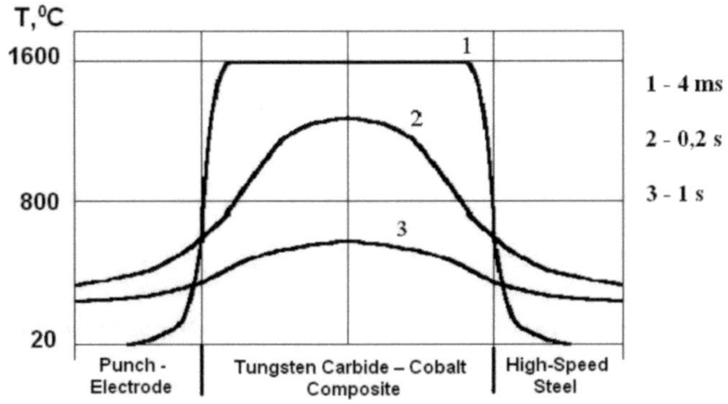

Fig. 6. The temperature distribution in the sample HVEDC process

Thus, consolidation of a powder material under HVEDC process occurs at a nearly constant temperature and constant mechanical pressure, created by the pressure system during all process of the high voltage electric discharge consolidation. Time dependences of mean specimen density during HVEDC process are shown in Fig. 4. These linear relationships reflect the movement of the punch at a constant rate. The constant velocity of the punch is from the fact that the densification process has a wave nature during HVEDC. Constant velocity of the wave compaction U depends on the speed of the punch v by the following relation (8).

$$U = v / (1 - \rho_0 / \rho) \quad (8)$$

Our analysis (Grigoryev, 2007) resulted in determination of the optimal parameters for high voltage electric discharge consolidation of powder materials with preset properties. Dimensionless parameters R, β (9) determine resulting density of tungsten carbide - cobalt composite layer after HVEDC

$$R = \frac{a}{\eta}\sqrt{\rho_m P}\,, \quad \beta = \frac{\sigma_T}{P}\,, \quad \Pi = (1 - \rho_0/\rho_m) * 100\% \quad (9)$$

where Π - initial porosity of a powder material. The diagram of dimensionless parameters R, β of optimum compaction modes is presented in Figure 7.

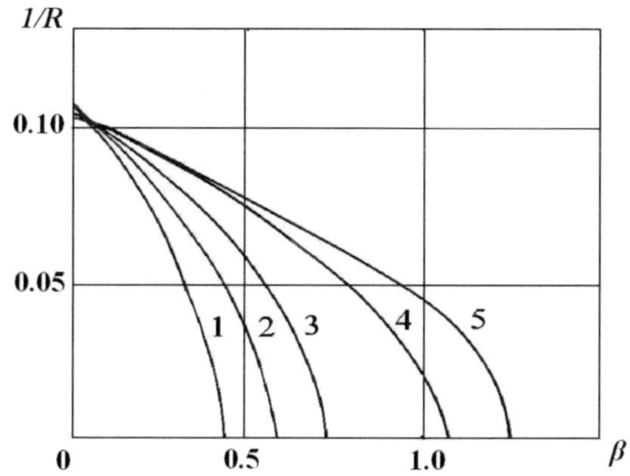

Fig. 7. The diagram of dimensionless parameters of optimum modes HVEDC process

Curves 1 - 5 (Fig. 7) determine the optimal parameters of high voltage electric discharge consolidation for different values of initial porosity: 1 - Π = 9%, 2 - Π = 20%, 3 - Π = 30%, 4 - Π = 50%, 5 - Π = 57%. The most important factor which determines the success of the HVEDS process is the current density amplitude in the powder specimen (Grigoryev, 2007). We studied the influence of pulse current density on kinetics of high voltage electric discharge consolidation of tungsten carbide - cobalt composite. Figure 8 shows the effect magnitude of pulse current on the resultant WC-Co composite densities after HVEDS (external pressure 200 MPa).

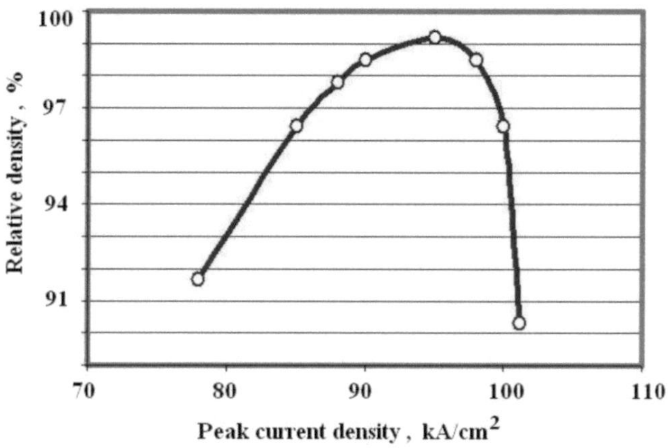

Fig. 8. WC-Co compact density dependence from pulse current density

This experimental dependence has a maximum on fixed peak current density (Fig. 8). The resultant WC-Co composite density increases within the current density region: from 80 kA/cm^2 to 95 kA/cm^2. The resultant density reaches the maximum value at ~95 kA/cm^2

and drastically decreases beyond 100 kA/cm². Representative micrographs of WC–Co compact samples were observed on the polished cross-sections using an optical and scanning electron microscopy. There is observed the difference between the microstructures of the specimens consolidated at the different pulse current amplitudes. In Figure 9 shows scanning electron micrograph of polished surface of the typical WC-Co composite consolidated by HVEDC at a current amplitude ~87 kA/cm².

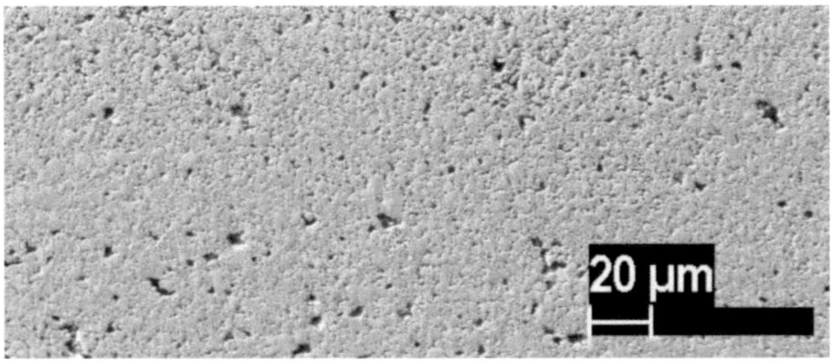

Fig. 9. Typical SEM image of WC–20Co cemented carbide consolidated by HVEDC (pulse current amplitude ~ 87 kA/cm²)

This specimen has small pores are uniformly distributed in the most of volume. Cemented carbide consolidated by HVEDC at a current amplitude ~95 kA/cm² has high density microstructure (Figure 10).

Fig. 10. SEM micrograph of the polished microstructure of the cemented carbide obtained by HVEDC (pulse current amplitude ~ 95 kA/cm²)

The grain size WC in the sample received by HVEDC corresponds to the initial size (<5µm) before the consolidation. The most of volume of this sample is pore less fully dense. The average relative density of this sample is >98%. The measurements of hardness of this sample show that the HRA value is 78±2. Near this level of pulse current the cobalt particles are in a fused condition and are redistributed in the compact volume due to magnetic pressure of discharge current pulse (Fig. 10). We determined that there is an upper level for the discharge peak current density beyond which the powder WC-Co composite material disintegrates like an exploding wire during HVEDC. Figure 11 shows the optical micrograph of the WC-Co compact sample after HVEDC (102 kA/cm²) beyond the upper level for the peak current density observed on the polished cross-section parallel to the direction of external pressure and pulse current.

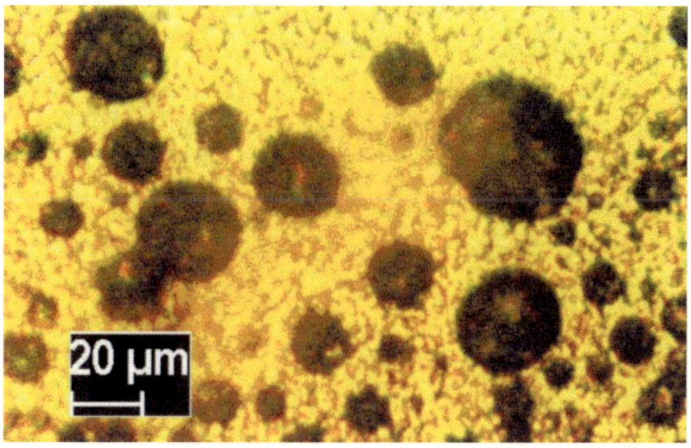

Fig. 11. Optical micrograph of the polished microstructure of WC-Co compact after HVEDC (pulse current amplitude ~ 102 kA/cm^2)

On this micrograph the dark spherical fields are pores that were arisen during HVEDC process. The measurements of average hardness of this sample show that the HRA value is 46±6.
There is a difference between microstructures at the edge and inside the sample depending on the distribution of magnetic pressure induced by the pulse current. The distribution of magnetic pressure (pinch effect) is defined by the distribution of a current density in the powder compact. The distribution of magnetic pressure has a parabolic profile with maximum inside sample (~ 18 MPa) when a current density has the homogeneous distribution in the powder column. Figure 12 shows the typical WC-Co compact structure with homogeneous distribution of a current density in the powder column along of diameter.

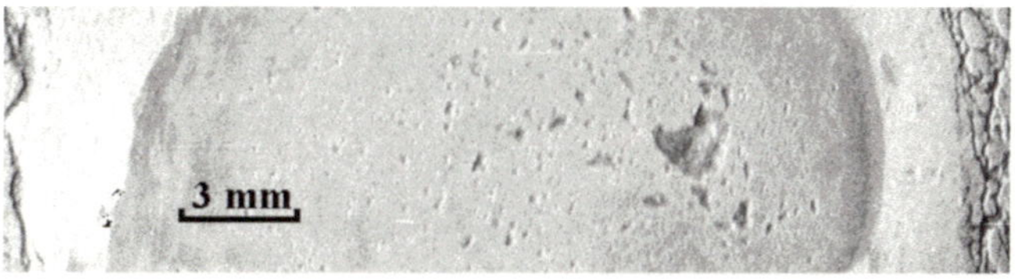

Fig. 12. WC-Co compact structure after HVEDC with homogeneous distribution of a current density (pulse current amplitude ~ 75 kA/cm^2, external pressure 120 MPa)

The magnetic pressure is more homogeneous in powder compact volume when the skin effect is strong. Figure 13 shows the typical WC-Co compact structure with the strong skin effect. In this case, the external compressed-air pressure was 200 MPa, and the peak pulse current density was 95 kA/cm^2.

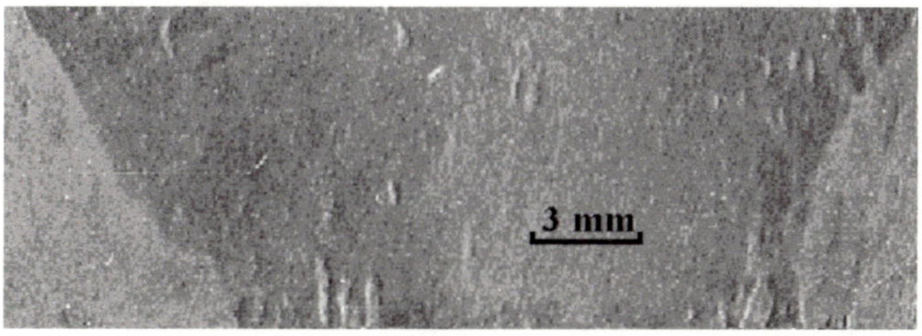

Fig. 13. WC-Co compact structure after HVEDC with a strong skin effect

XRD - phase analysis has displayed in WC–Co compact samples after HVEDC process the presence of W, α and β - phases Co. Lattice parameter of β - phase Co (HVEDC method) is equal 3,570 Å that corresponds with limit concentrations of W and C in Co (~10%). And lattice parameter of β -phase Co (conventional technology) is equal 3,555 Å (concentrations W и C ~ 3%). High voltage electric discharge consolidation retains balance of C under HVEDC-process and provides for receiving diphasic structure (WC + Co). XRD-results have displayed that the main phases of WC–Co powder were WC and α -Co (Fig. 14), but the main phases of WC–Co compact sample after HVEDC were WC и β -Co (Fig. 15). It concerned with a rapid cooling of WC–Co compact sample in the time of HVEDC process.

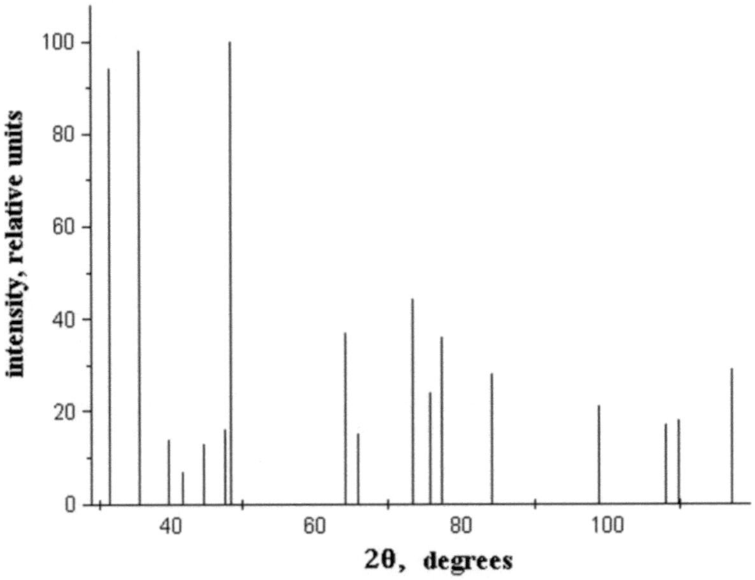

Fig. 14. XRD trait-diagram WC–Co powder before HVEDC

Correlation between intensities of XRD patterns for WC–Co powder and WC–Co compact samples after HVEDC depends on the homogeneity of the phase distribution and the texture of WC–Co compact induced by HVEDC process. Existence of η_1–phase in WC–Co compact samples after HVEDC is not detected.

Additionally we used HVEDC process for joining a cemented carbide layer to steel substrate. Short time forming a cemented carbide layer on a steel substrate reduces the heating of steel that reduces residual thermal stress magnitude in the resulting compound. It was ascertained that the high joint strength was obtained when the external pressure was 200 MPa and pulse current amplitude ~ 100 kA/cm^2. Figure 16a shows 1.2419 steel punch with coated cemented carbide layer for compacting powders of metal oxides. The mechanical properties of the joints are competitive to those of the conventional brazed steel-cemented carbide joints. Wear resistance of punches coated cemented carbide layer increases in 1,5 - 2 times.

Figure 16b shows neighbourhood of joining tungsten carbide - cobalt layer and 1.2419 steel by HVEDS. Our results reveal that HVEDC process produces the compound of consolidated

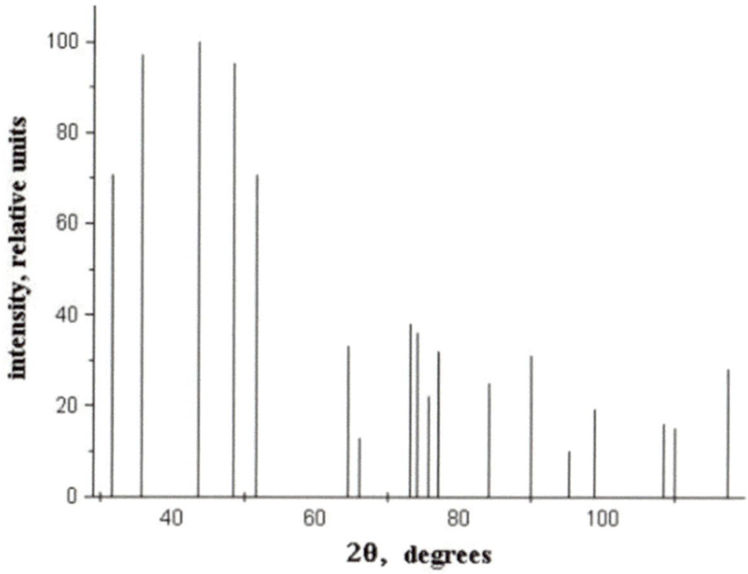

Fig. 15. XRD trait-diagram WC–Co compact sample after HVEDC

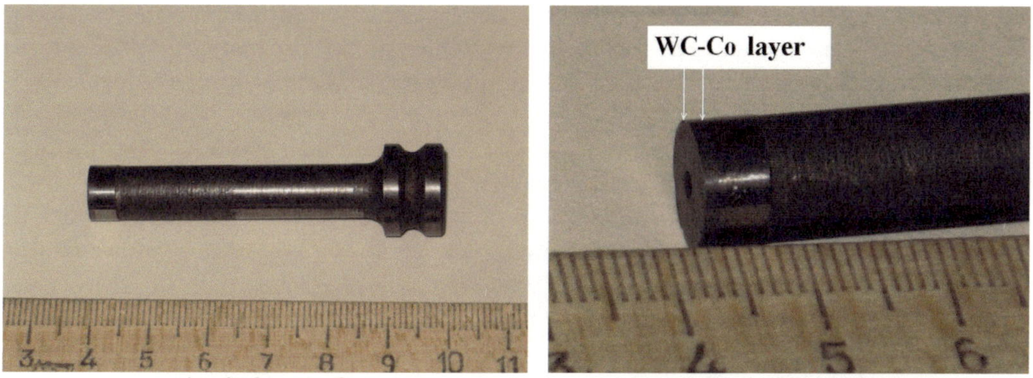

a) whole punch b) neighbourhood of joining

Fig. 16. The specimen of joining of cemented carbide with 1.2419 steel (EN) by HVEDS

cemented carbide layer and high-speed steel substrate without crack and pores. A main characteristic of the high voltage electric discharge consolidated cemented carbide/steel joints are their high strength, narrow and homogeneous interface. Figure 17 (a, b) shows micrographs of joining tungsten carbide - cobalt layer and 1.2419 steel by HVEDS.

The cobalt binder appears as the dark-gray network-like structure surrounding the hard WC phase (light-gray grains) in Fig. 17b. The average size of the WC grains does not exceed 5 µm in consolidated cemented carbide layer that corresponds to the initial WC grain size. The arrows in Fig. 17b indicate the contact surface between consolidated cemented carbide and 1.2419 steel substrate. The interface between cemented carbide layer and steel substrate has a sharp and clear boundary and no interdiffusion between the components of the cemented carbide layer and steel substrate.

a) interface WC-Co / steel b) arrows indicate the contact surface

Fig. 17. The contact area between 1.2419 steel substrate and cemented carbide layer

4. Conclusion

The response of WC–Co powder composites (loaded external pressure) to high energy electrical discharge has been described and understood in terms of peak current density and external pressure. It was found that the density and hardness of tungsten carbide - cobalt composite material reach its maximum values at certain magnitudes of applied pressure and high voltage electrical discharge parameters. There is an upper level for the high voltage pulse current amplitude beyond which the powder composite material disintegrates like an exploding wire. Near this level, the cobalt particles are in a fused condition and are redistributed in the compact volume due to magnetic pressure of discharge current pulse. The distribution of magnetic pressure is defined by the distribution of a current density in the powder compact. The magnetic pressure is more homogeneous in powder compact volume when the skin effect is strong. Densification to near theoretical density in a relatively short time can be accomplished with insignificant change in grain size by HVEDC. We have installed that the powder densification process has wave nature in high voltage electric discharge consolidation. We defined the velocity of wave front of densification process. Attempts to compact WC–Co powders by HVEDC process presage fruitful commercial results.

Additionally the high voltage electric discharge consolidation (HVEDC) allows the successful consolidation and simultaneous joining of cemented carbide to steel. HVEDC manufactures a high strength joints between the cemented carbide layer and steel substrate at any thickness of layer and substrate. Residual thermal stresses in the contact area of joints are significantly smaller magnitude after HVEDC process than for traditional methods of brazing and welding. In comparison to brazing, HVEDC process has the advantages of being faster, more precise and without the filler material. The high voltage electric discharge consolidation allows the successful joining of thick cemented carbide layers to steel substrates. The technique of electric discharge sintering is potentially widely applicable for joining composite materials to steel substrate.

5. Acknowledgment

The author would like to thank Tatiana Matveeva and Vladimir Grigoryev for their helpful discussions and invaluable assistance. The author is grateful to Pavel Dzhumayev for his help in electron microscopic studies.

6. References

Baranov, Yu.V., Troitskii, O.A., Avraamov, Yu.S. & Shlyapin, A.D. (2001). *Physical Bases of Electric-Pulse and Electroplastic Treatments and New Materials*, MGIU, ISBN 5-276-00177-4, Moscow

Bataronov, I.L. (1999), MECHANISMS OF ELECTROPLASTICITY, Sorovskii Obrazovat. Zhurnal, No. 10, (1999), pp. 93 - 99

Bazanov, V. N., Balankin S, A., Bashlikov, S. S., Grigoryev, E. G., Gunichev, V. V. & Yartsev, V. A. (1990). Application of electrical discharge for pulse heating and compaction of powder materials, *Proceedings of* Powder Metallurgy Conference & Exhibition (PM'90), London, UK, July 1990

Carroll, M.M., Kim, K.T. & Nesterenko, V.F. (1986). The effect of temperature on viscoplastic pore collapse. *Journal of Applied Physics*, Vol. 59, No. 6, (March 1986), pp. 1962-1967

Grasso1, S., Sakka1, Y. & Maizza, G. (2009). Electric current activated/assisted sintering (ECAS): a review of patents 1906–2008. Science and Technology of Advanced Materials, Vol. 10, No. 5, (October 2009), pp. 1-24

Crowson, A. & Chen, E. S. (Ed(s).). (1991). *Tungsten and Tungsten Alloys-Recent Advances*, Minerals, Metals and Materials Society, ISBN 0873391330, 9780873391337

Grigoryev, E.G. (2007). Modeling of physical processes of electro - discharge compaction of powder materials, *Proceedings of* Scientific Session of MEPhI-2007, Moscow, Russia, January 2007

Grigoryev, E. G. & Rosliakov A.V. (2007). Electro discharge compaction of WC-Co and W-Ni-Fe-Co composite materials. *Journal of Materials Processing Technology*, Vol. 191, No. 1-3, (August 2007), pp. 182-184

Grigoryev, E. G. (2010). Electric Discharge Sintering and Joining of Tungsten Carbide-Cobalt Composite with High-Speed Steel Substrate, *Proceedings of* INTERNATIONAL CONFERENCE ON ADVANCES IN MATERIALS AND PROCESSING TECHNOLOGIES (AMPT2010), Paris, France, October 2010

Groza, J.R. & Zavaliangos, A. (2003). Nanostructured bulk solids by field activated sintering. *Review Advance Materials Science*, Vol. 5, (2003), pp. 24-33

Kodasha, V.Y., Grozaa, J.R., Chob, K.C. , Klotzb, B.R. & Dowding, R.J. (2004). Field-assisted sintering of Ni nanopowders. *Materials Science and Engineering A*, Vol. 385, No. 1-2, (November 2004), pp. 367-371

Munir, Z.A., Anselmi-Tamburini, U. & Ohyanagi, M. (2006). The Effect of Electric Field and Pressure on the Synthesis and Consolidation of Materials A Review of the Spark Plasma Sintering Method. *Journal of Materials Science*, Vol. 41, No. 3, (2006), pp. 763-777

Olevsky, E. A. & Froyen, L. (2009). Impact of Thermal Diffusion on Densification During SPS. *Journal of the American Ceramic Society*, Vol. 92, No. S1, (2009), pp. 122-132

Orrù, R., Licheri R., Locci, A. M., Cincotti, A. & Cao, G. (2009). Consolidation/synthesis of materials by electric current activated/assisted sintering. *Materials Science and Engineering*, Vol. R 63, (February 2009), pp. 127–287

Raichenko, A.I. (1987). *Fundamentals of Powder Sintering by Passing an Electric Current*, Metallurgiya, ISBN , Moscow